Green
Education

Green
Education

An A-to-Z Guide

The SAGE Reference Series on
Green Society
Toward a Sustainable Future

JULIE NEWMAN, GENERAL EDITOR
Yale University

PAUL ROBBINS, SERIES EDITOR
University of Arizona

Los Angeles | London | New Delhi
Singapore | Washington DC

Los Angeles | London | New Delhi
Singapore | Washington DC

FOR INFORMATION:

SAGE Publications, Inc.
2455 Teller Road
Thousand Oaks, California 91320
E-mail: order@sagepub.com

SAGE Publications India Pvt. Ltd.
B 1/I 1 Mohan Cooperative Industrial Area
Mathura Road, New Delhi 110 044
India

SAGE Publications Ltd.
1 Oliver's Yard
55 City Road
London EC1Y 1SP
United Kingdom

SAGE Publications Asia-Pacific Pte. Ltd.
33 Pekin Street #02-01
Far East Square
Singapore 048763

Publisher: Rolf A. Janke
Assistant to the Publisher: Michele Thompson
Senior Editor: Jim Brace-Thompson
Production Editor: Kate Schroeder, Tracy Buyan
Reference Systems Manager: Leticia Gutierrez
Reference Systems Coordinator: Laura Notton
Typesetter: C&M Digitals (P) Ltd.
Proofreader: Rae-Ann Goodwin
Indexer: Joan Shapiro
Cover Designer: Gail Buschman
Marketing Manager: Kristi Ward

Golson Media
President and Editor: J. Geoffrey Golson
Author Manager: Ellen Ingber
Editors: Mary Jo Scibetta, Kenneth Heller
Copy Editors: Barbara Paris, Tricia Lawrence, Holli Fort

Copyright © 2011 by SAGE Publications, Inc.

Printed in the United States of America

Library of Congress Cataloging-in-Publication Data

Green education : an A-to-Z guide / Julie Newman, editor.

p. cm. — (The Sage reference series on green society: toward a sustainable future)
Includes bibliographical references and index.

ISBN 978-1-4129-9686-0 (cloth) — ISBN 978-1-4129-7461-5 (ebk)

1. Sustainable development—Study and teaching.
I. Newman, Julie.

HC79.E5G6898 2011 338.9'27071—dc22 2011006082

11 12 13 14 15 10 9 8 7 6 5 4 3 2 1

Contents

About the Editors

Green Series Editor: Paul Robbins

Paul Robbins is a professor and the director of the University of Arizona School of Geography and Development. He earned his Ph.D. in Geography in 1996 from Clark University. He is general editor of the *Encyclopedia of Environment and Society* (2007) and author of several books, including *Environment and Society: A Critical Introduction* (2010), *Lawn People: How Grasses, Weeds, and Chemicals Make Us Who We Are* (2007), and *Political Ecology: A Critical Introduction* (2004).

Robbins's research centers on the relationships between individuals (homeowners, hunters, professional foresters), environmental actors (lawns, elk, mesquite trees), and the institutions that connect them. He and his students seek to explain human environmental practices and knowledge, the influence nonhumans have on human behavior and organization, and the implications these interactions hold for ecosystem health, local community, and social justice. Past projects have examined chemical use in the suburban United States, elk management in Montana, forest product collection in New England, and wolf conservation in India.

Green Education General Editor: Julie Newman

Julie Newman, Ph.D., has worked in the field of sustainable development and campus sustainability since 1993. Her research has focused on the role of decision-making processes and organizational behavior in institutionalizing sustainability into higher education. In 2004, Newman was recruited to be the founding director of the Office of Sustainability for Yale University. At Yale, Newman also holds a lecturer appointment with the Yale School of Forestry and Environmental Studies, where she teaches an undergraduate course titled Sustainability: From Innovation to Transformation in Institutions. Prior to her work at Yale, Newman assisted with the establishment of the longest-standing sustainability office in the country at the University of New Hampshire, Office of Sustainability. Newman was a pioneer in the field of campus sustainability beginning in 1995, when she worked for University Leaders for a Sustainable Future (ULSF) while a graduate student at Tufts University. In 2004, Newman cofounded the Northeast Campus Sustainability Consortium to advance education and action for sustainable development on university campuses in the northeast and maritime region. This has led to a 10-year regional commitment and a set of annual meetings for sustainability professionals in the northeast and maritime region. She also co-coordinates a sustainability working group of the International Alliance of Research Universities as well as a Sustainability working group for the Council of Ivy Presidents. In addition, Newman is a member of the editorial board of *Sustainability: Journal of Record.* Newman lectures and consults for universities both nationally and internationally, participates on a variety of boards and advisory committees, and has contributed to a series of edited books and peer-reviewed journals. She holds a B.S. from the University of Michigan, an M.S. from Tufts University, and a Ph.D. from the University of New Hampshire.

Introduction

John Muir in 1911 acknowledged in his writings titled *My First Summer in the Sierra,* "When we tug at a single thing in nature, we find it attached to the rest of the world." A century ago, Muir ever so succinctly captured what now lies at the core of the dynamic discipline and inherent challenge of Green Education. Muir's quote is a potent reminder that we live in a world of interdependencies and interconnections. Nevertheless, we seem to find it easier to understand and interact with these complex and dynamic systems in pieces. The underlying challenge with this approach is that by focusing on the parts, we tend to neglect the whole. And by neglecting the whole, we continue to proliferate and stress the limitations on the Earth's systems, which manifest in drought, soil erosion, air, land and water pollution, global warming, loss of biodiversity and habitat, depleted fisheries, and correlated public health concerns. *Green Education,* a term that is often interchanged with *Education for a Sustainable Future,* provides an overarching multidisciplinary framework that calls upon the development of a broader and deeper understanding of the inherent connection between people, the environment, and the economy. Evoking Muir's vision, Green Education sets out to ensure that when a single thing in nature is tugged upon, we have an informed understanding as to how this relates to the rest of the world and the consequences of a single tug.

Why Green Education?

Every day, people make impactful decisions as consumers with respect to the use and acquisition of material goods or when preparing and eating every meal; as citizens and taxpayers when voting; as homeowners when heating, cooling, and building our homes; when traveling as commuters; and as employees in our workplace. The information that we have access to that informs these decisions and the consequences of many decisions tend to be too large or abstract to comprehend and therefore act upon, particularly at the level of the individual. Moreover, decisions tend to be driven by economic constraints forgoing externalities. Information is often available in pieces, which are easier to digest yet difficult to decipher what the greater impact upon the whole will be either locally or globally. Moreover, sustainability issues frequently represented in current events as captured by the media can be daunting while related scientific data may be inaccessible. We tend to have trouble as individuals comprehending how we can actually have an impact when the issues are global in scale, dynamic in content, and the results may take a decade or more to see and feel. Ultimately, Green Education seeks to find a balance in our ever-growing demand on our natural resource base while providing the tools to grapple with equitable use and distribution of these resources. The complexities that arise in this educational endeavor concern not only the physical attributes of resource use and depletion, but also the normative questions of how responsibility is taken or assigned.

Building a Foundation for Green Education

The field of Green Education dates back almost four decades to the 1972 Stockholm Conference on the Human Environment. This was the first international conference to collectively address people and the environment. The conference concluded with a call to broaden the basis of action by seeking engagement at the individual, business, and community level—as well as the local to global scale. The conference explicitly recognized the relationship between education and the emerging framework of what we now refer to as sustainable development. The document called for access to environmental education for all—from childhood through adulthood. The initial call to action was strengthened by the Tbilisi Declaration in 1978 that transpired from the Intergovernmental Conference on Environmental Education. This international declaration recognized the need for and proposed an interdisciplinary framework for environmental education. At that time, environmental education tended to target K–12 curriculum development exclusively, leaving higher education to focus on related research and teaching. Nine years later, in 1987, the Bruntland Commission created an overarching declaration for policy makers and educators alike defining sustainable development as a challenge to meet our present needs without diminishing the abilities of future generations to do the same.

In response to the call for action by the Bruntland Commission, yet preceding the 1992 Earth Summit in Rio, colleges and universities more formally became committed participants in the global push for sustainability through a sector-specific international declaration known as the Talloires Declaration. The Talloires Declaration, initiated and endorsed by a global group of leaders in higher education, is a 10-point action plan calling for the incorporation of sustainability principles into teaching, research, operations, and outreach. This call to action and framework set the course for the higher education sector to engage and take a leading role in shaping the breadth of what Green Education has to offer, from formal and informal curricula to applied technology and building design to interdisciplinary research. Since that time, higher education institutions have begun to establish sustainability action plans and committed resources to fulfill their potentials as sustainability leaders. Two decades later, it is common practice for colleges and universities to have designated staff devoted to ensuring sustainable campus practices that manifest in operational systems, student engagement, faculty research, and staff responsibility.

Campuses engaged in and committed to sustainability range from public and large private research institutions to small liberal arts colleges and community colleges. As the training ground for leaders in business, policy, international affairs, and every other sector links to issues of sustainability, higher education has the power to develop, study, and implement practicable models that can be adopted and shared by others. The inherent role of research and intellectual capital give higher education a respected and far-reaching voice. Now that the nation's ears are perked, colleges and universities can use this voice to drive ambitious transformation for sustainability.

Measuring Sustainability

One of the most notable trends that is on the rise in the field of Green Education is the creation of a measurement methodology for organizations to determine how they are performing today and how to establish targets for how they would like to perform in the future. Many colleges and universities and some secondary schools have begun to establish short- and long-term sustainability goals for their institutions. These goals may be quantitative or qualitative in nature, concerning everything from building design and performance to greenhouse gas emissions, landscape standards, water use, and transportation trends to curricular engagement opportunities. In an effort to determine how an organization or university is performing, third-party rating systems have been developed to compare and contrast and in some cases, incentivize others to excel.

The *Green Education* Volume

The entries in the *Green Education* volume seek to provide a historical perspective on the international and domestic treaties that have shaped the direction of the field today and expose you to current-day trends, examples, and challenges. The state of the field today will be explored in this volume via contemporary case examples from K–12, higher education, and community-based projects. It is not possible to cover every innovative example that has been identified. However, we encourage you as the reader to *tug* on one example and see how that will lead to additional learning outside of this volume. Green Education today distinguishes itself from other fields as programs and materials span curriculum development, research, operational systems, and outreach alike. In turn, these were adapted as the organizing principles for this volume.

Innovative Green Education efforts are continuously being defined, new and old knowledge is being integrated, new relationships are being built, and educational organizations are attempting to align sustainability behavior, professional capacity, and organizational systems to achieve the desired learning outcomes called for by Green Education. The National Research Council declared in 2002 that we are in the midst of "a transition to a world in which human populations are more crowded, more consuming, more connected, and in many parts of the world, more diverse, than at any time in history." Today, the type of change that is needed to reverse this trajectory is no doubt difficult. It's cultural—it's behavioral. It requires new perspectives—new knowledge—new mechanisms—new relationships. There is an inherent fragility in the system that lies in the alignment of desired behavior, citizen responsibility, professional capacity, and organizational systems. We need to reevaluate what we value—find new ways of accomplishing some of the same tasks and teach ourselves how to accomplish new tasks in new ways.

Julie Newman
Yale University

Reader's Guide

Community Education and Awareness

American Pastors' Creation Care Covenant
Earth Day Network
Global Green Day
Green Community-Based Learning
Greening of the Campus Conference
Green Week, National
Greens, The (PBS Website)
Sustainability on Television
Sustainability Websites

Educational Institutions Leading in Sustainability: Curriculum

Aquinas College
Arizona State University
Carnegie Mellon University
College of Menominee Nation
College of the Atlantic
Colorado State University
Dalhousie University
EARTH University (Costa Rica)
Green Mountain College
Lund University (Sweden)
Maastricht University (Netherlands)
Northland College
Rensselaer Polytechnic Institute
Stanford University (Global Ecology
    Research Center)
Stellenbosch University (South Africa)
Unity College
University of Michigan (Erb Institute)

Educational Institutions Leading in Sustainability: Operations

Australian National University
Berea College
Brown University
California State University, Chico
Carleton College
Cornell University
Dartmouth College
Dickinson College
Eastern Iowa Community College
Evergreen State College
Harvard University
Michigan State University
Middlebury College
Oberlin College
Santa Fe Community College
Tufts University
University of Copenhagen
University of
    Maryland, College Park
University of Pennsylvania
University of Vermont
University of Washington
Yale University

Educational Institutions Leading in Sustainability: Outreach

Allegheny College
Brandeis University

Drury University
Grand Valley State University
Tulane University
University of Chicago
University of Minnesota
University of New Hampshire
University of Oregon
University of Texas, Austin
    (Alley Flat Initiative)

Educational Institutions Leading in Sustainability: Research

Columbia University
Leeds University
Portland University
University of British Columbia
University of Glasgow
University of Tokyo
University of Virginia (ecoMOD)

History of Education for Sustainable Development

Agenda 21, "Promoting Education,
    Public Awareness, and Training"
    (Chapter 36)
Brundtland Report
CRE Copernicus Charter
Declaration of Thessaloniki
Environmental Education Debate
ESD to ESF (Education for Sustainable
    Development to Education for a
    Sustainable Future)
Geddes, Sir Patrick
Halifax Declaration
HESA: Higher Education
    Sustainability Act
Kyoto Declaration on Sustainable
    Development
Lüneburg Declaration
Swansea Declaration
Tbilisi Declaration
United Nations Conference on
    Environment and Development 1992
    (Earth Summit)

United Nations Conference
    on the Human Environment,
    Stockholm 1972
United Nations Decade of
    Education for Sustainable
    Development 2005–2014

Leaders in Green Education

Leadership in Sustainability Education

Measuring Sustainability

AASHE: STARS (Sustainability Tracking,
    Assessment & Rating System)
National Wildlife Federation
Sustainable Endowments Institute

Organizations for Sustainability Education

AAAS Forum on Science, Technology,
    and Sustainability
Association for the Advancement of
    Sustainability in Higher Education
Association of University Leaders for a
    Sustainable Future
Center for Environmental Education
Collaborative for High Performance
    Schools
Earth Charter International
Environmental Literacy Council
ERTHNXT
Higher Education Associations
    Sustainability Consortium
International Institute for
    Sustainable Development
NAEP Sustainability Institute
Second Nature: Education for
    Sustainability
Society for College and University
    Planning

Sustainability Education K–12

Audubon Adventures
Cloud Institute for Sustainability
    Education
Green Math

Sustainability Partnerships

List of Articles

List of Contributors

Agans, Lyndsay J.
University of Denver

Almeida, Sylvia Christine
Monash University

Bardaglio, Peter W.
Second Nature

Baumgarten, Martha S.
Knox College

Berkson, Emily K.
Knox College

Biggs, Lindy
Auburn University

Blagg, Lisa R.
Knox College

Boslaugh, Sarah
Washington University in St. Louis

Brouwer, Frank
*University of Sussex/London
RCE on ESD*

Chiu, Belinda H. Y.
BChiu Consulting

Coffman, Jennifer
James Madison University

Colucci-Gray, Laura
University of Aberdeen

Dahm, Cale T.
Knox College

Davis, Julie
*Queensland University of
Technology*

DeShasier, Bryan
University of Denver

Duffy, Lawrence K.
*American Association for the
Advancement of Science (AAAS)*

Ercoskun, Ozge Yalciner
Gazi University

Evans, Tina
Fort Lewis College

Ferber, Michael P.
The King's University College

Ferguson, Therese
Independent Scholar

Ferreira, Jo-Anne
Griffith University

Finley-Brook, Mary
University of Richmond

Flair, Levi W.
Knox College

Godfrey, Phoebe C.
University of Connecticut

Gonshorek, Daniel O.
Knox College

Gray, Donald
University of Aberdeen

Gunter, Jr., Michael M.
Rollins College

Harper, Gavin
Cardiff University

Helfer, Jason A.
Knox College

Henderson, DeAndre A.
Knox College

Kang, Leslie Y.
Knox College

Katzschner, Tania
Independent Scholar

Kopnina, Helen
University of Amsterdam

Kte'pi, Bill
Independent Scholar

Lanfair, Jordan K.
Knox College

LaRosa, Derek M.
Knox College

Lawrence, Kirk S.
University of California, Riverside

Mahone, Christian D.
Knox College

Marelli, Beatrice
University of Milan

Meek, David
University of Georgia

Miller, Justin
Ball State University

Newman, Julie
Yale University

O'Keefe, Sean G.
Knox College

O'Rourke, Caroline
University of Richmond

Panda, Sudhanshu Sekhar
Gainesville State College

Pietta, Antonella
University of Brescia

Purdy, Elizabeth Rholetter
Independent Scholar

Rands, Gordon
Western Illinois University

Rands, Pamela
Western Illinois University

Reid, Alan
University of Bath

Romano, Victoria M.
Knox College

Rosa, Angelina A.
Knox College

Ryan, Elizabeth
University of the Sunshine Coast

Sanya, Tom
University of Cape Town

Schmidt, David R.
Independent Scholar

Schroth, Stephen T.
Knox College

Slinger-Friedman, Vanessa
Kennesaw State University

Smiley Smith, Sara E.
Yale University

Snell, Carolyn
University of York

Stafford, Daniel T.
Knox College

Steffens, Ron
Green Mountain College

Swearingen White, Stacey
University of Kansas

Taylor, Whitney L.
Knox College

Trevino, Marcella Bush
Barry University

Turrell, Sophie
Independent Scholar

Ulloa, Sergio
Knox College

Vaca, Edel
Knox College

Van Grinsven, Rose M.
Knox College

Vedwan, Neeraj
Montclair State University

White, H. Courtney
Antioch University New England

Wilson, Sarah
University of York

Green Education Chronology

1848: The American Association for the Advancement of Science is founded. Its main priorities include publishing the journal *Science*.

1886: Aquinas College is founded in the Michigan city of Grand Rapids. The college would go on to implement a world-renowned Sustainable Business curriculum.

1946: The Scottish Centre for Ecology and the Natural Environment (SCENE) is established.

1960–2005: The percentage of Americans who participate in community recycling programs increases from 6.4 percent to 32 percent, a rise that is largely based on environmental education.

1962: Rachel Carson's *Silent Spring*, a book that educated the public about the dangers of the DDT pesticide, is published.

1965: The Society for College and University Planning (SCUP) is established. Among the responsibilities of the society include publishing the quarterly peer-reviewed journal *Planning for Higher Education*.

1967: CBS airs its documentary *The Poisoned Air*, educating the general public about the harmful effects of air pollution on humans.

1968: Bowling Green State University's Center for Environmental Programs is established.

1970: The University of Colorado becomes the first college to establish an Environmental Center.

1973: Washington University's Institute for Environmental Studies is founded.

1976: Brown University's Center for Environmental Studies (CES) is established.

1976: Colorado University establishes one of the first campus recycling programs in U.S. history.

1977: The First Intergovernmental Conference on Environmental Education is held in Tbilisi, Georgia.

1978: Dalhousie University adds Master of Environmental Studies to its list of available majors.

1983: The World Commission on Environment and Development, commonly known as the Brundtland Commission, is held.

1984: India's Centre of Environmental Education is established.

1987: The Brundtland Report releases its report after four years of research. The report is titled *Our Common Future.*

1990: In an effort to promote environmental education, Brown University establishes its Brown Is Green (BIG) program.

1991: Griffith University establishes its Master of Environment (Education for Sustainability) program.

1991: The Conference on University Action for Sustainable Development is held in the Nova Scotia city of Halifax.

1992: The Foundation of Environmental Education (FEE) establishes its Eco-Schools Program.

1993: The International Association of Universities signs a formal agreement called the Kyoto Declaration on Sustainable Development.

1995: New York City's Cloud Institute for Sustainability Education is established.

1995: Green Mountain College establishes its Environmental Liberal Arts curriculum.

1995: Penn State (Pennsylvania State University) founds its Center for Sustainability.

1996: Michigan University's Erb Institute, an organization focusing on sustainability, is established.

1997: Allegheny College's Center for Economic and Environmental Development is founded.

1997: Representatives from 83 countries congregate in the Greek city of Thessaloniki for the International Conference on Environment and Society.

1997: The University of New Hampshire forms its University Office of Sustainability.

2000: The Lund University Centre for Sustainability Studies is established.

2000: The University of Oregon establishes its Sustainable Development Plan.

2000: The Massachusetts Institute of Technology (MIT) forms its Green Building Task Force.

2000: Colby College's Environmental Advisory Group (EAG) is formed.

2001: The Education for Sustainability Western Network is established.

2001: The Center for Sustainable Communities and Civic Engagement is formed at Daemen College in the state of New York.

2001: Emory University elects its first campus environmental officer.

2002: Stanford University founds its Department of Global Ecology (DGE).

2003: The United States Partnership for Education for Sustainable Development (USPESD) is founded.

2004: Unity College in Maine announces it has become fully "green powered."

2004: Arizona State University establishes its Global Institute of Sustainability.

2004: Leeds University establishes its School of Earth and Environment.

2004: The University of Virginia establishes ecoMOD, a program designed to create environmentally sustainable dwellings for low-income people.

2004: The Northeast Campus Sustainability Consortium (NECSC) is established.

2005: The Sustainable Endowments Institute (SEI), a program designed to encourage sustainability on college campuses, is founded.

2006: The American Association for the Advancement of Sustainability in Higher Education (AASHE) holds its first conference in Arizona with approximately 600 people attending.

2006: Carleton College's Sustainability Website is launched.

2006: The Ohio University creates its Office of Sustainability.

2006: The University of British Columbia's Clean Energy Research Centre (CERC) is established.

2007: The College of the Atlantic announces that it will become the first "carbon-neutral" university.

2007: The University at Buffalo, State University of New York, (SUNY Buffalo) creates its Environmental Stewardship Committee.

2008: Dalhousie University establishes its College of Sustainability.

2008: The Green Education Foundation (GEF) is established.

2008: President George W. Bush signs the Higher Education Sustainability Act (HESA) into law.

2008: Luther College in the state of Iowa creates its Campus Sustainability Council.

2009: The American Association for the Advancement of Sustainability in Higher Education (AASHE) launches the STARS initiative, a program designed to award college campuses that teach and uphold the principles of environmentalism.

2009: The American Recovery and Reinvestment Act is passed. The legislation includes new opportunities for vocational training for "green jobs."

2009: Amherst College begins food composting its cafeteria waste as part of an effort to increase campus environmental standards.

2009: The University of Maryland's Sustainability Council is formed.

2009: Northland College adds Sustainable Community Development to its list of available majors.

2009: Ball State University begins construction on a geothermal energy system that, when completed, will reduce the university's carbon footprint by approximately 50 percent.

2010: The number of signatories pledging to uphold the principles of the American College and University Presidents Climate Commitment (ACUPCC) reaches 660.

2010: The College of the Atlantic ranks third out of 286 "green colleges" in *Princeton Review*'s list of the most environmentally friendly colleges.

2010: Santa Fe Community College's Sustainable Technologies Center (STC) is constructed at a cost of $11.4 million.

AAAS FORUM ON SCIENCE, TECHNOLOGY, AND SUSTAINABILITY

The AAAS Forum on Science, Technology, and Sustainability is supported by the American Association for the Advancement of Science (AAAS) Center for Science Technology and Sustainability as part of its international activities. Sustainability science is an emerging scientific field that discovers the knowledge that allows human societies to develop policies and strategies that will enable them to meet present demands without limiting future generations. The forum is a virtual meeting place for reporting on research advances and publishing events and activities for scientists interested in sustainability science; it is a partnership with the International Initiative on Science and Technology for Sustainability (ISTS). The forum, working in conjunction with AAAS affiliate organizations, is also tasked with educating and defining career paths for scientists and engineers who want to work in the field of sustainable development. The forum is a major vehicle to further AAAS's Center for Science Technology and Sustainability goals of promoting international scientific cooperation and building human capacity for the sustainable development workforce.

The forum discusses the challenges of climate change to both the developed and the developing world. Climate change will impact millions with loss of habitat, malnutrition, drought, and disease. For a "climate-smart" world, international cooperation and networking are needed to link scientific knowledge with cross-disciplinary innovative development strategies. The forum focuses on the ways in which science and innovation can be conducted and applied to meet human needs while preserving ecosystem services at local and global scales, that is, R. Buckminster Fuller's Spaceship Earth concept with an emphasis on life-support systems. Instead of broad sociopolitical discussions of the value of sustainability, the forum illustrates programs that analyze nature-society interactions and highlights people applying the resulting knowledge to create adaptations and transitions to healthier, sustainable communities. The forum supports AAAS's goals to enable centers of excellence that build capacity across public and private sectors and that develop innovative education programs, new technologies, and human sustainable management practices. These centers will produce a workforce that can develop solutions to the complex environmental and economic problems associated with current climate change projections. The forum's

Website (www.AAAS.org/programs/centers) allows the viewer to (1) download key documents that describe the underlying principles for sustainability science, (2) access a directory of individuals and institutions that are actively working in the field and a description of their projects, and (3) find links to content related to specific regions, research themes, or sustainability challenges. As Robert Chianese, a member of the AAAS Pacific Division Executive Committee, described in a 2009 sustainability symposium: "Sustainability is about how we live, wisely or carelessly, and the outcome of that will determine pleasantness and harmony—or ugliness and discord—of our future." He urged making sustainability "not just a household word . . . but a widespread household practice." The forum is a tool for accomplishing that vision.

Initiative on Science and Technology for Sustainability (ISTS)

Developing and advocating for sustainability science research and education, sustainable development policies, and sustainable lifestyles are challenges because of economic and political barriers. The forum Website informs about these challenges when encountered and announces meetings such as the November 2009 "Sustainability—Creating the Culture" meeting. AAAS's partner ISTS works to enhance access to knowledge about sustainable development from around the world and to evolve a vision for new sustainable technology. ISTS addresses the issue of unity in a nature–society system while maintaining a focus on the needs of the human condition in a world with a still-growing human population. Being integrative across science and multiple sectors of human activity, ISTS focuses on place-based activity at regional scales. At an intermediate scale, complexity is comprehensible, and innovative management is more politically possible at these local and regional scales. Transitions toward sustainability such as community gardens, have begun and can be reported on the forum.

ISTS aims to connect science and policy by developing an international open-ended network that is funded by the David and Lucille Packard Foundation, the National Oceanic and Atmospheric Administration (NOAA), and other regional institutions. The initiative's Web-based forum facilitates information exchange and engagement with a global virtual community that can link scholars, managers, and political leaders. This network can question the structures that perpetuate unsustainable practices and the ecological flaws in the early balanced approach theory of sustainable development.

Sustainability

Building on issues and international discussions by a select community in the 1970s, the concept of sustainable development grew in the 1980s and was promoted in 1987 by the Brundtland Commission (formally called the World Commission on Environment and Development). There are numerous definitions of sustainability, but the mainstream definitions emphasize the concept of conservation in which natural resources are used wisely and nonrenewable resources are not depleted. Fair distribution of the benefits for current and future generations is central to the conservation ethic. The definition of sustainable development has further evolved to assume a global ethic where current needs of the world population do not compromise the ability of future generations to meet their own needs. In this compromise between conservation and development as well as between present and future population needs, the repackaging of environmental management has led to inclusion of various values in the many current definitions of sustainability. In the industrialized world, there is a faith in technological progress and innovation, which leads to attempts to define sustainability for socioeconomic systems with the concepts of human capital and natural capital.

An early concept in sustainability was to solve the carrying capacity and unfair distribution of resources paradox of sustainable development by expanding the global economy. The decade of the 2000s has highlighted the problems with this idea and normal human behavior. A new animism, championed by an idealistic global media, has been included in some discussions of replacing sustainable development with ecological sustainability and reforming existing democratic structures and institutions. The scientific view of an ecologically sustainable development focuses on human ecosystem interaction that maintains the functional integrity of ecosystems, but also allows for continuation of ecological services. This view overlaps with the following concept of precautionary principles:

- Do not damage ecosystems so that they cannot provide services
- Watch for side effects of new technologies
- Do not overexploit renewable resources
- Take advantage of nature's organization principles—that is, industrial ecology
- Design resilient and adaptive human social systems
- Pay attention to national cycles that use waste and inhibit environmental diseases

Sustainability science is linked to the growing awareness of the impact human systems have on climate and disease. Research has demonstrated the importance of socioeconomic factors in producing disease. The ability of man to impact climate and hydrological cycles with subsequent impacts on agriculture and environmental diseases is leading to reevaluations of land use and resource practices. For example, chemical hazards related to resource development and energy use produce pollutants and influence human and wildlife exposure. Fire smoke transports metals such as mercury and fine particles that exacerbate cardiac and respiratory problems.

Career Paths

In an effort to develop career paths for a new workforce, the AAAS and the forum act as a bulletin board for career development activities. Postings of opportunities have included (1) Associate Research Scholar at the Center for Sustainable Urban Development by Columbia University, (2) Legality Assurance Expert at the European Forest Institute, (3) Short-Term Consultant to the Program on Forests at the World Bank, and (4) Graduate Fellowships in Sustainability Science at the University of Maine. In addition, Sarah Banas of AAAS has published a "Survey of University-Based Sustainability Science Programs." These activities are part of AAAS's missions to strengthen and diversify the science and technology workforce as well as to foster education in science and technology for everyone.

AAAS

The American Association for the Advancement of Science, commonly called "TripleA-S," is an international nonprofit organization and professional association whose members are dedicated to advancing science around the world by serving as educators, leaders, advocates, and spokespeople. Founded in 1848, AAAS serves 262 affiliated societies and academies of science, representing over 10 million individuals. AAAS publishes the journal *Science* and manages the premier science news Website "EurekAlert." It has four regional divisions and four primary program areas: (1) science and policy, (2) international activities, (3) education and human resources, and (4) Project 2061 (www.aaas.org).

See Also: Association for the Advancement of Sustainability in Higher Education; Sustainability Teacher Training; Sustainability Websites.

Further Readings

Dresner, S. *Principles of Sustainability.* London: Earthscan, 2002.
Duffy, L. K. "Diseases." In *Encyclopedia of Global Warming and Climate Change,* S. G. Philander, ed. Thousand Oaks, CA: Sage, 2008.
Marten, G. G. *Human Ecology.* London: Earthscan, 2001.
McManus, P. "Sustainability." In *Encyclopedia of Global Warming and Climate Change,* S. G. Philander, ed. Thousand Oaks, CA: Sage, 2008.
Myers, N. J. and C. Roffensperger. *Precautionary Tools for Reshaping Environmental Policy.* Cambridge, MA: MIT Press, 2006.

Lawrence K. Duffy
American Association for the Advancement of Science (AAAS)

AASHE: STARS (Sustainability Tracking, Assessment & Rating System)

As campus sustainability has become more prominent, so have efforts to assess what colleges and universities are doing to promote sustainable practices. Responding to a 2006 call from the Higher Education Associations Sustainability Consortium (HEASC), the Association for the Advancement of Sustainability in Higher Education (AASHE) developed the Sustainability Tracking, Assessment & Rating System (STARS) to evaluate and score individual campuses on a variety of measures. The STARS program is entirely voluntary, and campuses themselves compile the data needed to report publicly on their progress. After piloting the project with nearly 70 schools over a yearlong period, AASHE launched what it calls "STARS 1.0" in October 2009. By the start of 2010, over 120 U.S. colleges and universities had registered to participate in this initiative.

The STARS program contains three broad evaluation categories: Education and Research; Operations; and Planning Administration and Engagement. Participating institutions can earn up to 300 points (100 per category) based on their level of involvement with the various activities that comprise the different subcategories in each of these three broad groups. The overall score for an institution is an average of the percentage of the total points or credits it earns in each category. Thus, a university that earns half (50 percent) of the credits available in all categories has a final overall score of 50. A college that earns 40 credits in education and research, 50 operations credits, and 60 planning administration and engagement credits also has a final score of 50, once those three categories are averaged.

Similar to the evaluations in the Leadership in Energy and Environmental Design (LEED) rating program designed and managed by the U.S. Green Building Council, colleges and universities that participate in STARS may earn a bronze, silver, gold, or platinum rating, based on their final score, with platinum being the highest rating possible. Campuses that wish to participate in STARS without receiving a rating may also earn a STARS Reporter designation. Once assigned, a STARS rating lasts for three years, after which time institutions must resubmit their information.

STARS Categories and Credits

The three STARS categories and the credits contained within their subcategories reflect the fact that colleges and universities must embrace a wide variety of activities and policies across campus in order to pursue a sustainable future. At the same time, AASHE recognizes that institutions of higher education are highly diverse in their capacities to engage in these activities. As the AASHE Technical Manual for the program indicates, credits included in STARS reflect affirmative answers to four questions: (1) does the credit speak to improved environmental, social, and/or financial impacts, (2) is the credit relevant and meaningful for diverse institutions, (3) is the credit based on the presence of a strategy, and (4) is the credit measurable, objective, and actionable? The program includes an additional four possible "innovation credits" for activities that are not included in the three broad categories but that reflect new approaches to sustainability.

Further refinement of STARS credits occurs in the distinction between Tier 1 and Tier 2 activities. The former occur in every subcategory and are worth more points due to their respectively greater impacts on a campus. Tier 2 activities are available only under some subcategories. They reflect steps that, while laudable, may not have as large a measurable result. For example, a campus that has initiated a food procurement program that emphasizes local and organic foods in its dining halls earns more credits for this Tier 1 activity than a campus that offers a discount for reusable mugs, a Tier 2 activity.

With respect to the education and research category of credits, STARS participants examine three subcategories of activities related to cocurricular education, curriculum, and research. Cocurricular education credits reflect activities that colleges and universities offer to engage students in learning about sustainability outside the traditional curriculum. Examples of activities for which schools can earn points under this subcategory include student sustainability outreach campaigns and sustainability content in new student orientation programs. The second subcategory, curriculum credits, reflects sustainability content within formal courses and educational programs. Offering courses that are focused on or related to sustainability is among the ways that STARS participants can gain these types of credits. Third, colleges and universities for which research is a significant component of their mission can earn further credits when that research mission has a sustainability emphasis. Having faculty and/or departments conducting sustainability research are examples of specific credits schools may earn under this subcategory.

The operations category of the STARS program focuses broadly on the ways that institutions of higher education manage the nonacademic functions of their campuses. While the academic mission is an obvious centerpiece of colleges and universities, the policies and practices that support these academic functions have clear, measurable impacts on the environment. This broad focus is evident in the nine subcategories that provide the basis for operations evaluation: buildings, climate, dining services, energy, grounds, purchasing, transportation, waste, and water.

Within these subcategories, the three that contain the highest potential number of credits are buildings, climate, and energy. This emphasis reflects the significant impacts that campus building and energy policies and practices have on use of nonrenewable resources, air pollution, and greenhouse gas emissions. With respect to buildings, campuses earn credits by demonstrating efforts to address operation and maintenance, design and construction, and indoor air quality. One important aspect of building design and maintenance in this section is the application of LEED standards for existing campus buildings, particularly during major renovations, as well as for new construction. In general, institutions must be pursuing green building guidelines or policies to the maximum extent possible to achieve full credits. The credits for climate, on the other hand, center on steps these schools are taking to address greenhouse gas emissions. Following their development of a greenhouse gas inventory that accounts for such emissions, campuses can obtain credits for efforts to reduce them through offsets or other direct measures. Earning full points requires that a campus achieve climate neutrality, which as of early 2010 is an accomplishment of only one U.S. campus, the College of the Atlantic in Bar Harbor, Maine.

Additional credits in the operations category recognize numerous steps campuses have taken to make day-to-day activities more sustainable. These range from use of integrated pest management to efforts to promote bicycling and walking to reductions in water consumption.

The third and final broad category of evaluation within STARS recognizes that sustainability is also a focus in campus planning, administration, and engagement (PAE) activities. Five subcategories recognize accomplishments in the areas of coordination and planning, diversity and affordability, human resources, investment, and public engagement. Similar to the operations category, the PAE credits recognize a wide range of ways that colleges and universities can grapple with long-term sustainability goals. While an environmental focus is still apparent throughout this category, attention also focuses on the social equity aspects of sustainability, through efforts to promote diversity, affordability, and so on. Activities that can earn relatively large numbers of credits include the inclusion of sustainability in strategic campus plans (6 possible credits), providing sustainable compensation for the lowest-paid campus workers (8 possible credits), pursuit of sustainability-focused investments (9 possible points), and the provision of sustainability content in continuing education offerings (7 possible points).

STARS: What Lies Ahead

AASHE's STARS program presents U.S. colleges and universities with a comprehensive lens through which they can evaluate their sustainability progress. The credits contained in this rating program emphasize a full spectrum of sustainability activities. They also recognize that not all campuses are alike with respect to their size, emphases, and other characteristics. As the first institutions begin to complete their STARS assessments in 2010, participants and nonparticipants alike will be able to identify general areas of both strength and omission with respect to campus sustainability. That AASHE has labeled this iteration of its program "STARS 1.0" suggests that future versions are likely.

See Also: Association for the Advancement of Sustainability in Higher Education; Higher Education Associations Sustainability Consortium; Sustainability Officers.

Further Readings

AASHE Sustainability Tracking, Assessment & Rating System (STARS). http://www.aashe
.org/stars/index.php (Accessed January 2010).

Hermann, Michele. "Getting a Star for Sustainability." *University Business*, 11 (2008).

Sustainability Tracking, Assessment & Rating System (STARS) Version 1.0 Early Release
Technical Manual. http://www.aashe.org/files/STARS_1.0_Early_Release_Technical_
Manual.pdf (Accessed January 2010).

Stacey Swearingen White
University of Kansas

AGENDA 21: "PROMOTING EDUCATION, PUBLIC AWARENESS, AND TRAINING" (CHAPTER 36)

Women reading in an adult literacy class in India in 2002. Agenda 21 calls for increasing basic literacy levels, especially among women, as a foundation for improved environmental awareness.

Source: World Bank

The main topic of the United Nations Conference on Environment and Development (UNCED) Earth Summit was an action plan for the 21st century with the aim of sustainability of the planet. It is a reference publication including objectives, activities, and means of publication to be adopted at the local, national, and international levels. It was reviewed in 1997 by the United Nations General Assembly and its implementation was reaffirmed at the World Summit in Johannesburg, 2002. Briefly, Agenda 21 is divided into four sections: social and economic dimensions, conservation of resources, strengthening the role of groups, and means of implementation. Total number of chapters is 40. Chapter 36, "Promoting Education, Public Awareness, and Training" under Section IV, "Means of Implementation," is explained here and is related to all fields of Agenda 21. Chapter 36 of Agenda 21 defines education as a process by which societies reach their fullest potential. It is significant for promoting sustainable development and improving capacities of human beings on environmental issues. Education gives people the environmental and ethical awareness, values and attitudes, skills and behavior needed for sustainable development. Because sustainable development must ultimately involve everyone, access to education must be increased for all children, and adult illiteracy must be reduced. Chapter 36 emphasizes the importance of formal and informal education in making sustainable development central to the planning and conduct of activities in all spheres of life. As such, the chapter focuses on attempting to incorporate environmental training into formal, informal, and cross-disciplinary curricula, particularly public awareness and education programs on environmental issues.

36.1. This chapter includes broad proposals based on the Tbilisi Declaration 1977 and Jomtien World Conference on Education for All, and three main program areas are presented: (1) reorienting education toward sustainable development, (2) increasing public awareness, and (3) promoting training. These areas are explained under four subheadings giving the basis for action, objectives, activities, and financing issues.

Reorienting Education Toward Sustainable Development

Basis for Action

36.3. Education needs to explain not only the physical environment but the socioeconomic environment and human development. Education embraces all the ways in which societies develop values and sustainable lifestyles and global responsibilities for today and for next generations. Education, including formal and informal forms, raising public awareness, and training, is critical to improve environmental and ethical awareness, attitudes, skills, and behavior on sustainable development. To be effective, environmental education should be adapted in all disciplines and use formal and informal methods and effective communication.

Objectives

36.4. To improve sustainable development education, nations should seek to:

a) strive for universal access to basic education at the primary level at least
    80 percent of all girls and boys, through formal and nonformal schooling. Adult illiteracy should be cut to at least half of the 1990 level, and literacy levels of women should be brought into line with those of men;
b) encourage all sectors of society in environmental and development awareness on a worldwide scale;
c) universalize access to environmental and development education from primary school age through adulthood; and
d) promote environment and development concepts in educational programs in a local context, giving emphasis to knowledge and training decision makers at all levels.

Activities

36.5. In accordance with the countries' own needs, a total number of 15 activities is proposed on strategies, cooperation, advisory bodies, networks, and all stakeholders. These are:

a) All countries are urged to prepare national strategies and actions for meeting basic learning needs based on the Jomtien Conference, developing accessible and equal educational policies, and for strengthening international cooperation to decrease economic, social, and gender disparities.

b) Governments should update their policies, needs, costs, and schedules to integrate environment and development with all stakeholders. A review of curricula should be made with a multidisciplinary approach regarding different needs and sensitivities.

c) Countries are encouraged to build participatory advisory coordination committees or roundtables with all stakeholders for environmental education. They should assess community needs and mobilize different groups to implement their own environment.

d) Educational authorities should set up training programs for all educators and officials addressing the nature and methods of environment and development education and making use of them for nongovernmental organizations (NGOs).

e) Authorities should assist activity work plans at schools by involving children in environmental health, safe drinking water, sanitation, food, ecosystems, and ecological and economic impacts of resource use.

f) Educators should use proven educational methods and develop innovative teaching, also recognizing traditional systems of local communities.

g) The United Nations (UN) should review its educational programs in cooperation with UNESCO/UNEP, governments, and NGOs within two years. The UNESCO/ UNEP International Environmental Education Program should collaborate with the UN system, governments, and NGOs, and prepare a program for the needs of all educators. Regional and national authorities should also conduct similar programs to mobilize different sectors.

h) Information exchange by various technologies and capacities should be strengthened within five years. Countries should prepare learning materials and resources according to their needs.

i) Countries should support cross-disciplinary courses open to all university students. Existing regional and university networks should promote sustainable development research built upon new bridges with business and other sectors for technology, know-how, and exchange.

j) Countries could establish centers of excellence with an interdisciplinary approach; global institutions, universities, and existing networks could assist these centers.

k) Nonformal education is also a key instrument for delivering information and tapping new networks for global educational aims of the UN. Public and scholastic forums can also promote discussions and make suggestions on environmental issues and assist decision makers at local and national levels.

l) Educational authorities with relevant organizations should provide adult programs and encourage business, industrial, and agricultural schools for developing their curricula. Specific courses at the post-graduate level can be useful for training of decision makers as well.

m) Governments should foster opportunities for women to eliminate gender stereotyping in curricula. They should improve women's enrollment to programs as students or instructors. Priority should be given to young females.

n) Governments should bring indigenous peoples' rights, experience, and understanding of sustainable development into education and training.

o) The UN could monitor and evaluate decisions of UNCED on education and awareness through its agencies. It should disseminate decisions, ensure implementation, and review them through conferences and events.

Means of Implementation; Financing and Cost Evaluation

36.6. The conference secretariat has estimated the average total cost of $8 billion to implement this program. Half of this cost comes from the international community on grant or concessional terms. Actual costs depend upon specific programs of governments.

36.7. More support is needed for the governments through these measures such as:

a) giving priority to related sectors in budget allocations;
b) shifting allocations within existing budgets of primary education;
c) encouraging rich communities to assist poorer ones;
d) seeking private donors to shift their funds to poorest countries where rates of literacy are below 40 percent;
e) encouraging debt for education swaps;
f) increasing funds to private schools and NGOs;
g) promoting existing multi-school shifts, open universities, and e-learning;
h) using low-cost or no-cost mass media; and
i) setting up university twinning between developed and developing countries.

Increasing Public Awareness

Basis for Action

36.8. There is still a lack of awareness due to inaccurate or insufficient information about environmental issues. There is a need to foster environmental responsibility and commitment toward sustainable development.

Objective

36.9. The objective is to increase public awareness, including heightening sensitiveness and involvement, and devolving authority in decision making.

Activities

36.10. Actions include the following:

a) Countries should establish advisory bodies and should coordinate activities with the UN, NGOs, and media. They should encourage public discussions and networking on environmental policies.
b) The UN system should promote participation in public awareness activities and coordinate its information bodies and regional/country operations. For understanding the needs of communities, specific surveys should be made.
c) Countries should provide information services for raising awareness of all sectors.
d) Countries should support educational establishments and materials based on aesthetic and ethical dimensions of all sciences.
e) Countries and the UN system should work with the media, theater groups, and entertainment and advertising industries to promote a more active public debate on the environment. UNICEF should provide material to media for children and UNESCO/UNEP, and universities can give pre-service curricula for journalists.

f) Countries should use modern communication technologies, TV/radio programs in mobile units in rural parts involving interactive techniques.

g) Countries should encourage tourism programs of WTO/UNEP direct to museums, national parks, protected areas, etc.

h) NGOs should be supported for their high involvement and joint activities to highlight environmental problems.

i) Countries and the UN system should develop policies that emphasize listening and learning from indigenous people. Their values, culture, and sustainable living on Earth should be disseminated.

j) Young people and children should be involved in environmental programs developed by UNESCO/UNICEF/UNDP and NGOs.

k) Countries, the UN, and NGOs should encourage family activities in campaigns stressing women's contribution to transmission of social values.

l) Public awareness should be heightened regarding the impacts of violence in society.

Means of Implementation; Financing and Cost Evaluation

36.11. The conference secretariat has estimated the average total cost of $1.2 billion to implement this program between 1993 and 2000. This includes $110 million from the international community on grant or concessional terms. Actual costs depend upon specific programs of governments.

Promote Training

Basis for Action

36.12. Training is critical for promoting the transition to a more sustainable world. It is an important tool to improve skills to find employment and it is a two-way learning process to address environmental and development issues.

Objectives

36.13. The following four objectives are proposed:

a) to set up or improve vocational training programs on environment regardless of social status, age, gender, race, or religion;

b) to promote a flexible and adaptable workforce of various ages for the transition to a sustainable society;

c) to strengthen national capacities in training for employees and workers to meet their objectives and to facilitate the transfer of eco-technologies and know-how; and

d) to integrate environmental and ecological issues at all management areas of marketing, production, and finance.

Activities

36.13. The following 13 activities are proposed for training:

36.14. Countries should identify workforce training needs and assess measures to be taken to meet those needs.

36.15. National professional associations should set up their codes of ethics. They should sponsor programs on incorporation of skills and information on the implementation of sustainable development at policy and decision-making levels.

36.16. Countries should adapt environmental issues by exchanging methodologies in their training curricula.

36.17. Countries should encourage all sectors such as universities, government officials and employees, industry, and NGOs to train people in environmental management. Also training of trainers, new training approaches, and local resources are important.

36.18. Countries should set up training programs for school and university graduates to help them achieve sustainable livelihoods.

36.19. Governments are encouraged to consult geographically, culturally, or socially isolated people to enable them to contribute to sustainable lifestyles.

36.20. Government, trade unions, industry, and consumers should interrelate between good environment and good business practices.

36.21. Countries should provide locally trained and recruited environmental technicians to give local communities the service they need, starting with primary environmental care.

36.22. Special training programs should be given to special groups, and their impact on productivity, health, safety, and employment should be evaluated. Labor-market database and training guides should be created at all levels.

36.23. Aid agencies should assist the training programs.

36.24. Existing networks on the labor market, NGOs, and the business world should exchange the experience of training programs.

36.25. Governments should make preparedness plans against environmental threats and emergencies at all levels and give urgent practical training programs.

36.26. The UN system should extend environmental training and support for employers' and workers' organizations.

Means of Implementation; Financing and Cost Evaluation

36.27. The conference secretariat has estimated the average total cost of $5 billion to implement this program between 1993 and 2000. This includes $2 billion from the international community on grant or concessional terms. Actual costs depend upon specific programs of governments.

Chapter 36 of Agenda 21 lays out very broad proposals related to education. However, while the chapter emphasizes environmental and developmental education, it does not specify the particular educational resources and curriculum needed to attain sustainable development. Other chapters dealing with capacity building, human resources development, and changing consumer patterns help to reinforce the chapter's broad educational objectives. Taken as a whole, Agenda 21 provides the public with a strong understanding of the concepts behind environmental education. It is unfortunate that after 25 years, Agenda 21 still makes similar proposals as the 1977 Tbilisi Declaration, upon which it is based.

Although education, awareness, and training are the key instruments for a sustainable future, these are at risk of being a forgotten priority in financial and political agendas. Important efforts are usually focused on formal education, but employees, women, consumers, etc., are often forgotten. Adequate financial commitment by government and multilateral aid agencies continues to be challenging. Even with universal stakeholder engagement, ownership and action in educational initiatives are sometimes lacking.

It is a challenge for society to be sustainable. Education, public awareness, and training related to sustainable development gain more importance for society to be a transition culture. Many people, authorities, and all sectors should understand the close ties between human activities and the environment. Sustainable development education should offer not only the concepts and values but also the practical social actions for behavioral changes. This education for equitable sustainability is a continuous learning process based on respect for all life in interdependence and diversity. It requires individual and collective responsibility at the local, national, and global levels.

See Also: Brundtland Report; Millennium Development Goals; Tbilisi Declaration; United Nations Decade of Education for Sustainable Development 2005–2014.

Further Readings

Final Report of the World Conference on Education for All: Meeting Basic Learning Needs, Jomtien, Thailand, March 5–9, 1990 (New York, Inter-Agency Commission [UNDP, UNESCO, UNICEF, World Bank] for the World Conference on Education for All, 1990).

Goncz, Elzbieta. "Universities and Their Role in Enhancing Human and Social Capital for Sustainability." *Committing Universities to Sustainable Development Conference Proceedings* (2005).

Gorobets, Alexander. "An Eco-centric Approach to Sustainable Community Development." *Community Development Journal*, 41/1 (2006).

International Union for Conservation of Nature, The World Conservation Union Commission on Education and Communication. "Education: The Forgotten Priority of Rio?" Statement for Rio+5 on Chapter 36 Agenda 21, 1997.

Mader, Clemens. "UN Decade of Education for Sustainable Development—Students' Role and Contribution to Make a Difference." Committing Universities to Sustainable Development Conference Proceedings, 2005.

Projet de Société: Canada and Agenda 21.Winnipeg: International Institute for Sustainable Development (IISD), 1995. http://iisd.ca/worldsd/canada/projet/c36.htm (Accessed January 2010).

United Nations Educational, Scientific and Cultural Organization (UNESCO). "Educating for a Sustainable Future: A Trans-Disciplinary Vision for Concerted Action." EPD 97/CONF .401/CLD.l, 1997.

Ozge Yalciner Ercoskun
Gazi University

ALLEGHENY COLLEGE

Allegheny College, located in Meadville, Pennsylvania, is a private liberal arts college that enrolls approximately 2,100 undergraduates who are served by 162 faculty members. Allegheny College has emphasized sustainable practices, including outreach to its surrounding community, for decades. The environmentally friendly practices adopted by Allegheny College are many. In addition to the board of trustees adopting environmental guiding principles, several programs have been created on campus to impact the surrounding environment in positive ways. Many of these programs are operated through the Center for Economic and Environmental Development (CEED) founded in 1997 to coordinate projects that both faculty and students were implementing to address local and regional sustainability issues. CEED draws upon campus leadership to enhance the success of projects that address community challenges that have faced Meadville, Crawford County, and northwestern Pennsylvania for years. CEED is governed by a director, an executive committee, and a national advisory board. Several current projects are exemplary indicators of successful outreach to the community.

Greening the Gateway

One of the first projects undertaken by CEED was Greening the Gateway. This was a collaborative effort with the Pennsylvania Department of Transportation (PennDOT). A student competition was held for landscaping the area of highway coming into the town of Meadville with native flora. The winning student submission was then implemented, with volunteers and PennDOT workers digging and planting. It was discovered that salt from the PennDOT trucks used in winter was harmful to some of the shrubbery. Burlap bags were used successfully the next year to help maintain the health of the plants in the area. This early project quickly established Allegheny College as a college that was dedicated to and forward thinking about sustainability outreach issues.

Community-Focused Initiatives

CEED oversees multiple projects that have been implemented to support the local and regional communities of northwest Pennsylvania. The Local Foods Network (LFN) initiative connects local producers, local markets, and consumers. The LFN has established a relationship with the Meadville Area Local Growers (MALG) to hold farmers markets in Meadville. In addition, CEED has hosted events on campus to connect local producers to consumers in the community and on campus. The Taste the Bounty of Crawford County Dinner has provided college and local residents with connections to greater knowledge of the produce and foodstuffs available locally while building closer community connections and prosperity for local farmers and producers.

The Arts and Environments Initiative uses visual art projects to connect with the greater Meadville community. The Read Between the Signs (RBTS) project is supported by Allegheny College, the Pennsylvania Department of Transportation, and the Mid-Atlantic Arts Foundation. A sculptural fence created from discarded road signs covers the PennDOT storage lot that marks the entrance to Meadville. The sculpture portrays images of the Allegheny Mountains, the French Creek watershed, and surrounding landscapes of PennDOT workers, farms, and forests. This recycling project not only reuses discarded materials but states very clearly the community and college commitment to outreach with the sign placed at the entrance to town. Another project, Signs and Flowers, made it possible to construct large visual arts sculptures to beautify the PennDOT grounds and increase the presence of PennDOT in the community. Students from Allegheny College worked with PennDOT welders and crane operators to sculpt large creations that portrayed how humans impact the planet. The materials used were old road signs. Again a lasting visual presence was created in partnership with the Meadville community to provide a project created from used materials that acts as a reminder to all about how humans impact the environment.

Allegheny College also provides outreach to the Meadville community through the Green Room, which promotes sustainability. For the Green Room, Allegheny College collaborated with the Crawford County Industrial Park (CCIP). The project was supported by the CCIP, Allegheny College, and the Heinz Family Endowment. The CCIP is a restored industrial center and, after the restoration, a lobby was created as an entrance to the Industrial Park. To beautify and to reach out to this new structure in the community, student interns painted a mural that depicts the history of the site. The work went further to create a "green" break room. Tables and chairs were created with used materials when possible and an additional photo essay of the history of the park is displayed to link the importance of the park to the community.

Ecotourism Initiatives

Another project the CEED of Allegheny College has been able to implement is an ecotourism plan developed with the Crawford County Convention and Visitors Bureau. Allegheny College is situated in northwest Pennsylvania, 45 miles south of Erie, Pennsylvania, and 100 miles north of Pittsburgh, Pennsylvania. Being located less than a two-hour drive to these larger cities offers tourists a chance to spend even just one day with access to extensive areas of forest, farmland, lakes, and streams. In conjunction with its CEED, Allegheny College hosts a Website that describes in detail the resources the area has to offer tourists. In addition, information is available with one easy click on the option of touring the area in a car or in an organized bus tour. Convention and visitors bureaus have limited resources in both staffing and access to technology for continuing to provide updated information to the public. By maintaining updated information for tourists, Allegheny College is providing outreach to educate regional citizens about the flora and fauna in the area as well as providing an additional component of developing tourism for the Crawford County Convention and Visitors Bureau.

Ecotourism depends in part on an inviting and sustainable natural environment. To that end, the CEED at Allegheny College has entered into a community partnership with the Pennsylvania Sustainable Forestry Project. Again Allegheny is reaching out into the community to partner with local landowners and forestry professionals to ensure sustainability of the local environment. The project works in the community with landowners and other professionals to ensure that Forest Stewardship Council (FSC) standards are met. These standards work to educate and train citizens on how to manage socially and environmentally sound products that help sustain forests.

Commercial Initiatives

Sustainable practices can have a positive effect on commercial endeavors, supplying jobs and profits that assist a community to grow and to prosper. Allegheny College supports such economic growth through the Sustainable Manufacturing Initiative (SMI). Meadville is a town that depends on manufacturing for jobs. The SMI is a program in which faculty and students work to help manufacturers today meet the demands of the global economy. Manufacturers must meet standards for sustainability, but Allegheny College through CEED believes these are only minimal standards, and going beyond the minimums is a responsibility the college shares with its community.

One key community outreach component of the CEED at Allegheny College is its sustainable communities project, Meadville, PA: Not Your Run of the Mill Community. Several key aspects of this initiative are being developed to help with environmentally responsible economic development of the surrounding community. One component is the Main Street Program Grant Proposal. CEED developed an application to the Pennsylvania Department of Community and Economic Development for Main Street Program funding for the City of Meadville with input from community volunteers and the Meadville Redevelopment Authority (RDA). The funding will allow for a "Main Street manager" to help revitalize downtown businesses. Allegheny College has also provided support with a student intern to help with the revitalization efforts. Yet another section of the project is the City of Meadville Biodiesel Project. Students from Allegheny College focused a course project on developing a plan for City of Meadville vehicles to run on waste oil collected from local restaurants. The city council heard and voted to approve setting up a facility to convert the cooking oil collected from restaurants into biofuel. The biodiesel reactor is now converting fuel for part of the fleet of city vehicles. One other project is under development that artfully illustrates the exemplar outreach Allegheny College has through the CEED to develop sustainability projects in the community. A resource of Web links, video, audio, and print material is being collected to share with Allegheny College faculty and with local teachers in the Crawford Central School District. These materials will help faculty and teachers integrate environmental issues into the curriculum without causing disruption to the curriculum that must be covered.

Conclusion

Allegheny College proves that a small liberal arts college can have a tremendous influence on fostering sustainability. Through initiatives that focus on campus practices, community beautification, ecotourism, and sustainable commercial enterprises, Allegheny College has greatly influenced local, regional, and national perceptions of the value of green practices. Such practices provide students and faculty with meaningful learning experiences while also greatly benefitting the college, the community, and the region.

See Also: Education, Local Green Initiatives; Green Community-Based Learning; Social Learning.

Further Readings

Arnaud, B. S., L. Smarr, J. Sheehan, and T. DeFanti. "Campuses as Living Laboratories for the Greener." *Educause Review,* 44/6 (2009).

Elder, J. L. "Higher Education and the Clean Energy, Green Economy." *Educause Review,* 44/6 (2009).

Sherren, K. "A History of the Future of Higher Education for Sustainable Development." *Environmental Education Research,* 14/3 (2008).

Spencer, R. "Composting Helps Anchor University's Climate Commitment." *Biocycle,* 49/5 (2008).

"Sustainability and Accreditation." *University Business,* 12/5 (2009).

Stephen T. Schroth
Jason A. Helfer
Victoria M. Romano
Knox College

AMERICAN COLLEGE AND UNIVERSITY PRESIDENTS CLIMATE COMMITMENT

Arizona State University, one of the 660 signatories of the American College and University Presidents Climate Commitment, installed solar panels over this parking lot to increase the university's use of renewable energy.

Source: Wikipedia/Kevin Dooley

The American College and University Presidents Climate Commitment (ACUPCC) engages American institutions of higher education in efforts to address and to respond to climate change. Signatories to the ACUPCC pledge to develop comprehensive plans for their campuses to achieve no net greenhouse gas emissions. Working from this shared goal of climate neutrality, colleges and universities are free to develop plans and implementation timelines that best suit their individual institutional needs. As of early 2010, over 660 presidents and chancellors from institutions across the United States had signed the ACUPCC. While this number falls short of the stated goal of 1,000 signatories by 2009, it represents 35 percent of all American college students in all 50 states and reflects a significant trend in the level of attention climate change is receiving at the college and university level.

The ACUPCC developed out of conversations by college and university leaders concerned that higher education had a vital role to play in climate change action, research, and education. Sponsored by three organizations—Second Nature, ecoAmerica, and the Association for the Advancement of Sustainability in Higher Education (AASHE)—the ACUPCC emerged in 2006 with an initial commitment from 12 college and university presidents. Second Nature has become the lead support organization for the Commitment, while a steering committee of 22 leaders from different institutions conducts governance and fund-raising tasks. AASHE hosts the online reporting system on which signatories must post their progress.

Early institutional commitments to the ACUPCC were noteworthy in the breadth of the institutions they represented. Nearly 400 schools from 46 states had signed the Commitment in its charter signatory period, which ended on September 15, 2007. These schools represented the broad spectrum of higher education in the United States and included small and large, public and private, and research- and teaching-oriented institutions. The total number of charter ACUPCC signatories is larger than the number of worldwide signatories to the 1990 Talloires Declaration. Once the president or chancellor of a college or university signs the ACUPCC, that institution must conduct a greenhouse gas inventory of all campus emissions, with a target deadline of one year from the signing. Emissions calculators, such as the online tool developed by the organization Clean Air-Cool Planet, enable institutions to examine major sources of emissions, including purchased electricity, energy generated on campus, transportation, waste, agricultural practices, and landscaping. Among the broad institutional factors that contribute to comparatively larger greenhouse gas emissions are gross building area, the climate in which the institution is located, and, because of their larger research apparatus, doctorate-granting universities.

The results of these inventories inform the development of the institution's action plan for climate neutrality, which is to be finished two years following the inventory's completion. These plans include target dates and measures, descriptions of mechanisms to track progress, and evidence of steps to expand curricula and research emphases on climate change and sustainability. Additional requirements of the Commitment include the development of policies and programs that will address issues such as building standards, purchasing, transportation, and waste minimization. Upon completion of these action plans, institutions provide periodic progress reports to the ACUPCC.

In December 2007, the College of the Atlantic in Bar Harbor, Maine, became the first signatory institution to achieve its goal of climate neutrality. This small college, with just over 300 undergraduate students, was able to reach net zero emissions through steps such as relying on hydropower and a new wind turbine for its electricity, improving the efficiency of its building design standards, and relying on offsets for carbon emitted in certain campus travel activities. Other, larger institutions are likely to have even greater difficulties bringing their net emissions to zero, although several doctorate-granting institutions had completed their climate action plans by the end of 2009, including Arizona State University, Brandeis University, Cornell University, and the University of Florida. Future years will demonstrate the challenges and opportunities these and other signatory schools will face in reaching the goals of the ACUPCC.

See Also: Arizona State University; College of the Atlantic; Second Nature: Education for Sustainability.

Further Readings

American College and University Presidents Climate Commitment. http://www
.presidentsclimatecommitment.org (Accessed January 2010).

Fetcher, Ned. "Effects of Climate and Institution Size on Greenhouse Gas Emissions From
Colleges and Universities in the United States." *Sustainability: The Journal of Record*, 2
(2010).

Naditz, Alan. "In the Green: Making the Commitment." *Sustainability: The Journal of
Record*, 2 (2010).

Swearingen White, Stacey. "Early Participation in the American College and University
Presidents Climate Commitment." *International Journal of Sustainability in Higher
Education*, 10 (2009).

Stacey Swearingen White
University of Kansas

AMERICAN PASTORS' CREATION CARE COVENANT

Care for the environment among Christians in the United States, although contested, is a growing area of emphasis with a history of significant denominational and ecumenical statements and agreements. Most commonly called creation care, environmental stewardship in Christian circles is also known as missionary Earth keeping, ecojustice, ecomissiology, and evangelical environmentalism. The Website of the National Religious Partnership for the Environment (NRPE), an association of independent faith groups, including the Evangelical Environmental Network (EEN), the U.S. Conference of Catholic Bishops, the National Council of Churches (NCC), and the Coalition on the Environment and Jewish Life, hosts a robust collection of statements from many religious denominations and organizations. Such statements voice denominational goals to carry out certain programs or adopt policy positions concerning ecological issues. Additionally, they help to reorient members' attention to creation care as a religious responsibility.

Declarations and Initiatives

American mainline Protestant denominations have been making resolutions concerning the environment since at least 1954 when the Presbyterian Church (U.S.A.) drafted a resolution on natural resources. Since then, mainline denominations have produced numerous statements concerning environmental degradation on issues such as energy, toxic waste, consumption, agriculture, climate change, biotechnology, and other issues. Environmental stewardship did not begin to gain momentum until the limits-to-growth debate in the 1970s, which grew from pessimism of earlier confidence with progress in science and technology. Since then, mainline Protestant statements and declarations have included Perspectives on Christian Lifestyle and Ecology from the United Church of Christ (1975), Care for the Earth: Theology and Practice from the Reformed Church of America (1982), Resolution Concerning an Ecologically Responsible Lifestyle from the Disciples of Christ (1983), Policy Statement on Ecology from the American Baptist Churches (1989),

Statement on Environment from the Moravian Church (1990), Restoring Creation for Ecology and Justice from the Presbyterian Church (U.S.A.), Creation: Called to Care from the Church of the Brethren (1991), Affirm Environmental Responsibility and Establish an Environmental Stewardship Team from the Episcopal Church (1991), Environmental Justice for a Sustainable Future from the United Methodist Church (1992), Declaration on Environmental and Economic Justice from the Black Church (1993), Caring for Creation: Vision, Hope, and Justice from the Evangelical Lutheran Church in America (1993), and Walking Gently on the Earth from the Religious Society of Friends (1998). Additionally, the NCC, an umbrella organization including most mainline denominations, has drafted numerous statements on the environment.

The Roman Catholic Church has also been active in promoting concern for the welfare of the Earth. Roman Catholic teaching has historically encouraged the fostering of respect for human dignity, justice for the poor and marginalized, regard for the common good, and sustainable economic development. In light of these principles, Catholics encourage adherents to take an active role in promoting environmental justice in their parishes and communities. The two major statements that serve as the foundation for the Environmental Justice Program of the United States Conference of Catholic Bishops (USCCB) are Pope John Paul II's World Day of Peace (1990) and USCCB's 1991 pastoral letter Renewing the Earth. Additionally, popes have spoken on the environment often, as in John Paul II's 2001 General Address God Made Man the Steward of Creation.

For many decades, Evangelical denominations have also been active in promoting environmental sustainability, a fact evident in early resolutions from the National Association of Evangelicals (NAE) on Ecology (1970) and Environment and Ecology (1971). The primary statement representing Evangelical churches is the Evangelical Declaration of the Care of Creation (1993), which was signed by over 300 Evangelical leaders. Spearheaded by organizations such as the Au Sable Institute for Environmental Studies and the EEN, the declaration became the impetus for many creation care efforts. More recently, NAE and EEN sponsored a retreat for Evangelical leaders at Sandy Cove, Maryland, that led to signing the Sandy Cove Covenant (2004). Other Evangelical statements have included A Christian Land Ethic (1987), Message to Individuals and Churches (1990), Stewardship: All for God's Glory (1990), and Global Stewardship: The Christian Mandate (1997). Most recently, 86 leading U.S. Evangelical Christian leaders launched the Evangelical Climate Initiative (2006), a campaign for environmental reform calling on all Christians to advocate for federal legislation to limit carbon dioxide emissions and curtail global warming, and in March 2008, Southern Baptist leaders issued a similar statement confessing that their denomination had not done enough to slow global warming.

Contestation

Despite the significant statements and resolutions highlighted above, contestation remains regarding creation care, particularly among Evangelicals. For instance, in Evangelical circles an alternative to the Declaration for Creation Care called the Cornwall Declaration (2000), sponsored by the Cornwall Alliance, critiques the Declaration for Creation Care for viewing humans principally as consumers and polluters rather than as producers and stewards. The major issue of contestation revolves around the science of climate change. Following the resignation of climate change activist Richard Cizik as vice president of the

NAE, the Cornwall Alliance celebrated a statement by NAE President Leith Anderson asserting that the NAE has no formal position on climate change. Other Evangelicals reject the creation care movement altogether, arguing that God gave humans complete dominion over the Earth until the point at which Jesus will return and destroy it completely; therefore, humans can treat the Earth in any manner they desire. Hence, the creation care movement among Christians should not be interpreted as homogenous.

In spite of this contestation, many new denominational and parachurch organizations have recently ushered in support for creation care. For instance, Renewal: Students Caring for Creation is a growing movement of Christian college students dedicated to mobilizing and equipping their campuses to better steward the environment. Renewal recently released the "Green Awakenings Campus Report" chronicling environmental initiatives on 52 Christian college campuses across the United States and Canada. Other organizations include A Rocha, Au Sable, Blessed Earth, Catholic Conservation Center, Christian Environmental Association, Echo, Eco-Justice Ministries, Flourish, National Council of Churches Eco-Justice Working Group, National Religious Partnership for the Environment, North American Coalition for Christianity and Ecology, Plant with a Purpose (formerly Floresta), Quaker Earthcare Witness, Restoring Eden, Target Earth, the Regeneration Project, and Web of Creation.

See Also: Brundtland Report; Environmental Education Debate; Sustainability Websites.

Further Readings

Berry, R. J., ed. *The Care of Creation: Focusing Concern and Action.* Downers Grove, IL: InterVarsity Press, 2000.

DeWitt, Calvin B. and Ghillean T. Prance, eds. *Missionary Earthkeeping.* Macon, GA: Mercer University Press, 1992.

Ellingsen, Mark. *The Cutting Edge: How Churches Speak on Social Issues.* Strasbourg, France: Institute for Ecumenical Research, 1993.

Lowe, Ben, ed. *Green Awakenings: Students Caring for Creation.* http://www.renewing creation.org (Accessed February 2010).

Lowe, Ben. *Green Revolution: Coming Together to Care for Creation.* Downers Grove, IL: InterVarsity Press, 2009.

National Religious Partnership for the Environment. http://www.nrpe.org (Accessed February 2010).

Michael P. Ferber
The King's University College

AQUINAS COLLEGE

Aquinas College is a private undergraduate and graduate university located in the city of Grand Rapids in Kent County, Michigan. The Dominican Sisters of Grand Rapids founded Aquinas in 1886. Aquinas has a current enrollment of approximately 2,300 undergraduate and graduate students in 61 majors. Aquinas College is renowned for its Sustainable Business curriculum combining formal business training and environmental studies, implementing the nation's first undergraduate degree program in the field in 2003 and expanding it to the master's level in 2010. Aquinas also offers undergraduate degrees in Environmental Studies and Environmental Science. The sustainability curriculum is further enhanced through student groups such as Students Striving for Sustainability and student involvement in campus sustainability programs such as the Center for Sustainability. Aquinas is a member of the North Central Association and is accredited by the Higher Learning Commission.

Academic Degree Programs

Aquinas College is a leader in the field of sustainable business education. The school's Liberal Arts College introduced the nation's first undergraduate bachelor of science (B.S.) degree program in Sustainable Business in 2003. The degree combines traditional business courses with natural sciences courses in order to provide an understanding of the function of ecosystems and how to apply that knowledge in order to promote ecological and social awareness in the business decision-making process. There is an emphasis on the theories and practices necessary for businesses to achieve what is known as the triple bottom line: a combination of economic, environmental, and social capital.

The four-year curriculum is rooted in the theories of natural sciences writer Janine Benyus and ecologically conscious architect William McDonough. Benyus pioneered the field of biomimicry, which promotes the emulation of nature to solve human problems. It consists of a set of required courses in the four major areas of concentration: business, science, environmental studies, and sustainable business. Sustainable business students are also required to complete an internship with a local business through the Sustainable Business Innovations Lab in order to gain real-world experience in the development of sustainable business practices. Undergraduates may also minor or pursue a certificate in sustainable business. Sustainable business graduates will have the knowledge to aid businesses in their efforts to reduce negative environmental and community impacts while increasing long-term profitability.

Aquinas also offers undergraduate bachelor of science (B.S.) and bachelor of arts (B.A.) degree programs in Environmental Studies or Environmental Science. These interdisciplinary programs combine the fields of biology, chemistry, geology, mathematics, and physics in order to provide students with an understanding of the functioning of ecosystems. Students choose whether to concentrate in science in pursuit of a B.S. or to concentrate in human and cultural elements in pursuit of a B.A. Undergraduate Environmental Science majors also have the option of pursuing a double major in related fields such as biology, chemistry, geography, or sociology.

Required core courses include such topics as environmental studies, Earth environments, environmental biology, environmental chemistry, chemistry and society, human geography, cultural anthropology, applications of remote sensing and geographic information systems, and industrial ecology. Elective courses include special topics in environmental studies, environmental business management, geography of natural resources, and geography of water resources as well as field experience, readings, and independent projects.

The Liberal Arts College extended its Sustainable Business program with the introduction of a Master of Sustainable Business (M.S.B.) in 2010. This interdisciplinary master's degree will combine the fields of science, business, and environmental studies. The program consists of six conventional business courses offered in the Liberal Arts College's Master of Management program and six sustainable business courses. Graduate students in the Master of Management program have the option of pursuing a Sustainable Business concentration.

Student Organizations and Activities

Students may also gain environmental education and experience through a variety of clubs and organizations, both on and off campus, many of which have close ties to the academic curriculum. The Social Action Committee (SAC) promotes awareness and activism with regard to social injustices, many of which have environmental and sustainability connections such as nuclear power and world hunger. The SAC sponsors films, speakers, discussions, and rallies. The GEO Club centers on environmental and geographic issues and sponsors speakers, volunteer events, and field trips, and provides information for undergraduates interested in entering graduate school.

Students interested in sustainability and the environment run the Center for Sustainability at Aquinas College (C4S) under faculty direction. C4S was founded in 2005 out of the college's Sustainable Business curriculum focus. Associate Professor of Sustainable Business Deborah M. Steketee serves as the center's executive director. Students majoring in Sustainable Business may join Students Striving for Sustainability (S3), which plays a major role in C4S projects. S3 focuses on the proactive application of business, social, and environmental initiatives and the application of the sustainability paradigm to business through the development of organizations, communities, and activities. In addition to S3, students are involved in C4S projects through internships and employment opportunities.

C4S projects have included a full inventory of current Aquinas processes, practices, and curricula in the areas of social, environmental, and financial capital. The inventory was embedded into the sustainable curriculum through class assignments. C4S also maintains a comprehensive bibliography and collection of resource materials related to its sustainability mission and maintains an informative Website for use by consumers, nonprofit agencies, businesses, and governmental agencies as well as students. The center also hosts conferences and workshops.

Students interested in environmental issues are also active in many off-campus organizations and activities that allow them to put their educational foundation into real-world practice. A campus-based Habitat for Humanity group works with its Kent County counterparts to raise funds to build homes for residents who would otherwise not have access to home ownership. Students also participate in the EPIC organization, which works with the Grand Rapids community and local politicians for green energy and other environmental and sustainability issues. EPIC also sends students to lobby for green energy in Washington, D.C.

Campus Sustainability

Aquinas College's Sustainable Business and Environmental Science curriculum is also enhanced by administration efforts toward campus sustainability, which serve as student models of the implementation of sustainable practices in a real-world setting. Climate-friendly initiatives include the 2008 signing of the American College and University Presidents Climate Commitment and completion of a comprehensive, campus-wide greenhouse gas inventory in order to target future improvements. The Grace Hauenstein Library was the first building on campus to receive LEED Silver certification from the U.S. Green Building Council.

The school, its head chef, and its food-source companies are working to provide more locally grown and organic food on campus, to examine the sources and practices used at farms growing food served at the college, and to conserve resources through programs such as "Trayless Tuesdays" and controlled-use paper towel dispensers. Aquinas is joining 10 Grand Rapids city businesses, nongovernmental groups, and governmental organizations in a pilot project to test the feasibility of developing a full-scale urban organic waste composting facility in a city eco-industrial park.

Students actively participate in such campus sustainability efforts. The college administration actively seeks student as well as faculty and staff sustainability proposals for new green programs through the 2006 Campus Sustainability Initiative, which created three standing sustainability committees and a joint committee created by the three college governing bodies. The Sustainability Initiative will allow for greater coordination of sustainability efforts among staff, faculty, and students. Teams of Sustainable Business students work to develop sustainability suggestions for review by the provost. Examples of student projects in the process of implementation include a community bike system, an electronic waste recycling program, and the purchase of environmentally friendly cleaning products.

See Also: Green Business Education; Sustainability Officers; Sustainability Websites.

Further Readings

Barlett, Peggy F. and Geoffrey W. Chase. *Sustainability on Campus: Stories and Strategies for Change.* Cambridge, MA: MIT Press, 2004.

Maniates, Michael F. and John C. Whissel. "Environmental Studies: The Sky Is Not Falling." *BioScience,* 50/6 (2000).

M'Gonigle, R. Michael. *Planet U: Sustaining the World, Reinventing the University.* Gabriola Island, British Columbia, Canada: New Society Publishers, 2006.

Nance, Molly. "Converging Interests: Environmental and Sustainability Issues Have Permeated Every Level of Academia, From Art Instruction to Business Courses." *Diverse Issues in Higher Education,* 26/17 (2009).

Rappaport, Ann and Sarah Hammond Creighton. *Degrees That Matter: Climate Change and the University.* Cambridge, MA: MIT Press, 2007.

Rogers, Stephanie. "Green College Spotlight: Aquinas College in Grand Rapids, Michigan." EarthFirst. http://earthfirst.com/green-college-spotlight-aquinas-college-in-grand-rapids-michigan (Accessed June 2010).

Marcella Bush Trevino
Barry University

Arizona State University

Arizona State University's School of Sustainability, the first of its kind in the United States, offers both undergraduate and graduate degrees in sustainability. It is housed in this building, which generates energy from small wind turbines on the roof and from solar panels.

Source: Flickr/Kevin Dooley

Arizona State University (ASU), founded in 1885, is a public, state-controlled institution of higher education with campuses in Tempe, Phoenix, Glendale, and Mesa, Arizona. ASU is currently the largest public research university in the United States, with a Fall 2009 enrollment listed at 68,604 students under the instruction of over 2,800 faculty. Over the past decade, under the leadership of President Dr. Michael M. Crow, ASU has positioned itself at the forefront of teaching and research in sustainability, helping to train the next generation of leaders to help society become more sustainable. The university model envisioned by Crow is termed *The New American University,* with a focus on decision making for real-world solutions, including sustainability. This is perhaps most evident in examining ASU's approach to education in sustainability, spearheaded by the Global Institute of Sustainability, the School of Sustainability, and the Biodesign Institute.

Global Institute of Sustainability

With a focus on the urban environment, the Global Institute of Sustainability, established in 2004, is a product of ASU's dedication to environmental research. Focusing on sustainability research, education, and problem solving, and using the Phoenix metropolitan area as a laboratory, the institute brings together multidisciplinary teams from across the university as well as other institutions of higher education, K–12 schools, government agencies, and nonprofit organizations. In terms of research interests, the institute focuses on eight broad categories that bring together faculty, researchers, students, and external partners: economy and society; climate; energy, materials, technology; water; governance and policy; biodiversity and habitats; urbanization; and international development. A key tool featured at ASU and utilized by the institute for developing real-world solutions is the Decision Theater, which helps provide visualization to collaborative, complex, decision-making exercises surrounded by seven large connected screens. This environment allows for simulation and modeling in related sustainability topics from nanotechnology to urban planning.

The Global Institute of Sustainability also oversees sustainability initiatives on campus, including buildings and grounds, purchasing, transportation, and food service—all with the intent of reaching ASU's ultimate goal of climate neutrality. By far, however, ASU's dedication to curriculum development and green education has made the university an example of what sustainability education can become.

The Center for Sustainable Materials and Renewable Technology (SMART), located within the Global Institute and recognized as a National Center of Excellence, looks at sustainable systems—materials, climate, and energy—and works on innovations for business, industry (technology and materials), and policy makers. In addition to bringing together multidisciplinary researchers and students from across campus, SMART also works with industry partners such as Ford, AT&T, the United Nations, multiple U.S. federal agencies, and the cities of London, Beijing, Phoenix, and Chicago.

School of Sustainability

The School of Sustainability was established in 2007 to provide a degree-granting outlet to the Global Institute's transdisciplinary mission, bringing together economy, society, and science to address sustainability issues. With the connections already established by the institute, students in the various degree programs within the School of Sustainability are given the opportunity right from the start of their programs to take the classroom into the real world, working with schools, businesses, nonprofits, and government agencies to develop the next generation of sustainability leaders.

Under the leadership of over 60 faculty from more than 40 disciplines, the school offers degrees at all three levels of post-secondary education: bachelor's (B.S. and B.A.), master's (M.S. and M.A.), and has introduced the nation's first doctoral program in sustainability, which leads to a Ph.D. in the discipline. Students within the school come from strong academic backgrounds, with most possessing higher than average GRE scores and GPAs. Student backgrounds include degrees in disciplines such as journalism, economics, political science, engineering, and even theater.

For undergraduate students, the B.S. program focuses on environmental science and materials and resources management, with three difference tracks: Sustainable Energy, Materials and Technology; Economics of Sustainability; and Sustainable Ecosystems. Coursework includes classes from biology, chemistry, economics, engineering, earth sciences, and similar disciplines. The goal of the B.S. program is to help students understand how to sustainably use environmental resources, to apply the methods learned to environmental management of water, land, air, and human systems. Students also learn about environmental regulations and policy so they can better work with policy makers at any level of government and assist the private sector with compliance.

The B.A. program, in contrast, focuses on the societal impacts of sustainability, with tracks in Society and Sustainability; Policy and Governance in Sustainable Systems; International Development and Sustainability; and Sustainable Urban Dynamics with courses drawn from economics, sociology, anthropology, political science, ethics, and geography. In this program, students apply social methodologies to both environmental management and social institutions from the local to the global levels. They, like their B.S. peers, also learn how to develop more sustainable systems of water, air, land, and urban management, including work in policy regulations and environmental law.

Undergraduates are also given the opportunity to participate in the Community of Undergraduate Research Scholars, which puts students on research teams along with graduate students and faculty. In addition, the school is planning to develop a minor in sustainability, which can be used in coordination with a large number of undergraduate majors.

The M.S. in Sustainability program is geared for those students with backgrounds in the sciences and engineering who wish to apply those disciplines to improving sustainability. This 33-hour program consists of core courses, seminars, workshops, an optional thesis, and electives. The core courses help provide students with a foundation in sustainability methods and theory, providing a multidisciplinary background to further their studies. Examples of such courses include industrial ecology, human dimensions, quantitative methods, and perspectives on sustainability, with the latter two required for all students. The Challenge Seminars each focus on one main theme of the School of Sustainability, including water, energy, or ecosystems, and provide the foundation for their expertise building and research interests. The workshops, in contrast, are not theme based but problem based and give students the opportunity to develop real-world solutions to issues being researched by faculty and staff at the school at the time; each semester offers new workshop topics.

The M.A. degree has been developed for those students with backgrounds in the humanities and social sciences and follows the same structure as the M.S. degree: core courses, seminars, workshops, and an optional thesis. The core courses, while identical to the M.S. degree, allow students the opportunity to apply their interests and expertise in the social sciences and humanities to sustainability.

The Ph.D. program in Sustainability, the first terminal degree in sustainability in the United States, closely follows the same curriculum track as the two master's degrees, and the program admits students with either a bachelor's or master's degree. The total number of courses required for the degree varies depending on the student's background and previously earned degrees. Again, here are core courses (methods and theory), seminars (in the main themes of the school), and workshops (on real-world issues) as well as required research and dissertation hours. Graduates of the program will be able to assume leadership positions in research in sustainable solutions to environmental and human issues. The Global Institute offers a number of opportunities for graduate research, including the IGERT (Integrative Graduation Education and Research Training) program in Urban Ecology, research assistantships, and summer grants for research projects.

Each degree program also combines students and research interests into the Clustered Learning Network, which provides a learning opportunity to provide solutions using the methods learned in the classroom to sustainability problems. The Clustered Learning Network also utilizes an electronic network that allows collaboration with other disciplines and programs across campus.

Sustainability Education in Other Disciplines

For students from other disciplines with an interest in sustainability, the school has also coordinated concentrations within the B.A. program in Business, the School of Sustainable Engineering and the Built Environment (SSEBE) (offering B.S., B.S.E., M.S., M.S.E., and Ph.D. degrees), and the College of Law (Certificate in Law, Science, and Technology).

The Business concentration aims to address the growing need for sustainability experts who also understand business concepts. Coursework within the concentration includes

Environment and Society, Sustainable Energy and Materials Use, and International Development and Sustainability, offered through the School of Sustainability. This particular program aims to develop into a stand-alone major at some point, as well as an M.B.A. in Sustainability.

The SSEBE was created in 2009 within the Fulton Schools of Engineering to address sustainable engineering and construction, working with the Department of Civil, Environmental, and Sustainable Engineering (CESE) and the Del E. Webb School of Construction. In serving as the nexus for sustainable development within ASU's engineering arena, SSEBE regularly partners with a number of research centers on campus, including the Water and Environmental Technologies (WET) Center, the Center for Earth Systems Engineering and Management (CESEM), the Center for Environmental Biotechnology (CEB), and the Center for Sustainable Materials and Renewable Technology (SMART).

The WET Center is funded as a National Science Foundation Industry/University Cooperative Research Center (I/UCRC) and brings together a multidisciplinary team from the physical sciences, engineering, and the social sciences to address water-quality issues. By working with areas across campus, industry, and government agencies, students assist in research projects across a wide range of issues, including human health, renewable energy, water treatment, and urban planning.

CESEM helps train both engineers and managers in the design and management of systems that bring together the built, natural, and human environments. Coursework offered through CESEM includes Industrial Ecology, Earth Systems Management, and Climate Change Science. Research themes within CESEM concentrate on e-waste, sustainable network theory, urban systems, industrial ecology, and work with the Center for Sustainable Engineering, a consortium between ASU, University of Texas at Austin, and Carnegie Mellon University.

Fifty students are also enrolled in a Certificate Program in Sustainable Technology and Management, which offers tracks in Sustainable Engineering and Organizational Strategies. This program is offered by the Schools of Sustainability, Engineering, and Business; the certificate can also be earned as part of an M.B.A. program for Technology, Science, and Engineering.

Biodesign Institute

The Biodesign Institute focuses on a number of research areas, including improving healthcare services, infectious diseases, and renewable energy and bioremediation. The institute consists of a number of research centers that bring together researchers and students from a number of disciplines to solve problems involving human health and the environment. In the classroom, the institute offers a Ph.D. in Biological Design, a personalized program that combines coursework in the biological and physical sciences, as well as engineering. The Biodesign Institute building was the first in Arizona to receive LEED Platinum certification from the U.S. Green Building Council.

The Center for Environmental Biotechnology, one research center at the Biodesign Institute, examines sustainability at the micro level, creating renewable resource systems within microbiological environments. Bringing together biochemistry, genomics, and ecology, CEB works on projects such as renewable fuels (biofuels), water quality, and renewable energies from waste. CEB provides both graduate and postdoctoral researchers the opportunity to work in an emerging field in energy development.

The Center for Sustainable Health examines the intersection of human health and sustainability, bringing together medicine, psychology, economics, and sociology. The center aims to create health metrics that measure how the environment, and its manipulation through human activity, affects health. The center also aims to better educate policy makers and the public on how the environment relates to health and the health benefits related to a sustainable society.

Environmental Education Outreach

In addition to academic programs, both the Global Institute and School of Sustainability have also developed a number of programs at K–12 schools that provide researchers and students the opportunity to apply sustainability technologies and methods in the community, as well as to introduce environmental education to students. The Ecology Explorers program, based in metropolitan Phoenix, brings the university together with local schools to learn about environmental science while collecting data that is actually used by researchers, further expanding ASU's urban laboratory. The Service at Salado program helps local middle school students in south Phoenix work to revitalize an area of desert within the city.

A popular program that unites university researchers and high school students is the Southwest Center for Education and the Natural Environment (SCENE). This program allows high school students the opportunity to assist researchers on their projects while also getting the chance to begin their own research projects, which are featured at the Central Arizona Science and Engineering Fair. Upper-class students receive a scholarship for completing the program.

Finally, the Sustainable Schools program, funded by the National Science Foundation, puts together teams at three local school districts to create sustainable schools. These teams of administrators, high school teachers and students, and university faculty and students work to develop measurement methods of programs related to sustainability within their respective school systems, including facilities, community, and curriculum.

See Also: Education, State Green Initiatives; Leadership in Sustainability Education; Social Learning.

Further Readings

Arizona State University. "Biodesign Institute." http://biodesign.asu.edu (Accessed January 2010).

Arizona State University. "Global Institute of Sustainability." http://sustainability.asu.edu (Accessed December 2009).

Arizona State University. "School of Sustainability." http://sustainability.asu.edu (Accessed December 2009).

Blanchet, Kevin. "Sustainability Program Profile: Arizona State University School of Sustainability." *Sustainability,* 1 (2008).

National Wildlife Federation. "Campus Environment 2008: A National Report Card on Sustainability in Higher Education." http://www.nwf.org/Global-Warming/Campus -Solutions/Resources/Reports/~/media/PDFs/Global%20Warming/CampusReportFinal .ashx (Accessed November 2009).

Princeton Review. "Green Honor Roll." http://www.princetonreview.com/green-honor-roll
.aspx (Accessed November 2009).
Sierra Club. "Honor Roll: Our Top 20 Schools." http://www.sierraclub.org/sierra/200909/
coolschools/top20.aspx (Accessed November 2009).

Justin Miller
Ball State University

ASSOCIATION FOR THE ADVANCEMENT OF SUSTAINABILITY IN HIGHER EDUCATION

The Association for the Advancement of Sustainability in Higher Education (AASHE) is a membership-based organization consisting of colleges and universities in the United States and Canada whose main goal is to help create a sustainable society through the promotion of education for sustainability. Every individual in the participating institution becomes a member of AASHE and can take advantage of the resources provided by it. AASHE is a nonprofit, tax-exempt, 501 (c)(3) organization. Business organizations, government agencies, and nonprofits can become partners with AASHE in promoting sustainability in higher education. AASHE seeks to make sustainability a core consideration in the planning and execution of the entire gamut of functions and activities that characterize a college or university campus. It also serves as a clearinghouse for accumulated and rapidly expanding information on conceptualizing, adapting, and implementing sustainability practices and policies in the member institutions. In sum, AASHE works to advance sustainability and integrate appropriate practices across the entire spectrum of activities on campuses, ranging from teaching and research to governance and operations.

In keeping with its philosophy that institutions of higher education have a vital role to play in the drive to create a sustainable society, AASHE recognizes that important accomplishments, if they are to be emulated, need public visibility. To this end, AASHE gives annual awards for Campus Sustainability Leadership to institutions that have demonstrated outstanding achievements in advancing sustainability in the areas of teaching and research, operations, and administration and finance. Additionally, AASHE also recognizes student contributions toward promoting sustainability by giving out Student Sustainability Leadership Awards. For 2009, a team of students from Stanford University received the Student Sustainability Research Award for their paper describing a 25-year plan for providing solar photovoltaic recharging infrastructure to charge electric vehicles, resulting in sizable lowering of greenhouse gas emissions.

Sustainability represents a multifaceted challenge that requires that disciplinary and other constraints to knowledge sharing and collaboration be dismantled. Therefore, AASHE works with other nonprofits to advance the goal of achieving a sustainable society. In collaboration with the Society for College and University Planning, it promotes the Campus Sustainability Day, which is observed in the fall every year. AASHE provides a webcast that is available to all the participating institutions. AASHE also supports the efforts of the Campaign for Environmental Literacy in increasing the level of funding for environmental literacy by informing its membership and by helping to lobby the U.S. Congress. In keeping with the focus of AASHE to combat climate change, it works with the Energy Action Coalition to create plans and strategies for long-term efforts to lower greenhouse gas emissions from campuses.

Operationalizing Sustainability

The translation of sustainability as a theoretical concept into a set of measurable parameters represents an enormous challenge. One of the flagship AASHE programs is STARS (Sustainability Tracking, Assessment & Rating System), a voluntary, self-reporting framework for higher education institutions to monitor their progress toward sustainability and to be recognized for their leadership in this area. STARS differs from LEED (Leadership in Energy and Environmental Design) due to its encompassing nature and its applicability to the entire campus and not just individual facilities. There are three main categories in which campuses can earn points: Education and Research; Operations, Planning, Administration; and Engagement. Institutions that sign up for STARS become STARS Charter Participants and are recognized for their accomplishments in the field of sustainability.

STARS was created in response to a call by the Higher Education Associations Sustainability Consortium (HEASC) in 2006 to create a system that would provide a way to measure sustainability in everything that campuses do. The program is a proposed tool for higher education to attempt to evaluate and standardize how schools are performing. Seventy institutions participated in STARS Pilot, which was created with extensive input and feedback from members of the higher education community. The next phase of implementation will determine how successful this program will be.

Opportunities for Professional Development

The user forums on the AASHE Website allow experiences and insights gained in conceptualization and implementation of sustainable policies and practices to be shared widely. A July 2010 perusal of the forums found discussions of protocols concerning carbon offsets and the role that trees on campuses may play in lowering the net carbon footprint. Thus, AASHE enables best practices in a variety of areas related to sustainability on campuses to be documented and shared widely.

AASHE publishes a variety of newsletters, digests, and manuals that draw on the examples of sustainability practices being implemented around the nation. Surveys that document trends in salaries of campus sustainability officers and the differences in progress in improvement of sustainable practices across various units are also provided. Most of these publications are online and available through the AASHE Website.

As sustainability over the past few years has moved from a conceptual stage to the realm of measurement and benchmarks, AASHE has emerged as a leader in the "greening of the campus" movement that has gained popularity globally. The evolution of the concept of sustainability has created a pressing need for a robust framework to assess and compare performance across sectors and countries. AASHE has carved out its own niche in the arenas of both sustainability indicators and the process of developing them. AASHE differs from similar efforts to promote sustainability in higher education in other countries in terms of the comprehensiveness of its rating systems and its participatory approach. The stakeholders, including faculty, students, staff, and community members, are consulted at every stage of the formulation of the institutional plan. The plan therefore takes into account local priorities and contingencies, and the nonhierarchical decision making allows for a decentralized and democratized approach to sustainability. Another distinct feature of AASHE's approach to sustainability in higher education is its consideration of climate change and climate abatement as a focal issue. Lowering carbon emissions and striving for climate neutrality has thus become the pivot around which sustainability efforts are organized. AASHE attempts to harness the synergies between the different domains on campuses such as research and teaching and operations. Curricula can therefore be designed in a way that enables students to engage in hands-on learning with activities like waste management and recycling.

See Also: AASHE: STARS (Sustainability Tracking, Assessment & Rating System); Sustainability Officers; Sustainability Websites.

Further Readings

Beringer, Almut. "The Lüneburg Sustainable University Project in International Comparison: An Assessment Against North American Peers." *International Journal of Sustainability in Higher Education,* 8 (2008).

White, Stacey. "Early Participation in the American College and University Presidents Climate Commitment." *International Journal of Sustainability in Higher Education,* 10 (2009).

Neeraj Vedwan
Montclair State University

ASSOCIATION OF UNIVERSITY LEADERS FOR A SUSTAINABLE FUTURE

In 1990, 22 presidents and chancellors from colleges and universities around the world met together in Talloires, France, to develop strategies for dealing with rising concerns over sustainability issues. The result was the Talloires (pronounced Tal-Whar) Declaration, which was eventually signed by more than 270 institutions of higher learning from more than 40 countries. Member countries, which range from high to low/middle income and which constitute various levels of development, now include Australia, Canada, Finland, Hong Kong, India, Ireland, Japan, Mexico, New Zealand, Portugal, Spain, the United Kingdom, and the United States.

The Association of University Leaders for a Sustainable Future (ULSF) was created to serve as the secretariat of that group that signed the Talloires Declaration. After the creation of the Talloires Declaration, a number of other agreements followed suit, modeling themselves after the Talloires Declaration. Signed over the course of the following years, those agreements included the Halifax Action Plan of 1991, the Kyoto Declaration of the International Association of Universities of 1993, and the Student Charter for a Sustainable

Future, which was created by British student union representatives in 1995. ULSF is charged with providing motivation and support to all members through publications, research, and assessment, and ensuring that each institution continues to be "ecologically sound, socially just, and economically viable."

Growing from 40 members in 1990 to more than 1,000 in 2010, the Association of University Leaders for a Sustainable Future, which is based in Washington, D.C., continues to expand its agenda as well as its membership. The overall focus of the organization is on advancing global environmental literacy by encouraging institutions of higher education to develop comprehensive policies and programs that include the social and ethical elements of sustainability as well as those of sustainable development. Practical aspects of carrying out that mission include working with individual institutions to provide technical assistance and conducting training workshops that teach decision makers how to promote campus-wide sustainability not only through the education but also through slashing greenhouse emissions, developing sustainable food systems, and educating the public about sustainability issues. ULSF also provides referrals when necessary.

In addition to teacher training, ULSF's research and consulting efforts tend to focus on greening strategies and policies. ULSF also consults with member institutions about issues related to organizational change and social learning. Two of its goals in implementing the Talloires Declaration are to encourage strong commitment to greening campuses and engaging in remedial actions such as cutting down on carbon dioxide emissions. The organization also provides regular campus sustainability assessments. The efforts designed to encourage institutions to implement humane and sustainable food systems have often resulted in campus-wide involvement in sustainability projects ranging from recycling to participation in trayless dining. Considerable efforts at ALSU institutions are aimed at implementing the UN Decade of Education for Sustainable Development and participating in the creation and evaluation of UN Regional Centers of Excellence.

The Plan

The Talloires Declaration identified 10 actions that needed to be taken to adequately address the relationship between sustainability and education. Based on the assumption that colleges and universities play a major role in shaping the world views of future leaders of all societies, the Talloires Declaration established an agenda that incorporated those 10 points into all elements of campus life, encompassing teaching, research, operations, and outreach. The basic focus of the agenda was to increase environmental literacy across all disciplines to the point that sustainability became a way of life in the countries of all members.

Officially signed and agreed to by representatives from countries that included the United States, Japan, France, Nigeria, China, Lebanon, Thailand, and India, the 10-point action plan created by the Talloires Declaration commits all signatories to (1) increasing awareness of environmental sustainability development, (2) creating an institutional culture of sustainability, (3) educating the citizenry in environmental responsibility, (4) fostering environmental literacy, (5) practicing institutional ecology, (6) involving all shareholders in sustainability efforts, (7) collaborating in order to ensure an interdisciplinary approach, (8) enhancing sustainability capacities among primary and secondary schools, (9) increasing both national and international service and outreach capacities, and (10) maintaining the sustainability movement.

The Classroom

In order to meet guidelines established by the ULSF, the curriculum at member institutions must deal with social, economic, and environmental aspects of sustainable development, take a comprehensive view of sustainability, and be interdisciplinary to some degree. Among ULSF members, the sustainability curriculum has been incorporated into the fields of agriculture, applied science and technology, architecture and design, business, economics, education, engineering, environmental management, environmental studies, general degree and multidisciplinary studies, international studies, and law. In addition to those taught by particular departments and schools, many member institutions have also set up multi-departmental sustainability programs such as that taught through the Warner College of Natural Resources and the College of Natural Science at Colorado State University (CSU). This program, known as FEScUE (Flexible Extendable Scientific Undergraduate Experience) provides students with opportunities to work in small clusters that allow them to pursue sustainability training in ecology and evolution, bioinformation, or structural biology. The program employs the expertise of more than 100 faculty members who represent 17 CSU departments.

In addition to degree programs, many ULSF member institutions offer special nondegree and certificate programs designed to meet specific needs. At Clemson University, for example, the Center for Community Growth and Change and the Restoration Institute focus on providing South Carolina communities, particularly the Charleston community, with tools needed to practice sustainability in the natural and built environment and on offering lectures and conducting research on issues integral to a restoration economy. In British Columbia, the Campus Sustainability Office oversees a host of programs designed to promote sustainability on campus and throughout the area. The Student Electronic Network for Sustainability Education (SENSE) is an important element in achieving those goals. These special programs often have a global perspective. Such is the case at Virginia Tech, where the Projects in the Science Education Department operates the Malawi Project on Sustainability, Culture, and Education and Global Learning Communities in addition to projects such as the NASA-developed Earth Systems Connections: Elementary Curriculum Development Project and the NSF-supported Nano2Earth: Introducing Nanotechnology through Investigations of Groundwater Pollution.

Partnerships

The United Nations declared the years between 2005 and 2014 the Decade of Education for Sustainable Development and invited all participants in the sustainability community to come together to discuss sustainability issues and develop solutions for existing problems and circumvent future dilemmas through prevention action. Members of the Association of University Leaders for a Sustainable Future responded in 2000 by joining COPERNICUS-Campus, the International Association of Universities, and UNESCO in creating the Global Higher Education for Sustainability Partnership, a major global research project designed to last for several years. Its purpose is to develop sustainability networks at all levels from the local to the international and to provide methods of sharing research, resources, and other tools used in promoting sustainability. Special consultants on the project include Japan's United Nations University,

Mexico's Monterey Institute of Technology, South Africa's Rhodes University, and America's Goddard College.

See Also: Agenda 21, "Promoting Education, Public Awareness, and Training" (Chapter 36); Colorado State University; Lüneburg Declaration.

Further Readings

Barker, Tim. "Campuses Are Going Green." *St. Louis Post-Dispatch,* May 26, 2010.

Calder, Wynn and Richard Clugston. "Education for a Sustainable Future." *Journal of Geography in Higher Education,* 29/1 (March 1, 2005).

Dernbach, John C. *Agenda for a Sustainable America.* Washington, DC: ELI Press, Environmental Law Institute, 2009.

Franson, Melissa. *The Impact of Classroom Exposure to Sustainability, Course Content, and Ecological Footprint Analysis of Student Attitudes and Projected Behaviors.* Auburn University, AL: Unpublished Thesis, 2008.

Hignite, Karla, et al. *The Educational Facilities Professional's Practical Guide to Reducing the Campus Carbon Footprint.* Alexandria, VA: APPA, 2009.

Neapolitan, Jane E. and Terry R. Berkeley, eds. *Where Do We Go From Here? Issues in the Sustainability of Professional Development School Partnerships.* New York: Peter Lang, 2006.

Rockwood, Larry, et al., eds. *Foundations of Environmental Sustainability: The Coevolution of Science and Policy.* New York: Oxford University Press, 2008.

University Leaders for a Sustainable Future. "Association of University Leaders for a Sustainable Future." http://sustainability.nmsu.edu/documents/talloires.pdf (Accessed June 2010).

Elizabeth Rholetter Purdy
Independent Scholar

AUDUBON ADVENTURES

Over the past several decades, environmental education has become increasingly integrated into traditional scholastic settings—both in the classroom and on experiential-based field trips. Audubon Adventures is a major education program occurring at a national level within the United States. It was originally developed and is maintained by the National Audubon Society, which is a major environmental nongovernmental organization. The program materials that constitute the Audubon Adventures program are crafted by professional environmental educators and seek to promote basic scientific knowledge about birds, other wildlife, their habitats, and requirements. The program caters to children in the U.S. school system's grades three to nine. The name *Audubon Adventures* also applies to the title of the newspaper that is distributed to the program's students. Audubon Adventures is a curricular aid that can be integrated into preexisting curricula by both classroom teachers and those teaching in home-school settings. Any school, whether public or private, can participate in the program. At a general level, Audubon Adventures is exemplary of a larger move within the nation's educational system toward increasing environmental content. For example, Audubon Adventures curricula address various national and state educational standards, such as those related to science, math, and language. In terms of science standards, Audubon Adventures' program provides learning opportunities for students on the production, consumption, and application of energy. In facilitating these activities, the program advocates the pedagogical unity of reflection and action. Through integrating reflection and action in its curricula, Audubon Adventures synthesizes several major educational traditions, including service learning, experiential education, environmental education, and place-based education.

Teachers become involved with Audubon Adventures by enrolling in the program and by receiving an educational kit. The principal materials included in the Audubon Adventures kit include a newspaper about birds, wildlife, and habitat; a teacher's resource manual; access to the online National Audubon Society experts' forum, and a 20-minute video titled *Audubon Animals Adventure*. Through these materials, the program is designed to meet various educational standards within the United States' national and state educational standards. At a general level, the program seeks to generate understanding among students and to promote their learning and skills in language arts, science, and environmental literacy. The syllabus provides a supportive environment in which students learn, gain confidence, and have fun discovering nature in an exciting way. The syllabus has two major segments. In the first half, students focus on skill development, observation, and exploration. In the second half, students design and carry out a conservation project. Through English and Spanish editions, Audubon Adventures seeks to be a multicultural resource for the nation's students.

Topics covered by the Audubon Adventures curricula change annually, but primarily concern issues related to wetlands, migration, amphibians, and nature in the neighborhood. The program seeks to create a unity between what it characterizes as "issues" and "actions." Within the program, issues are environmental problems, current events, and policies. Examples of these issues include global warming, recent oil spills, and policies directed at sustainable development. Actions, within the program, are linked through curricular materials to the issues being presented. Proposed actions are directed to the individual level, such as ways to reduce one's carbon footprint, as well as to the policy level, such as how concerned students can make their voices heard. By connecting issues and action, Audubon Adventures draws on a rich tradition of pedagogical theory and education initiatives that have sought to link theory and practice together within educational contexts.

Audubon Adventures consists of both an in-school program directed toward third through fifth grade students and an after-school program designed for middle school students in grades six through nine. At a general level, the in-school program seeks to foster positive attitudes about nature among students and to provide curricular resources to teachers to help meet the national and state educational standards. The goal of the after-school program is to engage older students in a healthy, esteem-building, social, fun experience with peers that will produce positive environmental outcomes.

One of Audubon Adventures' after-school programs is exemplary of how an after-school initiative ties into theories of service learning, nonformal learning, and place-based education. In this program, students participated after school in a 16-week program. The program was divided into two sections. During the first section, students focused on developing their observation skills, engaging in activities such as hypothesis construction, data collection, journaling, field guide use, nature sketching, field note-taking, mapping, survey-ing, and species identification to practice the skills used by naturalists. The second section was focused on conservation outcomes. In this part, the students in the program identified a local environmental problem and then designed and implemented a project to address that problem. The students in this program chose to produce a video illustrating how waste management in New York City affects birds and other wildlife and their habitats. After producing this video, the group presented it to fourth graders at a local school.

Within this example, the students were involved in a type of service learning. Service learning comprises a set of teaching methods and philosophies that essentially seek to create learning opportunities through volunteer service within one's community. In this example, the students produced a video documenting an environmental problem as a way of addressing that problem and presented the results to younger students, thus extending the impact of the learning process. Learning through action as students did within this program is an example of experiential education. This subfield of education draws extensively on the works of John Dewey and Jean Piaget in advocating that learn-ers, of whatever age, make meaning from experience. Audubon's after-school program is also exemplary of place-based education, which is another major recent trend within education. Place-based education emphasizes the pedagogical value of students' local communities. This focus on "the local" takes innumerable forms. Although learning often focuses on a student's own "place," such as their immediate schoolyard, neighbor-hood, town, park, or community, in doing so, it simultaneously engages students in processes of reflection, thus making connections between local and larger regional, national, and global issues and processes. In this example, students focused on a local problem connected to various larger-scale issues and processes as the issue on which they wanted to engage in action.

Through an in-school and after-school program, Audubon Adventures seeks to pro-mote environmental literacy among students in grades three through nine. Its profession-ally designed curricula is meant to meet national- and state-level education standards as well as to provide safe and fun learning opportunities for students through which they can create meaningful lifelong relationships with nature.

See Also: Experiential Education; Outdoor Education; Place-Based Education.

Further Readings

Dewey, John. *Experience and Education.* New York: Free Press, 1938.

Fenwick, T. *Learning Through Experience: Troubling Orthodoxies and Intersecting Questions.* Adelaide, Australia: National Centre for Vocational Education Research (NCVER), 2009.

Gruenewald, D. "Foundations of Place: A Multidisciplinary Framework for Place-Conscious Education." *American Educational Research Journal*, 40/3 (2003).

Mezirow, J. "How Critical Reflection Triggers Transformative Learning." In *Fostering Critical Reflection in Adulthood: A Guide to Transformative and Emancipatory Learning.* San Francisco: Jossey-Bass, 1990.

David Meek
University of Georgia

Australian National University

Australian National University (ANU) is located in Canberra and is a leading international research and education institution in the field of sustainability. ANU has implemented an environmental policy and environmental management plan as well as an environmental planning committee, various campus sustainability initiatives, and student environmental groups to implement its commitment to becoming a leader in environmental best practices. All departments and individuals on campus contribute to ANU's commitment to sustainable operations. ANU's research and educational programs further enhance its sustainability goals. ANU has received funding from the Australian government to transform its Canberra campus into a model of sustainable practices.

Australian National University is using $1 million in government funding to transform its main campus, shown here, into a model of sustainability. New features will include a storm water harvesting system to irrigate campus landscaping.

Source: Wikipedia

Campus Sustainability Planning and Coordination

Australian National University implemented an environmental policy presided over by the vice chancellor to guide its campus sustainability efforts and ensure its compliance with regulatory requirements. ANU actively solicits input from the university community and expects that environmental awareness, responsibility, and sustainability goals will be integrated into all aspects of campus operations. ANU seeks to take a leadership role in defining environmental best practices for present and future generations both on and off campus. ANU also seeks a balanced approach that will fulfill its environmental mission while maintaining its role as a leading international education and research institution.

The Environmental Management Plan provides the guidelines for putting the university's environmental policy goals into practice, with an emphasis on the federal commitment to the principles of ecologically sustainable development (ESD). The plan includes goal specifications, environmental problems to be managed, and the necessary operational objectives, strategies, and policies to achieve these goals. The plan will also outline an environmental audit process to evaluate the university's progress every five years. Considerations in the prioritization of goals include the need to comply with or exceed statutory requirements, the scope of impact, the need to focus on causes of sustainability problems rather than symptoms, the impacts on the health and safety of the university and community populations, the availability of needed resources, and the levels of student and staff support.

The director of facilities and services will have primary responsibility for coordinating the plan and managing associated funding. An Environmental Management Planning Committee advises the university on environmental issues within the scope of the Environmental Policy and Management Plan. The committee will have a senior academic faculty member as chair and will consist of representatives from the faculty, staff, residential colleges, and student body as well as ex officio members from the Facilities and Services Division. The committee will provide an annual report on the university's performance in meeting plan objectives to the vice chancellor. The university instituted an award-winning environmental management and planning office known as ANUgreen. ANUgreen oversees the university's efforts to develop innovative environmental best practices throughout its research, teaching, and operations as well as community engagement and outreach efforts. ANUgreen provides guidance and support to faculty, staff, and students seeking to implement sustainability initiatives. ANUgreen welcomes student participation through a cooperative working relationship with the ANU Student Environment Collective, ANUgreen student internship positions, and student participation in ANUgreen events such as Celebrate Sustainability Day.

Campus Sustainability Groups and Operations

ANU promotes environmental and sustainability awareness among all students, faculty, and staff, who are actively encouraged to examine their environmental impact, to incorporate sustainability practices into their everyday lives and departmental operations, and to seek innovative new sustainability ideas that are environmentally and economically friendly. Strategies range from simple actions such as shorter showers or reusable mugs to complex operational changes such as the implementation of new energy-saving technologies.

The Social Environmental and Economic Sustainability at Work Program (SEE SAW) supports student, faculty, and staff sustainability and environmentally friendly practices through a variety of activities. The Green Labs program promotes environmental management practices in campus laboratories without compromising quality research capabilities. The Green Your Hall program supports environmentally friendly practices by both students and staff within the university's residence halls. ANU also has an environmental management plan for student residences, and many resident halls have their own green student groups. The residence hall student environmental group Fenner Green collects unwanted items from departing students and holds a reuse day at the beginning of each semester where remaining students may claim unwanted items for their own use.

The Environmental Collective is a semi-autonomous group housed within the Students' Association (ANUSA) and is the leading environment- and sustainability-based student group on campus. Although all ANU students are automatically enrolled in the collective, active participation is voluntary. The student body elects an environmental officer to manage the collective's meetings and activities. The environmental officer provides periodic reports to the ANUSA and represents the undergraduate student population on the ANU Environmental Management Planning Committee.

The Environmental Collective works with ANU's sustainability office and ANUgreen as well as with other on- and off-campus environment groups to promote environmental awareness, issues, and campaigns on campus. The collective's initiatives have included the organization of social events, facilitating student involvement in environmental conferences, working to improve ANU's environmental credentials, and organizing public forums and lectures.

The Australian National University Sustainability Learning Community (SLC) provides students, faculty, and staff members the opportunity to work together to support everyday campus sustainable living practices. The SLC operates a campus community organic garden, runs workshops on organic gardening and vermicomposting, organizes tree plantings and landscape rehabilitation efforts, and holds community-building social events and campus cleanup days. The SLC also runs the annual World Environment Day Great Green Debate at Bruce Hall as well as other residence hall green events and green competitions.

Academic and Research Programs

Australian National University's environmental policy includes the goal of maintaining its high level of environmental research and teaching, most notably in the area of ecologically sustainable development. The Fenner School of Environment and Society is the core of Australian National University's research and teaching sustainability initiatives at both the undergraduate and graduate levels. It is housed within the College of Medicine, Biology and Environment and is named after renowned Australian scientist and former foundation director of the Centre for Resource and Environmental Studies Professor Frank Fenner. The Fenner School is affiliated with the ANU Climate Change Institute, the National Institute for Rural and Regional Australia, the ANU Water Initiative, several National Climate Adaptation Research Networks, and the Australian Sustainable Cities and Regions Research Network.

The Fenner School oversees an integrative, interdisciplinary program focused on complex environment-society systems and challenges. The different disciplines and individual faculty and students who participate in the Fenner School work in dynamic teams in order to blend disciplinary foundations and integrative and applied skills. ANU believes that this integrative approach will better prepare its graduates to address the challenges in environment and sustainability they face both on campus and in their future lives and professions. The school also engages in outreach projects to benefit external communities, businesses, and organizations and to provide students with real-world experiences. The ANU Centre for Resource and Environmental Studies (CRES) pursues interdisciplinary research into a range of environmental, resource, and sustainability issues. CRES was absorbed into the Fenner School in 2007.

The ANU Climate Change Institute was founded to create a community of ANU researchers and teachers from the university's seven colleges who are dedicated to the interdisciplinary approaches to research and teaching in the field of climate change in order to contribute to climate change solutions at the local, national, and international levels. The institute also seeks to share its leading climate change research capabilities by serving as an intermediary between its researchers and outside governments, the private sector, and civil society. National collaborations include the Universities Climate Consortium, the Commonwealth Scientific and Industrial Research Organization (CSIRO), and the Bureau of Meteorology, and international collaborations include the International Alliance of Research Universities, the international global change research programs, and the Intergovernmental Panel on Climate Change.

The Economics and Environment Network (EEN) was established in 2002 as a replacement for the Ecological Economics Program. The EEN philosophy states that a sound understanding of economic principles can often help—and is sometimes crucial—in achieving more effective protection and management of our environment and natural resources. It provides greater collaboration, integration, and innovation in the fields of environmental, ecological, and resource economics. The network hosts activities such as seminars, workshops, working papers, virtual networking, teaching, and graduate student supervision. ANU's Environmental Economics Research Hub houses the network's activities. The EEN represents the largest group of researchers in the field within the Australian university community. The EEN works in association with the Crawford School of Economics and Government, the Fenner School of Environment and Society, the Research School of Pacific and Asian Studies, and the School of Economics.

Future Sustainability Initiatives

Australian National University recently received $1 million in Rudd Government funding to further its efforts to transform its Canberra campus into a model of sustainable operations. The funding will allow ANU to implement a variety of sustainable projects in research, education, and operations. Research- and education-based initiatives will include the Education Precincts for the Future project as well as green educational outreach events such as festivals, debates, seminars, training, and targeted advocacy workshops. The Education Precincts for the Future project is supported by the Rudd Government's Green Precincts Fund, which seeks better methods of community water and energy use in preparation for likely future shortages.

Operational initiatives will include the installation of photovoltaic solar panel arrays, the implementation of a trial for the solar split air conditioning system developed by ANU researchers, the introduction of a storm-water harvesting system for landscape irrigation, the conversion of one public sports field to synthetic turf to save water, an increase in the use of carbon-neutral transport on campus, the introduction of upgraded facilities for the benefit of long-distance bicycle commuters, and the introduction of water-saving initiatives to significantly reduce the university's use of potable water. The ANU Canberra campus will be transformed into a demonstration classroom of altered practices to reduce water and energy use and to provide climate change solutions. Their model will also inspire community sustainability mobilization.

See Also: International Alliance of Research Universities: Sustainability Partnership; Kyoto Declaration on Sustainable Development; Sustainability Officers.

Further Readings

ANUgreen. http://www.anu.edu.au/anugreen (Accessed June 2010).

Australian National University Facilities and Services Division Environmental Policy. http://facilities.anu.edu.au/index.php?pid=367 (Accessed June 2010).

Australian National University Sustainable Learning Community (SLC). http://slc.anu.edu.au (Accessed June 2010).

Barlett, Peggy F. and Geoffrey W. Chase. *Sustainability on Campus: Stories and Strategies for Change.* Cambridge, MA: MIT Press, 2004.

M'Gonigle, R. Michael and Justine Starke. *Planet U: Sustaining the World, Reinventing the University.* Gabriola Island, British Columbia, Canada: New Society Publishers, 2006.

Rappaport, Ann and Sarah Hammond Creighton. *Degrees That Matter: Climate Change and the University.* Cambridge, MA: MIT Press, 2007.

Marcella Bush Trevino
Barry University

B

BEREA COLLEGE

A small liberal arts institution located in Kentucky, Berea College has a long history of embracing progressive causes such as abolitionism and women's rights. Founded in 1855 by the abolitionist John Gregg Fee (1816–1901), Berea College was the first college in the U.S. south to admit both black and female students. It is one of the very few institutions of higher education in the country to offer every admitted student the equivalent of a four-year, full-tuition scholarship.

Seen in the light of these commitments, it is perhaps no surprise that Berea has been one of the earliest and strongest higher education advocates in the sustainability movement. Under the leadership of President Larry D. Shinn, the college has demonstrated in striking fashion how to transform the commitment to sustainability into a robust, multipronged strategic imperative. Dedicated to "a way of life characterized by plain living, pride in labor well done, zest for learning, high personal standards, and concern for the welfare of others," the college has sought to bring its campus facilities and operations, curriculum, and community outreach into alignment with this ideal. "Those of us who study the environment but don't incorporate what we know into how we operate as an institution," Shinn insists, "are failing in our educational task."

Shinn was one of the earliest signers of the American College and University Presidents Climate Commitment, which calls on higher education institutions to move toward carbon neutrality. Berea's aggressive climate action plan calls for a 65 percent reduction in greenhouse gas emissions by 2015. The college recently upgraded its heating plant from coal to natural gas, which has led to a sizable drop in its carbon footprint. Campus transportation policies contribute to a smaller footprint as well. Berea's fleet of vehicles includes three hybrids and 13 small electric carts. Carpoolers get top priority for parking, and a student-run program rents, fixes, and builds bicycles for students, faculty, and staff to use. In addition, the college operates a passenger van that transports students on regular runs from the college to local towns.

Berea has been equally active in other areas of campus life. With President Shinn's encouragement, students designed and implemented a recycling program on campus that quickly led the college to become the third-largest recycler in the state. On the green building front, besides the Ecovillage, Berea has undertaken $121 million of renovations to enhance energy efficiency and sustainability. Lincoln Hall, built in 1886 and listed on the

National Register of Historic Places, became the first LEED-certified renovation in Kentucky. This project reused many of the original materials, installed a roof made from recycled milk bottles, added motion sensors to minimize the amount of electricity used for lighting, and incorporated state-of-the-art monitoring systems for air quality, lighting, and physical comfort. Overall, the project cut energy use in Lincoln Hall by 45 percent while maintaining the historic character of the building.

The primary vehicle driving the campus sustainability initiative has been the Sustainability and Environmental Studies (SENS) program. SENS seeks to infuse the teaching of sustainability ideas and values throughout the curriculum "while guiding and supporting the efforts of the College to practice sustainability." Berea's 1,000 acres of farms and gardens and 8,000 acres of forests, including reservoirs, ponds, and streams, provide an outstanding ecological laboratory for SENS classes, internships, and projects.

Ecovillage

The most dramatic embodiment of the college's commitment to environmental stewardship is its $5.5 million living and learning community composed of 50 apartments for students who are married or single parents. The use of green design in this project, completed in 2003, has reduced energy and water use by about 60 to 65 percent. Occupied mostly by adult students with children, the Ecovillage incorporates a wide range of energy-conserving features, from dual-flush toilets and low-energy appliances to passive heating and cooling features. A Commons House provides a central laundry, recreation, and meeting space, and both community and individual gardens are scattered throughout the grounds. The complex includes a state-of-the art child-care facility, which provides early childhood education for the children of students, staff, and faculty. The child development center also serves as a learning laboratory for students in child and family studies, nursing, and related areas.

Another key feature of the Ecovillage is an environmental studies demonstration house for students interested in sustainability. The SENS House is almost entirely self-reliant in terms of energy, water, and waste treatment. Thick walls for increased insulation, concrete floors for ambient warming and cooling, and solar tubes for natural lighting all contribute to the building's small carbon footprint. The college forest supplied the timber used in construction, and the kitchens contain Energy Star appliances as well as cabinets and countertops made of recycled materials. A wood stove and solar panel combine to meet the house's energy needs, a cistern provides potable water, and all greywater produced by the house is diverted to a nearby greenhouse for irrigation.

Berea's ultimate goal, in the words of a 1998 faculty report, is to develop an education that furnishes "both a knowledge base and a model for sustainable living—economically, socially, ecologically, and spiritually—rather than serving as a ticket to hyper-consumption." The curriculum and programming reflect this shared ethos, and the institutional call for "plain living" carries over into campus operations. In exchange for free tuition, students earn a portion of their education costs by working 10 to 15 hours a week in one of 130 campus departments, including the college farm operation.

The Center for International Education's commitment to promoting environmental and social responsibility through education abroad is a good example of Berea's commitment to everyone becoming a model learner and teacher. While the air travel required to get to overseas program sites has a significant environmental impact, the center provides multiple opportunities for students to learn about sustainability firsthand from a variety of cultures and experiences. Furthermore, the study abroad orientation programs include discussions of the environmental impact of overseas education and information about more Earth-friendly travel. The Center for International Education, drawing upon Ithaca College as a model, provides Global Footprint Grants to facilitate research about sustainability and explore cross-cultural implications while abroad. Berea also encourages all education abroad participants to take the Green Passport Pledge to be environmentally and socially responsible both while abroad and after returning home.

In April 2008, Berea College joined with the University of Kentucky, the University of Louisville, and Centre College to address sustainability issues in Kentucky, a state where nearly all of the electricity is produced by coal and mountaintop removal is one of the main contributors to the state's environmental degradation. The four schools launched an initiative called Energizing Kentucky, which seeks to raise awareness of the energy and sustainability challenges that confront the state, nation, and world in the 21st century. A series of statewide meetings throughout 2008 kicked off this effort to foster greater collaboration and cooperation in the development of a more integrated state energy policy.

As a result of efforts like this, Berea has received many national accolades for being one of the greenest campuses in the United States. Among those that have praised Berea's sustainability efforts are the *Princeton Review, Newsweek, Sierra Magazine, Grist,* and Mother Nature Network. Despite this attention, campus members realize that there is still much work to be done. "Not a single campus is even close to achieving sustainability at this point," says Richard Olson, director of the Sustainability and Environmental Studies program at Berea. "Colleges need to get out ahead and model truly sustainable behavior to society."

See Also: American College and University Presidents Climate Commitment; Experiential Education; Place-Based Education.

Further Readings

Berea College. "Green Steps: Berea College's Movement Toward Sustainability." https://docs.google.com/viewer?url=http://www.berea.edu/sustainability/resourcedocs/GreenSteps Summary.pdf (Accessed July 2010).

Berea College. "Sustainability in Education Abroad." http://community.berea.edu/cie/abroad/green (Accessed July 2010).

Eilperen, Juliet. "Colleges Compete to Shrink Their Mark on the Environment." *Washington Post,* June 26, 2005. http://www.washingtonpost.com/wpdyn/content/article/2005/06/25/AR2005062501273.html (Accessed July 2010).

Jessie Ball duPont Fund. "The Revolution on Campus." *Notes From the Field,* 19 (Fall 2008). http://www.dupontfund.org/learning/revolution/campus-visits/berea-college/index.html (Accessed July 2010).

Scully, Malcolm G. "Berea College's 'Ecological About-Face.'" *Chronicle of Higher Education* (February 11, 2005). http://chronicle.com/weekly/v51/i23/23b01101.htm (Accessed July 2010).

Shinn, Larry D. "Sustainability: A Contemporary Berea Imperative." *Berea College Magazine* (Winter 2009).

Peter W. Bardaglio
Second Nature

BRANDEIS UNIVERSITY

The success of sustainability outreach at Brandeis University relates not only to the traditions of the institution in its efforts toward social justice, but is cultivated within an institutional environment of high student autonomy.

Jewish founded in 1948 but nonaffiliated, Brandeis University opened its doors to many who were turned away by other elite institutions. The founding of the institution sought to retain Jewish traditions of community and social justice while establishing a place for the "American ideal" of democracy through education. Although invited to join the Ivy League early on due to its academic reputation, the university declined membership citing the quota systems and other unequal practices the institutional leadership saw as unjust. A legacy of social justice persists and has stood to influence not only the mission and core values of the organization, but concepts of community and student development.

Commitment to Social Action

The university mission is undergirded by four pillars: dedication to academic excellence, nonsectarianism, commitment to social action, and sponsorship by the Jewish community.

Social action in particular has set a tone conducive to sustainability. This pillar asserts, "A complete education is not just the accumulation of knowledge. It is also the development of one's soul and individual ideology." Brandeis has made a commitment to social justice an integral part of its mission, as the commitment to social action also states, "We believe that the betterment of one's self is not just a personal endeavor but also a means through which to better serve society and humanity."

The university's commitment to social justice dates to Brandeis' earliest days, when the university counted among its faculty First Lady Eleanor Roosevelt and welcomed to campus a young civil rights leader named Martin Luther King, Jr. It continued through the 1960s and 1970s, with protests against South African apartheid. Brandeis remains a place for student activism that gives voice to the university's social justice mission; included prominently among these issues is environmental awareness, conservation, and activism.

As such, the Brandeis University campus is one that cultivates community partnerships, prioritizes contributions to the outside world, and empowers students to act as leaders within their communities. Sustainability efforts at Brandeis have sought to further this mission by first examining changes in thinking through educational outreach and academic efforts. Such attention to the relationship between the university and the external community efforts as supported by a long-standing tradition of justice and social action is unique to Brandeis.

Formal university efforts toward sustainability were established in 2007 with the creation and hiring of a sustainability coordinator intended to give support to accomplishing the goals of the ACUPCC signed onto by Brandeis in September 2007. The intention of sustainability efforts is an extension of the core values of the Division of Student Life, oriented toward social responsibility and justice, echoing the organizational saga of the university.

Outreach Efforts

Brandeis was one of only nine schools recognized for the Sustainable Endowments Institute's Champions of Sustainability in Communities Award. The competition honors colleges and universities that have collaborated with off-campus organizations on green projects and programs. Brandeis received honorable mention distinction for a class project in Professor Laura Goldin's Greening the Campus and Community course, in which a group of students partnered with the Prospect Hill Tenants Association to develop gardens for low-income families in Waltham.

Campus and community partnerships help create new sustainability programs and further mutual environmental goals. The Campus Sustainability Initiative is working with local community organizations, the Greening the Campus and Community class students, student clubs, and City of Waltham departments on many exciting projects. In addition to improving environmental impact and saving resources, many of these projects help train Brandeis students for future careers in community and environmental organizations.

Sustainability projects involving the local community include the following:

- Local community gardens
- Fall Farmers Market open to the public
- Donating students' unwanted books, clothes, food, and household items when they move out to local organizations
- Reducing local traffic congestion through shuttles and promoting public transportation options
- Pollution reduction in storm drains on campus and in Waltham through the Clean Streams Project
- Participating in the Waltham Greenhouse Gas Inventory
- Children's environmental education projects

Key examples of outreach efforts include the establishment of the student peer leadership group Eco-Reps, Students for Environmental Action (SEA), and service-learning opportunities within the Greening the Campus and Community course such as the Prospect Hill Tenant Project and Clean Streams Project.

Eco-Reps

Eco-Reps is an undergraduate peer-mentoring program modeled after similar programs taking place at more than 30 other universities nationwide. Brandeis Eco-Reps are student leaders implementing environmental peer education campaigns in residence halls. Eco-Reps is a program of Campus Sustainability Initiative with the goal to improve the climate change impact of the university through conserving energy and decreasing waste by increasing student awareness and engagement. Eco-Reps are responsible for coordination and implementation of the Green Room certification program that occurs within each quad (residence hall) at Brandeis. The Green Room certification program was conceptualized by a student leader and allows for education and awareness of sustainability efforts that students can voluntarily take part in. Students can be awarded the certification sticker by combating environmental waste in a number of ways, including using low-energy light bulbs or electronics, recycling or composting waste before trashing, and doing laundry with cold water and a simple drying rack. The list of the many possible ways to contribute is provided in a checklist at Brandeis sustainability's Green Room Website. By checking off 10 green activities, students can sign up for an inspection by their residence hall or quad Eco-Rep, who will award the green certification sticker. In addition, Eco-Reps coordinate with ongoing clubs and activities, such as the Students for Environmental Action (SEA) group.

Students for Environmental Action (SEA)

Brandeis Students for Environmental Action (SEA) is a student club aimed at promoting awareness and creating opportunities for students to take action toward creating a more sustainable campus and community. The activities of SEA include working with Eco-Reps to promote Green Room certification, promoting the use of bikes to reduce carbon emissions, and coordinating Earth Day events for the university. In addition, a major effort was undertaken by SEA to establish a "green fee" for students that would allow for a competitive grant fund coordinated by a separate board. The Brandeis Sustainability Fund was passed by a vote of the student body in 2010. The Brandeis Sustainability Fund will provide grants, advice, and support to any undergraduate student for projects promoting sustainability. Projects that receive funding could relate to energy efficiency, green buildings, waste management, renewable energy purchases, greening student events, and more.

Greening the Campus and Community

Greening the Campus and Community is a Community-Engaged Learning (CEL) course taught by Laura Goldin, in which students explore strategies for creating healthy, vigorous, environmentally sustainable communities in the face of increasingly challenging environmental problems. In this hands-on course, students collaborate with university partners and community organizations and agencies to design and implement sustainability projects creating measurable benefits to the campus and local community environment. Students work closely with the Campus Sustainability Initiative to achieve sustainability success stories. Two examples of projects undertaken by students of the course include the Prospect Hill Tenant Project and the Clean Streams Project.

Students at Brandeis University began their partnership with the Prospect Hill Tenants Association and other community groups with the intention of developing community gardens with low-income families. In the process, students learned that a community garden was not what the community needed, and subsequently redesigned the initiative to focus on smaller gardening projects closer to residents' homes. These efforts were integrated into an environmental education experience for the community's after-school program. The collaboration highlighted how community-guided sustainability efforts have the most potential for impact, and, as a result, the collaboration has the strength to grow in new directions.

In the fall of 2008, a group of students in the Greening the Campus and Community class created a project aimed at improvement of community storm water drainage, drinking water quality, and protection of wetland species on campus.

In conclusion, the success of the sustainability outreach initiatives are due in large part to the high level of student autonomy and leadership that is cultivated within and among the student body of Brandeis University. Students were responsible for pushing forward the university leadership in signing the Presidents Climate Commitment, and major outreach projects such as Green Room certification and the Sustainability Fund were student conceived and initiated. As the campus sustainability plan states, "Creating a sustainable campus is an exciting opportunity to continue Brandeis' history of connecting academic theory with social justice practice." Coupled with the work of the Campus Sustainability Initiative and the Greening the Campus and Community course, Brandeis University offers a range of sustainability outreach best practices.

See Also: American College and University Presidents Climate Commitment; Leadership in Sustainability Education; Sustainability Officers; Sustainability Websites; Whole-School Approaches to Sustainability.

Further Readings

Brandeis University Climate Action Plan. http://www.brandeis.edu/campussustainability/climate/actionplan.html (Accessed March 2010).

Sachar, A. L. *Brandeis University: A Host at Last.* Waltham, MA: Brandeis University Press, 1995.

Lyndsay J. Agans
University of Denver

BROWN UNIVERSITY

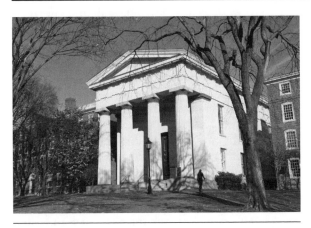

Most buildings on the Brown University campus have energy-efficient lighting, motion sensors, and light timers. Such efforts have helped Brown reduce its energy-related carbon emissions by 18 percent since 2007.

Source: Wikipedia

Brown University is a leading research and educational institution located in Providence, Rhode Island. Brown is a private university with an enrollment of over 8,000 undergraduate and graduate students. Brown serves as a model for environmental sustainability through its commitment to sustainable campus operations. Administrative divisions across the university have implemented a variety of projects dedicated to reducing the carbon footprint of Brown and the surrounding community. Brown's student groups and educational and research initiatives further its sustainable goals.

Campus Sustainability Initiatives

The Energy and Environmental Task Force assessed the university's current environmental impact and provided recommendations to improve sustainability in the future. Brown established the Energy and Environmental Advisory Committee (EEAC) in 2006. The EEAC developed short- and long-term goals and strategies for improved sustainability, releasing its first recommendations in 2007. The Energy and Environmental Office developed an energy action plan and provides an annual sustainability progress report. University President Ruth Simmons has signed several environmental initiatives, including a pledge sponsored by the New England Board of Higher Education and the Sustainable Campus Community Charter stemming from the International Campus Sustainability Network (ICSN). The university implemented the Brown Is Green (BIG) program in 1990 to promote environmental education and advocacy on campus and in the community and to serve as a national model for active learning. BIG projects have focused on energy conservation, pollution reduction, environmentally responsible design, energy efficiency, resource recovery, water conservation, and transportation. BIG also maintains a Website to provide information on the progress of campus sustainability initiatives as well as environmental topics, courses, research, student groups, and community projects.

Departmental Operations

Departments across the Brown campus have incorporated sustainable practices in their daily operations, including energy conservation, carbon emissions reduction, and recycling. Brown established a recycling program in the 1970s. The Brown Facilities Management Custodial Office manages the campus solid waste program, which diverted an average of 38 percent of collected waste into recycling during fiscal year 2008, the program's highest rate. Recycling collection containers are distributed across the campus. Facilities Management landscapers recycle yard waste. The Rockefeller Library, known as the "Rock," established its own electronics trash collection program and serves as a source of recycling information for the campus and broader communities. Brown participates in the annual RecycleMania competition. Purchasing seeks environmentally friendly materials, equipment, and suppliers, and departments seek alternate uses for old, unwanted furniture and equipment. Graphic Services has attained Forest Stewardship Council certification for its use of paper products from sustainable practicing forests. Various departments have gone paperless through the use of online forms.

Facilities Management's Planning Design and Construction Office is responsible for the implementation of sustainability initiatives in building construction and renovation as well as in campus energy use. All new buildings must achieve at least Silver certification in the U.S. Green Building Council's Leadership in Energy and Environmental Design (LEED) rating system. The Sidney E. Frank Life Sciences Building is LEED Silver certified, and seven more buildings under construction will seek the equivalent or higher upon completion.

Facilities Management's Sustainable Energy and Environmental Initiatives Office is working with the Energy and Environmental Advisory Committee to reduce greenhouse gas emissions for both existing buildings and new construction under the university's Greenhouse Gas Goals. The majority of buildings have been fitted with energy-efficient lighting, motion sensors, light timers, and steam trap systems that capture heat that the university's cogeneration plant converts to electricity. The Central Heat Plant now runs on natural gas, and Brown switched to a less carbon-intensive electricity supplier in 2009. Brown has lowered its energy-related carbon emissions by 18 percent since 2007, exceeding its goals.

Brown implemented the pilot program Community Carbon Use Reduction at Brown (CCURB) to further reduce greenhouse gas emissions both on campus and in the greater Providence area. The program is funded through the university with support from the Sidney E. Frank Foundation and is a collaborative effort with local community and civic groups. CCURB provides funding for student-led projects aimed at reducing the carbon output of the community outside Brown.

Brown created a Transportation Office in 2004 to oversee a comprehensive transportation demand management program. Green initiatives include incentives and information for carpools, subsidized UPASS bus fares in conjunction with the Rhode Island Public Transportation Authority (RIPTA), parking policies that discourage single-occupancy vehicles; use of high fuel-efficiency, hybrid, natural gas, or biodiesel-fueled vehicles; and bicycle-friendly measures, including a bicycle-sharing program. Brown maintains a partnership with the Internet service ZIPCAR, which provides rental cars for short periods ranging from an hour to a day.

Brown Dining Services has introduced a variety of sustainable food service and disposal programs, including the Local Food Initiative, an on-campus farmers market, and the Community Harvest and After the Harvest programs. Used fry oil is sent to Newport Biodiesel for conversion to fuel. Other green initiatives include the purchase of certified Fair Trade foods, composting, delivery of leftover food to local groups and food banks, the installation of grease reduction units, the elimination of Styrofoam, trayless dining, and the use of environmentally friendly napkins and containers. Dining Services received the 2009 Green Hospitality certification from the Rhode Island Hospitality Association and the Rhode Island Department of Environmental Management.

Student Groups and Organizations

Student organizations play a key role in Brown's sustainable campus operations. Student Eco-Reps work with Facilities Management to raise student environmental awareness on campus and the development and implementation of campus sustainability projects. The Brown Progress Initiative (BPI) is an interdisciplinary think tank that brings together individuals in the sciences, engineering, business and finance, and technology who are interested in sustainable product design and development. The umbrella organization emPower serves the various campus environmental groups.

Beyond the Bottle is a student group working to eliminate the use of bottled water on campus. The Brown Outing Club (BOC) helped implement the "bikes @ brown" and "bike share" programs. EcoFlow seeks ways to better manage and conserve water on campus. Real Food at Brown and the Sustainable Food Initiative are dedicated to the promotion of local, organic, and sustainable food. The Campus Climate group advises students on climate-related projects and helps them obtain MEEIO funding, provided for projects designed to reduce Brown's carbon footprint, and Supply Chain Sustainability funding, provided for projects designed to increase the efficiency and sustainability of the university's supply chain. Brown also features an environmental program house known as West House, where students can actively live a sustainable lifestyle.

Students also participate in a variety of off-campus sustainability initiatives. Brown students played an instrumental role in establishing Rhode Island's statewide mandatory recycling program. The student-led Sustainability Consulting Partnership (SCP) works with local Providence-area businesses and organizations to achieve sustainability goals without hurting profitability. Students also participate in local, state, and national organizations such as the Childhood Lead Action Project, the Environmental Justice League of Rhode Island, Project Get Ready Rhode Island (PGRRI), the Rhode Island Sierra Club, the Rhode Island Student Climate Coalition (RISCC), and Women in Science and Engineering (WiSE).

Beyond the Bag seeks to reduce on- and off-campus use of plastic bags through awareness programs. emPOWER contains a climate education group that makes educational presentations on climate issues at area high schools. The Outdoor Leadership Environmental

Education Project (OLEEP) mentors area high school students. Engineers Without Borders (EWB) consists of students and faculty dedicated to the social and environmental benefits of engineering. CCURB and Brown students work with the Rhode Island Environmental Justice League and the Capital Good Fund on the Providence Home Weatherization Team to make area homes more energy efficient. Pump It Up runs community awareness initiatives based on reducing automobile carbon emissions through proper tire inflation. Emerging Green Leaders (EGL) provides students interested in the green building movement opportunities to collaborate with and join similar community groups such as the Rhode Island Green Building Council and Emerging Green Builders Committee.

Academic and Research Programs

Brown University's Division of Engineering in cooperation with the Environmental Studies Program offers a bachelor of arts degree in Engineering with an Environmental Studies focus as well as a bachelor of science degree with an Environmental Science focus. These interdisciplinary degrees center on environmental policy, planning, and regulation. Brown offers a master of arts program in Environmental Studies. Brown has a partnership with the Marine Biology Laboratory at Woods Hole (MBL) to offer the Ph.D.-level Brown-MBL Graduate Program in Biological and Environmental Sciences. The Environmental Studies department is located in W. Duncan MacMillan Hall. The building features an environmentally responsible design and many energy-efficient measures. Brown also publishes *Watershed: Journal of Environment & Culture*.

Brown's Center for Environmental Studies (CES) was founded in 1978 to integrate environmental teaching, research, and service throughout the university and to implement an applied approach to environmental education and problem solving. CES also works with local and global governmental organizations, nongovernmental organizations (NGOs), and educational institutions. CES has a primary focus on urban environmental issues due to the university's Providence location. Issues studied at CES include solid and hazardous waste, environmental health, environmental justice, environmental accounting and insurance arrangements, and pollution prevention and toxics policies. The CES owns a Toyota Prius hybrid vehicle for faculty, student, and staff use.

CES is located in the Urban Environmental Laboratory (UEL) in a building renovated in the early 1980s with student participation through grants from the Richard King Mellon Foundation. Green features include superinsulation, passive solar heating, and a solar greenhouse. The building also serves as a community center and features a community garden dedicated to the growth of organic food in limited spaces. Students, staff, and community members may work plots. The garden also maintains public education activities, field trips for local primary and secondary school students, a newsletter and newsgroup, potluck dinners, and a Harvest Festival. Students participate in the garden through student coordinator positions as well as in a plant ecology class for experiential learning.

Brown houses a number of research and education facilities that promote on-campus and off-campus sustainability efforts. The Superfund Basic Research Program (SBRP) receives federal funds from the National Institute of Environmental Health Sciences and is dedicated to solving health and environmental issues associated with hazardous waste sites. The SBRP also participates in the interdisciplinary Reuse in Rhode Island: A State-Based Approach to Complex Exposures program dedicated to addressing Rhode Island's health and environmental issues.

The Watson Institute for International Studies houses the Global Environment Program, which utilizes natural and social science research and policy analysis to address the challenge of global environmental change. The Environmental Change Initiative (ECI) promotes on-campus and off-campus interdisciplinary collaborations on all levels. The Institute for Climate and Energy unites faculty, classes, and researchers from across disciplines to work together toward solutions to climate change and alternative energy. Brown University is teaming with Draper Laboratory in the creation of the Center for Energy Research (CER), dedicated to energy efficiency and smart grid research.

See Also: Sustainability Officers; Sustainability Websites; Whole-School Approaches to Sustainability.

Further Readings

Barlett, Peggy F. and Geoffrey W. Chase. *Sustainability on Campus: Stories and Strategies for Change.* Cambridge, MA: MIT Press, 2004.

Brown University. "Brown Is Green." http://www.brown.edu/Departments/Brown_Is_Green (Accessed June 2010).

M'Gonigle, R. Michael and Justine Starke. *Planet U: Sustaining the World, Reinventing the University.* Gabriola Island, British Columbia, Canada: New Society Publishers, 2006.

Powell, Christopher. "Brown University Sustainability Progress Report 2009." Office of Sustainable Energy and Environmental Initiatives, Facilities Management. http://www.brown.edu/Departments/Brown_Is_Green/documents/SustainabilityReport2009Update.pdf (Accessed June 2010).

Rappaport, Ann and Sarah Hammond Creighton. *Degrees That Matter: Climate Change and the University.* Cambridge, MA: MIT Press, 2007.

Marcella Bush Trevino
Barry University

Brundtland Report

The Brundtland Report, whose actual title is *Our Common Future,* is a seminal document in the global legitimization of the concept of sustainable development. The report was completed in 1987 by the UN-appointed World Commission on Environment and Development. Chaired by Gro Harlem Brundtland, prime minister of Norway at the time of publication, the stated mandate of the commission was to examine and propose remedial measures for the development and environmental problems facing the world. The commission's work was part of sustained efforts by the United Nations to galvanize common efforts toward sustainable development in acknowledgment that environmental problems are interlinked and can best be solved under a global framework. The interdisciplinary commission was composed of influential experts representing the developed and developing world. The commission incorporated views of public discussion panels and three advisory groups.

The Brundtland Report consistently argues that environment and development should be viewed as inextricably interlinked. Its stance is that development needs must be balanced with environmental conservation. This position is underpinned by the commission's conviction that present generations of humanity need to utilize the environment to maintain and improve their well-being (with particular emphasis on poverty alleviation) while preserving it as a resource base for future generations—which, in essence, constituted what the report termed *sustainable development.* Sustainable development, according to the Brundtland Report, does not set limits to growth but requires a change in the quality of economic growth to make it more equitably distributed and environmentally benign. Aiming toward kick-starting the sustainable development process in 2000, the report extensively examines humanity's problems and proposes solutions based on eight interlinked themes: population, energy, industry, food security, human settlements, international economic relations, decision support systems for environmental management, and international cooperation. The themes are illustrated with statistics and brief everyday anecdotes.

Specifically to education, the Brundtland Report dedicates only a few pages in chapter 4 as part of wider discussions on population and human resources. It argues that, together with improved health, education can improve the quality of a population, thereby changing it from a liability to an asset in the development process. Education can make a population more skillful, creative, productive, and better able to solve day-to-day problems. Additionally, the report points out, education can be useful in the process of changing values and attitudes toward environment and development. The report recommends inclusion of environmental education in curricula of formal education at all levels and in all disciplines. Definition of education in the report is not limited to formal education in schools, universities, and other institutions but also includes informal approaches through such means as religion, nongovernmental organizations (NGOs), tradition, and popular media. The report draws attention to the relatively poor enrollment of girls in education and advocates for their involvement in education, bearing in mind women's pivotal role in societal development. Vocational training to give people usable skills for self-sufficiency is favored.

Beyond what it specifically says on education, the Brundtland Report has wide-ranging ramifications for green education. The report's definition of sustainability contains two key aspects: poverty alleviation and environmental conservation. These central tenets of sustainable development have evolved into two main viewpoints referred to by some as the "Green Agenda" (focused on environment), contrasted against the "Brown Agenda" (focused on poverty alleviation). Green Agenda education (curricula, research, operations, outreach) is biased toward such concerns as global warming, pollution, and resource preservation. On the other hand, Brown Agenda education tends to center on issues such as community involvement and empowerment, affordability, and the rights of indigenous people. Broadly, although there are exceptions to the rule and bearing in mind that the two are not mutually exclusive, the Green Agenda is espoused by the developed countries while the Brown Agenda is mainly championed by the developing countries.

Another way to understand the Brundtland Report's implications for green education is by referring to David R. Krathwohl's four categories of knowledge: factual knowledge (basic elements students must know), conceptual knowledge (interrelationships between basic elements, including principles, generalizations, theories, and models), procedural knowledge (how to do something), and meta-cognitive knowledge (knowledge of cognition in general and awareness of self-learning processes). The report is not a primary source of quantitative factual information and, in any case, any such information that it contains is outdated. However, the report's qualitative categories and their scoping can aid in basic understanding of sustainability. Perhaps the strongest aspect of the Brundtland Report is conceptual on the basis that it analyzed a complex problem situation to come up with concepts and principles that define sustainable development. In this way, therefore, the report has directly or indirectly influenced what various players define as *green education*. In terms of procedural knowledge, the report does not go to the level of defining detailed operational tools. Instead, it sets out broad normative guidelines appropriate for national, regional, and international decision making. And finally, as a document giving an overview of the full range of the issues entailed in the concept of sustainable development as originally defined, the Brundtland Report is a veritable reference resource against which different approaches to green education can be checked to gauge where they are located, what they include, and what they exclude. In this way, the Brundtland Report can be used as a tool for meta-cognitive knowledge in green education.

See Also: EARTH University (Costa Rica); Millennium Development Goals; Stanford University (Global Ecology Research Center); University of Virginia (ecoMOD).

Further Readings

Hinrichsen, Don. *Our Common Future: A Readers Guide.* Washington, DC: Earthscan, 1987.
Krathwohl, D. R. "A Revision of Bloom's Taxonomy—An Overview." *Theory Into Practice,* 41/4 (2002).
World Commission on Environment and Development. *Our Common Future.* Oxford, UK: Oxford University Press, 1987.

Tom Sanya
University of Cape Town

California State University, Chico

California State University, Chico, is a public, state-funded university located in the city of Chico. The university's master and strategic plans call for the implementation of sustainable practices throughout its operational, educational, and research practices. The Institute for Sustainable Development, the Associated Students (AS) Environmental Affairs Council, campus and student Sustainability Coordinators, and the Green Campus Program lead the university's sustainability initiatives.

Campus Sustainability Initiatives

California State University (CSU), Chico, has master and strategic plans that include the incorporation of sustainable practices and design as well as a system-wide Sustainability Policy. Paid sustainability positions include the executive director of the Institute for Sustainable Development, the campus and AS sustainability coordinators, the Endowed Professor of Environmental Literacy, the AS recycling operations coordinator, and the Community Action Volunteers in Education (CAVE) assistant program manager. The provost hosts an informal monthly meeting of faculty, staff, students, and community members known as the Environmental Summit. CSU, Chico, has signed the American College and University Presidents Climate Commitment.

The various campus administrative departments have implemented a variety of sustainability initiatives, including the purchase of environmentally friendly goods and services, reuse and recycling programs, waste stream diversion, pollution prevention, and energy-saving measures such as the reconfiguration of campus computers to energy-saving settings. CSU, Chico, seeks compliance with state-mandated recycle content report (RCP) procurement goals and uses Green Seal–approved cleaning products. University Print Services has received Forest Stewardship Council certification for using paper products from sustainable practices forests. Waste diversion efforts include Diversion Excursion, America Recycles Day, RecycleMania, and Recycle Week. Environmental Health and Safety oversees hazardous and electronic waste disposal in conjunction with the student- and faculty-run Chemical Waste Identification and Reutilization Project (CWIRP).

The Grounds and Landscape Services Department and University Housing utilize mostly organic landscaping, native plant species, and environmentally friendly pest management techniques. The campus houses an arboretum with over 1,000 trees, dedicated in 1992. An Arboretum Committee oversees its educational use. The campus participates in the community Scour and Devour program, which coordinates individual trash collection efforts throughout the city of Chico.

CSU, Chico, has purchased three ecological reserves for land restoration, educational purposes, and community outreach. The Big Chico Creek Ecological Reserve (BCCER) is 4,000 acres and home to many diverse species of flora and fauna. Students and faculty have participated in its restoration. The Butte Creek Preserve (BCP) and the Eagle Lake Field Station are the other two reserves. The university signed a memorandum of understanding with the local Mechoopda tribe to ensure university development is in harmony with their views. The university also sponsors the On the Creek Lecture Series and self-guided creek-side tours.

The Campus Transportation Committee and the Transportation Department conducted a transportation demand management study to develop ways to improve the system. Sustainability initiatives include bike-friendly campus practices, the promotion of alternative transportation, and the use of alternative-fuel vehicles. CSU, Chico, subsidizes free student and staff Chico Area Transit and Butte County Transit bus passes through a partnership with the Butte County Area of Governments (BCAG).

The university will seek U.S. Green Building Council LEED certification or equivalent on all new construction and renovation. The residence hall Konkow House was retrofitted in partnership with the Strategic Energy Innovations (SEI) Green Campus Program. CSU, Chico, has passed a Resolution in Support of Green Building and Clean, Renewable Energy. Facilities management practices include reducing energy consumption, introduction of energy-efficient and renewable technologies such as the thermal energy storage system and solar panel arrays, and the retrofitting of existing buildings. The university also participates in the nonprofit North State Renewable Energy Group with local government agencies, businesses, and activists to promote further renewable energy projects on campus.

The Food Services Department has increased its use of reusable products and local, organic, and vegetarian and vegan food. CSU, Chico, hosts the Annual Organic Farming and Food Conference, which also includes a slow food dinner. Food Services has also signed a memorandum of agreement with the College of Agriculture's Organic Vegetable Project (OVP) to determine seasonal vegetable demands to guide University Farm plantings. The University Farm houses an organic dairy, which is used as a learning environment for agriculture students.

Sustainability-based campus committees include the Campus Conservation Committee, the Sustainability Fund Allocation Committee, and the Transportation Committee. CSU, Chico, is a member of sustainability-based organizations, including the Association for Advancement of Sustainability in Higher Education, Campus Compact, the City of Chico Sustainability Task Force, and the California Collegiate Recycling Council. The Center for Applied and Professional Ethics (CAPE) promotes ethical reflection about various issues of concern within and outside the university, including sustainability. CSU, Chico, also works with a variety of businesses, educational institutions, nonprofits, and government agencies. The Sustainable Business Partnership between the university and the Chico Chamber of Commerce links business students to real-world connections.

Student Clubs and Organizations

Students have played a key role in many campus sustainability initiatives. Student groups are housed with the Associated Students (AS). The Associated Students incorporates sustainability in its Strategic Plan, and its Environmental Affairs Council has passed several sustainability resolutions that became campus policies. AS funds the university's sustainability coordinator and a sustainability fund through student taxes and hosts approximately 50 sustainability interns each semester. AS also supports a recycling education coordinator. Students started the AS Chico Recycling Program (ASRP) in 1996 as well as a Recycling Education Outreach Program.

The Zero Waste Task Force is dedicated to the elimination of all campus waste streams on campus and is seeking a composting facility at the University Farm for pre- and post-food wastes from the residence halls. Students maintain the Compost Display Area and Garden (CDAG), and Slow Food Chico State teaches the campus the importance of local, healthy food options. The Take Back the Tap campaign installed a filtered-water system in the student union.

The Green Campus Program has completed several campus conservation projects, including the Sustainability House residence hall, residence hall energy competitions, the metering of campus buildings, the installation of EZ GPO software on campus computers, and the Greeks Going Green campaign. The AS Recycling Program collects nearly 600,000 pounds of recyclables annually and offers student internships. The AS Sustainability Program implements and funds sustainable practices throughout AS and offers internships and volunteer opportunities. Net Impact focuses on campus and community green business strategies. The Sustainable Consultation of Office Practices (SCOOP) is a collaboration of the Green Campus Program, AS Recycling, AS Sustainability, Net Impact, and the Institute for Sustainable Development. SCOOP conducts sustainability assessments and recommendation reports for campus offices.

Students also participate in off-campus sustainability initiatives. The Cause is an off-campus model sustainable living house. GRUB grows food on unutilized community land. Graduating students show their ongoing commitment to sustainability through the Sustainability Graduation Pledge. The Recycling and Rubbish Exhibit (RARE) provides educational opportunities for local elementary school students. Student Environmental Advocates promote environmental awareness and practices in Chico and surrounding areas. The student nonprofit Community Action Volunteers in Education (CAVE) provides numerous volunteer opportunities, including environmental projects. The student entrepreneurial group SEED conducted a SCORE Audit for Chico in collaboration with faculty and the local government members.

Academic and Research Programs

Sustainable development is integrated throughout the CSU, Chico, curriculum. The College of Natural Sciences offers bachelor of science (B.S.) and master of science (M.S.) degrees in Environmental Science. Students can also pursue undergraduate minors in Environmental Studies and Managing for Sustainability and a Professional Science master's degree. Business students can take linked sustainability courses. Students in English 130 created sustainability town hall meetings. Graduate students aided in the completion of a campus greenhouse gas inventory as part of a class project. The School of Graduate, International, and Interdisciplinary Studies offers international experiences and study-abroad opportunities such as the Environmental Literacy in Costa Rica program.

The Institute for Sustainable Development promotes the university's role in sustainable education, operations, and practices, and works with faculty to integrate the principles of sustainable development into the university curriculum. The institute also maintains the Our Sustainable Future informative Website and hosts on- and off-campus events.

The Office of Research and Sponsored Programs supports faculty projects related to sustainability. Other support comes from the CSU, Chico, Research Foundation, the Center for Ecosystem Research, the Geographic Information Center, the California Department of Fish and Game's Aquatic Bioassessment Laboratory, the Northstate Watershed Initiative, the Archaeological Research Program, and the Northeastern California Small Business Development Center.

CSU, Chico, hosts the annual This Way to Sustainability conference, Campus Sustainability Day, and Focus the Nation. The university offers the Rawlins Endowed Professorship of Environmental Literacy, the Rawlins Environmental Prize for student sustainability projects, and the Paul Persons Sustainability Award. The campus publication *Inside Chico* featured sustainability-related articles called Chico Verde. Campus Verde provides information about environmental activities occurring on campus, and the Environmental Action Resource Center (EARC) houses an environmental library for students, faculty, staff, and community members.

See Also: Outdoor Education; Sustainability Officers; Sustainability Websites.

Further Readings

Barlett, Peggy F. and Geoffrey W. Chase. *Sustainability on Campus: Stories and Strategies for Change.* Cambridge, MA: MIT Press, 2004.

California State University, Chico. "2008 Campus Sustainability Leadership Award Application." Association for the Advancement of Sustainability in Higher Education. http://www.aashe.org/resources/profiles/cat4_91.php (Accessed June 2010).

M'Gonigle, R. Michael and Justine Starke. *Planet U: Sustaining the World, Reinventing the University.* Gabriola Island, British Columbia, Canada: New Society Publishers, 2006.

Nance, Molly. "Converging Interests: Environmental and Sustainability Issues Have Permeated Every Level of Academia, From Art Instruction to Business Courses." *Diverse Issues in Higher Education,* 26/17 (2009).

Our Sustainable Future. "The Institute for Sustainable Development." http://www.csuchico .edu/sustainablefuture (Accessed June 2010).

Rappaport, Ann and Sarah Hammond Creighton. *Degrees That Matter: Climate Change and the University.* Cambridge, MA: MIT Press, 2007.

Marcella Bush Trevino
Barry University

CARLETON COLLEGE

Carleton College is a small, private institution located in Northfield, Minnesota, with an enrollment of approximately 2,000 students. Carleton offers an interdisciplinary Environmental and Technology Studies degree program, and members of the Carleton community are encouraged to take the Pledge of Sustainable Conduct. The school implements sustainability throughout its campus operations, with assistance from the Environmental Advisory Committee, the Sustainability Revolving Fund, and student sustainability assistants.

The 880 acres in the Carleton College Cowling Arboretum include restored tallgrass prairie, shown here in 2004. The college has also preserved and maintained the McKnight Prairie remnant's 33 acres of rare native prairie.

Source: Wikipedia/Mark Luterra

Campus Sustainability Initiatives

Carleton College has a master plan that includes sustainability goals and a climate action plan, and maintains an Environmental Advisory Committee. The school's administrators worked with students, faculty, and staff members to create the Pledge of Sustainable Conduct, committing the university community to achievable sustainable behavior. Carleton maintains a Sustainability Revolving Fund (SRF) for student-initiated projects that will reduce Carleton's carbon footprint while simultaneously reducing operating costs. The college's Sustainability Website was launched on Earth Day 2006. It provides information on environmentalism and sustainability as well as updates on campus initiatives. Carleton has also maintained the "Shrinking Footprints" sustainability blog since 2006, which features participation from students, faculty, and staff. Carleton's Purchasing Department requires the exclusive purchase of Energy Star appliances, green cleaning supplies, and paper with recycled content. There are campus-wide compost and recycling programs. One-stream recycling allows recyclable items to be placed into any bin, which simplifies the process and results in increased participation. Items are later sorted at recycling facilities. Unwanted student items are collected for sale to local families through the annual end-of-year yard sale in the West Gym. Many departments have implemented paperless forms and transactions, and campus printing implemented a campus-wide Go-Print program in 2008 to organize printing procedures, which saves paper and ink through the reduction of accidental printing.

All new construction projects seek, at a minimum, LEED Silver certification. The Carleton campus features sustainable residential houses, like the Green House and the Farm House. The school upgraded the campus electrical system to be more energy efficient and to run on a diverse array of energy types. All major buildings on campus feature computerized energy management and lighting control systems, are individually metered for electricity and water usage, and are being retrofitted with low-flow water fixtures. The campus installed a 1.65-megawatt wind turbine in 2004, which was the first college-owned utility-grade wind turbine in the country. The campus has reduced greenhouse gas emissions by 9 percent since 2004. The Plug Loads Database Project collects office and dormitory energy-consumption information for future energy-saving initiatives.

Carleton's Food Services Department seeks to increase the amounts of local, hormone- and antibiotic-free, organic, and Fair Trade foods served on campus. Dining halls compost pre- and postconsumer food waste, and composting bins are available campus-wide and in off-campus college housing. Students Organized for the Protection of the Environment and Food Truth are working to eliminate the use of trays during dining. They also replaced disposable ketchup packets with dispensers. The U.S. Environmental Protection Agency's P3: Prosperity, People, and the Planet program awarded Carleton a grant to purchase an Earth Tub, which will be used to compost waste from Farm House and reuse it in the house's garden. The Facilities Department is studying the possibility of installing a compost pulper in Burton Dining Hall.

Carleton's Transportation Department develops and promotes alternative forms of transportation. Carleton maintains a rideshare Website for interested carpoolers and participates in the federally sponsored Van-Go carpool program for students, faculty, and staff who commute to and from the Twin Cities. Carpooling faculty and staff may park in reserved spots. The campus fleet contains hybrid vehicles, and Carleton seeks to increase the fuel efficiency and reduce the size of its campus fleet. Carleton Facilities conducted a 2008 campus parking study to reduce the amount of land devoted to parking and to aid in the planning of future transportation sustainability initiatives. Alternative forms of transportation include free shuttles and a communal bike-sharing program.

Campus landscaping sustainability efforts include on-campus prairie and other native plantings, increased green space to reduce greenhouse gas emissions and soil erosion, use of corn-based fertilizer, and use of vegetated swales and storm water runoff ponds. A local company removes non-native wood waste for reuse as fuel.

Carleton's campus grounds house the Cowling Arboretum, known locally as "the Arb." The arboretum is a natural land preserve and native-species sanctuary that covers approximately 880 acres and includes pedestrian trails. It is composed of local habitats, including restored and remnant forests, the Cannon River floodplain, bur oak savannah, and tall grass prairie. Carleton also maintains the 33-acre McKnight Prairie remnant.

Student Academics and Organizations

Carleton College offers an interdisciplinary Environmental and Technology Studies program that combines the social and natural sciences, arts and literature, and humanities. Students within the major explore various dimensions of human interactions with environments at the local, regional, national, and global levels through coursework, hands-on laboratory and field research, research projects, seminars, internships and other work experiences, and off-campus studies. Available areas of specialization include Food and Agriculture, Conservation and Development, Landscapes and Perception, and Water Resources. Departmental activities are advertised through the ENTS Updates and Green Network e-mail lists.

Students receive sustainability education as part of their new student orientation. On-campus residents work with student sustainability dormitory assistants. Environmental Studies students, faculty, and staff serve on the Environmental Advisory Committee. Carleton also has offered paid student sustainability assistant positions since 2007. These assistants work with staff and faculty to make departmental operations more sustainable, write articles for the "Shrinking Footprints" sustainability blog, and publicize campus sustainability initiatives.

Other campus-based environmental groups include the Green Network of organizations and individuals interested in environmental advocacy, the student arm of the Minnesota Public Interest Research Group (MPIRG), and the Student Organization for Protection of the Environment (SOPE). Student sustainability projects have included a thrift store, a campus garden, and the Green Wars energy- and water-use dorm competition. The Yellow Bikes Club salvages abandoned student bikes to make them available for student use.

Other campus sustainability activities include having Green Bikes with attached milk crates available, and participating in the annual RecycleMania competition. The Food Truth group conducts educational events and sponsors a Clean Plate Club for students whose meals leave no waste.

See Also: Green Community-Based Learning; Social Learning; Sustainability Websites; Whole-School Approaches to Sustainability.

Further Readings

Barlett, Peggy F. and Geoffrey W. Chase. *Sustainability on Campus: Stories and Strategies for Change*. Cambridge, MA: MIT Press, 2004.

M'Gonigle, R. Michael and Justine Starke. *Planet U: Sustaining the World, Reinventing the University*. Gabriola Island, British Columbia, Canada: New Society Publishers, 2006.

Rappaport, Ann and Sarah Hammond Creighton. *Degrees That Matter: Climate Change and the University*. Cambridge, MA: MIT Press, 2007.

Sustainable Endowments Institute. "College Sustainability Report Card: Carleton College." http://www.greenreportcard.org/report-card-2010/schools/carleton-college (Accessed June 2010).

Marcella Bush Trevino
Barry University

CARNEGIE MELLON UNIVERSITY

Carnegie Mellon University was founded in 1965 through the merger of Carnegie Institute of Technology and the Mellon Institute of Science. Carnegie Mellon is a global research university with world-renowned arts and technology programs. Carnegie Mellon is located in Pittsburgh, Pennsylvania, and has over 11,000 undergraduate and graduate students and over 4,000 faculty and staff members. Students benefit from a low student–faculty ratio. The Carnegie Institute of Technology's Department of Civil and Environmental Engineering offers the Environmental Engineering, Science and Management program leading to a bachelor of science (B.S.) in Civil Engineering as well as interdisciplinary master's and Ph.D. programs. The university also houses renowned state-of-the-art research and computing facilities that allow students and faculty to pursue integrated and interdisciplinary field and laboratory studies across departments as well as with outside businesses and institutions. Carnegie Mellon's campus sustainability initiatives further enhance its curriculum and research activities.

Academic Degree Programs

The Environmental Engineering, Science and Management (EESM) concentration program is located within the Department of Civil and Environmental Engineering (CEE), part of the Carnegie Institute of Technology (CIT). CEE also offers a minor in Environmental Engineering. The program is accredited by the Accreditation Board for

Engineering and Technology (ABET). Core courses include the fields of math, basic science, civil and environmental engineering science, and engineering design projects. Electives include structural design, transportation engineering, and geotechnical engineering.

The interdisciplinary nature of the program allows students the opportunity to customize their course plans or follow predefined minors or double majors in their areas of interest. The program curriculum includes the identification, modeling, and solving of problems; critical and systems-level thinking; ethical reasoning; written, oral, and graphical communications; hands-on learning experiences; and individual and collaborative learning. Students who complete the program will graduate with a bachelor of science (B.S.) in civil engineering.

The EESM program offers an interdisciplinary course of graduate study at both the master's and Ph.D. levels in the areas of protection and restoration of air, soil, and water quality; green product and process design; and environmental sustainability. Students plan their course of study individually with the aid of faculty consultations and choose a specific area of specialization within fundamental engineering science or applied engineering design or development, but are also encouraged to pursue interdisciplinary and collaborative studies.

Specific programs of study at the master's level include water quality, air quality, sustainability, and green design. Graduate course and research topics include the areas of climate change, sustainable engineering, alternative energy, urban watersheds, fundamental physical, chemical, and biological processes in environmental systems and treatment processes, energy management, sustainable infrastructure, computational modeling on molecular and global scales, air quality, environmental microbiology, and biotechnology, nanoscience, and technology.

CEE faculty and graduate students often work on interdisciplinary research with faculty and graduate students from a variety of other campus departments, including the Institute of Complex Engineered Systems, the Tepper School of Business, the Robotics Institute, the Department of Engineering and Public Policy, and the School of Architecture. CEE also offers joint graduate programs, including the Architecture-Engineering Construction Management (AECM) M.S. and Ph.D. programs with the School of Architecture, the M.S./M.B.A. program with the School of Business, and the joint Ph.D. program with Engineering and Public Policy.

CEE and the Environmental Engineering Science and Management program house state-of-the-art group and individual laboratories, including an analytical laboratory. EESM students also have access to additional facilities located throughout the campus. EESM also encourages students to participate in research projects and to pursue study or work abroad. The CEE houses a state-of-the-art computing infrastructure, and students have access to the Pittsburgh Supercomputer Center, in which Carnegie Mellon is a lead collaborator.

Research Facilities and Programs

Carnegie Mellon students and faculty benefit from the university's state-of-the-art research facilities, including multiple interdisciplinary designated environmental research centers. Key research centers include the Institute for Complex Engineered Systems (ICES), the Steinbrenner Environmental Education and Research Institute (SEER), the Green Design Institute (GDI), the Center for Water Quality in Urban Environmental Systems (WaterQUEST), the Center for Sensed Critical Infrastructure Research (CenSCIR), the Center for Atmospheric Particle Studies (CAPS), the Center for Sustainable Engineering (CSE), the Western Pennsylvania

Brownfields Center, the Electricity Industry Center (CEIC), Institute for Green Science, and the Advanced Building Systems Integration Consortium. Much research involves collaborative efforts with other universities as well as industries, and many research results have been published.

SEER was established in 2004 and houses many research centers that explore the technological, economic, and social dimensions of environmental issues and building a sustainable future. CSE was formed in 2005 in partnership with the University of Texas at Austin and Arizona State University, and receives support from the National Science Foundation and the U.S. Environmental Protection Agency (EPA). The CEIC receives funding from the Sloan Foundation and the Electric Power Research Institute. The Western Pennsylvania Brownfields Center serves as a resource for local communities, small businesses, and property owners. The Advanced Building Systems Integration Consortium was founded in 1988 as a partnership with government and industry to study the electricity market, transmission, and infrastructure issues, including demand and security.

The Institute for Green Science is a research, education, and development center dedicated to a holistic approach to green or sustainable chemistry. Terry Collins, the Teresa Heinz Professor of Green Chemistry, is the director. The institute's work focuses on the contributions of chemists in the areas of pollution reduction, renewable energy technologies (especially solar technologies), and renewable sources of chemical feedstocks. Its research programs also utilize the TAML hydrogen peroxide activators invented in the Collins Group and patented and trademarked by Carnegie Mellon.

The Green Design Institute is dedicated to interdisciplinary education and research in the field of green design through partnerships with businesses, government agencies, and foundations. The institute houses the Design Decisions Laboratory and the Laboratory for Carbon Footprinting. Institute research is focused on sustainable infrastructures, energy and environment, life cycle assessment, and environmental management. Goals include the development and promotion of environmentally friendly and sustainable policy, design, management, manufacturing, and regulatory processes that are also economically beneficial. Students from a variety of degree programs participate in the institute through elective coursework. The institute offers community outreach through a Green Design Apprenticeship for high school students, courses and workshops for business executives and lifelong adult learners, memberships for industrial partners who wish to participate in its research programs, and the online availability of the Economic Input-Output Life Cycle Assessment (EIO-LCA) tool.

Campus Sustainability Initiatives

Carnegie Mellon's campus sustainability initiatives enhance the environmental curriculum by serving as a model for students and a place where they can apply what they have learned in the classroom. University President Jared L. Cohon is a leading expert in environmental and water systems analysis. He established environmental protection as a strategic priority for the university in 2001. The university's strategic and master plans contain sustainable goals and a sustainability report. The university also established the Green Practices Committee in 1998 to promote environmentally friendly, sustainable campus initiatives. Students are introduced to campus sustainability through a new-student orientation session on green practices. Students also serve as residence hall Eco-Reps and run the student sustainability group Net Impact.

Sustainability goals include increased use of renewable, pollution-free energy sources such as wind-generated electricity and solar panels; the reduction of toxic emissions and greenhouse gases; local, organic, and Fair Trade food use; green building design; and recycling programs. The campus features seven U.S. Green Building Council LEED-certified buildings, including two at the Gold level, and all future new buildings or renovations must meet or exceed LEED Silver certification standards. Current and future campus projects include the Intelligent Workplace (IW), the Collaborative Innovation Center, the Building as Power Plant to provide on-site energy needs, the use of Energy Star appliances, free bus passes for students and staff, a free car-sharing program, the use of biofuel and ethanol in campus vehicles, incentives for carpooling, composting, and the use of biodegradable to-go containers. Carnegie Mellon was named one of *Sierra* magazine's top 10 "coolest" schools in 2007.

See Also: Leadership in Sustainability Education; Social Learning; Sustainable Endowments Institute.

Further Readings

Barlett, Peggy F. and Geoffrey W. Chase. *Sustainability on Campus: Stories and Strategies for Change.* Cambridge, MA: MIT Press, 2004.

Carnegie Mellon University. "Green Design Institute." http://gdi.ce.cmu.edu (Accessed June 2010).

M'Gonigle, R. Michael and Justine Starke. *Planet U: Sustaining the World, Reinventing the University.* Gabriola Island, British Columbia, Canada: New Society Publishers, 2006.

Rappaport, Ann and Sarah Hammond Creighton. *Degrees That Matter: Climate Change and the University.* Cambridge, MA: MIT Press, 2007.

Sustainable Endowments Institute. "College Sustainability (Green) Report Card." http://www.greenreportcard.org/report-card-2010/schools/carnegie-mellon-university (Accessed June 2010).

Marcella Bush Trevino
Barry University

CENTER FOR ENVIRONMENTAL EDUCATION

The Center for Environment Education (CEE) is a national institution in India, engaged in developing programs and material to increase awareness about the environment and sustainable development. Protection of the environment and living harmoniously with the environment is a concept that is embedded in the constitution of India in articles 48A and 51G. In 1972, the then prime minister Indira Gandhi was the only head of state other than the head of the host nation to attend the Stockholm Convention—seen by many as a turning point and as a precursor to the heightened global awareness of environmental issues, and the stepping stone to international cooperation and commitment to resolve it. She formed the Department of Environment, which in 1980 was transformed into a separate ministry

The Center for Environment Education in India adapts educational materials about the environment to make them suitable for typical Indian classrooms such as this, which often do not have computers or other technology.

Source: U.S. Agency for International Development

dedicated exclusively to environment and forests. This ministry then created Centres of Excellence in partnership with nongovernmental organizations (NGOs) already working in the field to help further the cause of environmental education in India.

CEE was established in 1984 with support from the Ministry of Environment and Forests and was wholly funded by the government. It has been carved out of the Nehru Foundation for Development, its parent organization, which has been working in the field of science, environment, and education since 1966. CEE's main objectives are focused on trying to ensure a better environment and quality of life by means of improving awareness and understanding of environmental and developmental issues, and on promoting action for the conservation and sustainable use of nature and natural resources.

CEE's programs use a three-pronged approach: (1) creating awareness using existing educational opportunities, (2) targeting special groups not adequately covered in the existing system, and (3) capacity building for environmental education and educators. It works on 21 thrust areas, some of which are educating children, implementing higher education, communicating environment through the media, serving as a knowledge/resource center for key sustainable development issues, working with industry in various ways such as waste management and disposal of industrial effluents, developing programs for sustainable rural development, conservation of biodiversity of fragile areas, and training and capacity building of professionals. CEE has been designated as the national implementing agency for the United Nations Decade of Education for Sustainable Development (UNDESD) and is making efforts to promote the UNDESD goals of environmental education in India.

CEE has made efforts to bring international projects to India and then adapt them to the Indian context. One such example is the popular U.S. school magazine *NatureScope*. CEE has worked toward bringing out an Indian version of this magazine by adapting the language and changing the approach—keeping in mind the Indian classroom. For example, most classrooms in India are still equipped with only blackboards and chalk, and there is limited or no access to computers or any other form of technology. This has prompted the CEE to adapt the suggested activities in such a manner that there is little need for technology or any advanced material to conduct the activities. The flora and fauna that are highlighted in *NatureScope* are changed to ones that are more familiar to the classroom in Indian contexts.

One of the biggest advantages CEE has created for itself is that it is often able to act as a local player. With branches in different parts of India, the CEE is able to work closely with the local community and with other local NGOs. This puts the CEE in a better position to cater to the regional, cultural, and linguistic diversities of the region. The CEE has in the past used the "cluster method" to train teachers in environmental education. As part of this approach, it was engaged in capacity building of local NGOs and teachers, and has developed locale-specific materials in conjunction with these NGOs to train the teachers. This ensured that regional/local issues got preference in education and curriculum, and these needs were met in the local language—which provided the comfort of familiarity and faster assimilation among the teachers, students, and community members. The training was held in convenient locations so that the teachers did not have to travel long distances, thus encouraging them to attend. The sessions were also designed to provide networking opportunities for teachers so that they were able to meet others using the same programs and facing similar situations. This provided a forum for interchange of ideas and perhaps even resources in the future.

An Environmental Education Resource Bank

CEE has also helped develop an environmental education bank, which provides educational resources and kits to teach environmental education in different languages. These are referred to as "resource centers" and are run by nodal schools or NGOs. CEE plays a lead role in training local NGOs, which in turn train local teachers. The NGOs, equipped with a good understanding of the region, the languages, mindsets, and local environmental issues, are better placed to train the teachers. However, what is lacking is the skill and expertise to conduct training for environmental educational programs. The local district-level officials and state/central government officials have no expertise to train these teachers in environmental education. This situation naturally puts CEE in the role of a mediator—a combining force. It brings together the local public school teachers, the government officials, and the NGOs, and it starts the ball rolling for the environmental education programs. This involvement of various agencies brings about an environment of ownership and comfort among all stakeholders in this partnership.

Although CEE has an impressive program, it is essential that certain improvements are brought about in the system in order for it to continue further on the path envisaged. One of the primary problems of the situation lies in the Indian education system itself. The Indian education system is characterized by an extremely high reliance on textbook-based learning. This, at times, proves to be an impediment to hands-on learning, as it takes away from the children the practical aspects of learning. It must be noted here that this is more of an issue with the system than with CEE's approach; however, it is a factor that proves to be a stumbling block in CEE's work.

CEE has been able to provide some useful handouts, kits, and resources, but in conjunction with the above, it just adds to more material for the classroom. The impetus required to improve teaching methods is also something that needs to be a part of CEE's repertoire, something that at the moment is not. Further investment needs to be made in bringing in many more practical examples and in-class help. Teachers in these schools are not actually equipped to be able to teach environmental education with the effectiveness they have in other subjects, primarily because it is not their area of expertise. The support that they need to be able to venture out of their comfort zones needs to be provided by CEE.

CEE has conducted some very good workshops; however, data show that there seems to be an overreliance on one-day workshops. These are likely to turn into show-and-tell exercises for the teachers, rather than have any meaningful impact. For the workshops to be effective, they need to be part of a continuum that provides for contact with the teachers for a longer period of time than just an isolated one-day workshop. What may be required here are programs composed of longer durations, where the teachers attend a few sessions, return to their classrooms with the ideas discussed, and then reconvene in more workshops where they are able to analyze their experiences of trying to implement what they have imbibed. Such follow-up sessions provide a forum where the brainstorming is more effective, in the sense that teachers can collectively discuss what worked, what didn't, and how the integration of environmental education can be more rewarding to the class.

Green Teachers

The Green Teacher program is one initiative of CEE that tries to work around this by organizing five-day contact sessions. The problem remains the same; in a year, there are only two such contact sessions, and this causes the same lack of continuum as do the one-day sessions. The Green Teacher program also involves little practical training, which is the need of the hour to increase the efficacy of the workshop.

The resources and kits provided to the teachers in such sessions actually run the risk of being perceived as additional workloads to the already burdened teachers and their curriculum. The first step toward integrating environmental education in the classroom would be to change the value systems of the teachers, and then of the students. The teachers need to be motivated enough to want to learn about environmental sustainability, and then teach it in class. This appreciation of nature and of the environment is the primary aspect that needs to be inculcated in the teacher—something that is not ideally done in a one-day workshop.

The lack of adequate financial resources has adversely affected CEE's initiatives. There has been a gradual reduction in the funding that the CEE has received from the government. This naturally forces CEE to look for outside support to enable the projects to continue. The situation puts CEE under a lot of pressure and has forced it to spread itself thin by getting involved in a large number of activities quite diverse in nature. With this many activities in its schedule, there is a serious dearth of time or resources to be able to analyze completed projects for their efficacy or failure. It is important that CEE be able to thoroughly research the projects it has completed in order for future projects to be more effective. The absence of this aspect costs the CEE in its future projects.

Perhaps the financial situation outlined above also forced CEE to hire employees on a project basis. This hiring policy causes a lot of undue pressure on the staff, as they are always under pressure to look for continuing projects and to generate revenues to hold on to their jobs. The lack of job security has had its effects on the projects themselves.

India has a tradition of living in harmony with the environment. Over time, with industrialization, burgeoning population, and related issues, the environment has taken a back seat in the hearts of the populace. The CEE is an excellent concept and has sound moorings in its missions and values. However, it will take a lot more effort from all agencies concerned, especially the government, to realize the objectives of its very existence.

See Also: Experiential Education; Place-Based Education; Sustainability Teacher Training; United Nations Conference on the Human Environment, Stockholm 1972; United Nations Decade of Education for Sustainable Development 2005–2014.

Further Readings

Center for Environment Education. *Annual Report.* Ahmedabad, India: Center for Environment Education, 2008.

Government of India. "Constitution of India" (2008). http://india.gov.in/govt/constitutions_ of_india.php (Accessed November 2009).

Rangarajan, M., ed. *Environmental Issues in India: A Reader.* New Delhi: Dorling Kindersley, 2009.

Sylvia Christine Almeida
Monash University

Cloud Institute for Sustainability Education

Established in 1995, the Cloud Institute for Sustainability Education in New York City has played an important role in providing K–12 school systems and communities throughout the United States with the necessary tools and content to engage students in learning about sustainability. Originally a program of the American Forum for Global Education, the Cloud Institute became an independent organization in July 2002.

Jaime Cloud, founder and president of the nonprofit organization, is formally trained in organizational change. She taught global studies in New York City public schools for a decade previous to starting the Cloud Institute. Cloud's emphasis on not just incorporating new content into the curriculum but also transforming approaches to teaching and learning and providing students with hands-on experiences outside the classroom places her squarely in the tradition of John Dewey, the early 20th century democratic education reformer. This philosophy influences the operations of the Cloud Institute.

Although sustainability education in colleges and universities has received growing attention in recent years, not nearly as much work has been carried out at the K–12 level. Central to the approach of the Cloud Institute is the belief that K–12 education has a critical impact on American beliefs, attitudes, and behavior. Concerned that the dominant educational system in the United States does not adequately prepare future citizens to come to grips with the challenges and issues of the 21st century, the Cloud Institute helps prepare teachers to create a learning environment for students that will allow them to thrive in a complex, interdependent society. Cloud and her colleagues believe that by reshaping K–12 education in the United States so that it fosters sustainability, they can have a major impact on our ability to achieve a sustainable world.

In order to make the most of what Cloud refers to as this "huge leverage point in changing the consciousness of a nation," the institute offers curriculum, organizational, and professional development for teachers and administrators at schools around the country as well as in New York City. Courses and units developed by the Cloud Institute seek to foster a sense of place and purpose that empowers students to become effective change agents in their communities. Employing such pedagogical innovations as service learning, role-play simulations, and project-based learning, the Cloud Institute approach engages students in systems thinking, "win–win" negotiation, identifying problems upstream, and developing effective solutions.

Among the secondary school courses that Cloud has helped design and launch are Business and Entrepreneurship Education for the 21st Century, and Inventing the Future: Leadership and Participation for the 21st Century. Instructional units and programs from the Cloud Institute include Ecological Economics for Life; Introduction to Sustainability; Changing Consumption Patterns; Systems Thinking; Core Content and Habits of Mind of Education for Sustainability; and From Global Hunger to Sustainable Food Systems.

Besides city school systems, charter schools, and colleges, the institute serves a broad array of clients, including faith-based groups, banks, museums, corporations, government agencies, and libraries. Major funders of the Cloud Institute include the Ewing Marion Kauffman Foundation, Nathan Cummings Foundation, Geraldine R. Dodge Foundation, Bay and Paul Foundations, and Dr. Robert C. and Tina Sohn Foundation.

Much in demand as a speaker, Cloud likes to remind her audiences that "systems are perfectly designed for the results they achieve," a message that many people don't necessarily want to hear because it focuses on the need to dig below the surface and ask hard questions about underlying assumptions, goals, and intentions. Cloud points out, however, that such work is necessary if we are to build a sustainable future. "This kind of stuff—systems thinking and sustainability education—it is not instant orange juice," she says. "It's a way of thinking, a series of habits of mind; it's being able to transfer what you know and apply it in other areas; it's about asking better questions than the ones that have been asked of you—these are things that take time."

Among the "habits of mind" that the Cloud Institute seeks to cultivate is what it calls "protecting and enhancing the commons." Drawing on Fish Banks, an interactive role-playing simulation developed by Dennis Meadows, the institute has developed a simpler version of the game to help students understand the dynamics of the commons. The Fish Game challenges them to catch enough fish to support their families while others are trying to do the same. The students quickly learn that the only way to win is if everybody wins. If they ignore the limits on the fish stocks and fail to grasp the importance of such concepts as the replenishment rate and carrying capacity, they soon run out of fish to catch, and the economy and the community collapse.

Playing the Fish Game simulation, students explore sustainability in terms of the environment, social justice, equity, reciprocity, and intergenerational leadership, in the process excavating the mental models that have led to the kinds of unsustainable behavior so evident in the world around them. Such non–zero-sum exercises also underscore the key point that systems thinking and sustainability go hand in hand. As Cloud puts it, "Sustainability is mostly about evolving our economic and social systems in support of, and in keeping with, our ecological systems." The crucial distinction here is between "education *for* sustainability" and "education *about* sustainability"; the former encompasses but is not limited to the latter.

If there is one phrase that sums up the Cloud Institute's philosophy, it is "making connections." In a society where all too often, as Robert Putnam notes, we go "bowling alone," Cloud and her colleagues recognize the sense of disconnection that all too many students experience and the urgent need to address this situation. "If you're disconnected, then you can't care, can't be engaged," she observes. "You can't make any kind of positive change." The legacy of the Cloud Institute is clear: It helps connect students to their schools, families, and communities in ways that increase the likelihood we can create a future that is not only sustainable but perhaps even regenerative.

See Also: Early Childhood Education; Experiential Education; Integrating Sustainability Education Concepts Into K–12 Curriculum; Sustainability Teacher Training.

Further Readings

Cloud Institute. http://www.sustainabilityed.org (Accessed March 2010).

Federico, Carmela M., Jaimie P. Cloud, Jack Byrne, and Keith Wheeler. "Kindergarten Through Twelfth-Grade Education for Sustainability." In *Stumbling Toward Sustainability*, J. C. Dernbach, ed. Washington, DC: Environmental Law Institute, 2002.

Knapik, Michelle. "Go Fish . . . Within the Carrying Capacity." Geraldine R. Dodge Foundation Blog. http://blog.grdodge.org/2009/08/10/go-fish-within-the-carrying-capacity/#more-2399 (Accessed March 2010).

Long, Jane. "A Real Life: Jaimie Cloud." *Relevant Times: Fresh Ideas for Sustainable & Healthy Living* (September/October 2007).

"Not Instant Orange Juice—Educating for Sustainability." *Leverage Points* (September 19, 2006). http://www.pegasuscom.com/levpoints/lp78.html (Accessed March 2010).

Peter W. Bardaglio
Second Nature

COLLABORATIVE FOR HIGH PERFORMANCE SCHOOLS

The Collaborative for High Performance Schools (CHPS) movement seeks to improve the quality of students' educational experiences and help schools reduce energy needs and operate more sustainably by improving the nation's educational facilities. There is no cost to participate in the CHPS process; however, a resolution must be passed at the district level to begin the design process.

The CHPS process can be contextualized within the recent larger academic and corporate shift toward efficient design. As such, the CHPS process is similar to LEED certification. However, while similar, the LEED and CHPS programs are quite structurally and philosophically different, and interested educational communities must choose between one or the other in the planning process.

High-performance design can have a trickle-up impact, from the classroom to the boardroom. Among the primary benefits are higher test scores, increased average daily attendance, lower operation costs, improved teacher satisfaction and retention, decreased liability, and mitigated environmental impacts. Increasingly, studies are finding that there is a significant correlation between student performance and a school's physical structure. These studies confirm what teachers and students have long known, namely that a better school environment—one with better acoustics, lighting, indoor air quality—will deliver better student outcomes, reduce operating costs, and promote environmental sustainability.

The CHPS process is not complicated, but it does consist of a whole-building approach to design. Within this integrated framework, the interactions between technology and learning systems are considered together throughout the design process. The objective of this process is a school (whether retrofitted or new construction) that enhances teaching and learning, reduces operating costs, and helps reduce the school's environmental impact.

One of the primary objectives of CHPS is to improve and ensure student comfort and health through integrated design. Student comfort and health are seen as interrelated, and as closely tied to the quality of the learning as well as the learning process. For example, proper illumination has been shown to increase student learning, and good indoor air quality is essential for student, teacher, and staff health. These benefits, CHPS argues, are in turn tightly correlated with long-term cost savings for the school, such as in energy reduction.

A key element of the CHPS process is the design process. To ensure that the school's dreams become a reality, CHPS suggests that schools maintain focus on the five following key elements:

- *Set goals:* To help schools set attainable goals, CHPS has developed a set of criteria, known as the Collaborative for High Performance Schools (CHPS) Criteria. These criteria are a point system that helps the educational community to prioritize specific performance goals.
- *Communicate goals to designers:* Both early and sustained open communication of priorities between educational planners and architects and contractors is a must for timely and efficient attainment of performance priorities.
- *Pursue integrated design:* An integrated design team is essential to ensure that technological, architectural, and educational systems are tightly coupled, ensuring the optimization of student health and comfort while ensuring energy efficiency and sustainability.
- *Monitor construction:* Similar to any major project, milestones should be set, and the educational community should remain wary of any substitutions in either materials or design without open dialogue between the design team and the school community.
- *Verify goals:* The educational community should avail itself of the process of commissioning to ensure that the building has been designed as per specifications.

CHPS is largely focused in the states of California, Washington, Texas, Colorado, New York, and Massachusetts. Projects in each of these states generally follow specific zoning and construction requirements mandated at the state level. These criteria are elaborated in the state-specific documents on the CHPS Website. The process does not only deal with new construction; recently, a "major modernization" design process and priority setting system have been established. To initiate a CHPS process, it is necessary to pass a CHPS resolution. The organization's Website contains information for interested teachers, designers, facilities and maintenance staff, administrators, parents, and students who want to "go green" and raise awareness of the environmental, economic, and educational benefits associated with high performance. Of great value is the organization's informational booklet "Planning for High Performance Schools," which provides a road map of the process in addition to answering many common questions. The primary audience for this document is the educational community, defined within the CHPS process as superintendents, parents, teachers, school board members, administrators, facilities managers and directors, buildings and grounds committees, energy managers, and new construction project managers. For this diverse group, the document illustrates the importance of high-performance schools, and describes what elements comprise the design process and constitute the actual construction. It also details how to navigate the design and construction process to the school community to ensure that the project is realized according to the educational communities' priorities.

Once awareness has been generated within the educational community, and it is believed there is general consensus, the next step is for the district to pass a district resolution. The CHPS Website contains links to sample resolutions passed by other school districts as guidelines for schools interested in putting forth a resolution. For a resolution to be recognized as valid, the following criteria must be met: (1) at the trustee level, a formal commitment must be met supporting the CHPS process, (2) a commitment must be made that all new construction and major modernization projects meet the CHPS minimum qualifying point and prerequisite threshold, and (3) an electronic version of the resolution must be submitted online to the CHPS.

Once a resolution has been submitted, the project should be registered on the CHPS Website and the self-certification process completed. Upon registering, the school district will receive a scorecard to be used in this self-certification process. For each claimed prerequisite or point, the scorecard must be annotated. The CHPS committee reviews the scorecard for completeness, and at that point designates the project as CHPS-designed.

Throughout the design and construction process, schools have the option to have their construction documents scorecard and supporting documents independently reviewed. This review process is called commissioning, and is a means of ensuring project completeness.

By participating in the CHPS process, each school will receive its own Website, allowing for electronic submittal of forms and project tracking tools; project scorecards; and three CHPS criteria interpretations. Through being a part of the CHPS process, schools also gain access to a wide variety of valuable design resources, such as documents describing the benefits of high-performance schools and a database of high-performance products.

Through smart integrated design, the CHPS process seeks to build on the recent wave of private, governmental, and corporate interest in architectural efficiency and environmental sustainability. With projects initiated at the community level throughout the country, CHPS is changing not only the face of the nation's schools, but is also the learning experience that occurs within their walls.

See Also: Integrating Sustainability Education Concepts Into K–12 Curriculum; Leadership in Sustainability Education; Sustainability Officers; Sustainability Topics for K–12; Sustainability Websites.

Further Readings

Collaborative for High Performance Schools. http://www.chps.net/dev/Drupal/node (Accessed September 2010).

Fischer, J., P. E. Kirk Mescher, et al. "High-Performance Schools." *ASHRAE Journal,* 49/5 (2007).

Kats, G. *Greening America's Schools: Costs and Benefits.* Washington, DC: Capital E, 2006.

Kibert, C. J. *Sustainable Construction: Green Building Design and Delivery.* Hoboken, NJ: Wiley, 2007.

U.S. Green Building Council. "LEED Initiatives in Governments and Schools." (2008). https://www.usgbc.org/ShowFile.aspx?DocumentID=691 (Accessed September 2010).

Yudelson, J. and S. R. Fedrizzi. *The Green Building Revolution.* Washington, DC: Island Press, 2008.

David Meek
University of Georgia

COLLEGE OF MENOMINEE NATION

As one of 34 tribal institutions of higher learning operating in the United States, the College of Menominee Nation (CMN) is in a unique position to successfully promote environmental sustainability among its faculty and students and among Wisconsin's Native American population and the broader community. The Menominee have inhabited the surrounding land area for more than 150 years, and they have used that long tenancy to draw on both culture and ecology to develop an expanding body of sustainability knowledge. CMN's efforts toward promoting sustainability are guided by the sustainability coordinator and the Sustainable Development Institute (SDI), which has dedicated itself to using education, research, and community engagement to encourage greater environmental responsibility.

SDI works particularly closely with forestry officials to ensure sustainability of environmental resources and to maintain respect for Native American tribal lands and the Menominee culture. Attitudes toward sustainability are dictated by the belief in the sanctity of all forms of life that governs Menominee culture and by the recognition that successful sustainability is dependent on the proper balance between the needs of the Menominee people and the greater community. CMN reaches out to the international community through a series of Sharing Indigenous Wisdom International Conferences. The College of Menominee Nation has received considerable recognition for its work in developing an academic curriculum geared toward the promotion of sustainability and involving the student body in sustainability efforts.

Sustainability Development Institute

An original signatory of the American College and University Presidents Climate Commitment (2006), the College of Menominee Nation is actively engaged in generating a greenhouse gas emissions inventory and is working with the U.S. Environmental Protection Agency's Energy Star program to improve campus energy consumption. At its founding in 1993, the College of Menominee Nation established the Sustainability Development Institute in response to growing environmental concerns and the recognition that it had much to offer in the field of sustainability. Working with the Menominee Advisory Council on Sustainable Development, SDI focuses on generating relevant research, designing appropriate educational curricula, and identifying specific measures needed to achieve educational and community sustainability goals. The rubric under which SDI operates is three-pronged, concentrating on "Sustaining the Nation," which is defined as providing sustainability protection for all people of the United States in conjunction with protecting the lives and culture of the Menominee people and for all forms of life; "Sustaining the Spirit," which promotes and protects the knowledge, perspectives, and approaches of the indigenous community; and "Sustaining the Forest," which involves promoting social, ecological, and ethical protection of natural resources.

CMN officials identify five key areas of research and innovation: climate, health and wellness, ways of knowing, sustainable forestry, and sustainable development. In general, these efforts involve the Center for First American Forestlands, technical innovations carried out through the Menominee Media Center, the Apple Authorized Training Center, and iTunes University, and a host of research and innovation projects. In addition to providing research opportunities, SDI offers professional development courses and regularly holds workshops to enhance sustainability training. The institute also offers professional certifications in various sub-areas of sustainability.

Curriculum

Sustainability education is an integral part of academics at the College of Menominee Nation, and all members of CMN are required to enroll in the Introduction to Sustainable

Development course. The college also offers associate degrees in Sustainable Development and Sustainable Forestry, and many students pursue additional degrees in the field at other institutions. The number of students transferring to other institutions is likely to decline in the future because plans are under way to provide a four-year degree in the field. Fellowships and SDI internships are available to qualified students. In addition to mandatory classes in English, math, science, social science, and the humanities, students are required to take either Introduction to Native American Cultures or Introduction to Cultural Geography, Physical Geography, Geographical Information Systems, Introduction to Sustainable Development, and Implementing Sustainable Development.

While students in the Sustainable Development program approach sustainability from a broader perspective, those concentrating on Sustainable Forestry are required to focus their efforts on incorporating forest management with tribal beliefs and approaches. One project allows CMN faculty and students to work with those at Pennsylvania State University's Department of Rural Sociology and College of Forestry and the U.S. Department of Agriculture's Forest Service. Recognizing the vast knowledge of the Menominee people concerning forestry preservation, the U.S. Forest Service first established a partnership with CMN in 2003. A new six-year agreement was signed in 2009 to allow CMN's Center for First American Forestlands to continue work in progress. Through the Sustainable Development Institute, students are also involved in working with researchers at Iowa State University on the Sustainability Indicators Research Project.

Sustainability is further promoted at CMN by Strategies in Environmental Education, Development, and Sustainability (SEEDS), a student group that sells a special blend of organically grown and bird-friendly coffee to the student body. A portion of the project's proceeds is donated to local farmers. CMN students also regularly participate in national competitions such as the National Wildlife Federation's National Chill Out competition and RecycleMania. Other student efforts include participation in the Adopt-a-Highway program, campus-wide recycling projects, and involvement in efforts such as the Great Lakes Earth Day Challenge, in which CMN served as a drop-off point for electronic and solid waste. The Campus Sustainability Advisory Group, which is made up of 10 student volunteers, works closely with the sustainability coordinator on various policies and projects.

The success of CMN's sustainability efforts are evidence of the need for both government officials and the general public to recognize the advisability of drawing on the store of knowledge amassed by Native Americans, who have a long history of living in harmony with the world they inhabit. It also illustrates the need for education and public involvement in sustainability at all levels.

See Also: American College and University Presidents Climate Commitment; National Wildlife Federation; Sustainability Officers.

Further Readings

Dernbach, John C. *Agenda for a Sustainable America.* Washington, DC: ELI Press, Environmental Law Institute, 2009.

"Forest Service Renews Partnership with College of Menominee Nation in Wisconsin." *Indian Country Today* (December 2, 2009).

Franson, Melissa. *The Impact of Classroom Exposure to Sustainability, Course Content, and Ecological Footprint Analysis of Student Attitudes and Projected Behaviors.* Unpublished Thesis, Auburn University, AL, 2008.

Hignite, Karla, et al. *The Educational Facilities Professional's Practical Guide to Reducing the Campus Carbon Footprint.* Alexandria, VA: APPA, 2009.

Neapolitan, Jane E. and Terry R. Berkeley, eds. *Where Do We Go From Here? Issues in the Sustainability of Professional Development School Partnerships.* New York: Peter Lang, 2006.

Rockwood, Larry, et al., eds. *Foundations of Environmental Sustainability: The Coevolution of Science and Policy.* New York: Oxford University Press, 2008.

Sustainable Development Institute. http://www.sustainabledevelopmentinstitute.org (Accessed June 2010).

Elizabeth Rholetter Purdy
Independent Scholar

COLLEGE OF THE ATLANTIC

College of the Atlantic, chartered in 1969, was intended from the outset to have an academic emphasis that would set it apart. Focused on human ecology, or the interplay between humans and the environment, College of the Atlantic seeks to prepare students for paths that provide personal fulfillment while also enabling them to approach the complex sustainability challenges faced by communities around the world. Located on the coastline of Maine's Mount Desert Island—home to Acadia National Park—College of the Atlantic benefits from proximity to natural beauty as well as an ideal outdoor classroom.

In the first year of courses, 1972, the college had only four full-time faculty and enrolled only 32 students. Today, the college enrolls 300 students and boasts 30 faculty members. From the outset, learning at College of the Atlantic was an active experience, with early classes exploring the ocean and mountains surrounding their campus. Only one major, Human Ecology, is offered at College of the Atlantic; however, students are encouraged to customize their experience as their interests dictate. Students are encouraged to build their own courses, and nearly 29 percent of students do complete an independent study.

Also setting College of the Atlantic apart is its representative approach to governance. Students serve on formal college boards and committees and have full voting rights. All college meetings are held weekly, which enables the review of various committees' work by the entire college community.

Sustainability at College of the Atlantic is deeply embedded within the systems and structures of the institution. The college is used as a living laboratory, and it makes every effort to maintain transparency about the learning process. College of the Atlantic has undertaken a wide variety of innovative projects in an effort to address common institutional sustainability challenges.

Student housing is often one of the most difficult campus locations in which to institute changes to improve institutional sustainability. The Kathryn W. Davis Student Residence Village on the College of the Atlantic campus is an example of a truly exemplary sustainability effort, however. The buildings were built with superinsulation, wood-pellet stoves, and composting toilets. The building materials themselves are often recycled, including shredded newspaper insulation or specially certified material such as wood from old-growth forests.

Many colleges and universities throughout the world have established carbon emission reduction targets. In 2007, College of the Atlantic took an even greater stand by announcing its intention to become net zero, or the first carbon-neutral college. To achieve this goal, the college purchases its electricity from hydropower and builds only super-efficient structures. Campus heating and hot water is provided by using a wood-pellet boiler. General efficiency efforts have also been undertaken, including switching to compact fluorescent (CFL) light bulbs and stopping leaks. Finally, any remaining emissions are offset by purchasing credits from verifiable sources.

Procurement sustainability has also been a focus at College of the Atlantic, where 90 percent of purchased products are environmentally friendly. Products containing bleach, ammonia, and aerosols are avoided whenever possible, and paper products are 100 percent recycled.

In landscaping efforts, College of the Atlantic since its inception has chosen native plantings. Landscaping is maintained organically, without the use of pesticides or fungicides. The college also owns and operates an organic farm as well as an organic student garden.

Transportation is often a substantial challenge in developing a sustainable institution because of modern society's entrenched relationship with personal motorized vehicles as well as the tremendous variation in public transportation availability, depending on locale. At College of the Atlantic, however, transportation and its resulting impacts are eased. Students, faculty, and staff at College of the Atlantic can check a bicycle out of the library just as one would a book. This ready access to nonmotorized transportation helps to cut down on individual car use on campus, as does the nighttime bus shuttle operated by the school. The college also maintains a fleet of boats for research purposes. For these vessels, all diesel engines operate using E85 ethanol, a plant-derived diesel alternative.

College of the Atlantic provides an impressive model of the extent to which sustainability can be integrated into a high-quality educational institution. From the curricular to the operational, College of the Atlantic exemplifies what can be accomplished when leadership, staff, faculty, and students all work toward the goal of a sustainable institution driving their decision-making process.

See Also: Leadership in Sustainability Education; Outdoor Education; Whole-School Approaches to Sustainability.

Further Readings

College of the Atlantic. "About COA" (2010). http://www.coa.edu/about-coa.htm (Accessed June 2010).

M'Gonigle, R. Michael and Justine Starke. *Planet U: Sustaining the World, Reinventing the University*. Gabriola Island, British Columbia, Canada: New Society Publishers, 2006.

Rappaport, Ann and Sarah Hammond Creighton. *Degrees That Matter: Climate Change and the University*. Cambridge, MA: MIT Press, 2007.

Sara E. Smiley Smith
Yale University

COLORADO STATE UNIVERSITY

Colorado State University (CSU) is proud of its reputation as being "the Green University." Located in Fort Collins, CSU has built an international reputation for extensive research into sustainability, focusing on alternative fuels, clean engines, photovoltaics, smart-grid technologies, wind engineering, water resources, and the use of satellites in atmospheric monitoring and tracking systems. Sustainability efforts are interwoven throughout campus life, but promoting sustainability within the curriculum is considered key to achieving the goals of using education to further the sustainability agenda. Because of this focus, the faculty is ideally situated to create a curriculum that allows students to combine interests in sustainability with chosen fields of study. The existence of various centers, off-site locations, and partnerships provides students across the curriculum with a wide range of opportunities to see sustainability put into practice. The use of interdisciplinary and multidisciplinary programs permits students to approach sustainability from a range of perspectives, leading to solutions to global problems and identifying new fields of study. The curriculum includes opportunities to study sustainability at all levels. Undergraduates may take a 21-credit certificate program offered by the College of Liberal Arts, and graduate programs focused on sustainability cover a range of disciplines. In 2008, Colorado State established the School of Global Environmental Sustainability to serve as an umbrella organization for sustainability-related research and education. Its director works with various committees to improve the sustainability curriculum and design new courses.

Sustainability in the Curriculum

Each of CSU's eight schools offers a number of classes designed to incorporate sustainability into various professions. The Warner College of Natural Resources, for example, accomplishes this goal in the forestry geology and geography programs through such courses as Fishery and Wildlife Biology, Natural Resources Recreation and Tourism, Rangeland Ecology, and Water Resources. Other courses include Wildland Fire Measurements, Forest Ecology, Timber Harvesting and Environment, Wildlife Conservation, and Conservation Biology. A number of sustainability-focused classes are offered through a correspondence program, including Leopold's Ethic for Wildlife and Land, Wildlife Habitat on the Great Plains, Wildlife Habitat Evaluation for Educators, and Wildlife Policy, Law, and Administration. The Warner College is also involved in interdisciplinary sustainability courses.

Students taking courses in the Geology Department study sustainability in classes such as Blue Planet: Geology of Our Environment or Geology of Natural Resources. Geography students take a somewhat different approach, studying sustainability in classes such as Geography of Water Resources and Geography of Hazards. Students in the Natural Resources program may take Environmental Conservation and Global Environmental Systems at the lower level and progress to Integrated Ecosystem Management, Natural Resource Policy and Sustainability, Land Use Planning, and Wilderness Management. Other classes with a sustainability focus include Social Aspects of Natural Resource Management, Restoration Ecology, and Principles of Watershed Management.

Within the College of Liberal Arts, the Anthropology, English, Economics, History, Journalism, Technical Communications, Philosophy/Agricultural Sciences, Political Sciences, and Sociology programs all offer courses geared toward promoting sustainability. Specific courses include Humans in Global Health, Nature Writing, Economics of Energy Resources, Environmental Economics, American Environmental History, and Writing about Science, Health, and the Environment. The Philosophy Department, which has both a master's and a doctoral program geared toward sustainability, offers courses that concentrate on the ethics of either agriculture or the environment. The faculty of the Political Sciences Department offers courses on Environmental Politics and Policies, Politics of Environment and Sustainability, and Environmental Policy and Administration. Sociology professors teach undergraduate classes such as Agriculture and Global Society. Graduate courses include Sociology of Sustainable Development and Environmental Sociology.

The Practical Curriculum

In many departments, the sustainability focus has direct applications to the field of study. Such is the case with Animal Science, Agriculture, Agriculture and Resource Economics, Bioagricultural Sciences and Pest Management, Horticulture and Landscape Management, and Soil and Crops Sciences, which are all housed within the College of Agricultural Sciences. Likewise, there is a practical approach to sustainability in the College of Applied Human Sciences, which offers courses of study in Construction Management, Apparel Design and Merchandising, and Interior Design. Within the College of Veterinary Medicine and Biomedical Sciences, students may study Environmental and Radiological Health Sciences.

Within the College of Agricultural Sciences, students may begin their course of study by taking Plants and Civilization. Later, they progress to World Interdependence and Agricultural Ethics. At the graduate level, agriculture students take Understanding and Managing the Land. Pesticide Management is offered only as a correspondence course. Students combining the study of agriculture with that of resource economics examine sustainability in courses such as Agricultural Law, Economics of World Agriculture, and Economics of Water Resource Planning. Those involved in Bioagricultural Science and Pest Management become sustainability experts by taking such courses as Environmental Fate of Pesticides and Forest Health Issues. Those who concentrate on Horticulture and Landscape Architecture take such courses as Topics in Organic Agriculture and Landscape Irrigation and Water Conservation. Students of Soil and Crop Sciences study Environmental Issues in Agriculture early in their programs and go on to study Manure Management and the Environment as more advanced students.

Colleges such as Natural Science, Business, and Engineering offer particular courses in sustainability as well as incorporating sustainability into the overall curriculum. For instance, advanced botany and zoology students take Landscape Ecology; psychology students study Environmental Psychology; math students take Theory of Population and Evolutionary Ecology; and students studying statistics take Statistics for Environmental Monitoring. Students studying chemistry receive training in sustainability throughout their course of study.

There is a heavy sustainability concentration in much of the curriculum offered by the College of Veterinary Medicine and Biomedical Sciences. Students in this school take courses ranging from Environmental Health early in their programs to Environmental Health and Waste Management, Human Disease and the Environment, and Environmental Toxicology as they continue their academic careers. Students in this program may also serve as interns in Environmental Health. Graduate students in this field take classes such as Cancer Biology, Radiation Public Health, and Epidemiology of Infectious Diseases.

Special Programs and Certifications

Business students may enroll in an 18-month graduate program that leads to a master of science degree in Global Social and Sustainable Enterprise. As part of the program, students intern at institutions with established reputations in the field of environmental conservation and international development. Through the Human Dimensions of Natural Resources Department, students may pursue both undergraduate and graduate degrees with a focus on protected area management, global tourism, natural resource tourism, and environmental communication. Graduates from this program often go to work for state park systems. Others work as outdoor education specialists, tourism planners, recreation managers, and adventure guides.

Physics majors often receive hands-on experience in CSU's Little Shop of Physics, working with the Multiscale Modeling of Atmospheric Processes. Graduate students in the Mechanical Engineering program are involved in seeking solutions to global sustainability problems, working in areas such as biofuels, clean engines, photovoltaic solar cells, battery technologies, materials science, bioengineering, plasmas and space propulsion, dynamics and controls, and motorsport engineering. Students interested in fishery biology, forestry, natural resource management, rangeland ecology, and wildlife biology may participate in summer field programs on the Pingree Park Mountain campus high in the Rocky Mountains.

In response to the growing need to train students to design buildings that meet stringent LEED standards, CSU created the 12-week Green Building Certificate Program. They brought in some of the best-known experts in the field as guest lecturers. Those luminaries included Dave Nelson (daylighting), Alison Mason (solar engineering), and Bruce Hendee (landscape architecture). In 2004, the first students to enroll in the program graduated. Within a few months, two-thirds of them had passed the U.S. Green Building Council's LEED exam.

As a participant in the Colorado Master Gardener Program, Colorado State promotes green gardening. Through its extension program, CSU teaches the public about water conservation, gray water reuse, home irrigation systems, and rainwater harvesting. The 212 acres of the Environmental Learning Center provide students with opportunities to examine wetlands, riparian, cottonwood forest, and prairie ecosystems without leaving the area. The center is also available to the surrounding community, and both families and large groups take advantage of the unique experience provided.

CSU offers the Green Building Certificate Program through the Institute for the Built Environment to evening students. The classes offered through this program train students to meet LEED building standards.

Interdisciplinary and Multidepartmental Programs

Under the auspices of the Warner College of Natural Resources and the College of Natural Sciences, CSU offers an interdisciplinary graduate program in ecology that draws on the expertise of more than 100 faculty members representing 17 departments. Engineering graduate students are able to draw on the many projects sponsored by CSU that give them experience in studying aquatic habit assessment, hazards of land treatment, and storm water management.

CSU also offers the multidepartmental program known as FEScUE, which is funded by the National Science Foundation. It provides a forum for faculty and students from mathematics, statistics, and life sciences to work together in one of three research clusters: Ecology and Evolution, Bioinformatics, and Structural Biology.

PRIMES, the Program for Interdisciplinary Mathematics, Ecology, and Statistics, gives graduate students a chance to get involved in their chosen fields by working on cutting-edge research with experts in the field.

Outreach and Research

While the focus is on sustainability within the curriculum, Colorado State University is also committed to using outreach as a tool to promote national and global sustainability. The university is a member of the Association for the Advancement of Sustainability in Higher Education, and students had the opportunity to interact with experts in the field when Colorado State served as a host institution when the association's annual conference was held in Denver in 2010.

Partnerships are considered essential to sustainability outreach. Representatives from Colorado State worked with the San Diego–based Cannon Power Group to establish a wind farm at Maxwell Ranch, which the university established near the Colorado/Wyoming border. Under the guidance of Dr. W. S. Sampath, the university partnered with local industries to conduct solar research at Abound Solar, a research and development center. Students involved in Live Green spent their spring break in 2010 at the Great Sand Dunes National Park, helping park officials to evaluate the park's recycling program.

Because Colorado State is home to eight major research centers, students have unique opportunities to conduct research and gain experience at locations that include the Colorado State Forest Service, the Colorado Water Institute, and the Colorado Renewable Energy Collaboratory. Opportunities for in-depth research and practical experience often derive from participation in programs offered at various sites owned by the university or at those affiliated with it. Opportunities include working at the 15-acre, two-megawatt solar plant located on the Foothills Campus or at the Center for Multiscale Modeling of Atmospheric Processes, which is affiliated with the National Science Foundation. Students were also able to interact with sustainability experts in 2008 when CSU hosted the Colorado Global Climate Conference.

A number of CSU professors have been involved in groundbreaking sustainability research, and students are frequently involved in finding solutions to problems that threaten sustainability in the broader community. In two instances of outstanding research performed by CSU faculty, a member of the Department of Atmospheric Science was recognized in the prestigious *Geophysical Research Letters* in spring 2010 for developing new ways of examining the ways that aerosols evolve in the atmosphere, and a business professor received the 2010 Olympus Innovation Award in recognition of his establishing the Global Social and Sustainable Enterprise Program.

Student Involvement

Throughout the school year, students at Colorado State University participate in a number of activities designed to promote sustainability. The Green Warrior Campaign of 2010, for instance, encouraged students to conserve water, save electricity, and step up recycling activities. CSU's Live Green Team is made up of sustainability-conscious groups on campus, including the Coalition for Campus Sustainability, the Student Sustainability Center, the Live Green Community, the Student Environmental Leadership Network, the local chapter of Engineers Without Borders, the Environmental Action Collective, the Front Range Student Ecology Symposium, the Range Ecology Club, the Student Association of Fire Ecology, the local chapter of the Wildlife Society, the Environmental Learning Center, and the Emerging Green Builders of Northern Colorado.

Outside Evaluations

Despite the great strides in sustainability made by Colorado State, some outside evaluators believe more needs to be accomplished. In 2010, the Green Report Card awarded CSU an overall grade of B minus. The institution received solid Bs in administration, energy and climate change, and student involvement, and As in food and recycling, green building, and transportation, but evaluators gave the school Fs in the areas of endowment transparency and shareholder engagement because of the lack of public disclosure and public commitment to sustainability on the part of shareholders. The C given in the field of investment priorities was also in response to incomplete information on sustainability financial dealings.

Another assessment conducted by the *Princeton Review* that same year was more positive, citing CSU as one of the nation's top colleges in the field of sustainability education. The university received a score of 91 out of a possible 99 as a result of participation in the Talloires Declaration of 2001 and the American College and University Presidents Climate Commitment in 2008 and for continued efforts to expand sustainability efforts. Particular achievements of note included becoming the first institution in the world to receive LEED recognition for its Commercial Interiors certification program, weather forecasting data collection, adding green power to dormitories, the use of biomass boilers, and recycling efforts. Colorado State ranked 17th of 600 participants in the United States in 2010 in RecycleMania's Grand Champion competition because of successful recycling of more than half of all waste generated on campus.

Despite some criticism, Colorado State has clearly earned its title of the "Green University" because of its strong commitment to sustainability and the strength of its sustainability curriculum.

See Also: Association for the Advancement of Sustainability in Higher Education; Leadership in Sustainability Education; Place-Based Education.

Further Readings

"CSU Denver's Green Building Program Graduates Its First Class." *Colorado Construction,* 7/10 (2004).

Dernbach, John C. *Agenda for a Sustainable America.* Washington, DC: ELI Press, Environmental Law Institute, 2009.

"Forest Service Renews Partnership With College of Menominee Nation in Wisconsin." *Indian Country Today,* December 2, 2009.

Franson, Melissa. *The Impact of Classroom Exposure to Sustainability, Course Content, and Ecological Footprint Analysis of Student Attitudes and Projected Behaviors.* Unpublished Thesis. Auburn University, AL, 2008.

"Green Report Card: Colorado State University." http://www.greenreportcard.org/report -card-2009/schools/colorado-state-university (Accessed June 2010).

"The Green University." http://www.green.colostate.edu (Accessed June 2010).

Heald, Collette, et al. "A Simplified Description of the Evolution of Organic Aerosol Composition in the Atmosphere. *Geophysical Research Letters,* 37 (April 2010).

Hignite, Karla, et al. *The Educational Facilities Professional's Practical Guide to Reducing the Campus Carbon Footprint.* Alexandria, VA: APPA, 2009.

Neapolitan, Jane E. and Terry R. Berkeley, eds. *Where Do We Go From Here? Issues in the Sustainability of Professional Development School Partnerships.* New York: Peter Lang, 2006.

"The *Princeton Review*'s Guide to 286 Green Colleges." http://www.princetonreview.com/ green-guide.aspx (Accessed June 2010).

Rockwood, Larry, et al., eds. *Foundations of Environmental Sustainability: The Coevolution of Science and Policy.* New York: Oxford University Press, 2008.

"School of Global Environmental Sustainability." http://soges.colostate.edu (Accessed June 2010).

Elizabeth Rholetter Purdy
Independent Scholar

COLUMBIA UNIVERSITY

Columbia University, originally know as King's College, was founded in 1754 pursuant to a charter from King George III of England. The college's first classes were held in a school-house on what is now lower Broadway in Manhattan. After an eight-year suspension of instruction caused by the American Revolution, the revitalized college opened under a new name, Columbia College. The post-revolutionary Columbia continued to serve the needs of an urban population, but also welcomed a more economically, denominationally, and geographically diverse group of students and leaders. Today, Columbia continues to serve a diverse population in its distinctly urban setting. Columbia University prides itself on its forwarding-thinking, renowned research as well as its distinguished learning environment for both undergraduate and graduate students. Today, Columbia University counts approximately 3,500 faculty members and 26,400 students among its ranks and is based in two campuses, both in the Manhattan borough of New York City. Sustainability initia-tives and a wide variety of research are organized and overseen by the Earth Institute at Columbia University, a collaborative organization composed of many different centers and organizations within Columbia as well as many outside groups.

The Earth Institute: Solutions for Sustainable Development

The Earth Institute at Columbia University is charged with the overarching goal of achieving environmental sustainability in a world of continuous and constant environmental changes. Research is the foundation of the Earth Institute's work. The research projects of the Earth Institute involve more than 850 scientists, students, and fellows throughout more than 30 Columbia University centers and programs. Through the Earth Institute's Cross-Cutting

Noted economist Jeffrey Sachs directs Columbia University's Earth Institute, which organizes work on sustainability issues carried out by over 30 Columbia centers and programs and involving as many as 850 researchers.

Source: United Nations Development Programme/ Maureen Lynch

Initiative, Columbia encourages its researchers to approach problems of greater intricacy and to examine them with a holistic, interdisciplinary lens. These urgent and multifaceted issues of sustainability are approached with the goals of protecting Earth's ecosystems, oceans, and atmosphere, and broadening social and economic opportunities for Earth's population. The Earth Institute focuses on nine interconnected global issues: climate and society, water, energy, poverty, ecosystems, public health, food and nutrition, hazards, and urbanization.

Water

One of the greatest challenges of the 21st century is, and will continue to be, responsible, sustainable management of water resources. Rapidly growing populations, who are condensing in urban environments, combined with changing climates have begun to constrain freshwater availability in many regions. In order to help develop sustainable water policies and practices around the world, the Earth Institute aims to advance a multiscale predictive capability for water resource assessment at the local, regional, and global levels that takes into account shifting climates and demographics and, therefore, changing needs; to develop an ability to analyze the public and private investment in water resource development; to create technologies for the storage, treatment, and conveyance of water to improve access; to design policy instruments that facilitate incentives for efficient and equitable water use; to test both these developments in real-world settings to demonstrate their applicability; and to disseminate and continually develop the results of its research to support global water resource development and decision making. The Earth Institute hopes that these advances will help the Global Roundtable on Water (GROW) develop as a forum for international discussion of Earth's collective water future.

The Columbia Water Center is a newly developed center that seeks to create change in water usage and to help supply water to some of the most challenging settings in the world. Local partners in these extraordinary settings will assist with the creation and implementation of projects that emphasize the role of market forces and public-private partnerships. The Water Center has begun several projects in many locations throughout Asia, in Koraro, Ethiopia, in the transboundary river basins of the western United States, the Delaware River Basin, and the American Everglades.

Climate and Society

When examining the issues of sustainability and climate and society, the Earth Institute acknowledges that the people of Earth are both drivers of and responders to the changes in the planet's climate and that those same people create greenhouse gas emissions that affect the heat balance of the planet. These actions, reactions, and emissions as well as the direct exploitation of natural resources, considerably affect the climate of Earth. Conversely, the institute also believes that shifting weather patterns impact a society's ability to develop. In the 21st century, the scale of these interactions has reached epic proportions, and the Earth Institute intends to examine how sustainable development may be affected by a variety of stressors, both current and future.

The Columbia Climate Change Center focuses on the above issues. The goals of the Climate Change Center are to coordinate climate science research at Columbia University so that it is able to share information and to promote synergy between climate groups; to outline alleviation and adaption options so that groups are able to take initiative when reacting to anticipated effects of climate variability and change; and to create a framework at Columbia University to provide policy analysis and advice to stakeholders and policy makers.

Research projects conducted through the Earth Institute and its participating centers and groups are working on several issues, such as research of land use and climate change that resulted in the Dust Bowl of the 1930s. Researchers from the Lamont-Doherty Earth Observatory (LDEO) and the NASA Goddard Institute for Space Studies (GISS) investigated how the location and intensity of the epic drought in the United Sates and the dust blown into the atmosphere from eroded farmland interacted during the Dust Bowl. By examining the interrelationship of climate factors in the past and the accompanying health and social impacts, these researchers can gain improved understanding of how future climate variability and change will occur.

Columbia University graduate researchers supported by the Lenfest Center for Sustainable Energy are working toward the design of zero emission power plants by capturing carbon dioxide. Another Earth Institute–based project builds upon technology proposed by the Zero Emission Coal Alliance and investigates the proposed processes so that energy can be gained from coal without pollution. A team from the Center for Research on Environmental Decisions (CRED) has also done a great deal of research communicating the risks associated with climate change. After extensive research into human worry and reactions, the CRED team found that when scientific information is presented in concrete terms, in relation to recent situations, the decision maker has a better understanding of the possible experience.

Energy

Researchers concerned with this modern issue study in-depth energy and the ways it is connected with climate change as well as poverty and sustainable development. Several different Columbia centers and programs are doing forward-thinking research concerning energy, including the Lenfest Center for Sustainable Energy, which aims to develop technologies and institutions to guarantee a sufficient supply of sustainable energy. The center has chosen to focus on solar, nuclear, and fossil-fuel energy technologies and hopes, through carbon capture and storage, to be able to support the energy needs of the world's projected population in 2100 without increased carbon emissions. The Earth Engineering Center (EEC), although broader in general focus, also works to find solutions for the sustainable development of energy resources. A major project of the EEC is developing a process to recover renewable energy from municipal solid waste collections. Furthermore, the ECC has partnered with outside groups to research the most sustainable means for waste management.

Although not focused specifically on research projects, the Global Roundtable on Climate Change (GROCC) does important energy-focused work through the Earth Institute. Made up of stakeholders and decision makers from all over the world, from corporations and governmental and nongovernmental organizations, this group critically examines and attempts to build consensus on climate change issues. In addition to dialogue, GROCC also supports many hands-on research projects and the formation of the Global Task Force on Carbon Capture and Sequestration.

Urbanization

When focusing on urbanization, the Earth Institute works to understand not only city environments but also the interdependencies between those urban environments and rural and suburban landscapes. As Earth's populations quickly expand, they are also condensing in urban centers, which are currently suffering from acute stress. The backbone programs of the Earth Institute's urbanization studies are the Center for Sustainable Urban Development (CSUD), the Urban Design Lab, and the Millennium Cities Initiative. The CSUD focuses on integrating land use and transportation planning into the process of urban development. In addition to actively researching and disseminating information, CSUD is on the ground in developing cities such as Ruiru, Kenya, focusing on effective transportation planning and municipal land use. In addition, in 2007 CSUD organized the Rockefeller Foundation's Global Urban Summit to address issues of urban growth, especially in developing nations.

The Urban Design Lab (UDL) tackles the need for a comprehensive, design-based approach to achieve sustainable urbanism in the future, with New York City as its backdrop and test site. A great deal of the UDL focuses on the design of cities, but it also recognizes that a wide range of expertise and technology is needed to make its work relevant and usable. UDL has recently used the knowledge it has gained from work in New York to help plan and to develop Dongtan-2, a city in South Korea that is rapidly developing and contains abundant waterways, making it an ideal site for implementing new urban design strategies. UDL has also used its knowledge and resources to help reinvigorate the 145th Corridor in Manhattan, a neighborhood with a distinct identity and many opportunities for increased economic development, greater walkability, and improved air quality.

The Millennium Cities Initiative (MCI) works to assist mid-sized cities across sub-Saharan Africa in their efforts to achieve the Millennium Development Goals (end extreme poverty, create primary education opportunities for all, increase gender equality, decrease child mortality, improve maternal health, reduce the spread and effect of HIV/AIDS and other diseases, guarantee environmental sustainability for future generations, and develop a culture of global partnership to ensure development). MCI specifically works to strengthen the relationship between farmers and markets and between Millennium Villages and regional capitals, to increase investment in the Millennium cities and surrounding areas, and to help municipalities and other governing groups carefully use urban development strategies.

Hazards

As the population of the world increases and tends to cluster in crowded areas, the understanding of natural disasters becomes more and more important, and an understanding of how humans respond to them becomes vital to response. The Center for Hazards and Risk Research (CHRR) is the main arm of the Earth Institute that focuses on hazard issues. The mission of the CHRR is to further the predictive science of natural and environmental hazards and to help make risk management more scientific. CHRR acknowledges that one must also look at the social, political, and economic realities of the world, in addition to the scientific realities, when key decisions are being made. Important work of the CHRR includes preparation of natural disaster profiles for Indian Ocean countries that assess risk in relation to geographic and socioeconomic factors and the creation of the Center for International Earth Science Information Network (CIESIN), which works between the branches of the scientific community on online data and information management to create the best understanding possible of human-environment interaction.

Global Health

Global health is one of the focuses of the Earth Institute because poor health negatively impacts the human population and significantly decreases its ability to adapt to environmental change. The Center for Global Health and Economic Development (CGHED), one arm of the Earth Institute, works with low-resource countries to develop quality health systems for their citizens and to achieve the Millennium Development Goals, as well as sustainable economic development. The Access Project, which assists with projects supported by the Global Fund, also focuses on the broad goal of improving health systems for the poor.

The greatest effort focused on global health by Columbia University's Earth Institute has been put into the Millennium Villages. These 78 villages, in more than 10 countries and 12 geographic sites that house about 400,000 people are the focus of many groups' efforts to achieve the United Nations' Millennium Development Goals by 2015 through comprehensive, community-based, low-cost development strategies. Within many of these villages there is an especially high rate of disease within low-income populations, making economic growth nearly impossible.

Poverty

The Earth Institute strongly believes that extreme poverty and climate change are two interrelated issues, and that sustainable development can be planned to meet human needs and help spur economic growth. As Earth Institute researchers study extreme poverty, they keep in mind a variety of causes and symptoms, including hunger, malnutrition, lack of healthcare, lack of education, inadequate water access and sanitation, energy deficits, trade barriers, and gender inequality.

Many Earth Institute researchers tackle extreme poverty through the Millennium Villages and Access Projects described above. Other researchers, through the Center for International Earth Science Information Network (CIESIN), study the distribution of poverty, as well as the geographic and biophysical conditions of a community, through the Poverty Mapping Project. The Tropical Agriculture Program also gives Earth Institute's team members the opportunity to apply their research in a hands-on context in a variety of developing tropical countries—some of the most impoverished countries in the world. The researchers in this program use science expertise, recent technological advancements, management strategies, and policy tools to improve incomes and quality of life through sustainable farming while preserving the long-term integrity of the environment.

Food, Ecology, and Nutrition

Closely related to the concerns of extreme poverty in today's modern world are the dilemmas of food, ecology, and nutrition. Balanced diets, reliable food sources, clean drinking water, and stable agriculture create stability and well-being in a population. Unfortunately, most of the extremely poor areas in the world, often those most affected by climate change, do not have access to these things. The Earth Institute's researchers acknowledge that caloric intake is not the only indicator of an individual's or a society's health; instead, they believe that health environments and sufficient nutrients fulfill dietary needs. Most of the work of the Earth Institute concerning food, ecology, and nutrition is done through previously described projects such as the Millennium Villages and Cities project, the Tropical Agriculture Program, and much more.

Ecosystems, Health, and Monitoring

Ecosystems sustain invaluable services to the Earth such as air and water purification and food production, and support remarkable numbers of species. The Center for Environmental Research and Conservation (CERC) conducts a great deal of the Earth Institute's research on this global issue. CERC works to build environmental leadership and to solve complex problems in order to slow the loss of biological diversity and to achieve sustainable development. CERC conducts training courses to address the complex topics of climate change that affect society today. The Earth Institute is also a member of the Black Rock Forest Consortium, a group of education organizations that maintains a 4,000-acre Black Rock Forest field station that allows researchers from academia to research, hands-on, an ecosystem in New York State and provides K–12 students the opportunity to explore the area's many scientific wonders and to learn about the science of sustainability.

See Also: Education, State Green Initiatives; Experiential Education; Higher Education Associations Sustainability Consortium; University Employee Training: Private Partnerships.

Further Readings

Chichilnisky, G. and G. Heal. *Environmental Markets: Equity and Efficiency.* New York: Columbia University Press, 2000.

Columbia University. http://www.columbia.edu (Accessed February 2010).

McCaughey, R. A. *Stand Columbia: A History of Columbia University in the City of New York.* New York: Columbia University Press, 2003.

Stephen T. Schroth
Jason A. Helfer
Martha S. Baumgarten
Knox College

Constructivism

Constructivism puts the individual at the center of learning, forming meaning through experience. The definition of constructivism is multifaceted, depending on the context in which it is being applied, as a field or a practice in progressive education. The theoretical roots can be found most notably in the work of philosophers and psychologists. Among them were 18th-century philosopher Giambattista Vico, and his philosophy of learning belief that people can only understand what they have themselves constructed; John Dewey; Jean Piaget; and Lev Vygotsky—to name a few. Today, the relevancy of constructivism in green education is debatable. Controversial issues call into question whether advocacy of constructivism as an educational model can legitimately aid care of the natural world.

For green education, constructivism is both a blessing and a curse. The curse is that today's learners spend notably less time outside and are discovering degraded environments as their norm. Perceptions, values, and beliefs are being constructed within these experienced realities. Misconceptions can become deep seated in the mind of the learner. The construction of meaning is thus in danger of perpetuating erroneous beliefs and a narrow perspective of the world. The blessing is that educators have an opportunity to explicitly offer the individual experiences in which meaningful relationships and care of the natural world can be explored. Educators have an urgent and critical role in altering misconceptions and challenging the learners' existing framework of knowing about the natural world.

Traditional education has been dominated by lessons led by educators who impart knowledge, facts, and figures that are to be memorized with the expectation of fostering understanding. Yet common sense dictates that knowing about the world is embedded, in part, on individual experience. This is the basic premise of constructivism. Individual experience provides context and the basis for constructing a way of knowing about and making sense of the world. Concepts evolve over time as individuals reframe their understanding based on experiencing emerging conditions or disequilibrium. Understanding progresses as the learner confronts opposing ideas and reconstructs new ways of knowing.

For a term largely concerned with individual processes for making meaning, it is ironic that an agreed-upon definition for *constructivism* is difficult to pin down. The above, overly simplified explanation is not enough to elucidate the meaning of nuanced constructivism. Constructivism is referenced in relationship to philosophy, developmental psychology, education theory, and instructional methodology. While associations to constructivism are complicated, the individual's acquisition of new knowledge through experience is distinguishable by the shared aims of philosophers and psychologists John Dewey, Jean Piaget, and Lev Vygotsky. A deeper look at the work of each clarifies a few of the distinctions related to the various ways in which the term *constructivism* may be applied.

John Dewey advocated for education reform that ensured integrating individual experience in education. He saw experience that led to knowing about the world as increasing the individual's ability to function in society and an integral part of developing social responsibility.

The psychologist Jean Piaget believed individual learners have distinct developmental stages and mental patterns that guide behavior based on interpreted experiences regardless of culture. He also claimed that the learner would make accommodations when mental patterns were in what he called "disequilibrium." Disequilibrium is the point at which the learner is confronted with new information and existing knowledge no longer fits with experience. This allows the learner the flexibility to reconstruct the concept.

Lev Vygotsky recognized that individual meaning making can be deterred by the limits of one's own experience. He realized that educator-led instruction was important to provide scaffolding. In other words, he encouraged educators to understand the learner's existing knowledge and developmental stage, then to build upon this by adding new experiences. This could lead to the altering of understanding and contribute to new knowledge. Vygotsky maintained that learning was bound by linear developmental stages that he called the Zone of Proximal Development (ZPD). ZPD started with the basic naming of experience that ultimately led to more abstract thoughts, reaching higher metacognitive processing. Piaget claimed that thoughts develop language, whereas Vygotsky asserted that language develops thoughts and affects cognitive processing.

In the review of current constructivism literature, several controversial issues stand out. One issue argues that constructivism contributes to individuals' long-held misconceptions. Another concern suggests that constructivism disregards the value of cultural knowledge conveyed across multiple generations. And another infers that constructivism contributes to a capitalist paradigm focused on individualistic progress that is detrimental to the environment. While each of these concerns is legitimate, each is missing the important role of the educator in aiding adaptability and reframing the context of experiences based on the learner's preexisting references. Misconceptions can be corrected at the appropriate developmental stage in time. And social-cultural, as well as environmental, relevancy can be incorporated into learning experiences.

Constructivism provides meaningful context to the world and reality of the learner that cannot be found in facts alone. Neither radical constructivism nor an overemphasis of rote learning is productive. The interdependent balance of constructivism, alongside the accumulation of contextual facts, is essential to green education in the 21st century.

See Also: Environmental Education Debate; Experiential Education; Place-Based Education.

Further Readings

Cakir, Mustafa. "Constructivist Approaches to Learning in Science and Their Implications for Science Pedagogy: A Literature Review." *International Journal of Environmental & Science Education*, 3 (2008).

Gordon, Mordechai. "Between Constructivism and Connectedness." *Journal of Teacher Education*, 59 (2008).

Kruckeberg, Robert. "A Deweyan Perspective on Science Education: Constructivism, Experience, and Why We Learn Science." *Science & Education*, 15 (2006).

H. Courtney White
Antioch University New England

Cornell University

The account of Cornell University's path toward sustainable operations embodies a rich history extending back more than three decades and is characterized today by the type of university-wide integration and virility demonstrative of a leader among institutions of

Weather and environment sensors near the waters of Lake Cayuga, which Cornell uses for lake source cooling on campus. Because of this project, Cornell has achieved an 86 percent reduction in electricity use for cooling campus buildings since 2000.

Source: iStockphoto.com

higher education. The tradition of engaging the campus community in meaningful and practical ways, as Ezra Cornell originally did by employing students in constructing the first university buildings, remains intact today. While the New York state land grant institution's campus in Ithaca has grown to include over 250 buildings totaling 14 million square feet and is inhabited by over 20,000 students, a seemingly proportional number of student groups, administrative committees, and executive-level task forces are actively planning and implementing measures that reduce the environmental impact of the campus. Issues of energy supply, emissions of greenhouse gases (GHGs), material flows, engaging outside services, and long-term planning for the institution are deliberated over by these various committees with the goal of reducing the institution's impact on its ecological support systems.

Energy

Cornell has instituted numerous efficiency and conservation projects and programs across the Ithaca campus. The Cornell Combined Heat and Power Project is an example of a large-scale initiative, providing heat and electricity at an efficiency of 75 percent, compared with a 49 percent efficiency associated with generating each form of energy separately. At its core, the project involves an addition to the existing power plant of two newly installed natural gas turbines that are each paired with electric generators capable of producing a total of 30 megawatts of electricity. The heat produced during the combustion is captured and used to make steam. While the current performance of the system is not high enough to completely offset the use of coal, roughly 80 percent less coal was used when the plant came online in January 2010. Coal will be phased out by mid-2011. At that point, the $82 million project is estimated to reduce the university's GHG emissions by 28 percent. In addition to reducing environmental impacts, the new infrastructure and a fuel flexibility that includes biofuels were sought to help reduce the costs of providing energy to existing buildings and to account for the increasing demand for electricity and heat needed for future construction. Similar projects were undertaken at Princeton University, Massachusetts Institute of Technology, Yale University, and the University of Cincinnati, among others.

Lake source cooling is another of Cornell's large energy projects recognized for increasing the sustainability of the school's operations and energy supply. When the project came online during the summer of 2000, it reduced the electricity used for cooling in campus buildings by 86 percent, or 10 percent of the total electrical use on campus. Rather than sourcing cooling from mechanical chillers driven by electricity, lake source cooling employs the cold waters of nearby Lake Cayuga to condition the air. An inlet located 240 feet below the surface of the lake draws in water that is 39 degrees Fahrenheit. The cold water is pumped through a heat exchanger through which a separate water loop is chilled and then routed through the campus air conditioning. The heat is transferred from the buildings to the lake and dissipates during the winter. This innovative process reduces the emission of GHGs and other pollutants resulting from the traditional generation of cooling that relies on the use of electricity and refrigerants. In addition to lake source cooling, the university has added both hydroelectric and solar power into its renewable energy portfolio. Roughly 2 percent of the total electric demand is met by a local hydroelectric power plant at Fall Creek, supplied by water from Beebe Lake. Although providing significantly less power than the hydro plant, installations of photovoltaic panels on a few buildings as well as highly visible locations such as atop bus shelters, serve as outreach and teaching tools available to the campus community. The university also investigated the potential for developing a wind farm consisting of eight turbines. A pre-feasibility study highlighted issues and documented concerns of local residents and ultimately led administrators to suspend further study.

Cornell's tradition of energy conservation has generated $7 million in savings through projects implemented and operating during the two decades between 1980 and 2000. Such success was made possible through the vision and work of the utilities department as well as numerous campus committees. Since 1980, for example, a committee of representatives from Campus Life has engaged students through education on the issue of energy use in the residence halls. Similar task forces and groups continue to drive sustainable operations through the present century. The Kyoto Task Team, a group of students, faculty, and staff, was formed by the Utilities Department in 2001. The objective of the group was to position the university to achieve a reduction of GHG emissions of 7 percent below 1990 levels by 2008, as outlined in the Kyoto Protocol. The team developed energy conservation projects and programs, vetted paths toward greater energy efficiency, oversaw studies of renewable energy options, and established the auxiliary working group the Energy Conservation Initiative. The initiative was charged with reducing energy demand in campus buildings by 20 percent and was supported in that effort by the addition of university staff responsible for retrocommissioning, or "tuning-up, " building systems. Additional groups with a related focus on energy include the Cornell University Renewable Energy Society, a student group advocating alternative energy, and the Facilities Services Sustainability Computing Group, a committee of employees conducting outreach and education on reducing the power consumed by computers and office peripherals.

Buildings

The U.S. Green Building Council's Leadership in Energy and Environmental Design (LEED) standards and certification process has gained prominence in the arena of sustainable construction, renovation, and maintenance. Surrounding issues such as building site, materials, construction practices, and heating, ventilating, and air conditioning, LEED standards target environmentally friendly options. The Green Buildings Oversight Committee at Cornell has taken steps to ensure that these standards are reflected in the university's approach to its built environment. The committee, made up of administrators and faculty, instituted a green building policy for projects that surpass the $5 million mark.

Buildings in this category must meet two criteria. First, they must be built to LEED Silver standards. Second, they must achieve at least 30 percent energy savings relative to the nationally recognized baseline for energy efficiency established by the American Society of Heating, Refrigerating and Air-Conditioning Engineers (ASHRAE). Presently, Cornell's LEED buildings include Weill Hall and the Alice H. Cook House. Weill Hall, built in 2008 and measuring 263,000 square feet, is the largest life science teaching and research facility in the state of New York. It achieved LEED Gold. The Alice H. Cook House, built in 2004, earned a similar accolade—it was the first LEED-certified green residence hall in the state. Numerous other building projects on campus are aspiring toward LEED certification.

Beyond LEED standards, the Cornell University Planning Office and the Campus Planning Committee target long-term sustainability goals that will continue to meet the needs of the campus community into the future. Among their considerations are issues of land use and maintaining a balance with natural landscapes with respect to building sites as well as decisions regarding infrastructure developments associated with university operations forecast over a 30-to-60-year time horizon. The Department of Building Care, together with the Environmental Health and Safety Office, strives to preserve the health of the campus community by maintaining the health and safety of the building stock. In support of this mission, these offices identify and vet the least invasive cleaning products, equipment, and custodial protocol. Chemical-free products, including numerous Green Seal Certified products, are vetted and used whenever possible. The employees who use the products complete specific trainings with the goal of ensuring the health and safety of the staff as well as the buildings. Such deliberations and thoughtful controls improve indoor air quality through the reduction of airborne contaminants in addition to extending the life of the buildings and infrastructure.

Climate Action Plan

In June 2007, President David Skorton signed the American College and University Presidents Climate Commitment (ACUPCC). In addition to various tangible action items and near-term emission reduction goals, the commitment prescribed that the university develop a Climate Action Plan that presented a path toward climate neutrality, or net-neutral emissions. Cornell's plan targets 2050 as the date by which it will achieve neutrality. The process of developing the plan was intentionally designed to engage the campus community as well as the surrounding community. The design also carried the goal of serving as a step-by-step model for other organizations, including municipalities and larger governing bodies, to review and adopt for their own use. The first step involved developing an inventory of GHG emissions. Cornell estimates it produced roughly 319,000 metric tons of CO_2 equivalent (MTCO2e) per year, 55 percent from on-site combustion, 27 percent from purchased electricity, and the remaining 18 percent from commuting and air travel. Next, over 700 ideas for reducing emissions were gathered from campus and local community stakeholders. Ultimately, the ideas were organized into 19 unique initiatives outlining mitigations under five headings: Green Development, Energy Conservation, Alternative Transportation, Fuel Mix and Renewable Energy, and Offsetting Actions. The various committees tasked with developing the specific actions nested within these headings bolstered the emissions reduction measures by simultaneously identifying complementary educational and research opportunities as well as appropriate vehicles for targeted outreach.

Transportation

Cornell's award-winning Transportation Demand Management (TDM) program is the model among institutions of higher education. The program focuses on reducing the parking demand of faculty and staff through various incentive structures. Efficient and environmentally friendly alternatives to single-occupancy trips include subsidized bus passes for the regional public bus system and rideshare incentives. Strict regulations and enforcement for on-campus parking and the adoption of a fee structure for individual parking permits that sets higher costs for parking spots closer to the center of campus are both supported by shuttle services that connect parking lots along the campus periphery. Within a year of implementation, the program effectively reduced the number of parking permits issued by 25 percent. Participation in the ridesharing enjoyed a 10 percent increase. Taken together, the elements of the TDM program resulted in about 33 percent participation among faculty and staff and 2,600 fewer cars on campus. Cornell also promotes bicycling and pedestrian activities as an additional measure that secures green spaces from the construction of parking lots. Web resources include state bike laws and maps offering information on routes and bike rack locations. The campus fleet incorporates alternative fuel vehicles as well. Diesel vehicles operate on fuel with at least a 5 percent blend of biodiesel, and some farm vehicles use as much as 20 percent in the fuel mix. The fleet also uses about 10 electric cars to handle lighter-duty operations.

Waste Management

Solid Waste Management at Cornell falls under the purview of the Grounds Department. Roughly 11,000 tons of material is discarded each year across the campus; roughly 66 percent of this is diverted from landfills and incineration through composting or recycling. The Farm Services Department manages an eight-acre composting facility. It is the second-largest facility in the county and was recognized by the U.S. Environmental Protection Agency, receiving its Environmental Quality Award in 2009. The facility collects waste on a fee-for-service basis from campus animals, greenhouses, food scraps, and agricultural activities. The compost produced is used on campus in place of chemical fertilizers, thereby closing the loop on material flows and turning wastes into a useful soil amendment. Recycling bins on campus collect mixed office paper, corrugated cardboard, scrap metal, glass, and plastic. Universal wastes such as florescent lamps and electronics are collected and recycled as well. In addition to recycling and reusing, Cornell also focuses on the reduction of waste through outreach efforts such as trayless dining and reducing paper usage by making double-sided copies when printing.

Purchasing

The Green Purchasing Task Force at Cornell leads the institution through a culture shift in procurement protocols and contract changes that increase the use of products and services recognized as having a lower impact on human health and the environment. Composed of student, faculty, and staff volunteers, this group approaches purchasing on two mutually reinforcing fronts. Utilizing the Central Purchasing Department's Website, the task force provides resources to the campus that facilitate the identification of green suppliers and products. Such resources include detailed lists of environmentally preferred suppliers that rate vendors according to their production methods, transportation system, and life cycle performance. The availability of a green certification logo key increases the ease of identifying a preferred vendor by cataloging and explaining logos such as Energy Star, Green Seal, and the Forest Stewardship Council. In addition to providing the Web-based resources, the task force actively engages the campus community on issues of green purchasing by advising and attempts to set new normative behavior through publicly acknowledging and reinforcing the adoption of a green procurement protocols.

Dining Services

From food supply to managing food wastes, Cornell's Dining Services offers a multitude of examples for how to incorporate sustainability into university operations. The Cornell Dining Local Food Advisory Council, which emerged from the effort of students to increase the amount of locally grown food and is now staffed by chefs, food distributors, and faculty, guides education and outreach initiatives across the campus on issues of sustainable food and agriculture. Through the work of the council, roughly 23 percent of the produce consumed on campus is sourced from local and regional growers, including purchases from the student-run organic farm located on campus. In addition to farm tours, themed dinners connect students to local growers and issues of sustainable food production. The 30 dining units across campus all offer Fair Trade coffee, and numerous locations operate a trayless dining program that significantly reduces the amount of food and water waste. Reusable mugs offset the use of disposal drinking vessels, biodegradable packing is purchased and used for to-go menu items, and all the pre- and postconsumer food scraps are composted.

See Also: American College and University Presidents Climate Commitment; Kyoto Declaration on Sustainable Development; Yale University.

Further Readings

Cornell University. "Cornell Sustainable Campus." http://www.sustainablecampus.cornell.edu (Accessed June 2010).

Cortese, Anthony D. "The Critical Role of Higher Education in Creating a Sustainable Future." *Planning for Higher Education*, 31/3 (2003).

Keniry, Julian. *Ecodemia: Campus Environmental Stewardship at the Turn of the 21st Century.* Washington, DC: National Wildlife Federation, 1995.

Koyanagi, Dean. "From Advocating to Institutionalizing Sustainability at Cornell University." In *The Green Campus: Meeting the Challenge of Environmental Sustainability,* Walter Simpson, ed. Alexandria, VA: APPA, 2008.

Sharp, Leith. "Green Campuses: The Road From Little Victories to Systemic Transformation." *International Journal of Sustainability in Higher Education*, 3/2 (2002).

David R. Schmidt
Independent Scholar

CRE COPERNICUS CHARTER

In the fall of 1993, the Copernicus Charter was first presented to the Association of European Universities (CRE) at its conference in Barcelona, Spain. Drafted by COPERNICUS (the Co-Operation Programme in Europe for Research on Nature and Industry Through Coordinated University Studies), the University Charter for Sustainable Development was intended to galvanize university action around environmental sustainability in Europe. As of 2009, the charter has been signed by 305 university leaders in 37 European nations.

The Copernicus Charter first acknowledges the urgent need for humanity to shift to more sustainable practices, citing the Brundtland Commission's definition of sustainable development, or, caring for the needs of the present without compromising the ability of future generations to meet their own needs. It then asserts the need for a clear plan of action that has as its focus the preparation of decision makers and teachers to become empowered with knowledge about sustainable development.

Given this broad call to action, the charter proposes the following 10 principles of action:

- *Institutional Commitment:* Universities shall demonstrate real commitment to the principle and practice of environmental protection and sustainable development within the academic milieu.
- *Environmental Ethics:* Universities shall promote among teaching staff, students, and the public at large sustainable consumption patterns and an ecological lifestyle, while fostering programs to develop the capacities of the academic staff to teach environmental literacy.
- *Education of University Employees:* Universities shall provide education, training, and encouragement to their employees on environmental issues, so that they can pursue their work in an environmentally responsible manner.
- *Programs in Environmental Education:* Universities shall incorporate an environmental perspective in all their work and set up environmental education programs involving teachers, researchers, and students—all of whom should be exposed to the global challenges of environment and development, irrespective of their field of study.
- *Interdisciplinarity:* Universities shall encourage interdisciplinary and collaborative education and research programs related to sustainable development as part of the institution's central mission. Universities shall also seek to overcome competitive instincts between disciplines and departments.
- *Dissemination of Knowledge:* Universities shall support efforts to fill in the gaps in the present literature available for students, professionals, decision makers, and the general public by preparing informative didactic material, organizing public lectures, and establishing training programs. They should also be prepared to participate in environmental audits.
- *Networking:* Universities shall promote interdisciplinary networks of environmental experts at the local, national, regional, and international levels, with the aim of collaborating on common environmental projects in both research and education. For this, the mobility of students and scholars should be encouraged.
- *Partnerships:* Universities shall take the initiative in forging partnerships with other concerned sectors of society in order to design and implement coordinated approaches, strategies, and action plans.

- *Continuing Education Programs:* Universities shall devise environmental educational programs on these issues for different target groups; for example, business, governmental agencies, nongovernmental organizations, and the media.
- *Technology Transfer:* Universities shall contribute to educational programs designed to transfer educationally sound and innovative technologies and advanced management methods.

Although these 10 principles offer significant guidance, they do not frame or explicitly direct action for sustainable development, leaving universities ample room for interpretation. While this flexibility can lead to greater creativity or to setting appropriate action, it can also result in inaction. Criticisms of these nonbinding environmental agreements cite lax or little change among signatory institutions. The Copernicus Charter attempted to circumvent such struggles by requiring that signatories be high-level university administrators, in the hopes that the decision to participate would involve prior thoughtful consideration.

See Also: Brundtland Report; International Alliance of Research Universities: Sustainability Partnership; Sustainability Officers; United Nations Decade of Education for Sustainable Development 2005–2014.

Further Readings

Calder, Wynn and Richard M. Clugston. "Progress Toward Sustainability in Higher Education." *The Environmental Law Reporter: News & Analysis,* 33/1 (January 2003).
International Institute for Sustainable Development. "Declarations for Sustainable Development: The Response of Universities." (2010). http://www.iisd.org/educate/declare.htm (Accessed August 2010).
United Nations Educational, Scientific and Cultural Organization. Education for Sustainable Development, United Nations Decade of Education for Sustainable Development. "Copernicus" (2010). http://portal.unesco.org/education (Accessed August 2010).

Sara E. Smiley Smith
Yale University

Dalhousie University

Dalhousie University, one of Canada's leading higher education institutions, maintains three campuses in downtown Halifax, Nova Scotia. Dalhousie was established in 1818 and currently has an international student body of over 15,000 students. Dalhousie has been a leader in environmental and sustainability education for over 20 years, often working in collaboration with governmental, nongovernmental, educational, private-sector, and community groups at the local, national, and international levels. Dalhousie offers degrees at both the bachelor's and master's levels through the School for Resource and Environmental Studies. The College of Sustainability was founded in 2008 and houses Canada's first degree-granting program in Environment, Sustainability, and Society. Dalhousie's commitment to campus sustainability further enhances its curriculum.

Academic Degree Programs

Dalhousie University's interdisciplinary environmental studies curriculum is based on the themes of environmental and resource management, science, and policy. Environmental and sustainability courses and degrees are housed within the School for Resource and Environmental Studies (SRES). SRES is one of four departments housed within the Faculty of Management. SRES was formerly the Institute for Environmental Studies, founded in 1973. SRES collaborates with other campus programs, including the Marine Affairs Program and the Lester Pearson Institute for International Development as well as with other institutions, both in Canada and internationally, including the Nova Scotia Agricultural College.

The school's interdisciplinary approach to natural resource management, environmental studies, and sustainable development explores many different perspectives and draws from a variety of diverse disciplines, such as the natural and social sciences, law, and policy and public administration. Specific topics of interest include climate change, depletion of the ozone layer, long-range transport of air pollutants, contamination and depletion of freshwater and marine resources, deforestation, soil degradation and desertification, loss of biological diversity, mismanagement of chemicals and wastes, overpopulation, rural decline, and urban decay. The school also seeks to broaden student education through the promotion of community service activities.

Dalhousie University offers a four-year undergraduate Bachelor of Management program in which students may chose environmental management, among other areas of concentration. The program is a joint collaboration between the four schools within the Faculty of Management: Business Administration, Information Management, Public Administration, and Resource and Environmental Studies. The program emphasizes management in the public and nonprofit sectors, although the private sector is also covered. Dalhousie also offers a Cooperative Education bachelor of science program that combines academic coursework with alternating periods of career-related work-experience terms. Dalhousie's campus features Ocean Pond, a man-made ecosystem that serves as a learning center for biology students.

Dalhousie is also a leader in graduate environmental studies. The university has offered a two-year Master of Environmental Studies (MES) program since 1978. The MES program is part of the School for Resource and Environmental Studies. Students complete coursework and a thesis emphasizing natural resource and environmental issues. Dalhousie has offered a 16-month Master of Resource and Environmental Management (MREM) program since 2004. Students complete coursework and an applied internship in order to gain the techniques and scientific knowledge to solve resource and environmental problems. Although the School for Resource and Environmental Studies does not offer a Ph.D. program, it participates in the interdisciplinary Ph.D. program. Graduate students include both those seeking a professional career and workers seeking to enhance their job skills.

Dalhousie established the College of Sustainability in 2008 to provide a comprehensive, collegial base for the interdisciplinary study of sustainability-based problems. The college houses Canada's first degree-granting program in Environment, Sustainability, and Society (ESS). In addition to the ESS major, ESS students may also pursue a double major or combined honors program in a variety of other disciplines. Participating faculty members are drawn from dozens of academic disciplines both to teach and to work on campus and in community-based sustainability projects. Graduates leave as globally aware citizens with critical thinking and environmental leadership capabilities. The college also hosts lectures, seminars, and activities for students, faculty, and the greater community.

Dalhousie students further their environmental education and gain firsthand experience in the implementation of sustainability practices through participation in student organizations and activities related to sustainability. Dalhousie maintains an Office of Sustainability to coordinate and oversee campus-wide faculty, staff, and student sustainability initiatives. The office annually directly employs between four to seven students and involves many others through its various sustainability projects. The Dalhousie Student Union houses its own student-run Sustainability Office and directly communicates its concerns and ideas to the Office of Sustainability as well as to the campus administration.

Dalhousie students have the opportunity to join a wide variety of campus organizations, many with a primary or secondary environmental focus. Student societies that promote sustainability education and activism include SustainDal, the Dal Environmental Law Society, the Student Society, the Dal/Kings Oxfam Society, and the College of Sustainability Student Society. SustainDal holds campus water and energy conservation and waste reduction campaigns. Student-led sustainability events include Green Week and an annual Dump and Run, a community-wide garage sale held as an environmentally friendly alternative to the annual campus ritual in which departing students dispose of unwanted items.

Campus Sustainability Initiatives

Dalhousie's campus sustainability initiatives enhance the environmental curriculum by serving as a model for students and a place where they can apply what they have learned in the classroom. The President's Advisory Council on Sustainability is composed of student as well as faculty and community representatives. The council passed the Sustainability Policy and Statement of Principles, and the university released a Sustainability Plan for Dalhousie in 2010 with provisions for the publication of annual reports thereafter to measure progress toward the achievement of sustainability goals. The President's Advisory Council also provides a forum for the discussion of sustainability issues. Dalhousie has also completed a greenhouse gas inventory and a Sustainable Transportation Plan.

Dalhousie has implemented a campus-wide effort to incorporate sustainability into planning, assessment, and policy development and implementation. Dalhousie administration seeks to engage students, faculty, and staff members into its sustainable initiatives and community and business partnerships. Dalhousie opened its university Office of Sustainability in 2008 to help implement individual and organizational change and educational programs. The director of Sustainability and a team of student assistants run the Office of Sustainability. The office presents an annual Dalhousie Sustainability Trophy and other awards to those residence and university groups that implement the most effective campus sustainability initiatives.

Campus sustainability initiatives include resource conservation; the use of phosphate-free cleaning products; the use of recycled paper and other sustainable products; recycling and disposal of hazardous waste; alternative and sustainable transportation methods; travel avoidance goals; increased bike friendliness; green building design and system and building-wide sustainability upgrades; local, organic, and Fair Trade food purchasing; trayless cafeteria service; campus community gardens; energy-efficiency standards; and climate strategies. Dalhousie is seeking LEED and other green certification for all new and renovated buildings, such as the Kenneth Rowe Building, Life Science Research Institute, Life Science Center, and New Academic Building. Past and ongoing projects include the One Million Acts of Green campaign, the Paper Cut campaign, the opening of a Campus Bike Centre, the ChemEx Chemical Exchange Program, and ReThink! Sustainability on Campus.

Dalhousie participates in a number of broader sustainability initiatives as well. Dalhousie is a member of the Association for Advancement of Sustainability in Higher Education (AASHE) and the Canada Green Building Council, and became an Energy Star Partner in 2009. The university has signed three international declarations related to environment and sustainability: the Halifax Declaration, the Talloires Declaration, and the UNEP International Declaration on Cleaner Production as well as the University and College's Climate Change Statement for Canada. The Sustainable Endowments Institute awarded Dalhousie an overall grade of B on the College Sustainability (Green) Report Card.

See Also: Association for the Advancement of Sustainability in Higher Education; Green Week, National; Whole-School Approaches to Sustainability.

Further Readings

Barlett, Peggy F. and Geoffrey W. Chase. *Sustainability on Campus: Stories and Strategies for Change*. Cambridge, MA: MIT Press, 2004.

Dalhousie University. "Office of Sustainability." http://office.sustainability.dal.ca (Accessed June 2010).

M'Gonigle, R. Michael and Justine Starke. *Planet U: Sustaining the World, Reinventing the University.* Gabriola Island, British Columbia, Canada: New Society Publishers, 2006.

Rappaport, Ann and Sarah Hammond Creighton. *Degrees That Matter: Climate Change and the University.* Cambridge, MA: MIT Press, 2007.

Sustainable Endowments Institute. "College Sustainability Report Card (Green Report Card)." http://www.greenreportcard.org/report-card-2010/schools/dalhousie-university (Accessed June 2010).

Marcella Bush Trevino
Barry University

DARTMOUTH COLLEGE

Founded in 1769, Dartmouth College is the smallest school in the Ivy League, with an enrollment of under 6,000. Originally founded to educate and Christianize Native Americans, it now boasts a nationally ranked undergraduate program and business, engineering, medical, and other graduate programs. Named one of the world's most enduring institutions in 2004, Dartmouth is located in Hanover, New Hampshire. With a 269-acre campus, Dartmouth owns the 27,000-acre tract of land, the Second College Grant, which is sustainably forested and maintained. Dartmouth has a long history of environmental sustainability, research, and educational initiatives, many of which have been student initiated and grassroots with strong support and leadership from the faculty and administration.

The history of student initiatives began in 1909 when a student founded the Dartmouth Outing Club to increase interest in winter sports. It is now the nation's oldest and largest college outing club, with programs that include maintaining cabins and 75 miles of the Appalachian Trail, canoeing and kayaking, rock and ice climbing and mountaineering expeditions, and hunting and fishing. Dartmouth's practical approach to environmental conservation also led to early faculty involvement and institutional commitment. Dartmouth's first three Environmental Studies courses were offered in 1970, coinciding with the first Earth Day, held that year. Among the oldest in the country,

While Dartmouth's main campus is only 269 acres, the college also owns this 27,000-acre expanse of land, called the Second College Grant, which is used for wilderness activities and research and to demonstrate sustainable forestry.

Source: Wikipedia

the college's Environmental Studies Program was interdisciplinary from the beginning, engaging the biology, geology, government, and engineering departments. Its first director was Gordon MacDonald, a member of President Richard Nixon's Council on Environmental Quality. With a $100,000 grant from the National Science Foundation and a grant from the Henry R. Luce Foundation, the fledgling program was broadened to include applied and theoretical aspects of environmental science and policy. In 1972, Donella Meadows began her 29-year tenure with the department. A MacArthur Fellow, she remains one of the most influential environmental thinkers of the 20th century, authoring nine books, including the groundbreaking *Limits to Growth,* publishing the biweekly *The Global Citizen,* and founding the Sustainability Institute. By 1980, Dartmouth's Environmental Studies Program passed its probationary review period and, like other departments, has since had full-time, tenure-track faculty.

In early 1980s, student interest in environmentalism waned and enrollment in the program decreased. While other institutions dropped their programs altogether, Dartmouth faculty globalized the environmental science courses, which helped to renew student interest in the latter half of the decade. In 1989, Dartmouth alumni created the Dartmouth Environmental Network to increase awareness of and professional opportunities for Environmental Studies students. On Earth Day 1996, the college officially announced the offering of a major in Environmental Studies. The program now also includes a minor in Environmental Science, Environmental Studies, modified major options, and an off-campus program in South Africa. Dartmouth College's efforts went beyond the classroom. In 1988, recommendations to make the college's food system more sustainable resulted in a campus-wide composting system and an organic farm. The subsequent campus recycling system became one of the nation's most successful campaigns, with student volunteers taking charge of dormitory recycling. The farm initiative started to gain momentum when a group of Environmental Studies students and members of the Dartmouth Outing Club built a small organic vegetable plot three miles north of the college. In this educational working garden, students continue to engage in independent research, projects, and experiential learning in sustainable food and energy systems. The organic farm also maintains an interdisciplinary approach, with departments ranging from biology to religion utilizing it in lectures, workshops, and class projects.

Meanwhile, student-initiated grassroots efforts became a major impetus for Dartmouth's sustainability efforts. One of the first in the 1990s was the Environmental Conservation Organization (ECO), which focused on recycling in the dorms. The ECO continues today, broadening its reach to advocate for climate change awareness and energy efficiency. Several first-year students also worked to pilot a sustainable living floor in one dormitory cluster, known as East Wheelock. This living experiment eventually evolved into the creation of the Sustainable Living Center (SLC) in 2007, an on-campus residence educational and experiential community with 19 beds. Students run and lead the living center as a freestanding building that aims for independence from outside sources of energy and minimizes the output of waste products. Future plans for the SLC include physical retrofits as well as an organic garden and a root cellar to allow residents to eat as locally and sustainability as possible.

One of the most highly visible student initiatives, the Big Green Bus, began in 2004. Students from the environmental studies and engineering departments converted a retired school bus to run on waste vegetable oil and retrofitted it with recycled, sustainably produced materials, energy-efficient appliances, and solar panels. It is now a coach bus, and every summer since 2004, a group of about 15 students travels across the United States raising awareness about the interconnections between people, economy, and the environment, and promoting sustainable living practices.

Members of Dartmouth's Greek system also work to make the fraternities, sororities, and coeducational social houses more sustainable. These efforts became formally recognized and coordinated under the Green Greeks Program in 2006. This formalization included a new position of a Greeks sustainability intern, whose key role is to create environmental programs within the Greek system.

In the spring of 2006, students launched the Sustainable Move-Out to reduce the waste that occurs during spring move-out and fall move-in. Students clean and sort reusable appliances and room decorations that others historically tossed, selling them to new and returning students and community members each fall. The proceeds are used to support environmental initiatives on and off campus. Other student-led efforts include the Dartmouth Animal Welfare Group, focused on raising awareness about issues of animal ethics on campus, increasing students' interactions with animals, and supporting national movements that also promote animal welfare, especially as it relates to the food industry, product testing, and scientific laboratories. Meanwhile, the student group Ecovores raises campus awareness about food and the economic, social, and environmental trade-offs between various modes of agricultural production. Ecovores works with local farmers and producers to encourage and develop a stronger local, sustainable food supply with the college's dining services, which, as of 2010, receives 4 percent of its food from local sources. Green Lite Dartmouth was started to provide real-time feedback on energy usage in dormitories by displaying an animated polar bear.

Dartmouth's environmental efforts were not only student-run, however. In an effort to coordinate both the institutional and student-led environmental initiatives, the college created the Dartmouth Sustainability Initiative in 2005 to focus on "reducing the environmental footprint of the College by integrating sustainability principles and practices into campus operations and working with student organizations to increase campus awareness of sustainability issues." In 2006, the Environmental Protection Agency (EPA) and the Department of Transportation recognized it as one of the best places for commuters, with its transportation initiatives such as Zipcars, carpooling, and free shower passes for bikers and walkers. Dartmouth is also one of the seven institutions that comprise a consortium supporting the Hubbard Brook Ecosystem Study, the longest-running, most comprehensive ecosystem study in the world at the 7,600-acre Hubbard Brook Experimental Forest in New Hampshire. In 2008, Kathy Lambert, an alumna of Dartmouth who had started the Science Links Program with the Hubbard Brook Research Foundation to bridge science and public policy, became the college's sustainability manager. The college also completed an audit of 25 percent of its main buildings that use 70 percent of its energy. As a result, the trustees of the college allocated $12.5 million in loan funds for renewable energy projects around campus, including efficient lighting upgrades and daylighting controls.

In 2009, Dartmouth began a three-year project funded by the Northeast Sustainable Agriculture Research and Education Program to create a long-term, profitable relationship between local farmers, the college, and the surrounding community. That same year, Dartmouth received a $330,000 grant from the New Hampshire Green House Gas Reduction Fund to support its commitment to reduce greenhouse gas emissions by 30 percent below 2005 levels by 2030. The Energy Conservation Campaign and Energy Pledge Drive encourage students, staff, and faculty to commit to 8 to 12 actions that conserve energy and that use energy more efficiently. Over 2,000 have signed the Energy Pledge as of early 2010, with measurable reductions in energy use across campus.

Building on its interdisciplinary history, the Dickey Center for International Understanding funds the Dartmouth Council on Climate Change on how global warming and climate change have social, environmental, economic, and political impacts at the local and international levels. A delegation of students extends their advocacy efforts by lobbying for climate change legislation in the U.S. Congress as part of Powershift, a nationwide grassroots effort. The Dickey Center also houses the Institute for Arctic Studies, founded in 1989 to lead polar studies and advocate for the study of Arctic climate change. Other international environmental efforts include the Global Security Fellows Institute, a partnership with the University of Cambridge to study the environment, economic development, ethnic identity, and migration across transnational boundaries.

Whether international or local, faculty driven or student driven, Dartmouth College has continued to stay at the forefront of ecological research and conservation.

See Also: Greening of the Campus Conference; HESA: Higher Education Sustainability Act; Leadership in Sustainability Education.

Further Readings

Dartmouth College. "Dartmouth Outing Club." http://www.dartmouth.edu/~doc (Accessed November 2009).

Dartmouth College. "Dartmouth Sustainability Initiative." http://www.dartmouth.edu/~sustain (Accessed November 2009).

Energy Action Coalition. http://energyactioncoalition.org (Accessed November 2009).

Hubbard Brook Ecosystem Study. http://www.hubbardbrook.org (Accessed November 2009).

Meadows, Donella. *Beyond the Limits: Confronting Global Collapse, Envisioning a Sustainable Future*. White River, VT: Chelsea Green, 1992.

Meadows, Donella. "The Global Citizen, 1996–2001." http://www.pcdf.org/meadows (Accessed November 2009).

Meadows, Donella. *The Limits to Growth*. White River, VT: Chelsea Green, 1972.

Meadows, Donella. *The Limits to Growth: The 30 Year Update*. White River, VT: Chelsea Green, 2004.

Northeast Sustainable Agriculture Research & Education. http://nesare.org (Accessed November 2009).

Belinda H. Y. Chiu
BChiu Consulting

DECLARATION OF THESSALONIKI

In 1997, participants from nongovernmental organizations, as well as intergovernmental and governmental representatives of over 83 countries met in Thessaloniki, Greece, for the International Conference on Environment and Society: Education and Public Awareness for Sustainability. Out of this conference came the unanimously supported Declaration of Thessaloniki, which put forth the critical idea of education for sustainable development.

The Thessaloniki Declaration argued that the concept of environmental sustainability must be linked with poverty, population, human rights, and health. In addition, Thessaloniki was critical in advancing the connection to and importance of education for sustainability. With regard to formal education, this declaration affirmed that all subject disciplines must address issues related to the environment and sustainable development and that university curricula should be reoriented toward a holistic approach to education.

In 1995, a major United Nations Educational, Scientific and Cultural Organization (UNESCO) International Workshop was held in Athens on Reorienting of Environmental Education (EE) to Sustainable Development, which recommended the organization of an international conference to examine progress 20 years after Tbilisi (1977), which had formally introduced environmental education. The focus of the conference at Tbilisi was the importance of substantive social change to precipitate necessary environmental change and impact. The declaration also recognized that sustainability initiatives must take place at all levels of society and must be transdisciplinary in nature, although the idea of education for sustainable development (ESD) has roots in the first United Nations (UN) Conference on the Human Environment (1972) and was more fully articulated in Chapter 36 of the Rio Declaration (1992), which recognized education as the prerequisite for sustainable development and described the needed provisions for such education. The workshop in Athens also recommended a five-year follow-up on the Rio Declaration of 1992.

Participants at Thessaloniki reinforced the Tbilisi platform of social change and interconnectedness and argued that the concept of environmental sustainability must be clearly linked with poverty, population, food security, democracy, human rights, peace, health, and a respect for traditional culture. By turning the focus to systems and structures of education, the Thessaloniki Declaration affirmed that all subject disciplines must address issues related to the environment and sustainable development and that university curricula is a critical location for creating a more sustainable future. Finally, the declaration called for governments and leaders in education to honor the commitments they had already made in signing past declarations (e.g., Talloires) of environmental sustainability and to take action to that end.

Sustainable development is presented as the objective to which education should devote itself as an instrument of choice. Education is seen as an ongoing process aimed at developing the capability of adapting to rapid changes in the world, but first and foremost, it is seen as a process of transmitting knowledge and information to make the public understand the problems and to stimulate awareness around sustainability and the environment. The declaration states: "Education is an indispensable means to give to all women and men in the world the capacity to own their own lives, to exercise personal choice and responsibility, to learn throughout life without frontiers, be they geographical, political, cultural, religious, linguistic or gender." (Article 9)

The conference itself was the result of preparatory work during the numerous regional workshops that were held throughout the world in the months preceding the meetings at Thessaloniki. The conference opened the door to continuing discussions aimed at clarifying the concepts associated with "environmental education" as well as laying a framework for understanding the concept of education for sustainable development. One of the key documents to come out of the conference was the work based on the continuing activities under Agenda 21, Educating for a Viable Future: A Multidisciplinary Vision for Concerted Action, which proposes an analysis of the concept of sustainability linked to the concept of sustainable development. This repositioned the concept of sustainability within economic, environmental, and social sustainability with a clear connection to matters of human rights, equality, and education. It states:

The reorientation of education as a whole towards sustainability involves all levels of formal, nonformal and informal education in all countries. The concept of sustainability encompasses not only environment but also poverty, population, health, food security, democracy, human rights and peace. Sustainability is, in the final analysis, a moral and ethical imperative in which cultural diversity and traditional knowledge need to be respected. (Article 10)

While acknowledging the relevance of previous recommendations about environmental education (Belgrade, 1975; Tbilisi, 1977; and Moscow, 1987) and the importance of continuing the important work of these declarations, the Declaration of Thessaloniki has been criticized by some and supported by others for the shift away from "environmental education" toward "education for sustainable development."

The declaration emphasizes the respect for the capital constituted by natural and cultural resources and recommends introducing the concept of sustainability. Environmental education for sustainable development is viewed as an area of participation in civil society and as an instrument that will enhance understanding of how natural biogeochemical mechanisms work and their impact on socioeconomic choices.

A critical and lasting implication of the Thessaloniki Declaration is not only the holistic concept that sustainability must incorporate social and economic sustainability in addition to environmental sustainability, but recognition of the critical place of education in creating sustainable development. The work is also cited as a catalyst for the development of the Lüneburg Declaration of 2001, which called for the recognition of higher education as a critical location for education on sustainability. This declaration prioritized the role of education for a sustainable future and made way for the United Nations' establishing 2005–2014 as the Decade of Education for Sustainable Development.

See Also: Agenda 21: "Promoting Education, Public Awareness, and Training" (Chapter 36); Lüneburg Declaration; Tbilisi Declaration; United Nations Decade of Education for Sustainable Development 2005–2014.

Further Readings

International Conference, Environment and Society: Education and Public Awareness for Sustainability. "Declaration of Thessaloniki." (December 1997). http://www.unesdoc .unesco.org/images/0011/001177/117772eo.pdf (Accessed June 2010).

Meadows, Donella. *Beyond the Limits: Confronting Global Collapse, Envisioning a Sustainable Future.* White River, VT: Chelsea Green, 1992.

United Nations Educational, Scientific and Cultural Organization (UNESCO). http://www .unesco.org/new/en/unesco (Accessed June 2010).

Lyndsay J. Agans
University of Denver

DICKINSON COLLEGE

One of the oldest institutions of higher learning in the United States, Dickinson College is located in Carlisle, Pennsylvania. A private liberal arts college, it was founded by visionary Benjamin Rush in 1783 and named after John Dickinson, one of the major players in the debate surrounding the Declaration of Independence. Dickinson grew quickly, along with the new nation that had successfully broken its bonds with the mother country.

With around 2,400 students, modern-day Dickinson, which has pledged to become carbon neutral by 2020, is kept intentionally small. In the 21st century, the focus at Dickinson College is on integrating sustainability into all facets of college life. All departments and members of the campus community are held accountable for actions adding to the college's carbon footprint, and all are involved in working to ensure that sustainability goals are met. Both curriculum and support programs are designed to promote sustainability. The Office of Campus Sustainability and the Center for Environmental and Sustainability Education oversee activities dealing with sustainability at Dickinson. Students are given the opportunity to intern with either Facilities Management or the Center for Environmental and Sustainability Education to gain hands-on experience in the field. Special programs include the Biodiesel Shop, the College Farm, and the Alliance for Aquatic Resource Monitoring (ALLARM).

Various committees have been created to deal with particular aspects of sustainability at Dickinson. The President's Commission of Environmental Sustainability (PCES) draws on the expertise of faculty and staff. Meeting twice each semester, its membership is made up of campus administrators and representatives from each college division. The Society Advocating Environmental Sustainability (SAVES) and the Socially Responsible Investment Committee are open to all students, faculty, and staff. Other groups include the Sustainable Foods Committee, Students for Social Action, and EarthNow! Dickinson alumni continue to play a role in sustainability activities through Alumni for Sustainable Dickinson.

Office of Campus Sustainability

The Office of Campus Sustainability, which oversees campus operations, is responsible for many of the day-to-day activities involved in promoting sustainability at Dickinson. The sustainability coordinator serves as the liaison between operations personnel and faculty, students, and staff. In addition to operations carried out by the Office of Campus Sustainability involving waste management, recycling, ride sharing (Zipcars), and student residential programming, the office encourages students to participate in special programs and campaigns such as RecycleMania, Earth Day, and the Green Devil Challenge (a program aimed at reducing energy and water consumption by 10 percent). As part of Dickinson's Trash on the Plaza program, each day volunteers go through trash in order to encourage recycling. Under the Red Bike program, which aims to cut down on carbon dioxide emissions, Dickinson students may borrow bicycles free of charge. As each school year ends, volunteers collect unwanted items and hold a huge yard sale. All proceeds are donated to the United Way.

Dickinson's sustainability operations, carried out by the Office of Campus Sustainability, have received international acclaim. In 2007, Dickinson instituted the Interior Space Temperature Policy for both heating and cooling systems. The campus uses the central energy plant to provide steam heat and chilled water to most of the campus. A system of dual-fuel boilers utilizes both gas and oil, depending on cost and availability. Facilities Management has the capability to monitor heating systems on much of the campus in order to employ greater fuel efficiency. Renewable Energy Certificates are used to offset electric distribution. To reduce carbon dioxide emissions, Dickinson purchases only electricity generated from wind power and operates hybrid vehicles.

Dickinson is extremely proud of its Biodiesel Shop, a student-led initiative. Using only vegetable-oil waste generated by local restaurants, the shop produces 50 gallons of biodiesel at a time. Dickinson has also used composting to promote sustainability, reducing the food waste formerly sent to landfills by 700 pounds a day. Student workers take on the responsibility of collecting composted waste each day and transporting it to the College Farm, where it is used as fertilizer.

All new buildings are constructed according to sustainability standards. Two recent buildings won the Leadership in Energy and Environmental Design (LEED) Gold certification. New buildings are designed to meet the standards of Silver certification, and existing buildings are being renovated to meet LEED certification standards. Much attention has been paid to Dickinson's Center for Sustainable Living, generally referred to as the Treehouse. The Treehouse, which utilizes a corn-pellet stove, dual-flush toilets, faucet aerators, a gray water system, and Energy Star appliances, serves as a model for sustainable living.

Center for Environmental and Sustainability Education

While the Office of Campus Sustainability is involved in carrying out sustainability operations, the Center for Environmental and Sustainability Education is engaged in developing a curriculum designed to promote sustainability. This office also offers sustainability courses, faculty development activities, and sustainability modules. It is also engaged in both community and global research. The office is heavily involved in all elements of institutional planning in the field of sustainability and considers communication and outreach integral to the goal of using education to promote sustainability. Activities in that endeavor include hosting conferences, developing living laboratory initiatives, and providing education on climate change.

Outside Evaluations

In 2010, Dickinson College received an A– from the Green Report Card. The college was awarded solid A grades for its commitment to green building, sustainable landscaping, energy efficiency, water conservation, waste reduction, and responsible financing and purchasing decisions. Dickinson was also praised for having two committees and eight full-time staff devoted to advancing sustainability and for its extensive use of sustainability internships. Particular attention was paid to the fact that by 2030, Dickinson will try to reduce greenhouse gas emissions by 75 percent from 2008 levels, and for its positive efforts toward achieving food system sustainability. Despite strong efforts at construction and transportation sustainability, Dickinson received Bs, indicating a need for improvement in these areas.

In 2010, the *Princeton Review* evaluated the sustainability efforts of American colleges in the *Green Honor Roll*. Dickinson was one of only 15 schools to achieve the highest score of 99. The chief reasons for this high score were the creation of the Center for Environmental and Sustainability Education in 2008 and Dickinson's efforts at integrating sustainability into all aspects of college life. The report paid particular attention to efforts toward food sustainability, noting that by foregoing nonbiodegradable material in its dining hall, Dickinson has been able to turn 800 pounds of organic material and compostable tableware into compost each week. Dickinson was also praised for collecting used fryer oil to be turned into biodiesel for use by college vehicles. Construction sustainability efforts also contributed to Dickinson's high rating.

Dickinson College's commitment to integrated sustainability serves as a model for other educational institutions and supports the advisability of promoting sustainability through education.

See Also: Leadership in Sustainability Education; Sustainability Officers; Whole-School Approaches to Sustainability.

Further Readings

Dernbach, John C. *Agenda for a Sustainable America*. Washington, DC: ELI Press, Environmental Law Institute, 2009.

Franson, Melissa. *The Impact of Classroom Exposure to Sustainability, Course Content, and Ecological Footprint Analysis of Student Attitudes and Projected Behaviors*. Unpublished Thesis. Auburn University, 2008.

"Green Report Card 2010: Dickinson College." http://www.greenreportcard.org/report-card-2010/schools/dickinson-college (Accessed June 2010).

Hignite, Karla, et al. *The Educational Facilities Professional's Practical Guide to Reducing The Campus Carbon Footprint*. Alexandria, VA: APPA, 2009.

Neapolitan, Jane E. and Terry R. Berkeley, eds. *Where Do We Go From Here? Issues in the Sustainability of Professional Development School Partnerships*. New York: Peter Lang, 2006.

"The *Princeton Review*'s Guide to 286 Green Colleges." http://www.princetonreview.com/green-guide.aspx (Accessed June 2010).

Rockwood, Larry, et al., eds. *Foundations of Environmental Sustainability: The Coevolution of Science and Policy*. New York: Oxford University Press, 2008.

"Sustainability at Dickinson College." http://www.dickinson.edu/about/sustainability (Accessed June 2010).

Elizabeth Rholetter Purdy
Independent Scholar

DRURY UNIVERSITY

Drury University is a private liberal arts school with approximately 1,600 students, located in Springfield, Missouri. The university was founded by Congregational ministers in 1873. Drury adheres to the Native American belief that humans inherit the Earth from their ancestors and borrow it from their children, and the university is strongly committed to the concept of sustainability outreach. Drury is one of more than 500 American schools that belong to the Association for the Advancement of Sustainability in Higher Education.

In 2005, Drury University began a campus-wide effort to become a greener university, and designated it a Year of Sustainability. Throughout the 2005–2006 school year, the university hosted environmentalist speakers and groups, offered programs on sustainability, and encouraged students to explore how they could make the world a greener place.

The Drury curriculum is designed to teach students about the environment from an interdisciplinary perspective. In addition to environmental elements incorporated into biology, architecture, history, psychology, and other subjects, classes such as Sustainable Development are designed solely to promote sustainability. Students enrolled in this class receive practical experience, engaging in projects such as conducting water assessments on campus and researching recycling. First-year students at Drury often take Global Perspectives for the 21st Century to learn about climate change, habitat loss, overpopulation, urbanization, and globalization. Instead of taking a general ethics course, Drury students may opt for Environmental Ethics. Many of the activities involving sustainability at Drury University are overseen by the director of campus sustainability, who works under the auspices of the

Office of Campus Operations and Sustainability. Operations activities at Drury that promote sustainability include installing a ground source heat exchange system at Stone Chapel, installing hybrid photovoltaic/solar thermal systems in some dormitories, converting light fixtures, bathroom fixtures, and exit signs to make them more energy efficient, and replacing older campus vehicles with electric vehicles. Despite its success in the areas of curriculum and operations, it is in the area of sustainability outreach that Drury has been most successful. The university is constantly involved in partnerships with other schools and with the business community in an effort to educate the public about sustainability.

Outreach and Interdisciplinary Sustainability

While an interdisciplinary approach is used to coordinate sustainability efforts at Drury, many of the schools within the university also offer special classes designed to promote sustainability. Such classes frequently offer students opportunities for outreach, a cornerstone of Drury's sustainability agenda. For example, the Hammons School of Architecture offers studios and theory courses that examine relevant elements of sustainability. In 2008, the Design/Build Studio gave students hands-on experience when they built a Habitat for Humanity residence that received LEED Platinum recognition. This residence was one of only 44 edifices in the United States to receive that recognition, and it was the first Habitat for Humanity dwelling to be designed to meet such high sustainability standards. Students in the Chemistry Department and fifth-grade students from local schools also learned about sustainability as they worked as volunteers on the home. The residence was designed to take advantage of natural resources for both heating and cooling. Solar energy was used, and a storm and water management system was employed outdoors. Only drought-resistant plants were used. Drury architecture students have also worked with the *Extreme Makeover: Home Edition* (ABC) team to incorporate sustainability into the designs of three residences built for families with particular needs. Since 2008, the architecture curriculum has included a Sustainable Design track for students who choose to concentrate their efforts on sustainable architecture.

Other programs at Drury University also combine curriculum instruction with sustainability outreach. Students involved in Environmental Studies, Environmental Science, and Environmental Health Science partnered with the Water Shed Center, a Springfield–based nongovernmental organization (NGO), to establish a field station that is engaged in ecological and water quality research. Students within Drury's Communication Department regularly participate in campus and community projects dealing with public relations, advertising, communication, and broadcasting. Students in the Business Department are able to use actual case studies to increase their knowledge of green business practices. Drury created the Ozarks Center for Sustainable Solutions with funds provided by the Department of Natural Resources. As part of the program, student interns work with local schools and businesses on pollution prevention, finding solutions to problems that threaten their sustainability. One such project involved retrofitting a large number of school buses with diesel-emission reduction equipment to cut down on carbon dioxide emissions.

Partnerships and Finances

Drury University partnered with Ozarks Technical Community College, an adjacent campus, to begin building a new shared student life complex in 2010. The garage has park-and-plug capacity and storage for bicycles. A bike shop is also available on the premises.

Additionally, the complex offers greenway trail access. All student housing was designed to meet green standards. A retail center offers Fair Trade coffee and an organic grocery store and cafe, which are locally owned.

Remaining committed to administrative and financial sustainability is an integral element of incorporating sustainability into all facets of campus life and engaging in continued sustainability outreach. At Drury, the 40-member Sustainability Council works closely with other relevant individuals and departments to oversee such activities. In 2007, Drury increased its sustainability planning by adding new full-time staff devoted to financial sustainability. Each Drury student pays a Sustainability Fee, and the Sustainability Council has an annual budget of approximately $20,000.

Outreach is also an important element in promoting administrative and financial sustainability, and Drury annually partners with Missouri State University, the City of Springfield, Green County, the National Biomass Producers Association, and local manufacturers to sponsor the Ozarks New Energy Conference to teach the public about sustainability.

The Pitt Family and Sustainability

At any institution of higher learning, celebrity donors help to call attention to the cause of sustainability. In 2009, Hollywood actor Brad Pitt and his younger brother Doug, both of whom grew up in Springfield, contributed $600,000 in honor of their father to aid efforts at obtaining LEED Gold certification for the O'Reilly Family Event Center being built at Drury University. To show their gratitude, officials named the playing field of the basketball arena the William A. Pitt Court after Pitt senior. In addition to the $6 million donated by the O'Reilly family, the Mabee Foundation and various donors contributed an additional $1 million to construct the 3,100-seat arena, which became the first arena in the United States to be constructed on such stringent sustainability guidelines. Ecofriendly features of the center include low-flow water fixtures and a recycling area. In addition, both the roof and the pavement are made from solar-reflective materials, and renewable energy is generated as a result of using environmentally friendly energy systems.

Outside Evaluations

Drury University students have often won recognition for their sustainability efforts, and these undertakings frequently have an outreach focus. In 2009, Drury's Students in Free Enterprise team placed second in the United States in recognition of its development of the Ozarks Carbon Exchange Fund, which allows them to partner with local schools and NGOs to promote sustainability.

Drury has received considerable attention from outside sources in recognition of its sustainability efforts. In 2009, Drury University was one of only four schools to earn an Honorable Mention from the Green Report Card. Drury's inclusion in this prestigious list was based largely on its achieving LEED Platinum status on the Habitat for Humanity home and on plans to retrofit existing low-income houses to make them more environmentally sound. It was noted that Drury had major potential for becoming one of the powerhouses in sustainability among American institutions of higher learning. In 2010, the local press recognized Drury's efforts at food-service sustainability when trays were removed from the dining area as an energy-saving endeavor. That same year, Drury University was the only institution of higher learning in the state of Missouri to be named as a green college in "The *Princeton Review*'s Guide to 286 Green Colleges."

True to its mission to improve the planet and pass on a healthier and more sustainable world to the children of the future, Drury University has managed to make great strides in incorporating sustainability into higher education and in educating its students and the wider community about the environment in which they live.

See Also: Association for the Advancement of Sustainability in Higher Education; Place-Based Education; Whole-School Approaches to Sustainability.

Further Readings

Association for the Advancement of Sustainability in Higher Education. "Drury University." http://www.aashe.org/resources/profiles/2009/drury-university (Accessed June 2010).

Barker, Tom. "Campuses Are Going Green." *St. Louis Post-Dispatch,* May 26, 2010.

"Champions of Sustainability in Communities." http://www.greenreportcard.org/report -card-2009/awards/champions-of-sustainability (Accessed June 2010).

Dernbach, John C. *Agenda for a Sustainable America.* Washington, DC: ELI Press, Environmental Law Institute, 2009.

Franson, Melissa. *The Impact of Classroom Exposure to Sustainability, Course Content, and Ecological Footprint Analysis of Student Attitudes and Projected Behaviors.* Unpublished Thesis. Auburn University, Alabama, 2008.

Hignite, Karla, et al. *The Educational Facilities Professional's Practical Guide to Reducing the Campus Carbon Footprint.* Alexandria, VA: APPA, 2009.

Lee, Evelyn. "Habitat for Humanity and Drury University LEED Platinum Home." http:// greenweb.federatedmedia.net/archives/526 (Accessed June 2010).

Neapolitan, Jane E. and Terry R. Berkeley, eds. *Where Do We Go From Here? Issues in the Sustainability of Professional Development School Partnerships.* New York: Peter Lang, 2006.

"President's Council on Sustainability." http://www.drury.edu/section/section.cfm?sid=303 (Accessed June 2010).

"The *Princeton Review*'s Guide to 286 Green Colleges." http://www.princetonreview.com/ green-guide.aspx (Accessed June 2010).

Rockwood, Larry, et al., eds. *Foundations of Environmental Sustainability: The Coevolution of Science and Policy.* New York: Oxford University Press, 2008.

Rogers, Stephanie. "Brad Pitt Helps Missouri's Drury University Go Green." http://earth first.com/brad-pitt-helps-missouri%E2%80%99s-drury-university-go-green (Accessed June 2010).

Elizabeth Rholetter Purdy
Independent Scholar

E

EARLY CHILDHOOD EDUCATION

Children taking part in a conservation education project in Roma, Texas, in 1997. Early childhood education for sustainability augments outdoor learning and nature education by introducing a greater focus on environmental and sustainability topics.

Source: U.S. Department of Agriculture, Natural Resources Conservation Service

Young children are the most vulnerable and most at risk of environmental challenges, current and future. Yet early learning about environment and sustainability issues and topics has been neglected and underrated in early childhood education, even though there is an expanding body of research literature—from economics, neuroscience, sociology, and health—that shows that early investments in human capital offer substantial returns for individuals and for communities and have a long reach into the future. Early childhood education for sustainability (ECEfS)—a synthesis of early childhood education (ECE) and education for sustainability (EfS)—builds on groundings in play, outdoor learning, and nature education, but takes a stronger focus on learning about, and engagement with, environmental and sustainability issues. Child participation and agency is central to ECEfS and can relate, for example, to local environmental problem solving such as water and energy conservation or waste reduction in a child care center, kindergarten, or preschool, or young children's social learning for indigenous reconciliation and cultural inclusivity. While the ECE field has been much slower than other educational sectors to take up the challenges of sustainability, this situation is rapidly changing as early childhood practitioners begin to engage—it is fast moving from the margins of early childhood curriculum and pedagogic decision making into the mainstream.

This presents challenges, however, as ECEfS is somewhat misunderstood and misrepresented and, as a new field, is under-researched and under-theorized.

Despite its slow start, there is now a groundswell of interest in ECEfS, especially with the impetus of the United Nations Decade of Education for Sustainable Development 2005–2014. This contrasts with the "patches of green" that characterized the sector previously. In Australasia, for example, the first ECEfS conference was held in New Zealand as recently as late 2006. International interest in ECEfS has mirrored Australian and New Zealand developments, with an international coalition now beginning to emerge. The inaugural international gathering focused on ECEfS was only in 2007. This led to the first international report on ECEfS published by UNESCO (2008). A follow-up meeting in November 2008 resulted in *The Gothenburg Recommendations on Education for Sustainable Development*. This document identifies early childhood as a "natural starting point" for all ongoing EfS within a framework of lifelong learning. This is an important inclusion as this is the first time that ECEfS has been so recognized as contributing to education for sustainability. Some media commentators have portrayed early childhood education for sustainability as greenwashing that scares preschoolers and sets them on pathways that may lead to fear of the future, disempowerment, and disengagement and that might escalate environmental worries into anxiety disorders or depression. They argue that ECEfS simply transfers environmental problems created by adults over to children to resolve. Education for sustainability in early childhood that contributes to such pathologies is clearly inappropriate; indeed, it is unethical. Child participation and child agency are central ideas in ECEfS—the field is based in children's rights. These rights are not only bound to the present, however; they extend to the rights of children to be involved in shaping and creating healthy and sustainable futures. A rights-based approach to ECEfS recognizes that young children are current and future citizens who are already affected by environmental decision making. This kind of ECEfS also recognizes young children as already having capabilities and capacities for shaping and creating their worlds. They are not simply passive recipients of adult agendas.

Others argue that before-school learning about environmental and sustainability issues is inappropriate as preschoolers are too young to understand the complex ideas and relationships involved in sustainability and, further, that they are incapable of actively engaging with sustainability issues and topics. Yet there is increasing evidence derived from childhood studies and children's rights literature that demonstrates children's capability to actively engage with, and transform, practices that relate directly to them. Furthermore, there are a small number of studies from the emerging ECEfS field that show that preschoolers are capable of learning about topics such as water and energy conservation and waste management, and that they can be activists for creating change. Furthermore, their actions have been shown to positively influence parental behaviors such that there are intergenerational learning benefits from children to adults resulting in parents changing behaviors at home toward more environmentally friendly practices such as conserving water and using public transport. The early evidence is compelling that ECEfS has the potential to make real contributions to sustainability.

There are two key challenges for this fledgling field, however. The first concerns teacher/practitioner education for sustainability. Without the engagement of their teachers and caregivers, young children are unlikely to have substantial and ongoing opportunities to participate in and learn through environmentally focused activities and projects. There needs to be significant uptake by higher education and technical institutions and colleges where early childhood teachers and other early childhood practitioners are educated and trained. There is a clear role for early childhood policy developers, professional associations, and teacher educators and trainers—both pre-service and in-service—to lobby for EfS to be integrated into early childhood teacher education.

The second challenge for ECEfS relates to research. While rising interest by early childhood practitioners in education for sustainability is heartening, the ECEfS field has been largely neglected by researchers. At a time when teacher interest is on the rise, there is only a small evidence base—and the associated theorizing that comes with research—on which to grow the field. As teachers engage, it is vital that an evidence base emerge in parallel. This is important because good research adds legitimacy to practitioners' work, helps to theorize what is largely an atheoretical field of educational endeavor, informs program improvement and minimizes the replication of mistakes, and enhances prospects for funding to support new and ongoing initiatives.

Early childhood education for sustainability has energy and much to contribute, both to early childhood education and to education for sustainability. As the challenges described above are addressed, ECEfS is poised to make a difference in the lives of young children now and into the future.

See Also: Sustainability Teacher Training; Sustainability Topics Correlated to State Standards for K–12; Sustainability Topics for K–12; United Nations Decade of Education for Sustainable Development 2005–2014.

Further Readings

Centre for Environment and Sustainability. *The Gothenburg Recommendations on Education for Sustainable Development*. Gothenburg, Sweden: Chalmers University of Technology and University of Gothenburg, 2009.

Davis, J. "Revealing the Research 'Hole' of Early Childhood Education for Sustainability: A Preliminary Survey of the Literature." *Environmental Education Research*, 15/2 (2009).

Pramling Samuelsson, I. and Y. Koga. *The Contribution of Early Childhood to a Sustainable Society*. Paris: UNESCO, 2008.

Julie Davis
Queensland University of Technology

EARTH CHARTER INTERNATIONAL

The Earth Charter is an international declaration of 16 principles, or ethics, that were collectively created to try to ensure a global commitment to sustainability, justice, and peace in the 21st century. The Earth Charter seeks to act as a universally binding document that calls for a global commitment "to bring forth a sustainable global society founded on respect for nature, universal human rights, economic justice, and a culture of peace. Towards this end, it is imperative that we, the peoples of Earth, declare our responsibility to one another, to the greater community of life, and to future generations." The Earth Charter is a quintessentially liberal document in that it seeks to honor the rights of individuals while uniquely extending those rights to the Earth and all living beings. Similar to the Universal Declaration of Human Rights in terms of proposing global ethical standards, the Earth Charter was created by individuals around the world, as opposed to international governing bodies.

The development of the Earth Charter derived from the 1987 United Nations World Commission on Environment and Development (the Earth Summit) report, which called for a new charter to address the issue of sustainable development. In 1994, building on this earlier idea, Maurice F. Strong (chairman of the Earth Summit), Mikhail Gorbachev, and others worked to launch the Earth Charter as a civil society initiative that included not just humans but all life on Earth. By early 1997, an Earth Charter Commission was formed and from 1997 to 2000, 42 national Earth Charter committees around the world were established, each adding input to the creation of the charter. Finally, it was approved at a meeting of the Earth Charter Commission at the United Nations Educational, Scientific and Cultural Organization (UNESCO) headquarters in March 2000, and the official launch was on June 29, 2000, in The Hague, Netherlands. Since then, many endorsements have occurred around the world by political, religious, educational, nongovernmental, and activist organizations representing millions of people, while others have rejected it on a number of differing grounds.

Charter Topics

The charter opens with a preamble that states,

> "We stand at a critical moment in Earth's history, a time when humanity must choose its future. As the world becomes increasingly interdependent and fragile, the future at once holds great peril and great promise. To move forward we must recognize that in the midst of a magnificent diversity of cultures and life forms we are one human family and one Earth community with a common destiny."

It further affirms the need for a global commitment, as quoted above, and sees the 16 principles as the way to ensure the desired outcomes. The 16 principles are organized into four pillars: (1) Respect and Care for the Community of Life, (2) Ecological Integrity, (3) Social and Economic Justice, and (4) Democracy, Nonviolence, and Peace. The 16 principles each have a key sentence followed by supporting details. For example, the key sentence for one of the principles under the first pillar is: "1. Respect Earth and life in all its diversity." A principle from the third pillar is: "9. Eradicate poverty as an ethical, social and environmental imperative." Another from fourth and last pillar is: "16. Promote a culture of tolerance, nonviolence and peace." Thus, it can be said that the principles affirm the ethics of equality, democracy, and justice for the world's people and other living beings, without any concrete commitment to implementation. However, the charter ends with a section titled "The Way Forward" that aims to inspire action. It begins by stating: "As never before in history, common destiny beckons us to seek a new beginning. Such renewal is the promise of these Earth Charter principles. To fulfill this promise, we must commit ourselves to adopt and promote the values and objectives of the Charter."

Charter Supporters

The Earth Charter is the first collectively created "people's" charter to challenge and question what many see as global capitalism's unequal and destructive use of the Earth's resources. It is a unique declaration of people's global commitment to liberal values promoting the common good. As such, liberal institutions at all levels from local governments, nongovernmental organizations (NGOs), educational institutions, religious and professional groups, as well as socially responsible private companies have signed on in support. However, as of 2010, world governments have not fully endorsed it, nor has the United Nations. As a result, it is not legally binding but acts as goals to be achieved. Supporters also commend its unique understanding of the interconnections of oppressions and its recognition that all human needs are inseparable from the Earth. The charter therefore has the potential to eventually legally challenge global capitalism's emphasis on private profit over the needs of all living beings. In addition, a major emphasis of the charter has been its adoption in schools around the world as a tool in teaching issues of democracy, ecology, and sustainability. To help in its educational use a guide was created by the Earth Charter Initiative in conjunction with UNESCO and was updated in 2010.

Charter Critics

Critics of the Earth Charter are conservative religious, political, and economic groups that see it as being opposed to their more sectarian religious, nationalist, and procapitalist views. Religious conservatives, for example, see it as an ecumenical threat to biblical/church authority and one that, if made legally binding, could hold them to universal Earth-based ethics rather than to the ethics of their religion. In addition, many see it as a form of global governance that would superimpose its "laws" onto their nation, their businesses, and subsequently onto them. Other critics are radical anticapitalist and feminist groups who see it as lacking issues of accountability. For example, the charter fails to pinpoint who specifically is responsible for the causes of our ecological problems and, consequently, who should be held accountable to remediate them. Therefore, it seemingly holds all equally responsible for climate change and global inequality and assumes all are equally empowered to create the proposed changes. Thus, from more critical perspectives, the charter lacks the ability to promote any real structural change along the lines that it proposes.

Conclusion

The Earth Charter is a unique global document that brings together the hopes and dreams of many of the world's people for a more just, equitable, sustainable, and peaceful future. Based on the belief that we can change our relationships to each other, Earth, and all living beings, the charter serves as a guide that has yet to be implemented.

See Also: Kyoto Declaration on Sustainable Development; Millennium Development Goals; United Nations Decade of Education for Sustainable Development 2005–2014.

Further Readings

Bosselmann, K. "In Search of Global Law: The Significance of the Earth Charter." *Worldviews*, 8 (2004).

Corcoran, P. B., Mirian Vilela, and Alide Roerink. *Earth Charter in Action: Towards a Sustainable World*. Amsterdam: KIT Publishers, 2005.

Davion, V. "The Earth Charter and Militarism and Ecological Feminist Analysis." *Worldviews*, 8 (2004).

Henderson, H. "Beyond Economism: Toward Earth Ethics." *Ecological Integrity*. In *Earth Charter in Action: Towards a Sustainable World*. Amsterdam: KIT Publishers, 2005.

Holland, J. and Elizabeth Ferrero. *The Earth Charter: A Study Book of Reflection for Action.* El Segundo, CA: Redwoods Press, 2005.

Rast, J. "The Earth Charter's Spiritual Ministry." http://www.contenderministries.org/articles/earthcharter.php (Accessed January 2010).

Phoebe C. Godfrey
University of Connecticut

EARTH DAY NETWORK

U.S. Forest Service staff meeting the public at an Earth Day information booth in Washington, D.C., on April 22, 2010. As many as 17,000 organizations worldwide participate in the Earth Day Network.

Source: U.S. Department of Agriculture

Earth Day Network (EDN) promotes a healthy and sustainable environment for present and future generations, diversifying the environmental movement worldwide through a combination of education, environmental awareness, public policy, and consumer activism campaigns and programs. EDN promotes bottom-up initiatives to provide opportunities for citizens from the local to the national and global levels. Its programs and activities aim to encourage civic engagement, expand the definition of environment, mobilize communities, implement environmental education programs, help bring clean water and sanitation, help students become environmental leaders, and support and encourage Earth Day events and actions around the world. EDN is an organization founded by the organizers of the first Earth Day in 1970; it is located in Washington, D.C., and promotes actions domestically and internationally. The network is composed of more than 17,000 partners and organizations in 174 countries, and more than 1 billion people participate in its development and environmental protection activities to help impact their communities and create positive changes. EDN is a member of Earth Share, a network that represents environmental charities in many workplaces, offering annual employer-sponsored programs that let employees contribute a few dollars per paycheck as a charitable donation.

Earth Day

The only event celebrated simultaneously around the globe by people of all backgrounds, faiths, and nationalities is Earth Day, held annually on April 22 to commemorate the anniversary of the birth of the modern environmental movement in 1970. Many communities also celebrate Earth Week or Earth Month by organizing a series of environmental activities throughout the month of April.

Earth Day founder Gaylord Nelson, then a U.S. senator from Wisconsin, proposed the first nationwide environmental protest to thrust the environment onto the national agenda. In 1970, 20 million people poured into the streets in the United States to protest against the health and environmental impacts of uncontrolled industrialization. The first Earth Day also had participants and celebrants in thousands of schools, colleges, universities, and local communities across the United States. It also led to the creation of the U.S. Environmental Protection Agency, and the passage of the first comprehensive environmental laws: the Clean Air, Clean Water, and Endangered Species Acts.

Recent Earth Days, observed by more than 500 million people and several national governments each year, continue to renew the commitment to improve the environment worldwide, focusing on many environmental problems, and particularly on key issues such as fighting global warming and developing a new green economy.

Earth Day Network's Programs and Campaigns

EDN coordinates large- and small-scale projects sustaining networks, programs, partnerships, and campaigns to lower carbon footprints, promote civic participation, and develop a sense of environmental responsibility among citizens, communities, corporations, and governments to affect change at the local, national, and international levels.

These projects sustain communities to reach sustainable, clean-energy economies through tangible and implementable steps. Program guides, fact sheets, environmental education curricula, toolkits, lesson plans and action plans can be downloaded from the EDN Website. These resources allow the public to learn how EDN's programs and campaigns affect the world, how crucial the issues are to human health, and how to get involved and promote healthier, more sustainable communities.

The Global Water Network supports projects that raise awareness about water and sanitation issues, and seeks to increase water and sanitation access to communities in need. It also allows individuals, organizations, or businesses to help fund water projects in rural areas around the globe. Through this project, water sector stakeholders are linked on a global scale, enabling them to share information and strategies to promote ecologically and socially sustainable water and sanitation services.

The Climate Change Solutions campaign is a three-year global campaign to educate and motivate all levels of the global community, from K–12 to college students, governments, organizations, and businesses, to reduce their contribution to global warming.

The Water for Life campaign aims to increase awareness and stewardship of local, national, regional, and global water resources by providing a clear overview of the most important water-related issues, such as the relation between climate change and the availability of freshwater resources, water demand and supply, and water quality and health. It also focuses on the importance of the Integrated Water Resource Management (IWRM) to coordinate the development and management of water, land, and related resources to maximize economic and social welfare in a sustainable way.

EDN is part of Campaign for Communities, a joint coalition of the Earth Day Network, the NAACP National Voter Fund, the Southwest Voter Registration Education Project, and Project Vote/ACORN. Its mission is to broaden the environmental movement and to see environmental equality, justice, and the right of all people to live in healthy communities by including nontraditional environmental issues that face underrepresented communities, including low-income, youth, and minority communities. Main objectives include issues not typically associated with the environment and consist of educating and mobilizing these communities, expanding participatory democracy, and training and promoting community leaders on how to empower themselves and their communities through civic participation.

EDN's Communities of Faith Climate campaign educates leaders and members of religious communities in order to bring climate change education and opportunities for civic activism to those involved with a religious life.

The Offset Your Carbon Emissions campaign helps people to make their Earth Day events, business events, or family events carbon neutral. In order for an event to be carbon neutral, people must reduce; for example, meet locally, walk, use public transportation, or carpool; reuse or repair products; and offset what cannot be reduced by supporting carbon-reducing projects that displace emissions from energy used for the event.

Some EDN networks, programs, partnerships, and campaigns promote a more democratically active citizenry at school. They show the connections between civic and environmental education, allowing students to understand how their actions influence environmental health at local and global geographical scales. These projects also seek to improve the daily lives of students. The Educators' Network is one of the most innovative and successful programs in the United States, providing lesson plans to help teachers integrate environmental issues into core subjects at all grade levels, and coordinating thousands of environmental events and activities throughout the year. The most important activities are the Green Schools Campaign and Education Grant Programs.

The National Civic Education Project presents solutions to the lack of civic and environmental education in today's society, creating personal responsibility for the environment among students around the world.

EDN's Green Schools program also promotes green schools, creating many benefits regarding not only environmental topics, but also socioeconomic ones, in the short and long term. Green schools' most important advantages range from reductions in air pollution emissions and in energy and material costs, to improvements in energy efficiency, resources conservation, recycling, waste reduction, an enhancement of social equity and students' health, as well as increasing student motivation and performance. This method also presents a cost-saving opportunity for schools because the savings derived from reducing energy and material consumption are higher than the cost of implementing the project.

Earth Day Network's Tools

The EDN Footprint Calculator is a tool that estimates the size of the user's personal impact, and can help evaluate EDN project and campaign results. The Ecological Footprint quiz estimates the area of land and ocean required to support the user's personal consumption of food, goods, services, housing, energy, and the absorption of personal wastes. The quiz is based on national consumption averages, and gives an idea of a personal ecological footprint relative to other people living in the same country. It also compares a personal ecological footprint in relation to a sustainable one, represented by Earth's biological carrying capacity expressed in global hectares or global acres, standardized units that take into account the differences in biological productivity of some ecosystems. A personal footprint is subdivided into four consumption categories: carbon usage (including home energy and transportation), food, housing, and goods and services, and is also subdivided into four ecosystem types of cropland, pastureland, forestland, and marine fisheries. At the end of the quiz, the calculator indicates the number of planets necessary to provide enough resources to support these consumption habits for the world population. If the number of planets needed to sustain consumption is less than one, it implies that people using the calculator are living a sustainable lifestyle, while if the number of planets is more than one, it implies that they are living an unsustainable lifestyle that would require the biological capacity of more than one Earth. In the later case, suggestions to reduce the user's personal footprint by implementing simple actions are provided.

See Also: Educating for Environmental Justice; Education, Local Green Initiatives; Education, State Green Initiatives; Global Green Day.

Further Readings

Earth Day Network. http://network.earthday.net (Accessed July 2010).
EarthShare. http://www.earthshare.org (Accessed July 2010).
Environmental News Network. http://www.enn.com (Accessed July 2010).
Global Footprint Network. http://www.footprintnetwork.org (Accessed July 2010).
University of Michigan, Center for Sustainable Systems. "Sustainability: Learn it—Live it." http://css.snre.umich.edu/facts/index.html (Accessed July 2010).

Antonella Pietta
University of Brescia

EARTH University (Costa Rica)

EARTH is a private nonprofit international university located in the town of Guácimo in the province of Limón, a lowland region in the east of Costa Rica. EARTH derives its acronym from the Spanish title *Escuela de Agricultura de la Región Tropical Húmeda* (Agricultural School of the Humid Tropical Region). The W. K. Kellogg Foundation, the U.S. Agency for International Development, and the Costa Rican government worked with other national and international agencies to create this unique university in the mid-1980s. The higher education initiative emerged from recognition that unsustainable agricultural practices were damaging and straining soil, water, forest, biological, and other natural resources across Central America.

EARTH opened in 1990 when it began to offer a four-year degree in agronomy that incorporated classroom lectures, laboratory work, field practice, and community service. In spite of having less than 1,500 total alumni and existing for just a relatively short time, EARTH has already garnered respect from international research and educational institutions. In keeping with the university's mission to affect change in low-income areas, 80 percent of students are granted full or partial scholarships. EARTH University Foundation, a nonprofit organization based in Atlanta, Georgia, helps fundraise to support this mission; however, the sales of environmentally friendly goods and services produced on campus are increasingly able to finance the university's programs. University-run businesses show that profitable ventures can be environmentally sustainable.

EARTH's core objective is to prepare students to return after graduation to communities in the humid tropics and to promote sustainable development. Another core objective of EARTH University is to create a positive community impact. The university contributes to management and water quality monitoring in the surrounding Parismina watershed. Food that is not produced on campus is purchased under contracts with local farmers. Students spend one day per week working with a local family or school; they have installed over 1,000 biodigesters that utilize livestock manure to produce cooking fuel.

With a very low faculty-to-student ratio, EARTH's approximately 40 faculty members and just over 400 students come from over two dozen countries. EARTH was founded under the direction of President José Zaglul, who was born in Costa Rica and studied abroad until he received a doctorate from the University of Florida. Zaglul returned to Costa Rica to work with various institutions, including the Centro Agrícola Tropical de Investigación y Enseñanza (CATIE), an international center for tropical research, before founding EARTH. Today, Zaglul travels extensively, promoting educational models such as EARTH that are hands on, community oriented, and ecologically sustainable. In 2009, Zaglul addressed the American College and University Presidents Climate Commitment (ACUPCC) Leadership Summit.

In 2007, *Grist* magazine voted EARTH the third-greenest university in the world. EARTH provides an unusual model for higher education that is accessible, holistic, practical, and focused on sustainability. Classes provide entrepreneurial and field experience, including a community service internship. The 8,200-acre campus includes a working banana farm, rainforest reserve, and cattle farm. Students help run these operations as they research ways that the university can develop or expand sustainable resource management and green entrepreneurship. Innovations from EARTH students have been disseminated around the world, such as the example of recycling plastic bags placed around bananas during transport from the field. EARTH has been able to minimize the use of agrochemicals during banana production with organic fungicide. Chemical-free production allows EARTH to reuse water and compost plant waste, in contrast to conventional banana plantations.

The university's research focuses broadly on agricultural systems, ecosystem management, carbon forestry, and biodiversity conservation. The university calls its research agenda "applied, guided, and beneficial." Facilities include a food processing lab and an ethnobotanical garden, demonstrating the range from high- to low-technology initiatives. The university expanded in 2005 to include a new campus called EARTH–La Flor. This satellite site will host a green conference center and a sustainable technology center created with support from the Ad Astra Rocket Company, which intends to conduct plasma research.

In 2007, EARTH University declared itself carbon neutral. Its analysis of carbon balance included three installations—the campus in Guácimo, the La Flor campus in Guanacaste, and the EARTH University Foundation office in Atlanta. EARTH analyzed its carbon mitigation and sequestration sources such as its forests, banana, oil palm, and heart of palm plantations and other crop and livestock production systems and found it had the capability to capture 16,324 tons of carbon dioxide annually. In fact, EARTH captures 15,170 tons more carbon dioxide annually than it emits. The university has several initiatives to reduce emissions, including the use of electric vehicles for on-campus transportation. A biodigester collects wastewater from the cafeteria and dormitories and uses it to produce energy. Students live on campus and are not permitted to keep cars: many rely on a bicycle for transportation. Three times a year, EARTH has a "day without a car" when all vehicles must stop at the university gates, which are located several kilometers from most buildings. This symbolic action receives local, national, and international attention.

EARTH University has a relatively long history of selling carbon credits to offset greenhouse gas emissions released from other locations. Emitters pay to plant trees on the Costa Rican campus. The university established carbon trade with the Port of Rotterdam even before the Kyoto Protocol popularized international exchange of carbon credits between industrialized and developing countries. EARTH has since become an international certifier of carbon neutrality and works with banks, airlines, and others to verify emissions reduction.

EARTH receives an endowment from the U.S. government and extensive support from EARTH University Foundation, a nonprofit organization based in Atlanta. The foundation organizes fund-raising campaigns using online marketing and social networking tools. "I plant . . ." is a program in which supporters buy a tree that is planted on campus—plaques with the names of tree donors are visible across campus. International sales of banana paper, a value-added product made on campus from recycled objects mixed with fiber made from banana stock, help the university become self-financing. In 2007, EARTH partnered with the giant health-food chain Whole Foods to sell Earth-bananas: this was the first product to receive the Whole Foods Whole Trade guarantee. The program represents the most stringent Whole Foods certification system and focuses on social and ecological sustainability in addition to promoting a high-quality product that helps to alleviate poverty.

Affiliated U.S. educational institutions include the University of Florida, Ohio State, and Michigan State in addition to partners in Germany, Chile, Argentina, and Uruguay. Regional EARTH networks are impressive: graduates make up a critical number of trained agronomists in some countries. Alumni associations, called AGEARTHs, exist in 16 Latin American countries. EARTH recently started a newsletter called *Impacto*, dedicated to reporting the impacts of campus events and student work as well as the accomplishments of alumni and donors. The university publishes a journal twice a year called *Tierra Tropical: Sostenibilidad, Ambiente, y Sociedad.*

See Also: American College and University Presidents Climate Commitment; Michigan State University; Sustainability Officers.

Further Readings

EARTH. "Universidad EARTH." http://www.earth.ac.cr/ing/index.php (Accessed January 2010).

EARTH University Foundation. "EARTH University." http://www.earth-usa.org/Page5334.aspx (Accessed January 2010).

Mary Finley-Brook
Caroline O'Rourke
University of Richmond

EASTERN IOWA COMMUNITY COLLEGE

Eastern Iowa Community College is a nationally recognized leader in environmental and sustainability issues located in the rural Midwest. It is the only higher education institution in a six-county service area and is known for its environmental and energy technology education. The college offers both traditional associate degree and certification programs as well as adult, business, and community courses and programs. It has adopted a sustainability policy as part of the Chancellor's Sustainability Initiative to ensure its sustainable operations and educational goals are achieved.

Campus Sustainability Initiatives

The Eastern Iowa Community College District (EICCD) covers Clinton, Jackson, Muscatine, and Scott Counties and parts of Cedar and Louisa Counties. The district borders the Mississippi River and includes Clinton Community College to the north, Muscatine Community College to the south, and Scott Community College in Bettendorf in the middle. Its mission statement includes a commitment to global sustainability both in the present and in the future as well as campus-based sustainability education, research, and operations.

EICCD seeks to be a Midwestern Premier Center of Excellence for Sustainability Education. The Eastern Iowa Community College District board of trustees adopted a sustainability policy for business and operational procedures in 2008 as part of the Chancellor's Sustainability Initiative (CSI). The policy and its implementation were guided by the American College and University Presidents Climate Commitment, which the EICCD signed in 2007. The signature commits EICCD to the achievement of climate neutrality as well as the promotion of greenhouse gas reduction research and education. The EICCD held a retreat for interested faculty, staff, and students soon after signing to explain its sustainability goals and to gain input.

After creating a sustainability policy, district administration formed district-wide assessment teams to discuss sustainability goals, determine priorities, and develop action plans for the implementation of both immediate and long-term actions. Teams covered the areas of business practices and transportation, community leadership, curriculum, facilities, recycling, and measurement and celebration. District administration also named an oversight team for the development and implementation of sustainability-related goals and projects.

Other sustainability initiatives include the maintenance of a speaker's bureau on sustainability topics; the maintenance of a sustainability section on the EICCD Website featuring information and tips, a feedback form, a carbon footprint calculator, and other useful links; ongoing energy-efficiency workshops for homeowners; Eagle Watch; Migratory Bird Day; Energy and Our Environment day camps for kindergarten through fourth grade students; and the Scott Community College Environmental Club. EICCD also participates in the Scott County GREEN (Guidance for Residential Energy Efficiency Neighborhoods) program as well as other sustainability outreach initiatives.

Academic and Research Programs

The Eastern Iowa Community College District offers both classroom-based and online academic programs for traditional and adult students, business professionals, and community members. Students interested in pursuing their education beyond the associate degree opt for the College Transfer track. Its schools have an enrollment of over 8,000 students in its associate of arts and associate of science degrees and College Transfer programs, over 3,000 students in its 40-plus career technology programs, and over 30,000 students in its business and industry, continuing education, and adult education classes. EICCD receives accreditation from the Higher Learning Commission. Academic partnerships include the Department of Recreation, Park & Tourism Administration, and the Western Illinois University–Quad Cities College of Education & Human Services.

Eastern Iowa Community College is known for its environmental and energy technology education and is a national sustainability leader in the field of green building technologies. The state-of-the-art Advanced Technology Environmental and Energy Center (ATEEC) located on the Scott Community College campus is LEED Platinum–certified according to the standards of the U.S. Green Building Council. The center serves as an educational facility and research laboratory for students and faculty as well as for industry and government officials at the local and national levels. It was an Incentive Grant winner in 2008.

EICC offers a Green Building Technologies degree program. Core courses include Green Construction Technologies, which covers the construction of sustainable residential housing that is also high quality and affordable. Faculty members have the flexibility to add their own resources to the curriculum. Course content topics have included reducing construction impact, sustainable materials, energy efficiency, greening HVAC systems, and LEED-compliant interior design. The Green Building Technologies program houses the Student-Built Homes program, through which high school students can receive dual enrollment college credit in green technology as well as hands-on construction experience.

Eastern Iowa Community College offers an online Health, Safety, and Environmental Technology (HSET) certificate and degree program, which was instituted over 20 years ago with U.S. Environmental Protection Agency funding. The online program attracts a diverse international student base. Students gain knowledge in health, safety, and environmental regulations and compliance; managing hazardous materials and wastes; protecting workers and the environment; and conservation technology. The Eastern Iowa Community College District in conjunction with the Western Illinois University–Quad Cities offers the accelerated Natural Resources Management Track. Degree options are an associate of science in Conservation Technology and a bachelor of science in Recreation, Park, and Tourism Administration with a minor in Environmental Studies.

Classes are held at both the Western Illinois University's Moline campus and the Nahant Marsh Education Center in Davenport, Iowa, located on a 262-acre nature preserve on the Mississippi River. Course topics cover knowledge of the natural and social sciences. Students also participate in labs, fieldwork, and internship experiences. This interdisciplinary program prepares students for careers in outdoor recreation and natural resources management such as conservation scientist, forester, naturalist and environmental educator, park ranger, and wildlife biologist.

Eastern Iowa Community College implemented its popular Renewable Energy Systems Specialist program in 2009 to train technicians in sustainable energy such as wind power, solar power, and biofuel energy. This Associate of Applied Science degree program with certificate career options is a cooperative program with Western Illinois University. The training emphasizes electromechanical knowledge and includes real-world and experiential learning opportunities. Career opportunities include installing, maintaining, and troubleshooting photovoltaic systems and wind turbines, electrical wiring, mechanical installation, and site analysis. Related occupations include renewable energy technician, solar systems installer, windsmith, and wind turbine manufacturing specialist.

See Also: Experiential Education; Green Business Education; Place-Based Education.

Further Readings

Barlett, Peggy F. and Geoffrey W. Chase. *Sustainability on Campus: Stories and Strategies for Change.* Cambridge, MA: MIT Press, 2004.

M'Gonigle, R. Michael and Justine Starke. *Planet U: Sustaining the World, Reinventing the University.* Gabriola Island, British Columbia, Canada: New Society Publishers, 2006.

Rappaport, Ann and Sarah Hammond Creighton. *Degrees That Matter: Climate Change and the University.* Cambridge, MA: MIT Press, 2007.

Marcella Bush Trevino
Barry University

EDUCATING FOR ENVIRONMENTAL JUSTICE

Environmental justice refers to equitable distribution of environmental goods, such as natural resources and clean air and water, among human populations as well as between species. The ethical dilemmas embedded within the concept of environmental justice are fourfold. On the one hand, proponents of environmental justice seek to redress inequitable distribution of environmental burdens, such as hazardous and polluting industries affecting vulnerable groups like ethnic minorities or economically disadvantaged populations. Second, environmental justice refers to the developed and developing countries' unequal exposure to environmental risks like the consequences of climate change. In both cases, environmental justice entails equitable spatial distribution of burdens and benefits to different nations or social groups. Third, temporal environmental justice refers to the issues associated with intergenerational justice (or concern for future generations of humans). The final issue involves the so-called biospheric egalitarianism (i.e., concern with other species and their exclusion from anthropocentric priorities).

The first ethical dilemma is illustrated by the definition of the participants of the Central and Eastern European Workshop on Environmental Justice held in Budapest in 2003, where environmental justice was referred to as a condition where members of disadvantaged, ethnic, minority, or other groups suffer disproportionately at the local, regional, or national levels from environmental risks or hazards, and/or suffer disproportionately from violations of fundamental human rights as a result of environmental factors. The U.S. Environmental Protection Agency (EPA) defines environmental justice as the fair treatment and meaningful involvement of all people regardless of race, color, national origin, or income with respect to the development, implementation, and enforcement of environmental laws, regulations, and policies. Oxfam publications indicate that the poor suffer most from the effects of globalization and environmental degradation. Global warming is believed to affect the world's poor—those least able to protect themselves against crop failures and rising sea levels—far more severely than the more affluent, as it is assumed that the poorest people almost always live in the poorest environments, since in the rural areas of the developing world, the poor have been forced onto marginal areas by the process of enclosure, leading to deforestation, soil erosion, and agricultural failure.

Educational programs addressing the first ethical dilemma span the range of the high school curriculum, particularly history and the introduction to politics, as well as higher vocational and university courses addressing international politics and global justice areas. While school curricula do not include specific courses addressing inequitable distribution of environmental burdens, but rather integrate the subject within existing departments, some universities have developed specialized courses on global inequality and environmental threats, particularly addressing the first ethical dilemma.

The second dilemma, associated with unequal international exposure to environmental risks, reflects the current deadlock between developed and developing countries in regard to issues like the international Kyoto Protocol negotiations. While developing countries recognize environmental issues as global, they ask that developed countries pay for solutions. Poor nations fear that international agreements will limit their attempt for economic growth, whereas economically powerful nations refuse to substantially reduce their greenhouse gas (GHG) emissions if developing countries do not make similar sacrifices. While developing countries often point out that developed countries are largely responsible for the present environmental problems, such as global warming and high emissions, developed countries argue that growing economies increasingly contribute to this problem themselves. Developing countries ask whether developed countries have the right to ask them to curb their economic growth while they themselves are enjoying the benefits of it. As the 1987 report of the World Commission on Environment and Development (WCED) points out, both the poor and the affluent contribute to environmental degradation as the failures that need correction arise both from poverty and from the shortsighted ways that we have often pursued prosperity. In this view, many parts of the world are seen as being caught in a vicious downward spiral, where poor people are forced to overuse environmental resources to survive from day to day, and their impoverishment of their environment further impoverishes them. The communication of the International Development Research Center in 2009 reflects on these concerns by asserting that many of development paths followed by industrialized countries of the north are not viable, while the economies of the south will have to continue to grow to redress some of the wretched disparities between the south and north and to offset the erosion of economic gain as a result of the rapid increase in world population. This leads to the conclusion that for this growth to occur without causing irreversible damage to the planet's ecology, the north must dramatically reduce its current level of consumption.

High school history and political courses addressing the differences between developed and developing countries' politics start by addressing the issue of the political interdependency between countries and regions. University courses addressing the second dilemma are similar to those addressing the first one, namely in courses focused on the issues of global inequality and addressing the underlying causes of environmental change. Additionally, courses of global environmental politics and environmental studies integrating sociological and political insights may specifically address the political deadlock between developed and developing countries in regard to environmental problems.

The temporal concept of environmental justice associated with intergenerational justice reflects concerns for future generations of humans, entailed by the definition of sustainable development encompassed by the Brundtland Report—meeting the needs of the present without compromising the ability of future generations to meet their own needs. The fate of future generations may be uncertain due to present growth and consumption patterns. Children in (high) school may be introduced to the concept of sustainable development, while higher education courses involving ethics and philosophy may address the moral implications of intergenerational justice.

Related to the concern about future generations is the final issue of biospheric egalitarianism (concern with other species and their exclusion from anthropocentric priorities). From a conservationist point of view, the increasing human population and the effects of greater consumption will have a negative effect on the conservation of pristine nature areas and further decrease in biodiversity. High school courses that address the effect of human activity on the environment range from biology to history. New educational programs address the effect of human industrial activity on the welfare and survival of other species. Higher education courses within the fields of conservationist biology, climate studies, environmental protection, and the like, convey factual knowledge about the effect of human activities on other species. Ethics and philosophy courses address the moral implications of anthropocentrism.

Balanced courses on environmental justice take all four types of ethical concerns of environmental justice into account.

See Also: Brundtland Report; Kyoto Declaration on Sustainable Development; Leadership in Sustainability Education; Sustainability Topics for K–12.

Further Readings

Brundtland Report. *Our Common Future: The World Commission on Environment and Development.* United Nations World Commission on Environment and Development, 1987.

Carter, Neil. *The Politics of the Environment: Ideas, Activism, Policy.* New York: Cambridge University Press, 2007.

Elliot, Lorreine. *The Global Politics of the Environment.* New York: Palgrave Macmillan, 2004.

Lidskog, Rolf and Ingemar Elander. "Addressing Climate Change Democratically. Multi-Level Governance, Transnational Networks and Governmental Structures." *Sustainable Development* (2009).

Rhodes, Edwardo Lao. *Environmental Justice in America.* Bloomington: Indiana University Press, 2003.

Helen Kopnina
University of Amsterdam

Education, Federal Green Initiatives

The U. S. federal government has a variety of "green initiatives" across agencies and within the administration that include public awareness campaigns, executive orders, legislation, and funding. In contemporary terms, the catalyst and origins of modern federal green initiatives

The American Recovery and Reinvestment Act of 2009 funds infrastructure improvements, such as this roadwork in Rhode Island, while also offering grants for projects to increase energy efficiency and establish sources of renewable energy, among many other projects.

Source: Wikipedia

can be traced to the establishment of the Environmental Protection Agency (EPA) in 1970. More than 40 years later, numerous policies, programs, and financial incentives comprise a diverse and growing platform of federal green initiatives.

Federal Antecedents

President Benjamin Harrison's Forest Reserve Act of 1891 set aside 13 million acres of land as protected national forest in the public domain. Later, President Grover Cleveland added 25 million acres, and President William McKinley added an additional 7 million acres during his administration. Additionally, President Theodore "Teddy" Roosevelt's work from 1901 to 1909 preserved a vast amount of U.S. national forests and parks. Roosevelt's administration established numerous areas of protection and conservation, including 150 national forests, 51 federal bird reservations, 4 national game preserves, 5 national parks, 18 national monuments, and 24 reclamation projects. Another important historical federal initiative was the Civilian Conservation Corps (CCC), which ran from 1933 until 1942. On April 5, 1933, President Franklin D. Roosevelt (FDR) issued Executive Order 6101, bringing about the CCC, and implementing a general natural resource conservation program on public lands in every U.S. state and the territories of Alaska, Hawaii, Puerto Rico, and the U.S. Virgin Islands. As part of the New Deal legislation proposed by FDR, the CCC was designed to provide relief for unemployed youth during the Great Depression, and aimed to preserve forests, prevent soil erosion, and provide flood control, among other projects.

Progression of Federal Green Initiatives (1970–2010)

Since the inception of the EPA in 1970, the federal government has made great strides in its focus on green initiatives. As the largest energy consumer in the United States, the federal government has both an opportunity and a responsibility to lead by example. While the inception of the EPA can be considered the first significant step toward a greener government and sustainable on a civic level, the Office of the Federal Environment Executive has a mission to promote sustainability and environmental stewardship throughout federal government operations.

1970s

The EPA was created under President Richard Nixon; the U.S. Congress passed what at the time were considered to be modern environmental statutes, such as the Clean Water and Clean Air Acts. William Ruckelshaus was appointed the administrator of the agency and served in that role until 1973. The focus of the 1970s centered on identifying remedies for air and water pollution. Through these early environmental efforts, jobs were created and provided for welfare recipients. In an EPA press release dated January 27, 1976, the EPA's administrator Russell Train announced that the EPA had helped 719 individuals on welfare in seven states to prepare for and assume environmentally related jobs. A $1 million fund and a timeline of 14 months for this training were established under an agreement issued by the Department of Labor and the EPA on October 3, 1973.

1980s

While the 1970s focused on the agency's establishment and the identification and reduction of air and water pollution, in the 1980s, significant steps were taken to clean up abandoned hazardous wastes sites. This included the development of an emergency response plan for environmental accidents, after the largest environmental accident to date—the Three Mile Island partial nuclear core meltdown in 1979. In June 1979, President Jimmy Carter proposed the concept of a Superfund to provide funding for the cleanup of abandoned hazardous waste dumpsites across the nation. In addition, it was proposed to fund an immediate cleanup of hazardous substances in waterways and on land annually. In 1981, the EPA identified the first 114 top priority Superfund sites. This hazardous waste focus in the 1980s included the cleanup of asbestos in schools, the establishment of the Hazardous and Solid Waste Amendment of 1984, and the Safe Drinking Water Act Amendments in 1986.

In the wake of the *Exxon Valdez* oil spill in 1989, the end of the 1980s marked the end of a series of aggressive environmental cleanup efforts and significant legislation.

1990s

The 1990s ushered in a more proactive approach in federal green initiatives: to preemptively stop pollution. In September 1990, the EPA Science Advisory Board proposed a comprehensive environmental risk strategy. This long-term strategy broadened the already far-reaching scope of the EPA. As a result, in October 1991, the EPA announced the National Environmental Education Act of 1990, a move to spur environmental support and understanding through consumer education initiatives. While the education movement was still in its infancy, additional acts, amendments to acts, Superfund and cleanup initiatives, and responses to international disasters comprised the day-to-day functions of the EPA. The National Environmental Education Act of 1990 (NEEA) established an Office of Environmental Education in EPA's headquarters, and a position was instituted in each of EPA's regional offices to provide leadership and coordination, and to develop and support a variety of programs. The national environmental education (EE) grant program, the President's Environmental Youth Award (PEYA), the Environmental Education Training Partnership (EETAP), and the National Network for Environmental Management Fellows (NNEMS) are among the better-known programs implemented under the legislation. The National Environmental Education Foundation (NEEF) was established by the National Environmental Education Act, and receives a portion of the EPA's appropriation each year to leverage support for environmental education from the private sector.

One of the more significant efforts in consumer education was the establishment of the EPA's Energy Star program in 1992. The program was established in an effort to reduce energy consumption and greenhouse gas emissions by power plants. The Energy Star logo appears on electronics, appliances, buildings, and other products that use 20 to 30 percent less energy than required by established federal standards. Since 1992, this program has been launched internationally in Australia, Canada, Japan, New Zealand, Taiwan, and the European Union.

Another movement in the 1990s was the development of the Leadership in Energy and Environmental Design (LEED) program. The LEED program was a recognized green building certification system to provide third-party verification that a building was designed and built using strategies to improve performance in energy savings, water efficiency, carbon emissions reduction, improved outdoor environmental quality, stewardship of resources, and sensitivity to their impacts.

It is important to note that the attention, commitment, reactive and proactive actions, and educational initiatives by the U.S. federal government since the 1970s reduced the risk of future environmental accidents. While the 1970s and 1980s were marred by severe environment accidents, the end of the 1990s brought EPA plans for cleaner cars and fuel standards, Energy Star building awards, and demonstrating that the Clean Air Act benefits far outweighed the costs.

2000s

In early 2000, the EPA endorsed a cleaner fuel plan and program in an effort to reduce car, truck, off-road equipment, and school bus emissions. In response to the terrorist attacks of September 11, 2001, on the World Trade Center, the EPA led and participated in the various environmental monitoring and cleanup actions in New York City, including the following:

- Monitored air, water, and dust for potential environmental hazards
- Vacuumed debris and dust from streets and other outdoor spaces
- Disposed of hazardous waste from the World Trade Center site
- Created an online database to report monitoring results to the public and press
- Set up wash stations and provided protective equipment for recovery workers
- Developed indoor cleaning and testing programs for residences in Lower Manhattan

As part of this effort, and in an effort to further support its continued education initiative, the EPA collected and disseminated as much information as possible on its outdoor and resource monitoring, including air, dust, water, and river sediment sample quality. This information is made available on the EPA Website.

In 2003, 167 solar panels were installed on the White House; this was the first solar electric project of its kind on the White House grounds, though not the first time the White House has had solar panels. In 1977, President Jimmy Carter, in response to the energy crisis in the 1970s, had a solar-powered water system installed. It was later removed by President Ronald Reagan because of persistent water leakages.

As part of the water quality movement, the EPA established the WaterSense program in 2006. This program was established to preserve the U.S. water supply by encouraging water efficiency through the use of a special label, similar to the Energy Star program for consumer products. WaterSense was unique in that it was not a regulatory, enforced program, but a voluntary program. If a manufacturer's products passed the program specifications, the manufacturer was awarded the right to use the WaterSense label. Like Energy Star, WaterSense made it easy for consumers to identify environmentally friendly products.

Housed within the White House Council on Environmental Quality, the Office of the Federal Environmental Executive was created by executive order in 1993. The office serves as steward for the implementation of President Barack Obama's Executive Order on Federal Sustainability (EO 13514) and the GreenGov initiative, working collaboratively with the Office of Management and Budget and each of the agencies through the Steering Committee on Federal Sustainability.

President Obama's Executive Order 13514

On October 5, 2009, Executive Order 13514, Federal Leadership in Environmental, Energy, and Economic Performance, was set into motion by President Obama. This executive order sets sustainability goals for an integrated strategy in the federal government and to make reduction of greenhouse gas emissions a priority for federal agencies. This particular executive order is but one of many green executive orders that demonstrate the federal government's commitment to a "best practice" of leading by example. Past green executive orders include the following:

- EO 13150, Federal Workforce Transportation
- EO 13211, Actions Concerning Regulations That Significantly Affect Energy Supply, Distribution or Use
- EO 13212, Actions to Expedite Energy-Related Projects
- EO 13221, Energy Efficient Standby Power Devices
- EO 13302, Amending Executive Order 13212, Actions to Expedite Energy-Related Projects
- EO 13327, Federal Real Property Asset Management
- EO 13352, Facilitation of Cooperative Conservation

Green technology advocates estimate the economic stimulus plan of the late 2000s will spend approximately a half-trillion dollars over two years on tax rebates and various projects, including the construction of renewable energy plants and the installation of "smart" meters. These new green projects are not only environmentally friendly but also socially responsible and economically influential.

As an example of the impact of President Obama's stimulus plan, in April 2009, the EPA announced an award of just over $20 million to the Nebraska Department of Environmental Quality. This award is for the improvement of a rural and urban aging wastewater infrastructure that, in turn, will protect human health and will create jobs as well as boost local economies. The funds were provided by the American Recovery and Reinvestment Act of 2009.

American Recovery and Reinvestment Act of 2009

The purpose of the American Reinvestment and Recovery Act of 2009 is to create and save jobs, jump-start the U.S. economy, and, perhaps most importantly, build the foundation for long-term economic growth. The main purpose of this act is to identify and target projects that will modernize the nation's critical infrastructure, encourage America's energy independence, and expand educational opportunities, increase access to healthcare, provide tax relief, and protect those in greatest need.

The Recovery Act provides $7.22 billion for specific programs administered by the EPA. This act contains six Program Plans that represent the foundation of the EPA's contribution to the nation's economic stimulus. The six Program Plans involve the following:

- *Clean Water State Revolving Fund Recovery Act Plan:* Investing in construction of water quality protection and wastewater treatment infrastructure
- *Drinking Water State Revolving Fund Recovery Act Plan:* Ensuring clean drinking water
- *Brownfields Recovery Act Plan:* Cleaning up former industrial sites for new commercial or community use, and training and placing persons in environmental careers
- *Underground Storage Tank Recovery Act Plan:* Cleaning up petroleum leaks from underground storage tanks
- *Superfund Recovery Act:* Cleaning up uncontrolled hazardous waste sites
- *Clean Diesel Recovery Act Plan:* Supporting the use, development, and commercialization of strategies to reduce diesel emissions

Funding for these programs protect and increase "green" jobs, sustain communities, restore and preserve the economic viability of property, promote scientific advances and technological innovation, and ensure a safer, healthier environment. These programs were chosen carefully both for their ability to put people to work now and for their environmental value. Progress and results are monitored in detail to ensure that American workers and taxpayers are reaping the economic and social benefits of these investments.

As of 2010, numerous economic incentives for energy efficiency and renewable energy were offered at the federal level to both corporations and individual taxpayers. For individual taxpayers, a number of incentives, deductions, and credits were available, including the Residential Energy Conservation Subsidy Exclusion, the Residential Energy Efficiency Tax Credit, and the Residential Renewable Energy Tax Credit.

One of the most popular of these measures is the Residential Energy Efficiency Tax Credit. The federal tax credit for energy-efficient home improvements was established by the Energy Policy Act of 2005. After expiring December 31, 2007, the credit was extended and expanded by the Energy Improvement and Extension Act of 2008 (H.R. 1424: Div. B, Sec. 302) and the American Recovery and Reinvestment Act of 2009 (H.R. 1: Div. B, Sec. 1121). In addition to extending the credit, H.R. 1424 and H.R. 1 strengthened the efficiency requirements for most equipment, extended the credit to stoves that use biomass fuel and asphalt roofs with appropriate cooling granules, raised the cap for the credit, and redesigned the way the credit is calculated. The credit applied to energy efficiency improvements in the building envelope of existing homes and to the purchase of high-efficiency heating, cooling, and water heating equipment. Efficiency improvements or equipment must serve a dwelling in the United States that is owned and used by the taxpayer as a primary residence.

The federal government has also provided corporate tax incentives, deductions, and credits for energy efficiency and renewal, including the Energy-Efficient Commercial Buildings Tax Deduction, the Residential Energy Conservation Subsidy Exclusion, the Business Energy Investment Tax Credit (ITC), the Renewable Electricity Production Tax Credit (PTC), and the Energy-Efficient New Homes Tax Credit for Home Builders. One of the first of these was the Energy-Efficient Commercial Buildings Tax Deduction. The federal Energy Policy Act of 2005 established a tax deduction for energy-efficient commercial buildings applicable to qualifying systems and buildings placed in service from January 1, 2006, through December 31, 2007. This deduction was subsequently extended through 2008 and then again through 2013 by Section 303 of the federal Energy Improvement and Extension Act of 2008 (H.R. 1424, Division B), enacted in October 2008.

In addition to tax-related incentives, the federal government also made loans and grants available for the purposes of increasing energy efficiency and establishing sources of renewable energy. The federal grant programs include the Tribal Energy Program Grant, the U.S. Department of Treasury Renewable Energy Grants, and the U.S. Department of Agriculture (USDA) Rural Energy for America Program (REAP) Grants. Federal loan programs include Clean Renewable Energy Bonds (CREBs), Energy-Efficient Mortgages, Qualified Energy Conservation Bonds (QECBs), U.S. Department of Energy Loan Guarantee Program, and USDA Rural Energy for America Program (REAP) Loan Guarantees.

The Food, Conservation, and Energy Act of 2008 (H.R. 2419), enacted by Congress in May 2008, converted the federal Renewable Energy Systems and Energy Efficiency Improvements Program into the Rural Energy for America Program (REAP). Similar to its predecessor, REAP promotes energy efficiency and renewable energy for agricultural producers and rural small businesses through the use of (1) grants and loan guarantees for energy efficiency improvements and renewable energy systems, and (2) grants for energy audits and renewable energy development assistance. REAP is administered by the USDA. In addition to these mandatory funding levels, discretionary funding may also be issued each year. Of the total REAP funding available, approximately 88 percent is dedicated to competitive grants and loan guarantees for energy efficiency improvements and renewable energy systems. These incentives are available to agricultural producers and rural small businesses to purchase renewable energy systems (including systems that may be used to produce and sell electricity) and to make energy efficiency improvements. Eligible renewable energy projects include wind, solar, biomass, and geothermal; and hydrogen derived from biomass or water using wind, solar, or geothermal energy sources. The USDA also gives competitive grants to eligible entities to provide assistance to agricultural producers and rural small businesses "to become more energy efficient" and "to use renewable energy technologies and resources."

The U.S. government continues to move forward across environmental, social, and economic areas in the creation and implementation of federal green initiatives. Alternative and renewable energy sources and energy efficiency are seen as more than "green" issues but also as a priority of national interest and security.

See Also: Education, Local Green Initiatives; Education, State Green Initiatives; Sustainability Officers.

Further Readings

Energy Star. U.S. Environmental Protection Agency. http://www.energystar.gov (Accessed July 2010).

U.S. Department of Agriculture. "Business and Cooperative Programs." http://www.rurdev .usda.gov/rbs/busp/9006loan.htm (Accessed July 2010).

U.S. Environmental Protection Agency. http://www.epa.gov (Accessed July 2010).

Lyndsay J. Agans
Bryan DeShasier
University of Denver

Education, Local Green Initiatives

In the past decade or so, the field of environmental education has increasingly come to rely on using local environment and environmental issues as resources for learning. The focus on the local, and in many cases lived, environment has marked a departure from the previous emphasis on learning abstract and seemingly remote concepts and issues that students often found difficult to relate to. Local environment, at once meaningful to students and parents alike, provides better opportunities for engagement and participatory learning. Often the political and social issues surrounding local environment appeal to students and motivate them to understand the problem and its possible solutions in all their complexity.

The focus on local environmental problems in environmental education has been traced by some scholars to the 1977 United Nations Educational, Scientific and Cultural Organization (UNESCO) conference in Tbilisi. This conference recognized that the roots of the modern environmental crisis lay in the dominance of cultural values, often specific to the West, focused on unlimited growth and unchecked consumerism. This diagnosis of environmental problems constituted a powerful critique of not only the blind faith in markets that drove environmental degradation but also the dominance of the technical approaches to managing the environment. An alternative perspective emerged from the conference that emphasized a local approach to environmental problem solving. The implications of this new approach, however, amounted to more than just a switch in scale of analysis; it also called for a participatory process in which community participation would become integral to the understanding of and possible solutions to the local environmental issues. Equity and efficiency were therefore placed on an equal footing in analyzing environmental problems and their possible solutions.

The value of local green initiatives in education has over time been widely recognized. The pedagogy of locally based environmental education is based on several interconnected principles. The central feature of this philosophy is the belief that learning can be enhanced by engaging students in environmental concepts, knowledge, and skills through the specificities of the place. The tangible emotional ties to a place and its concreteness can facilitate an open-ended learning environment where students pursue questions that they find personally meaningful, in contrast to a traditionally structured and top-down approach to learning.

By focusing on local issues, students are encouraged to make connections between individual actions and broad outcomes. The students can also trace the linkages between local economic and environmental aspects of their lives and the regional, national, and global forces that affect them directly and indirectly. The element of interconnectedness, which binds places and people together in ways that often lead to unpredictable outcomes, is also underscored. Ultimately, the importance of scale in the analysis of environmental and social systems becomes clear. For instance, riparian issues and upstream and downstream relations that are not immediately evident highlight the necessity for an integrated and systemic view of environmental issues. Scale is not only critical to understanding the nature of the problems but also in considering appropriate policy and management options that are available to planners and policy makers. Students learn about the potential of and limits to the individual action, subject to the constraints and opportunities generated by the broader system. Perhaps most importantly, they become aware of the powerful idea that individuals, through their choices, can have a positive or negative impact on the environment.

Another important aspect of local green initiatives in education pertains to the centrality accorded the participatory approach. The involvement of the community as a whole in learning, including curriculum and assessment, locates schools within the broader social and political context. Learning is therefore reconceptualized in an unorthodox way, as a semiautonomous activity that takes place in diverse situations, in and out of the classroom. Pro-environmental attitudes and environmental knowledge is thus broadly enhanced, and it is theorized that this will ultimately feed into a grassroots environmental movement. It is posited that a general increase in environmental knowledge and change in attitudes will drive people to think about sustainability in their lives and the ways in which it can be applied to their specific needs and goals.

Schools at the local level have increasingly incorporated environmental considerations at various levels of their operations. In efforts involving students, teachers, administrators, and policy makers, changes have been introduced to school environments, including the areas of food, buildings, spaces, and resource use. These improvements include reducing exposure to toxics, creating healthy spaces and buildings, and offering healthy food choices in schools. The impetus for these changes has often come from concerned parents, who are becoming increasingly aware of the connection between individual health and the state of the environment. Greater media focus on the linkages between exposure to hazardous substances, including toxins and diseases, has also fueled these efforts.

Studies of environmental literacy levels in the United States have generally found them to be inadequate. Dominant among the approaches suggested to increase environmental literacy is the one that emphasizes establishing a connection with the environment. Especially for younger children, environmental literature, including picture books with their vivid characters and age-appropriate narrative, can help develop an appreciation for nature. This literature usually anthropomorphizes animals but does not detract from the goal of building environmental awareness. More general ideas related to the interdependence of culture and the environment can also be conveyed through literature. These ideas can then be reinforced in students' minds by their personal observations of their immediate environment.

Some universities have introduced innovative environmental education curriculum that is based on immersion in the natural environment. Arcadia, an academic program run by Lawrence University, involves students living and taking classes in a natural setting, away from the campus. The emphasis is on coursework that is integrated with field experiences, while the students live an austere lifestyle in a close-knit community. The program, in the classical tradition of liberal education, seeks to bring together a small group of like-minded learners in a natural setting that seeks to provide an embodied environmental education.

Conceptualizing Local Green Education

Environmental literacy comprises more than knowledge, skills, and attitudes and their transfer to students. The setting in which environmental education takes place has important implications for how environmental education is conceptualized and how learning takes place. Environmental education, when considered holistically, involves becoming members of specific environmental communities and achieving proficiency in their culture and practices. The goal of environmental education, in other words, should be more than environmental sensitization—it should be to become environmental practitioners. Depending upon whether environmental education takes place in a classroom or in a community through experiential learning, students will learn different ways of being environmentally proficient. In a classroom, students may learn more about the technical aspect of environmental science and about the web of interconnections between the environment and other parts of society. In an organizational setting, students may focus on becoming members of a community, with its characteristic culture, practices, and social dynamics. The former may involve becoming conversant with the specialized language and vocabulary of environmental science, whereas the latter may emphasize informality and collaboration.

Some environmental education initiatives are conceptualized in terms of creating sustainable empowered communities. Going beyond the confines of the classroom model, a community-school partnership is forged to facilitate lifelong environmental education in an autonomous setting. Community School Movement in Israel seeks to improve the education mission as well as general social well-being by establishing partnerships between schools and neighboring communities. Education is thus seen as more than a product, and students are not mere customers, but the community and its values are incorporated in curriculum, assessment, and other aspects of the learning process. The community is involved in formulating the programs and activities, which are chosen in accordance with its priorities. One of the goals of community schools is to break down the barrier between formal and informal education and use events and activities in the community as a part of the curriculum. In the absence of national guidelines, in Israel, on environmental education, communities exercise substantial control over the curriculum. Typically, historical, archeological, and cultural issues are included in the category of the environmental. Both the community and the school, especially teachers involved in the program, have positive perceptions of collaboration and the impact on educational quality. The negative judgments appear to be minor and center on the perceptions of interference that teachers and community members occasionally have about each other. Partly, the success of the community school seems to be built on the cultural homogeneity that characterizes the participants in the program.

Local issues have been used in environmental education to make abstract issues concrete. Especially for younger students, who may have difficulty grasping seemingly distant and unrelated environmental issues such as biodiversity conservation, the local environment can provide a convenient entry into more general themes and topics. Students are imparted relevant knowledge, skills, and understanding as they learn to collect and weigh evidence, examine values and differences therein, and, in the process of thinking through local problems, develop their own positions and attitudes on environmental issues. Students often communicate their findings to the community, and, in the process, become familiar with the politics surrounding environmental problems.

Local Environmental Issues, Methods, and Strategies in Education

In general, environmental education has moved away from an emphasis on the experience of wilderness and pristine environments in shaping environmental consciousness to the recognition that lived environments, including urban environments, can form the basis of an enduring environmental awareness. Not all communities are, however, equally conducive for environmental education to occur. Community characteristics such as income and social status may affect the relationship to the environment of the students residing there. Important insights into how students view the environment of their communities have been gained using participatory techniques such as ethnography and participant action research. In lower-income, marginalized neighborhoods, concerns about security and lack of basic amenities may inhibit the appreciation of the environmental aspects among students from these areas. In contrast to the attitudes, however, the differences in environmental knowledge between students from communities of differing socioeconomic status are not as pronounced. The lesson here is that environmental relationships and perceptions of one's environment are shaped by the mediating effect of the community of residence. If environmental education is to be effective in forging students' relationship to their immediate environments, the role that place of residence plays in shaping environmental education has to be factored in.

The pedagogy of place-based learning provides support for basing environmental education in the concerns and context that are familiar to students. The place, in this case the community in its physical, social, and political dimensions, becomes a microcosm of the world and provides a test case for examining the broader forces at work. The transformative approach to education, involving critical pedagogy, is utilized to organize learning of scientific knowledge of environmental issues. Under this approach, the student worldview, as it relates to socioeconomic issues, especially disparities and inequality, becomes a gateway for learning scientific concepts and knowledge. Instead of encountering scientific facts and knowledge as disembodied subject matter, students learn to think about the specific place they live and work in, its salient characteristics and features, and use science as a methodology and process to address questions meaningful to them. The aim of place-based pedagogy for environmental education is not merely achieving environmental literacy but helping students challenge the dominant environmental discourses, rooted in cultural preferences, and fashion more harmonious ways of living in their environment.

The concept of service learning has been utilized to increase student awareness of environmental problems, for instance, water quality issues and strategies to ameliorate the problems. In urban settings, it is not uncommon for students to have poor awareness of water quality problems, especially where cities are part of large watersheds. Efforts have been made to combine ecological restoration research in wetlands, undertaken by scientists in state agencies, with encouraging civic participation in carrying out various research and data collection activities. Not only do volunteers provide valuable time and labor, their knowledge of and attitudes toward the environment are also positively influenced. The notion of community participation in both defining the problems of interest and possible pathways for their resolution is a critical component of the environmental service learning programs.

Environmental problem solving undertaken as part of coursework can help educate students about the legal, technical, and social aspects involved in environmental management and their interrelationship. The rationale for this approach also includes the potential for independent learning through increased motivation. Problem solving allows students a launching pad into the relevant regulations and for collecting technical data. Additionally, students learn data presentation skills and how to use evidence to build consensus on required courses of action.

Among the issues that have been utilized for participatory environmental learning and planning are assessing contamination and level of toxics of local sites, watershed management, ecological restoration, and recycling. All of them have aspects—for example, related to air and water quality—that affect the residents directly. These issues are also more open to public input and feedback because of the inherent uncertainties involved in defining their scale, scope, and impact. A participatory approach based on dialogue, sharing of knowledge, and shared problem definition can go a long way in building consensus and producing broadly satisfactory outcomes. The process by which diverse stakeholders are brought together for a sustained and mutually beneficial interaction leading to enhanced knowledge of the system is often referred to as a social learning approach. The choice of problem, in other words, can lead to a reconceptualization of learning to a collective and iterative process rather than one simply involving a unidimensional transfer of information.

The choice of appropriate educational and problem-solving tools used to promote understanding of the local system is important. Collective envisioning exercises, for instance, involving a group engaged in creating a resource map, are useful for eliciting knowledge of landscape and environmental relationships. Cognitive mapping provides a look into how individuals perceive their environment. Newer computer-based mapping programs and software, such as geographic information systems (GIS), help improve the understanding of spatial relationships and patterns. GIS maps and analysis can also be used to facilitate a common understanding of the natural system.

Local Green Initiatives in a Cross-Cultural Perspective

In many developing countries, education is seen as a tool for social change and for inculcating values favorable to fostering a strong sense of citizenship and economic growth. As environmental conditions have become exacerbated—often due to flawed economic policies and lack of environmental laws and regulations—the role that education can play in generating sustainable development has attracted greater attention.

Environmental education in developing countries may focus on indigenous knowledge and practices. This is both because of better cultural fit with the indigenous worldview and the conservationist objective of reviving endangered traditions and knowledge. Emphasis on indigenous environmental beliefs and practices may also arise out of a desire to undo the damage done to these cultures by colonialism. The categories and content of indigenous environmental knowledge enable learning to be more relevant and meaningful to the participants, therefore producing a profound emancipatory impact.

A key objective of the local environmental education efforts outside the first world context is to contribute to the nation-building project. Not only are these countries confronted with economic and environmental problems, the sense of citizenship and civic responsibility are also often found to be lacking. Environmental education thus is seen as part of a broader attempt to shape notions of active citizenship through empowering individuals and communities. In doing so, the goal is to invert the top-down development model, primarily by building local capacity and injecting institutional accountability. In certain countries, where institutional capacities are limited, nongovernmental organizations may play an important role in providing participatory environmental education.

In sum, the local green initiatives in education are achieving a number of interrelated objectives: making environmental education less abstract and more specific and relevant, diversifying the channels of learning, dissolving the boundary between formal and nonformal education, involving communities in problem definition and analyzing possible solutions, and fusing citizenship with environmental stewardship. A lifelong learning approach is taken as opposed to learning as an instrumental and limited activity. Local environments, including their physical, cultural, economic, and political dimensions, are seen as interconnected, and a robust understanding is recognized as being based on acknowledging and understanding this reality. A defining feature of local environmental education is its resistance to imposition of a one-size-fits-all approach. The objectives, priorities, content, and process to be followed will depend on the community characteristics and the outcomes of a democratic, participatory model of public engagement. In doing so, local green initiatives in education celebrate diverse ways of knowing and being in the world.

See Also: Education, State Green Initiatives; Environmental Education Debate; Experiential Education; Outdoor Education; Place-Based Education; Tbilisi Declaration.

Further Readings

Buxton, Cory. "Social Problem Solving Through Science: An Approach to Critical, Place-Based, Science Teaching and Learning." *Equity & Excellence in Education,* 43/1 (2010).

Fisman, Lianne. "The Effects of Local Learning on Environmental Awareness in Children: An Empirical Investigation." *Journal of Environmental Education,* 36/3 (2005).

Hoga, Kathleen. "A Sociocultural Analysis of School and Community Settings as Sites for Developing Environmental Practitioners." *Environmental Education Research,* 8/4 (2002).

Langen, Tom and Rick Welsh. "Effects of a Problem-Based Learning Approach on Attitude Change and Science and Policy Content Knowledge." *Conservation Biology,* 20/3 (2006).

Scrivener, Chris. "Getting Your Voice Heard and Making a Difference: Using Local Environmental Issues in a Primary School as a Context for Action-Oriented Learning." *Support for Learning,* 18/3 (2003).

Tal, Revitali Tali. "Community-Based Environmental Education: A Case-Study of Teacher-Parent Collaboration." *Environmental Education Research,* 10/4 (2004).

Tedesco, Lenore and Kara Salazar. "Using Environmental Service Learning in an Urban Environment to Address Water Quality Issues." *Journal of Geoscience Education,* 54/2 (2006).

Upitis, Rena. "Four Strong Schools: Developing a Sense of Place Through School Architecture." *International Journal of Education & the Arts,* 8 (Interlude 1, 2007). http://ijea.asu.edu/v8i1 (Accessed February 2010).

Wells, Rachael and Pauline Zeece. "My Place in My World: Literature for Place-Based Environmental Education." *Early Childhood Education Journal,* 35/3 (2007).

Neeraj Vedwan
Montclair State University

EDUCATION, STATE GREEN INITIATIVES

In response to growing concerns about global warming and climate change, in 1990, the U.S. Congress passed the National Environmental Act, which required the national Environmental Protection Agency to create the Office of Environmental Education within the Office of Public Affairs. Because the United States is a federal system, each state was given responsibility for implementing the act according to specified federal guidelines. Each of the 50 states set out to comply with the act based on varying degrees of commitment to the concept of sustainability. State green initiatives were implemented across the

Bird-watching at a wildlife refuge in California in 2008. The state has been very active in creating green initiatives, including the California Student Sustainability Coalition, the Green California Initiative, the Green Building Initiative, and the Green Building Action Plan.

Source: U.S. Fish & Wildlife Service

United States under pressure from individuals, environmental groups, state legislators, and educators. Within each state, many educational institutions used state green initiatives to carry out their own sustainability agendas. More than 600 American colleges and universities signed the American College and University Presidents Climate Commitment, and 378 became members of the Association of University Leaders for a Sustainable Future. More than 600 institutions of higher learning committed themselves to striving for carbon neutrality. In 2007, the College of the Atlantic, located in Bar Harbor, Maine, became the first college to achieve that carbon net zero goal. In general, private schools have proven able to move at a faster pace in their efforts to achieve sustainability goals. This is partly because they tend to have more control than public institutions over their efforts, and they may also have greater access to financial resources. Statewide green initiatives are at varying stages of development, but the fact that they exist at all is clear evidence that state governments, educational institutions, and business and community leaders are responding to calls for greater efforts toward sustainability in the United States.

National Environmental Act of 1990

Faced with a growing body of knowledge about threats to the environment, Congress responded in 1990 by acknowledging the reality of global warming, ocean pollution, threats to species diversity, and the existence of toxic contaminants in the air, water, and land of the United States. The congressional solution to these problems was to enact the National Environmental Act to provide a means of using the resources of government at all levels to work with educators and local political, business, and community leaders to address sustainability issues designed to reverse existing damage and to prevent future environmental damage. Because creating a "well-educated and trained professional workforce" was considered essential to accomplishing these goals, American schools became the focal points for achieving the new agenda. Grants, fellowships, internships, awards, and other forms of support were provided to encourage schools to initiate curriculum reform, special projects, and activities that promoted sustainability and to engage in green building, recycling, and energy-saving measures. The act called on state and local governments, nongovernmental organizations (NGOs), the media, and the private sector to provide educational institutions with their full support.

The Office of Environmental Education was created to provide support at the federal level. Providing education and training became the chief responsibility of the Environmental Education and Training Program, which was created to furnish schools with model curricula, resource materials, technologies, teacher training programs, educational materials, and audio-visual materials. Certain educational materials were designed to target particular groups, such as elementary and secondary students or senior citizens. The office was also required to offer environmental education through seminars, teleconferences, and workshops. Another duty was to create assessment tools designed to evaluate the success of various efforts toward sustainability.

State and Regional Campus Responses

In the field of higher education, various state green campus initiatives led to the formation of a plethora of state and regional campus sustainability organizations designed to put particular initiatives into effect. The northeastern states created the Campus Consortium for Environmental Excellence (C2E2), the Environmental Consortium of Hudson Valley Colleges and Universities, and the Northeast Campus Sustainability Consortium. C2E2, which emphasizes professional networking and development, implemented separate initiatives dealing with regulatory compliance, environmental management, and sustainability. Membership in the Environmental Consortium of Hudson Valley Colleges and Universities is restricted to the 43 institutions located in the area. This organization, which focuses on both regional and global sustainability, is able to utilize the vast resources of the Hudson Valley to provide practical experience for area students at all levels. With members in both the northeastern United States and eastern Canada, NECSC concentrates on a range of sustainability issues.

Institutions in the south have established the Environmental Initiative of the Associated Colleges of the South (ACS), the Southeast Higher Education Sustainability Network (SHES), and the Texas Regional Alliance for Campus Sustainability (TRACS). Membership in ACS is restricted to independent schools. The focus within this group is public sustainability education and addressing academic and extracurricular sustainability issues on member campuses. SHES serves chiefly as a discussion forum, but it does hold periodic teleconferences to promote networking. TRACS exists to unify Texas-area schools around the issue of sustainability and climate change and to encourage the implementation of environmental issues on member campuses.

Institutions in Minnesota, Iowa, and Wisconsin created the Upper Midwest Association for Campus Sustainability (UMACS) in the hope that neighboring states would join the organization at a later time. Sponsors include the College of St. Benedict and St. John's University. The Minnesota Office of Environmental Assistance frequently provides support.

A number of states have also created their own sustainability organizations for the purpose of promoting statewide sustainability. California created the California Student Sustainability Coalition (CSSC) in 2003 to motivate students to tackle sustainability initiatives. In the Illinois Community College Sustainability Network, founded in 2007, the focus is on professional training and development. The Maine Green Campus Consortium has been highly effective in encouraging sustainability action throughout the state and on individual campuses. The Michigan Higher Education Partnership for Sustainability (MiHEPS) serves mainly as an information resource. The New Jersey Higher Education Partnership for Sustainability (NJHEPS), however, is more action oriented. Its various projects include the NJDEP

Greenhouse Plan, which gathers energy reports from member campuses, and the creation of teams that promote sustainability through design, energy use, and the curriculum. Sharing information is also a focus of the South Carolina Sustainable Universities Initiative (SUI), but SUI is also engaged in implementing sustainability curricula and operations on member campuses. The focus of the Pennsylvania Consortium for Interdisciplinary Environmental Policy (PCIEP) is on bringing together policy makers and educational institutions to promote sustainability in both policy and practice. In addition to a focus on education and the exchange of information, the Vermont Campus Sustainability Network is involved in working with various communities to implement sustainability projects and initiatives. With more than two decades of experience in sustainability in education, the Washington Center for Improving the Quality of Undergraduate Education has a comprehensive sustainability agenda that includes cultural pluralism and community outreach.

Statewide Building Initiatives

Virtually every state in the United States now mandates that all state-funded buildings be designed to meet Leadership in Energy and Environmental Design (LEED) standards, and any renovations must meet specific sustainability guidelines. American colleges and universities have been at the forefront of the sustainability building movement. Many green campuses are involved in sustainability from a twofold perspective: they attempt to combine sustainability education with sustainable building practices. Attempts toward building sustainability generally take place through residential sustainability, green campus housing, or sustainability-themed living-learning communities. Pennsylvania's Dickinson College, for instance, built two structures that received LEED Gold certification, and older buildings were restructured according to LEED Silver standards. Dickinson also erected the Center for Sustainable Living, known locally as the Treehouse, which serves as a model for sustainable living by utilizing a corn-pellet stove, dual-flush toilets, faucet aerators, a greywater system, and Energy Star appliances.

State initiatives requiring educational institutions to meet LEED standards are dependent on specific laws and governmental commitment to sustainability. In 2005, by executive order, Arizona issued a mandate for all state-funded buildings to meet LEED Silver certificate standards. That same year, the state legislature passed the Energy and Natural Resources Conservation Act, which established an Office of Sustainability and a state task force to work with various entities on sustainable building. As is often the case, California leads the nation in sustainability efforts in higher education. In addition to requiring LEED Silver certification in all new buildings, the California Green Building Standards Code mandates sustainability in building through landscaping, appliance efficiency, building design, and utilization of recycled materials. Colorado's sustainability requirements in building are carried out according to its high-performance green building standard, which requires all projects to strive to achieve the highest certification possible.

While project administrators in Florida may strive for LEED certification, certification may also also be obtained from agencies such as Green Globe and the Florida Green Building Coalition. The same is true in New York, where the Office of General Services accepts green certifications from LEED, Green Globes, and the American National Standard Institute as stipulated by the Green Building Construction Act. When Maryland passed the High Performance Building Act in 2008, legislators specifically required all public schools using state funds to comply with either LEED Silver certification or to earn two Green Globes. The legislature committed itself to paying half of the extra costs entailed in meeting these requirements. Also in 2008, New Jersey passed a new law requiring LEED Silver certification for all new state-owned buildings 15,000 square feet or larger. Six years earlier, an executive order had mandated the use of LEED guidelines in the design of all new schools built in the state. Through the Sustainable Building Tax Credit established in 2009, New Mexico grants tax credits to offset the cost of requiring LEED certification for all new buildings. Credits range from $3.50 per square foot for buildings achieving Silver certification to $6.25 per square foot for those that achieve Gold certification. Credits are adjusted for residential structures. According to an executive order signed in 2006, all public buildings over 15,000 square feet are required to obtain LEED Silver certification.

While much attention has been paid to large state universities, other colleges have also been involved in sustainability deriving from state initiatives. At Nicolet Area Technical College located in Rhinelander, Wisconsin, the entire campus community has embraced the concept of sustainability. One of the school's initiatives involves increasing awareness of sustainability throughout the region and implementing sustainable living on the Nicolet campus. Well aware that the state imports 95 percent of its energy requirements, Nicolet has chosen to focus on renewable energy. Using state grant money and assistance from the Wisconsin Technical College System Foundation, Nicolet installed a wind turbine and two solar electric panels in order to generate both wind and electric energy for use at the Renewable Energy Center, which is used for data collection and analysis. Because the systems are connected to the public service system, the school receives renewable energy credits. Teachers across the curriculum use the equipment to provide their students practical experience.

Outside assistance is often necessary to help colleges and universities meet state green initiatives while accomplishing something unique to each campus. Those involved in the Association of College and University Housing Officers-International's (ACUHO-I) 21st Century Project, for instance, will benefit from the project goal of erecting campus residences that "reflect the ever-changing roles" that residential halls play in campus life today. From the 21st Century Project viewpoint, technology is a major element in building residences that are both functional and attractive to students. Colorado College, Indiana University, and Baylor University have all applied to have prototype residence halls built on their campuses as part of the first phase of the multiyear 21st Century Project.

California

While there are certain commonalities among statewide initiatives, each state has implemented initiatives based on its own political, geographical, and cultural climate. Fulfilling its role as a leader in innovation among the 50 states, the State of California has created the Green California Initiative, which is designed to promote environmental literacy throughout the state. Early in the 21st century, Governor Arnold Schwarzenegger acknowledged that the state's valuable resources needed to be preserved. By signing the Green Building Initiative and implementing the Green Building Action Plan, Schwarzenegger committed California to green building. Other initiatives dealt with promoting energy efficiency and conservation and instituting green purchasing procedures throughout the state. One of the most successful efforts of the Green Building Initiative was obtaining LEED Platinum certification for the California Department of Education Building. It is the first existing building in the state to receive such recognition. Erected in 2003, the building initially received Gold certification and scored 95 out of a possible 100 on the U.S. Environmental Protection Agency's Energy Star rating, but subsequent renovations brought the building up to Platinum status. Renovations included improvements in air ventilation and distribution, plumbing, recycling of waste, environmentally friendly purchasing, utilization of mass transit for employees, and employing environmentally friendly roofing and paving materials, cleaning supplies, and pest management.

California's Education and the Environment Initiative, which was implemented in 2003, targets classrooms in the state, incorporating sustainability education into the K–12 curricula in science, history, social science, English, language arts, and mathematics, and involving government agencies and community leaders in partnerships with schools to ensure the success of California Environmental Principles and Concepts.

Maryland

Within the State of Maryland, statewide initiatives in higher education have centered on green building practices on both public and private college campuses and on developing curricula that are responsive to the growing need to prepare graduates for living and working in a world that is becoming ever more conscious of global warming and climate change. In addition to STEM (science, technology, engineering, and math) initiatives already in place, new efforts have concentrated on incorporating sustainability education into the humanities and social sciences.

In line with Maryland's goal of reducing overall energy consumption, college and university campuses have begun experimenting with new ways of saving energy. Green building practices have been put in place, and state-funded buildings are renovated, built, operated, and maintained according to LEED standards. All new buildings over 7,500 square feet on four-year public campuses are required to meet LEED Silver standards or higher. Maryland institutions are also developing new methods of conserving valuable resources, expanding waste reduction, and engaging in antipollution practices.

Oklahoma

In 1992, the Oklahoma legislature passed a bill creating the Oklahoma Environmental Education Coordinating Committee (OKEECC) under the auspices of the Oklahoma Conservation Commission, bringing together various state agencies involved in promoting sustainable education. In order to bring NGOs onto the playing field, the Kirkpatrick Foundation lobbied the state government to create the Conservation Education Initiative, which allowed state and private groups to work together on promoting sustainability in education. The role of NGOs has thus been integral to successfully creating a viable sustainability agenda in Oklahoma schools. Agreeing to serve as home base for the Conservation Education Initiative, the Oklahoma branch of The Nature Conservancy became deeply involved. One of its most successful efforts involved the creation of a Website with a searchable database designed to provide easily obtainable sustainability resources for educators, students, parents, government agencies, and the public. The site also provides teachers with advice on incorporating sustainability across the curriculum and contains a nature locator. Martin Park Nature Center is also involved in the initiative and encourages full use of its 140-acre park.

New York

Many states are in the process of creating new programs designed to supplement existing responses to statewide green initiatives. New York, for instance, announced in spring 2010 that the New York State Outdoor Education Association (NYSOEA) was responding to recent No Child Left Inside legislation by establishing an Environmental Literacy Committee that was charged with developing a statewide environmental literacy plan. Funding is derived from federal grants that have been made available to states that have passed such legislation. Consisting of 30 members representing various sectors of the state, the committee began holding meetings and discussion groups to address the issue of environmental literacy. A separate grant was earmarked for developing a statewide environmental resource directory and extending environmental literacy beyond the classroom.

Statewide green initiatives have done much to advance the cause of sustainability in education throughout the United States. The most successful efforts have been in states where the public is generally supportive of such efforts and where decision makers are highly committed to meeting the challenges presented by global warming and climate change. Many educational institutions have been transformed by these initiatives, and they, in turn, are transforming the worldviews and living habits of broader communities in their efforts to create a more sustainable environment for everyone.

See Also: American College and University Presidents Climate Commitment; Association of University Leaders for a Sustainable Future; Dickinson College; Maine Green Campus Initiative.

Further Readings

American Association of Sustainability in Higher Education. "State and Campus Sustainability Organizations." http://www.aashe.org/resources/regional_orgs.php (Accessed July 2010).

Dernbach, John C. *Agenda for a Sustainable America*. Washington, DC: ELI Press, Environmental Law Institute, 2009.

Dunkel, Norbert W. "Green Residence Halls Are Coming: Current Trends in Sustainable Campus Housing." *Journal of College and University Student Housing*, 36/1 (2009).

EEAC. "Environmental Literacy Initiative in New York State." http://www.eeac-nyc.org/news_detail.asp?id=17 (Accessed July 2010).

Environmental Protection Agency. "National Environmental Act of 1990." http://www.epa.gov/enviroed/whatis.html (Accessed July 2010).

Franson, Melissa. *The Impact of Classroom Exposure to Sustainability, Course Content, and Ecological Footprint Analysis of Student Attitudes and Projected Behaviors*. Unpublished Thesis, Auburn University, AL, 2008.

Hignite, Karla, et al. *The Educational Facilities Professional's Practical Guide to Reducing the Campus Carbon Footprint*. Alexandria, VA: APPA, 2009.

Jorgensen, Haley. "A Green Campus Culture in Wisconsin." *Techniques: Connecting Education and Careers*, 81/4 (2006).

Neapolitan, Jane E. and Terry R. Berkeley, eds. *Where Do We Go From Here? Issues in the Sustainability of Professional Development School Partnerships*. New York: Peter Lang, 2006.

O'Keefe, Kevin M., et al. "2009 Maryland State Plan for Postsecondary Education." http://mhec.maryland.gov/highered/2004plan/2009marylandstateplan_initial_working_draftnov19.pdf (Accessed June 2010).

Peterson, Jenny Coon. "Environmental Initiative Brings Together State Agencies, Private Organizations." *Oklahoma Gazette,* April 29, 2009.

Rockwood, Larry, et al., eds. *Foundations of Environmental Sustainability: The Coevolution of Science and Policy.* New York: Oxford University Press, 2008.

State of California. "Welcome to Green California!" http://www.green.ca.gov/default.htm (Accessed July 2010).

21st Century Project. http://www.21stcenturyproject.com (Accessed July 2010).

U.S. Green Building Council. "LEED Public Policy." http://www.usgbc.org/DisplayPage .aspx?CMSPageID=1852#state (Accessed July 2010).

Elizabeth Rholetter Purdy
Independent Scholar

ENVIRONMENTAL EDUCATION DEBATE

This article explores the main areas and features of debate in environmental education (EE). It surveys contemporary and historic debates to illustrate how EE goals, activities, and approaches are variously conceptualized, contextualized, developed, and contested. It also traces the principles and conditions that afford a purposeful and generative culture of debate, considering the case for and substance of educational reform as an exemplar to underscore the value of debate for developing and furthering the field. Key forums for EE debate include (1) the meetings, conferences, and networks of the field and cognate areas, for example, as represented by North American Association for Environmental Education, World Environmental Education Congress, and American Educational Research Association; (2) associated and transdisciplinary professional and scholarly groupings; and (3) the field's publications and wider related literatures, for example, as listed at www.eelink.net.

Key EE Debates

Following initial bursts of practical and scholarly activity about the philosophy and practice of EE during the 1970s, the grounding, quality, development, and achievements of EE have been widely debated, particularly since the Tbilisi Declaration (1977) on the international objectives and characteristics of the field. Key areas of debate have required developing and appraising arguments about the following:

- How diverse aspects of and approaches to education take up local to global environmental issues and conflicts of interest
- How environmental learning and teaching can usefully be conceived and practiced in and across a variety of settings, contexts, interests, and tensions
- How human-environment relationships are appropriately construed through (1) and (2)

Such areas of debate have been fed largely by deliberations in three dynamic and contested fields of inquiry: education, environment, and (to some extent) their intersections in EE. Thus, on the one hand, debates have had to examine a diverse set of perspectives on the varied and complex roots and drivers of what informs and counts as EE, and on the other, the specific patterning, interpretation, and implementation of EE goals, approaches, and programs.

As Stephen Toulmin notes in *The Uses of Argument* (1958), some aspects of arguments are the same throughout all fields and hence characterized as field-invariant, while others vary from field to field and are thus field-dependent. How the worth of an argument is assessed in an EE debate may lead to an over- or underemphasis of these aspects as well as generate an important spark for further deliberation, for example, disputes about the origins, trajectories, and efficacy of EE given different readings of each field. To illustrate, presented with the claim that all education should be construed as EE, debaters may wish to examine on what grounds and with what effects might this be advanced and contested, how it plays out in all educational practices and settings (e.g., in which contexts, now or at other times, here and elsewhere), must the debate reach a resolution regarding the claim, and how it relates to other factors and features of debate in both this and other fields of endeavor relevant to the proposition.

Those EE debates that have received most critical engagement have been those that have marshaled an evolving and sometimes elaborate set of arguments, for example, in relation to the form and focus of educational reform given a particular constellation of environmental conditions and issues, the rationale for championing some approaches to EE over others, as well as tensions and shifts in curriculum priorities and practices. Equally, EE debates take place in different ways in diverse settings and may be simultaneous, in sequence, or with interruption. Thus, local and wider debates about EE can vary considerably in terms of their structure, energy, vigor, and duration as well as the particular inflections, participants, and treatment of core and peripheral arguments and issues in and across each field, for example, those taking place largely in relation to a particular worldview, milieu, body politic, or framework, in contrast to others possible or impinging on the EE debate at hand.

Constituting the Fields of Inquiry

Debate drawing upon the field of education often requires mastery of established, alternative, and emerging positions on (1) the purposes, histories, and ideals of education and educational systems; (2) how and where we learn, and accordingly, what counts as worthwhile teaching and curriculum, for example, in a range of formal and nonformal educational settings, individually and collectively, or at different stages in and across a life course; (3) the role of local and wider conditions and contexts in shaping the politics, culture, remit, and priorities of education and schooling in particular settings, for example, in debating the social, economic, cultural, philosophical, pedagogical, and policy cases for or against having a mandated or standardized curriculum, teaching standards, sequencing and progression of learning, and public and private education; (4) the links between education and the wider world, such as the educative aspects of societal goals promoting participation in education, vocationalism, a neoliberal economy, nation building, or the realization of a "better world"; and (5) how education is studied and developed, and how insights from these are brought to bear in debate as well as in reconstituting the debate about education and its reform.

EE debates drawing on arguments and evidence in the environmental field tend to differ in terms of form and focus. Those addressing the natural science aspects of environmental dimensions, problems, and issues are usually less equivocal than those engaging aspects of the social sciences and humanities. This is particularly so when it is assumed an evidence base must primarily structure environmental debate and then implications are read off from that, rather than, say, require a contest of premise, argument, rebuttal, and conclusion. Similarly, when much of the physical science of the environment is characterized as reasonably documentable and can thus be presented and qualified as to which theories or methodologies are in play and which data must be sought, key areas of debate then tend to be about how objective and impartial, trustworthy, or uncertain the current science or its warrants are. This may be offset against other perspectives and claims about the environment, such as phenomenologically based interests, aspects of which may be overtly or covertly values-based or values-driven, for example, by ethical, social, political, cultural, or aesthetic concerns. However, such distinctions can become muddied or collapse, as in the case of debates addressing anthropogenic climate change. These are often characterized by contestation of its probable causes and effects and their significance for the human political economy—including education—alongside the variability, risks, and uncertainties issuing from its impacts on human and nonhuman communities and habitats around the globe.

Other aspects of environmental debate often addressed in relation to EE include the modeling, impacts, and experience of threats to the quality of air, land, and water; population levels, carrying capacities, and globalization in relation to lifestyles, natural resources, and the biosphere; grounds for maintaining the diversity and integrity of ecosystems, particularly biodiversity and habitats; and the links between the aforementioned and other matters, for example, degraded environmental quality and sociocultural forms of inequality (concerning well-being, poverty, diseases, community development pathways, environmental injustices, etc.).

From this it can be adduced that to ensure positions and deliberations are transparent and accessible, attention should be given to examining the sources, construction, representation, and priority of issues and arguments. Moreover, media, governments, nongovernmental and community-based organizations, lobbyists, advocates, businesses, academics, etc., all bring a perspective to bear that may require recognition and contestation, for example, in terms of an implied or explicit narrative of the issue or the normativity of the position or perspective advanced.

William Scott and Stephen Gough's landmark collections *Key Issues in Sustainable Development and Learning: A Critical Review* and *Sustainable Development and Learning: Framing the Issues* are noteworthy in this regard because they attempt to bring all three fields of inquiry into play when debating the interconnections between contexts, interpretations, and priorities ascribed to learning and sustainability. Their work uses contrasting readings and vignettes to invite and structure debates, focusing on the grounding, relevance, rigor, and impact of historic and contemporary accounts of (1) values and valuing, and risk and complexity; (2) local to international sustainability and education contexts for policy and practice; (3) the language and meaning of key concepts, for example, humans and nature, ecology and economy, learning and behavior, sustainability and capacity, science and technology, self and society; and (4) various ideologies, understandings, and practices of pedagogy and evaluation in environmental and sustainability education in global northern and southern contexts.

Other prominent debates about the substance, grounding, development, parameters, and efficacy of work in EE also draw across the aforementioned fields but in different ways to reinvigorate debate both between and beyond the fields, for example, by advocating either a disciplinary, interdisciplinary, or transdisciplinary perspective. They can be differentiated by how they engage and appraise the various merits and limitations of arguments as to the following:

- Distinctions made between theories or practices of education couched in terms of *about, in, through, for being with, by,* and *for* the environment, and more recently, "education *as* sustainability"—that is, debated in terms of their continued priority and usefulness to either furthering, broadening, or focusing EE approaches and methodological innovation
- Identification of "currents" with a longer tradition in the development and shaping of EE (i.e., the Naturalist, Conservationist/Resourcist, Problem-Solving, Systemic, Scientific, Humanist, and Value-centered) and more recently emerging currents (Holistic, Bioregionalist, Praxic, Socially Critical, Feminist, Ethnographic, Eco-Education, Sustainable Development/Sustainability)—that is, debated in terms of either their fit with current local and wider approaches to framing and practicing EE approaches, or their potential and shortcomings as a framework for learning that stimulates or reinvigorates approaches to EE
- Possible (re)conceptualizations of EE based directly on (1) a "philosophy," for example, of deep ecology, conservation biology, socially critical forms of analysis such as ecofeminism or posthumanism; (2) a "context for 'intervention,'" for example, based on a particular principle, issue, program, subject discipline, technology or technique, praxis, or educational setting (such as a museum, schoolyard, home, classroom); or (3) an "inspiration or category of interest," for example, pedagogies of socioecological justice, place, decolonization, stewardship, ecological citizenship, ecocritical literacy, experience, consciousness—that is, debated in terms of what is expected in the "rhetoric and reality" of standard, diverse, and converging approaches to EE, including what is or ought to be reproduced, contested, rejected, transformed, or introduced within the field
- Characterizing paradigms in EE goals, practices, and approaches as positivist, interpretative, or socially critical, and more recent commentaries on "post" paradigmatic approaches to EE—that is, debated in terms of the field's capacities to elaborate and articulate simple, complex, and critical ways of understanding and perhaps reworking EE
- Perspectives on the links between EE and other fields, typified by examining similarities and differences with Education for Sustainable Development (ESD) or its variants, for example, whether EE is a part of ESD, ESD is a part of EE, ESD and EE partly overlap, or ESD is a stage in the evolution of EE—that is, debated in terms of the frameworks within which EE is currently and might otherwise be directed, imagined, developed, and challenged

Principles for Engaging and Reconstituting EE Debate

To further and transcend arguments about such themes, as well as appreciate their dependencies and commonalities, and strengths and limitations, it is important to establish how a debate is configured and prosecuted. To progress, a debate requires an appreciation of the origins and foundations of arguments, the role of ongoing research and evidence, and from where arguments, new insights, and issues emanate in addition to how they are engaged effectively. This is because debate is typically expected to offer an advancement of human knowledge and a spur to progress and flourishing. Thus, good debate is reasoned to be that which prompts us to live life forward, in fresh or novel ways. It may also serve to deparochialize viewpoints about EE, illuminate ignorance and muddle, combat cant and hubris, help (re)imagine and (re)constitute a democratic, ecologically sustainable, and open society, and so forth. Equally, poor EE debate might be marked by trite, stale, or trivial subject matter, dull or blocked process, disparagement, or be judged to support violence toward others (human and nonhuman), intentionally or otherwise.

A customary way of putting this is that there is an art and craft to debate. To be judged generative and appealing, the form and practice—and thus, aesthetics—of debate should not be ignored. Attention to how an EE debate is contextualized highlights how recurrent or cross-cutting themes and questions in and out with the field are deliberately engaged as necessary and sufficient for debate, and thus, how corresponding argument is advanced with purpose. Equally, examining how a debate is innovative invites consideration as to whether, on the one hand, it is understood to have reached stalemate, stagnated, or run its course; or, on the other, been refueled, reframed, or redirected by a breach of its current conditions and practice, for example, through fresh insights, resources, twists, perspectives, challenges, or approaches to its terms, substance, practice, and progress within or beyond the field. Thus, exploring how an EE debate is contextualized and innovative prompts reflection as to how particular lines of argument and their warrants are variously engaged and how a debate relates to wider matters discussed elsewhere, at other times, or in other ways. Consequently, criteria for evaluating an EE debate may include examining its particular and common appeals, contents, structure, form, warrants, evidentiary bases, authorities, clichés, special pleading, contingencies, qualifiers, reflexivity, influence, sophistication, distinctiveness, completeness, and transferability.

A key challenge in this regard is whether a field champions and demonstrates robust debate regarding how the most pressing problems of the day are addressed. For the social theorist Bruno Latour, this requires that debate engages: How many, who, and what are we (a question of metaphysical multiplicity) and how can we live together (a question of ontological unity)? Both entail (environmental) educational predicates and implications, in that the questions presume that all people have the capacity—and arguably, duty—to form, articulate, and review their beliefs and positions on such questions, but also that they are curious, intelligent, and no matter what their standing, stage, identity, or course in life, are willing and able to question, challenge, contradict, and criticize not only the cherished views, assertions, and positions of others but also their own or those of their community or tradition—in good faith and without coercion.

Arguably, decline in or obstacles to a lively culture of debate should be lamented in any field, not just EE. The risks associated with a dominant discourse are widely appreciated, while those of a cultural climate of "anything goes" or celebration of difference as an end rather than a means to exploring some commonality of purpose and value in a field are increasingly recognized in relation to EE. Indeed, specific concerns about the stifling or silencing of EE debate have raised whether a critical and generous disposition is displayed toward others and their views (e.g., avoiding ad hominem attacks and fallacious arguments) and how criticality and reflexivity are advanced and practiced in EE debates in the various forums.

Insider and Outsider Perspectives in EE Debate

Given the above, it should be noted that as with many small fields, those in EE have not always sought out conversation partners from outside the field, in part owing to sensitivities about bias, partisanship, or conflict, but also lack of opportunity and status. Thus, while there may be an expectation and welcoming of EE debate, it still tends to proceed through "insiders" focusing on the relative merits of different education ideals, movements, or imperatives in developing or evaluating EE and thus, returns to questions of who or what sets the agenda for EE. In this regard, typical foci of debate become (1) the merits, contributions of, or alternatives to local and international perspectives on nature study, conservation education, outdoor education, values education, progressive education, and science education; or (2) whether EE should favor a particular model of knowledge, attitudes, values, and/or behavior change; curriculum development; educational mainstreaming or critique—such as via critical or participatory approaches, for example, action research.

However, as noted above, recent thematics for debate have broadened the range of viewpoints and arguments on offer and can thus foment other possibilities. For example, some arguments focus on the degree to which EE was or is construed to be part of either a wider hegemonic or counterhegemonic project (usually by drawing on political and social theory and critiques of the dominant social paradigm). Others focus on the possibilities and case for situating human processes, principles, and practices in ecologically informed aspirations and paradigms. These include (1) examining the benefits and limitations of providing indoor and outdoor learning experiences from ecocentric rather than anthropocentric interpretations of EE; (2) changing human behavior or empowering people by combating nature deficit disorder, addressing social and ecological injustice, developing arts-based responses to nature, etc.; (3) balancing ameliorative and anticipatory forms of learning or individualized and social approaches to learning; or (4) transforming, challenging, reforming, or mollifying various norms and institutions of modern ways of living, as in through contrast with indigenous cosmologies and practices.

A key debate that can smudge insider and outsider interests is that on "Stevenson's Gap" and the case for reform in both EE and education more widely. Robert Stevenson's analysis of EE during the 1980s argued there is a gap between policy rhetoric and school reality for EE. He highlights the socially critical and political action goals of EE since Tbilisi and the uncritical role of much schooling and curriculum inquiry and development in maintaining and reproducing the present social order. Revisiting this analysis 20 years later, he argues the gap persists.

While Stevenson explains this in terms of the structural organization of schools, the primacy of demands on teachers to maintain order and control, and teachers' presuppositions about knowledge and teaching, commentators and critics also suggest other and deeper issues to contend with. For example, one line of argument is to trace the inadequacies and malfunctions of the education system, social inequalities, the agendas of modernity and globalization, and the hybridization of the EE field (e.g., with ESD, adult education, emancipatory educational and political traditions) and how these continue to fissure the field's configuration and possibilities for reform. Others argue that the No Child Left Behind Act has come to dominate the discourse and practice of schooling in the United States through a focus on achievement and accountability. Consequently, it is claimed environmental educators have either accommodated this by playing the game or resisted it by attempting to change the rules and reworking the constraining regularities of public schooling. In this regard, place-based education has been advocated as an alternative route to curriculum development and school-community relations because it gives priority to local cultural, environmental, economic, and political concerns. (Iteratively, of course, while this may be formative for EE, it too requires debate.) Another strategy is to redirect attention away from a focus on the passionate and motivated EE teacher toward fostering wider engagement with the adequacies of curriculum transformation processes, normative frameworks for EE, and situated engagements with environmental learning by students and communities. Stevenson concedes some and challenges other such points. He reformulates his argument to suggest that the focus of the debate about the gap should now be on teachers' practical theories and the contexts shaping their practices, in particular, the power of professional communities to contribute to teacher learning, and how environmental educators might build their normative and technical capacity, both individually and collectively, to shape and develop effective EE practice.

Conclusion

While many justifiably subscribe to the view that there are equally legitimate and incommensurable ways of practicing EE, an end to debate is no less in sight. Examining the culture and depth of purposeful and generative debate reveals not just consensus and dissensus but continuing ambivalence and pockets of ambiguity about the enduring objectives and characteristics of the field. Equally, the forums noted earlier afford ample opportunity for engaging debate that is historically attentive and progressive in orientation. Looking across the arguments and themes, then, it is hoped that EE debate will continue to nurture questions that engage in the following:

- What is understood to be the field's history, conditions, and horizons?
- What is now necessary and sufficient as a definition and defense of a case either for or against particular forms of EE?
- What is minimal as EE in theory and practice, and how does this relate to maximal and optimal interpretations, for example, to overcome gaps in rhetoric and reality, to avoid capture by the latest buzzword or fashion, or to be efficacious in addressing ongoing and emerging socioecological problems?
- Are there particular kinds of ethics/experiences/processes/contexts/politics/understandings/ conditions—or for that matter, intersections of these—that are essential to an effective EE worthy of the name?

See Also: Outdoor Education; Place-Based Education; Social Learning; Sustainability Topics for K–12.

Further Readings

Robottom, Ian and Paul Hart. *Research in Environmental Education: Engaging the Debate.* Geelong, Australia: Deakin University Press, 1993.

Sauvé, Lucie. "Currents in Environmental Education: Mapping a Complex and Evolving Pedagogical Field." *Canadian Journal of Environmental Education,* 10 (2005).

Scott, William and Stephen Gough, eds. *Key Issues in Sustainable Development and Learning: A Critical Review.* London: Routledge, 2004.

Scott, William and Stephen Gough, eds. *Sustainable Development and Learning: Framing the Issues.* London: Routledge, 2004.

Stevenson, Robert, ed. "Revisiting Schooling and Environmental Education: Contradictions in Purpose and Practice." *Environmental Education Research,* Special Issue, 13/2 (2007).

Alan Reid
University of Bath

Environmental Literacy Council

The Environmental Literacy Council was created in an effort to assist teachers, students, policy makers, and the general public to investigate issues of sustainability and to find resources that assist them in making informed decisions related to green concerns. A non-profit organization, the Environmental Literacy Council provides free background information on many common environmental science concepts—resources that reflect a variety of perspectives—as well as curricular materials that provide teachers with the tools necessary to craft instruction that is appropriate for their classes. As environmental studies became an increasingly important part of the curriculum at many schools, the Environmental Literacy Council offered assistance and support to teachers who wished to center their instruction more on green issues. The Environmental Literacy Council was heavily criticized by some, however, as being biased and not above misrepresenting scientific findings to question global warming. As of 2009, the Environmental Literacy Council ceased operations, although its Website still remains online.

Background

The Environmental Literacy Council was founded by Jeffrey Salmon, a former speechwriter for Vice President Richard "Dick" Cheney and the director of the George C. Marshall Institute (GCMI) from 1991 until 2001. A predecessor organization, the Independent Commission on Environmental Education, was also funded by GCMI. The Environmental Literacy Council's Website first appeared in 1997 and stemmed from a perceived need for a group to provide resources to teachers and the general public. The Environmental Literacy Council shared space with the GCMI and received funding from a variety of sources, including the Sarah Scaife Foundation, the Earhart Foundation, the Charles G. Koch Charitable Foundation, and the John M. Olin Foundation. All of these foundations are known for supporting a variety of conservative causes, although the Environmental Literacy Council also received funding from 2001 until 2003 from a variety of federal agencies, including the National Science Foundation, the National Endowment for the Humanities, and the U.S. Department of Education.

Resources

The Environmental Literacy Council's Website provides a host of resources for classroom teachers, parents, and students. The Website provides content with regard to specific areas, including air and climate, land, water, ecosystems, energy, food, and environment and society. Each of these content areas provides an icon that links to a page with content about the subject. The air and climate page, for example, contains information about air, climate, and weather, and provides information regarding the thermosphere, mesosphere, stratosphere, and troposphere. Links are also included to some selected Websites including information from the National Aeronautics and Space Administration (NASA) on Earth's atmosphere and from the Environmental Protection Agency (EPA) on pollution issues around the globe. Also included are teaching modules developed by the University Corporation for Atmospheric Research and the University of South Dakota. Similar content is included for each of the other content areas.

Also on the Environmental Literacy Council's Website are resources regarding a host of other environmental topics, such as biodiversity, biogeochemical cycles, climate, weather, economics, and assorted other topics. Each of these also had groups of resources that supported the topic and prepared activities that could be utilized in the classroom or at home. Resources were also grouped by state, so that those who live in a particular geographic region would be able to access statutes, governmental agencies, and nature preserves that were topical and local. In addition to links to the pertinent state's agencies, links to maps, geographic surveys, and historical documents related to the environment were available. For younger users, links to Creature Features provided information about specific endangered animals, while older students could engage in geography quizzes and other games. For teachers, there were review materials that were geared to the Advanced Placement (AP) environmental studies test and assessment tools that could be used for evaluation purposes. An area also existed where teachers could share lessons they had developed with their peers by posting this work on the Website for others to access.

The information and resources provided by the Environmental Literacy Council attempted to allow users to view the environment from multiple perspectives. Many of the supporters of the Environmental Literacy Council had a more conservative approach to sustainability than many other supporters of green education. As a result, its Website tried to present an environmental issue and then explore that issue from the perspective of multiple stakeholders, including those who were more skeptical regarding claims made about the need for sustainability. Among other instructional strategies, students were frequently encouraged to role-play, that is, to assume different personas with opposing perspectives, to gain a better understanding of how different people view things. During the administration of President George W. Bush, the Environmental Literacy Council partnered with the National Science Teachers Association (NSTA) to prepare NIH-funded professional development modules that explored environmental science issues. These were available to the public free of charge.

While the Environmental Literacy Council prepared materials that were topical, attractive, and accessible to a wide range of educators and parents, its materials were not without critics. Many advocates of incorporating sustainability studies in the K–12 classroom charged that the Environmental Literacy Council's materials were biased, deceptive, and based upon faulty science. For example, when the Environmental Literacy Council stated that a study suggested that global warming was a naturally occurring phenomenon that had arisen periodically on a 1,500-year cycle, the author of that study accused it of misinterpreting her findings. Those who favored stronger advocacy on the part of green education believed that the Environmental Literacy Council's materials were a deliberate attempt to obfuscate and obstruct environmental studies. Despite this controversy, the Environmental Literacy Council's materials were widely used and often cited by state departments of education and other teacher-oriented groups.

See Also: Association for the Advancement of Sustainability in Higher Education; Association of University Leaders for a Sustainable Future; Environmental Education Debate; Halifax Declaration.

Further Readings

Chepesiuk, R. "Environmental Literacy: Knowledge for a Healthier Public." *Environmental Health Perspectives*, 115/10 (2007).

Environmental Literacy Council. "What Is Environmental Literacy?" (2010). http://www
.enviroliteracy.org (Accessed June 2010).

The Halifax Declaration. http://www.iisd.org/educate/declarat/halifax.htm (Accessed August
2010).

Stephen T. Schroth
Jason A. Helfer
Sean G. O'Keefe
Knox College

ERTHNXT

ERTHNXT is a domestic nonprofit organization whose mission is to safeguard all species by instilling through education an ethic of environmental stewardship in the
nation's youth. It seeks to achieve this mission through a curriculum of scientific-based
ecological knowledge. Through this curriculum, its goal is to engage students in actions
that are transformative for both themselves and the environment. The organization
believes that education is the best means by which all life can be safeguarded, because
activities can be diffused within a community as knowledge is disseminated. The organization does not function within traditional educational institutions, but rather is more
a form of nonformal education, providing adult leaders with a "toolkit." The toolkit is
designed to educate young people to learn about the importance of trees in the environment, how to plant and take care of trees, and share the information they have learned
with others.

The organization was originally called Future of All Life, Inc., and changed its name
to ERTHNXT in 2007. The original name refers to the seminal book about biodiversity,
The Future of Life, written by advisory board member Dr. Edward O. Wilson. In this
book, Wilson offers a plan for preserving Earth's biological diversity. The founding members of ERTHNXT were inspired by this book, and developed the organization to help
inform children and youth about the environment and their ability to make environmentally conscious decisions. However, the founding members found that the meaning of the
original name confused those who were unfamiliar with the book or topic. The new name,
ERTHNXT, is meant to clearly communicate the organization's perception of the priority
of preserving all life, and to imply a sense of urgency for preserving biodiversity and the
planet. In the age of text messaging, the organization's name is spelled to reach out to
youth and support the organization's call to action "Send the Message."

One of ERTHNXT's flagship programs is titled "Trees for the 21st Century." This
initiative was inspired by Kenyan Nobel Peace Prize winner Wangari Maathai. This program involves an active campaign of tree planting. Since its inception, the program has
engaged over 43,000 students in tree-planting actions, planting approximately 41,000
trees. To date, the organization's activities have been focused in the mid-Atlantic region,
principally in the states of Delaware and Pennsylvania. ERTHNXT is engaged in a wide
variety of partnerships to help further its effectiveness. These organizations include both
environmental nongovernmental organizations, such as the National Wildlife Federation,
as well as private corporations, including Delmarva Power. The organization's noncorporate

"Program Partners" help the organization by increasing the number of students who participate in the programs. In 2007, ERTHNXT partnered with the Girl Scouts of America, announcing a nationwide tree-planting campaign. Through its collaboration with the Girl Scouts, it provided each of the 148 councils in 46 states with 95 trees, involving thousands of girls in planting more than 14,000 trees. The corporate partners assist ERTHNXT financially, a "win-win" situation that helps ERTHNXT achieve its goal while greening the corporation.

ERTHNXT employs a variety of age-appropriate curricula to help foster stewardship and environmental consciousness among students. Students in its youngest grade level (second and third grades) are engaged in fun environmentally themed games, watch educational environmental videos, and learn about stewardship by keeping a terrarium—picking out appropriate plants and caring for them. The students also learn about the story of Wangari Maathai, and create connections to this environmental movement by learning about Kenyan children and the environmental challenges they face. Students in this age group also learn biological information about tree structures, and gain an appreciation of the values of trees by learning about the diverse products that are produced from them.

For the students in the next age bracket (fourth through sixth grades), the curriculum begins placing more of a focus on biological knowledge. For example, at this level students learn about invasive species, and also about tree identification, getting hands-on experience with dichotomous keys. In addition to the tree-planting exercises, they continue to build on their general knowledge of Kenya and learn more about the environmental issues youth face.

Students in the next higher age group (sixth through eighth grades) begin developing critical thinking skills, which are generally promoted across educational theories. Students learn to think critically by exploring causes of stream pollution, and advance their problem-solving abilities by discussing potential mediating actions. They build on this critical knowledge by learning more about the ecological features of riparian zones, and specifically the roles trees play in those habitats. For students in the oldest age group (9th through 12th grades), critical thinking is once again emphasized; at this level, students explore the carbon footprint concept, map out their carbon footprint, and discuss ways that they can green their lives through simple activities, such as recycling. Building on biological knowledge of trees gained in previous activities, students learn about native regional species by comparing detailed descriptions with photographs. They also learn about the biology of seed dispersal.

Although ERTHNXT engages in diverse forms of environmental education, its objective of instilling an ethic of environmental stewardship remains principally grounded in its tree-planting projects. The organization and its partners select native coniferous tree species that are accustomed to the specific hardiness zone of the region. The tree kits available for sale by the organization include a grade-appropriate activity guide for facilitators, a tree planting and stewardship guide, as well as 6- to 12-inch-tall seedling trees.

ERTHNXT is exemplary of a new breed of educational nongovernmental organizations. These organizations are employing nonformal learning methodologies, such as disseminating curricula through after-school activities and environmental clubs, to help foster an ethic of environmental stewardship in the nation's youth.

See Also: Early Childhood Education; Educating for Environmental Justice; Outdoor Education; Sustainability Topics for K–12.

Further Readings

Dewey, John. *Experience and Education.* New York: Free Press, 1938.

Fenwick, T. *Learning Through Experience: Troubling Orthodoxies and Intersecting Questions.* Adelaide, Australia: National Centre for Vocational Education Research, 2009.

Gruenewald, D. "Foundations of Place: A Multidisciplinary Framework for Place-Conscious Education." *American Educational Research Journal,* 40/3 (2003).

Mezirow, J. *How Critical Reflection Triggers Transformative Learning. Fostering Critical Reflection in Adulthood: A Guide to Transformative and Emancipatory Learning.* San Francisco: Jossey-Bass Publishers, 1990.

David Meek
University of Georgia

ESD to ESF (Education for Sustainable Development to Education for a Sustainable Future)

Education for Sustainable Development currently enjoys huge momentum and is attracting exponential adherence. It is the educational manifestation of the concept of sustainable development that became well known in the World Conservation Strategy and that soared into prominence with the publication of the Brundtland Commission's report, *Our Common Future,* a landmark statement on sustainable development. Their definition of the term is probably better known than the report itself: "Sustainable development is development that meets the needs of the present without compromising the ability of future generations to meet their own needs."

Education for Sustainable Development (ESD) has a long history associated with a partnership between the United Nations Environment Programme (UNEP) and the United Nations Educational, Scientific and Cultural Organization (UNESCO), established in the early 1970s to strengthen environmental education internationally. In the past 30 years, environmental education has expanded its reach to encompass a wider engagement with development issues under the banner of Education for Sustainable Development. A key outcome of the Rio Earth Summit's Agenda 21 in 1992 was widespread international commitment to strengthen the role of education, training, and public awareness in achieving sustainable development. The Johannesburg Plan of Implementation at the World Summit in 2002, in seeking to further strengthen the contribution of education, training, and public awareness of sustainable development, specifically called for the declaration of a Decade of Education for Sustainable Development. This was launched by the United Nations (UN) in 2005 (2005–2014) and proposes that all levels of the education and training system need to be reoriented toward a more sustainable model. It is the author's view that ESD is paved with good intentions; however, they can prove futile, or even dangerous, if they are devoid of critical understanding of socioecological systems, an ecological and social conscience—love and respect—an ethical relationship.

Sustainability has evaded consensual definition for almost 30 years, and although the term may be recent, the concept is not. The term, coined in the late 1980s, suggests something sustained, held up, stable, durable, and lasting. As with many well-intended concepts and principles, *sustainable development* as a term has suffered as a catch-all buzzword that can be filled at will by different users to hold their own meanings and intentions to which one might argue there has never been any real commitment.

As a "brand," sustainable development carries a great deal of goodwill but also much baggage and has generally been regarded in policy and planning circles as marginal to mainstream decision making. Some more thoughtful people recognize it as tuned to a world of declining resources that will require adaptive strategies quite different from those being pursued currently. Early calls for sustainable development emphasized resource limits and used rhetoric of disaster, threat, and scarcity.

The concept of sustainable development is strongly associated with an environmental and green agenda. The environment, environmentalism, and green issues seen as separate by and from the mainstream is an enduring and pervasive perception and has for a long time been marginalized as a contained body of thought. There thus exist enduring connotations and associations of sustainable development being concerned around environmental issues and an unconcern about the plight of the poor.

For some authors, this represents a "mislabeling of various societal messes as environmental problems." The absence of a clear view of what "environment" really is renders it open to all manner of interpretation and legitimization. The flexibility of the concept resides exactly in the process of separating things natural from things social, which permits a discursive reading of what environment or nature is to serve specific social ends, while ignoring the inevitable mediations between environment and society. The paradigm thus of the relationship between humans and the environment is considered key.

Compounding confusion, sustainability is used in a narrow as well as in a broad sense. Used narrowly, it emphasizes only environmental and ecological concerns as they manifest both locally and globally. Hugely important as they are, addressing them alone could never deliver sustainability. Sustainability, in its broad sense, also requires economic viability and greater social equity.

A shift is required from the goal of standard of living to that of quality of life, transforming the drive to simply get and consume into the profoundly different one of pursuing deep psychic fulfillment—a step forward, not backward, as it is too often portrayed. Achieving sustainability then requires attention to psychology and even to spiritual issues, to satisfy values deeper than advertising-induced desire. Sustainability is not only about curbing environmental abuse—it is more about enjoying a more sane and more just way of life. The universe is not a dead clockwork mechanism but a living process, constantly unfolding and creative. A profound psychological impact of such a shift could help us to no longer feel alienated from the world, nor compelled to defend against this feeling through acquisitive consumption, but can instead disencumber ourselves to open up and feel an integral part of this astounding and benevolent planet.

In recent years, there has been a major shift in approaches to sustainable development—partly in response to the limitations of traditional models and partly in response to a global trend among scientists, economists, and environmentalists away from narrow determinism toward developing a worldview that embraces the complexity of our natural and social system. A fundamental change in human thinking is considered key to sustainability.

Education in a Changing World

Education for Sustainable Development or its younger offsprings, such as education for sustainability, education for a sustainable future, and education for the development of responsible societies, are valid and urgent responses to social and planetary trends that threaten the global environmental and human condition with disaster.

Many authors and thinkers are engaged in and working at developing a new language for explaining and talking about human learning. It is a period of major revisions, and the discussion above has illustrated that ESD carries much baggage. While the sustainable development discourse has been mainstreamed and has had enormous canvassing and heuristic capacity, many people continue to emphasize very different meanings and dimensions. As many all-inclusive definitions hide real conflicts between various dimensions, there are authors who have avoided the term *sustainable development* altogether in search of a new umbrella word.

Some thinkers have sought to mollify or mount an alternative conceptualization of sustainable development and use alternative terminology that takes the development out of sustainability, such as education for sustainability or education for a sustainable future or education for the development of responsible societies.

Education for Sustainable Development, holistic conception though it is intended to be, has not attracted peace, social justice, and antidiscriminatory educators to its conferences, programs, educational policy, and training initiatives to the same extent. This is why the author, much like many other thinkers, feels more resonance with the more value-free term *Education for a Sustainable Future* (its younger sister). It is potentially a more neutral, embracing term for more holistic connective integrative thinking that will further the discourse from single- or separate-issue environmentalism to mainstreaming sustainable development in the 21st century.

The term *resilience* also is appearing more frequently in discussions about environmental and societal concerns, and it has a strong claim to actually being a more useful concept than that of sustainability. Resilience thinking offers a different way of understanding the world around us and of managing our natural resources. It makes an important distinction between the amount of knowledge and the kind of knowledge we pursue and acquire. The philosophy of resilience emphasizes an accurate understanding of socio-ecological systems and how they function—it conceives resource systems and people as part of them.

From Learning of Ecology to the Ecology of Learning

A powerful case for sustainable education has been made by Stephen Sterling (2003) who argues that education can only contribute to social transformation if it relinquishes the modernist agenda characterized by managerial, mechanistic, and transmissive approaches and comes to be informed by an ecological paradigm characterized by "whole-system thinking," participation, empowerment, and self-organization. In this way, education can play a crucial role in helping foment a sustainable future.

There is no question that we are living in an increasingly fragmented world. Problems lie very deep within our fundamental approach to the world. The world today is in the grip of multiple crises. We are witnessing simultaneous transitions in terms of our economy, urbanization, and our ecological life support systems. We live in a complex world with complex problems where we can no longer single out any one "BIG" problem, as our problems are interconnected global-local problems with irreversible long-term consequences.

Education is the key to steering the world on a more sustainable path. Unless people have the tools to analyze and understand the world around them, they will not be able to address the challenges that face our society and the environment. Our current generation requires leaders and citizens who can think ecologically, understand the interconnectedness of human and natural systems, and have the will, ability, and courage to act. This period we are living in has been referred to as the "Great Turning" requiring all our best qualities of leadership and collective compassion and ingenuity to avoid it becoming the end of human civilization altogether.

Much like resilience thinking, S. Sterling argues that a fundamental change in human thinking is key to sustainability, and ESD is far more than getting ecological concepts into the curriculum and education—it is less about the "learning of ecology" and more about the "ecology of learning." A fundamental change in the way we view the world is required rather than just the development of ecological understanding.

Throughout this article, the author has suggested that making space for interdisciplinarity in education will help develop a more integrative way of seeing the world. Confronted by the need to act given the increasingly evident confluence of multiple environmental, social, and economic crises, there is arguably an enormous opportunity, possibility, and willingness perhaps to learn to see the world anew and to sink and fudge differences.

As an example, universities produce conservation professionals with excellent skills to describe the current environmental decline, but without the skills to stem it. More appreciation is required that conservation courses must embody consilience—the fusion of knowledge traditions.

In my own profession, planning professional bodies have agreed that future planners will need to be able to go beyond the basics to be leaders and innovators in promoting sustainability. At recent planning conferences, a consensus is emerging with regard to what our next generation of city and regional planners will need to know, such as, for example, being able to identify and interact with diverse interests, mediate differences, and undertake consensus building to help different constituencies reach agreement in the face of new global energy and climate challenges. These are all tall orders that assume interdisciplinarity and promoting transformative agendas for sustainability.

Issues of climate change and sustainable development also offer a profoundly instructive example of the gulf between the worldview informing most educational institutions and the worldview needed to engage with the complexities of the contemporary world. Climate change confronts education with interconnectedness, chaos complexity uncertainty, and the controversial. Our future remains intrinsically unknowable. An adequate response on the part of universities to climate change would require a shift of significant proportions. Science is reluctant to address the aesthetic, economic, emotional, political, social, and spiritual dimensions of climate change and its impacts and implications. Climate change for the most part is conceived as a topic to be located within the science curriculum.

Deeper questions as to what happened at Copenhagen in December 2009 and whether Copenhagen talks opened doors to a new global order also show how the highly complex climate negotiations were too political for the technicians and too technical for the politicians, and the process has perhaps not yet caught up with new dynamics and new integrative skills—therefore a dysfunctional process and a gap between the art of the possible and the requirements of science. There appeared to be no common base of exploration and explanation.

The Discomfort and Hope of Interdisciplinarity

The author is of the conviction that it is the role of all education to help society appreciate and understand humanity's current critical position with respect to global trends and patterns such as globalization, poverty and inequity, energy and resource use, loss of biodiversity, and climate change. It must be the mission to educate for life and to address the challenges facing society. We want to nurture a foundation of skills, knowledge, and versatility that will last a lifetime despite a changing environment.

All education should be mindful of these trends and plan to prepare its students for a preferred future. There is urgency to bring new depth, clarity, and compassion to every level of human endeavor—from unlocking individual potential to finding new approaches to global-scale problems. We need to focus on research, education, and leadership for humanity's most pressing problems. This raises the questions of whether our way of educating provides adequate tools and is able to generate these new conceptual resources.

This entry proposes that by valuing and mainstreaming interdisciplinarity in all educational endeavors, we can perhaps contribute to more effectively engage and reverse trends and deepen understanding. The author believes that interdisciplinary teaching and research represent the future. We still have much to learn, including how to learn. We all still have work to do to help us honor each other's contributions to the rich aesthetic and intellectual resources and to help our students and learners understand their value.

The dawn of the 21st century sees humanity in a bind: our most pressing problems are complex and interdisciplinary, yet in the face of this complexity, we tend to seek deeper and narrower expertise for these solutions. Our schools and universities, as long-lived institutions, reach the 21st century having to face deep, unresolved questions. This crisis is more serious in some institutions than in others but concerns all of them.

To a large extent, we train our children, students, academics, and scientific and artistic leaders to be deeply functional experts in one area, while the problems they face spill over every imaginable boundary. That it is a time for boundary pushing within all of the modes and methods of creating knowledge is increasingly becoming apparent. The current acceleration in the demand for knowledge in general cannot be faced alone by departments in rigid institutional frameworks as in the case of most of our universities and training institutions.

One needs to ask: how do we bring diverse people together to think beyond their normal boundaries? How do we aggregate rather than segregate? How do we think, communicate, and develop shared understanding across disciplinary boundaries when for more than three centuries fundamental differences emerged over how we have conceptualized/used nature, science and society. In "natural" science we are concerned with so-called hard facts, the observable, the measurable, what is known objectively, and we explain things. In social sciences, on the other hand, we are concerned with "soft," more messy, and intangible issues—human values, human needs, human interests, human senses—and we interpret things.

As the natural scientists are perceived to be dealing with hard objective truths and less messy, wicked, social problems, within our current dominant positivistic outlook, it appears that more and more areas of life are "scientized" and the natural is no longer available naturally. Science, social behavior, and economics have been "siloized" over more than three centuries.

The word *interdisciplinarity* has gotten quite a workout lately; the term/concept is peppered somewhat meaninglessly through corporate, academic, and business prose to make proposals palatable, relevant, cutting-edge, fashionable, and contemporary. Although there is willingness and eagerness to jump on the interdisciplinary bandwagon, the term is readily used in universities to describe programs and courses. However, many difficult issues, challenges, paradoxes, and complexities are often ignored, not explicitly addressed, recognized, or understood.

There appear to be very few spaces and resources and very little time to explore and surface them. I would argue that there is an important need to stimulate much more discussion on current bold aspirations if we are serious about achieving better interdisciplinarity. There needs to be much more openness and discussion and raising awareness in terms of perceptions, pitfalls, and misunderstandings. Skills that have been refined and practiced for more than three centuries are deep and entrenched, and new ways would need practice. One can't assume that new practices just happen on their own.

See Also: Agenda 21: "Promoting Education, Public Awareness, and Training (Chapter 36); Green Community-Based Learning; United Nations Decade of Education for Sustainable Development 2005–2014.

Further Readings

Birch, E. L. and C. Silver. "One Hundred Years of City Planning's Enduring and Evolving Connections." *Journal of the American Planning Association,* 75/2 (2009).

Colby, M. E. "Environmental Management in Development: The Evolution of Paradigms." *Ecological Economics,* 3 (1991).

Hawken, P. *Blessed Unrest—How the Largest Movement in the World Came Into Being and Why No One Saw It Coming.* New York: Penguin Group, 2007.

Knight, A. T., R. M. Cowling, M. Rouget, A. Balmford, A. T. Lombard, and B. M. Campbell. "Knowing But Not Doing: Selecting Priority Conservation Areas and the Research-Implementation Gap." *Conservation Biology,* 22/3 (2008).

Orr, D. *Down to the Wire.* New York: Oxford University Press, 2009.

Selby, D. "The Firm and Shaky Ground of Education for Sustainable Development." In *Green Frontiers: Environmental Educators Dancing Away From Mechanism,* J. Gray-Donald and D. Selby, eds. Rotterdam and Taipei: Sense Publishers, 2008.

Sterling, S. "Whole System Thinking as a Basis for Paradigm Change in Education: Explorations in the Context of Sustainability." Ph.D. Dissertation, University of Bath, 2003. http://www.bath.ac.uk/cree/sterling/sterlingthesis.pdf (Accessed July 2006).

Swyngedouw, E. and M. Kaike. "The Environment of the City . . . or the Urbanization of Nature." In *A Companion to the City,* G. Bridge and S. Watson, eds. Oxford, UK: Blackwell Publishers, 2000.

Walker, B. and D. Salt. *Resilience Thinking Sustaining Ecosystems and People in a Changing World—How Can Landscapes and Communities Absorb Disturbance and Maintain Function?* Washington, DC: Island Press, 2006.

World Commission on Environment and Development. *Our Common Future.* New York: Oxford University Press, 1987.

Worldwatch Institute. *State of the World: Transforming Cultures From Consumerism to Sustainability.* Worldwatch Institute Report on Progress Toward a Sustainable Society. New York: W. W. Norton, 2010. http://blogs.worldwatch.org/transformingcultures/wp-content/uploads/2009/11/SOW10-PreviewVersion.pdf (Accessed January 2010).

Tania Katzschner
Independent Scholar

EVERGREEN STATE COLLEGE

From Earth Day in 1970 to climate-change actions in 2010, colleges and college students have played a key role in environmental and sustainability movements. When the State of Washington sought to meet a growing student population, the enabling legislation in 1967 hinted at the potential of a new college for changing times. The legislators specifically identified the need for a large campus in an underserved region and the goal of reaching these students with innovative teaching techniques, which led to the 1,000-acre campus near Olympia, the state capital, and the college's approach to learning. When Evergreen State College opened its doors to 1,000 students in fall 1970, it entered an era in higher education when innovations in teaching and mission-focused education were being explored at existing colleges and at other newly launched institutions, such as the University of California, Santa Cruz. These two new state-funded institutions—rooted in innovative curriculum and the faculty, staff, and student bodies attracted by such curriculum—served as prototypes and applied demonstrations in delivering innovative teaching methods. At Evergreen, this innovative focus was also shaped by the environmental initiatives that have been identified with the school throughout its history.

Integrated Learning Model

Today, with nearly 5,000 students on its wooded campus, Evergreen State College is renowned as a liberal arts college whose mission and success is based on integrative learning practices. The college is among few institutions that have been crafted from the ground up with a focus on learning communities that are often organized around the built and natural environment of the campus, the region, and beyond. Rather than enrolling in individual courses, students join faculty each term to focus on a unified thematic study, culminated not by grades but by a narrative assessment. The focus on learning communities in general education and discipline-specific curriculum has shaped the entire school, with environmental and social-change groups emphasized in student life, residential halls, and dining services.

While any field of study can benefit from such an integrated learning model, this emphasis is well suited to interdisciplinary fields such as environmental studies and sustainability. The integrated nature of the formal studies also supports greater academic focus on campus life than at most colleges. Themed dorms tend to identify with green learning and sustainability, and students staff a forest garden and an organic farm that both serves as a learning laboratory and provides locally grown food for the school's dining services. The campus's built environment serves as a living laboratory where studies on energy use and sustainable practices occur; the forest and tidal wetlands offer a landscape for both study and outdoor recreation.

Academics and Sustainability

Areas of study in which the college excels include education and environmental studies, fields that the college offers at the undergraduate and graduate levels. The college's largest graduate program is the Master of Public Administration, which reflects the public service aspect of the college's mission and its hosting of the Washington State Institute for Public Policy. An additional option features a Master of Environmental Studies/Master of Public Administration joint degree.

At the undergraduate level—and for a state college of its size—Evergreen students demonstrate notable success in continuing on to graduate studies. The integrated studies model of education has also garnered Evergreen numerous commendations as both a best value and as simply one of the best liberal arts institutions, with the college's accomplishments noted by *US News and World Report*'s recent college rankings as first in the West for undergraduate teaching at master's universities. Other citations of excellence note the college's student engagement, and the college received high ranks from the *Princeton Review* (including a listing as one of 15 colleges in its "Green Rating Honor Roll") and in the recent book "Colleges That Change Lives." Evergreen has been honored repeatedly by *Sierra Magazine*'s "Cool Schools" list of "eco-enlightened U.S. universities" and as one of Planet Green's Top 5 Green Colleges.

The college's attention to student assessment is also applied to its own narrative assessment of programs and of college-wide goals. In a recent assessment of sustainability in college programs, Evergreen defined such curricular elements as living within limits, equitable resource distribution, and understanding social, economic, and environmental interconnections. Overall, slightly less than 60 percent of programs noted some level of sustainability during their program review process, with less integration of sustainability in core programs but a substantial focus on sustainability concepts in environmental studies, evening and weekend studies, and inter-area studies.

Community and College

Although Evergreen is recognized nationally, it draws 75 percent of its students from Washington and remains a state institution, with coursework and research entwined with local and regional issues. The "green" college is also an urban college, with programming and courses integrated with the nearby city of Olympia, which supports urban studies as well as work on state legislative and policy issues. In 1982, Evergreen launched a satellite campus in Tacoma to serve urban-based students who are not able to commit to the residential campus in Olympia. Additionally, the college is home to at least seven public service centers, including the Longhouse Education and Service Center serving on-campus students and surrounding Native American communities and Northwest Indian Applied Research Center; the Labor Education and Research Center; the Evergreen Center for Educational Improvement; and the Center for Community-Based Learning in Action.

Educational Leadership

Evergreen's footprint in educational leadership builds from its successful tradition in shaping its curriculum to meet learning goals. The college supports regional, national, and international innovations and training at the Washington Center for Improving the Quality of Undergraduate Education. One initiative hosted by the center is the

Curriculum for the Bioregion, with the center leading the region in developing sustainability-focused curriculum. A center-initiated survey of 18 regional colleges observed that key sustainability issues—such as global warming, climate change, and energy—were underrepresented in regional curriculum, which echoed survey results by the National Wildlife Federation 2008 National Report Card on Sustainability in Higher Education. To address this lack, the Bioregion program created a learning community of faculty from various disciplines and campuses, with initial focus on introductory courses in biology, composition, chemistry, philosophy, and religious studies. The program seeks eventually to develop interdisciplinary curriculum models for sustainability education that asks students and disciplines to apply their learning in ways that support and shape healthy communities.

Such interdisciplinary learning is key to the overall mission of the Washington Center, with two of its core programs focused on instructor training, curriculum sharing, and assessment of learning communities in higher education.

The college's Sustainability Council also notes that the college's emphasis on community-based learning and whole-systems thinking supports an active curricular approach to the social and environmental aspects of sustainability. The council's key goals focus on alternative transportation, clean energy systems, sustainable food practices, and waste reduction and feature a "Greener Living" program to engage the college community in sustainability initiatives. Evergreen's early leadership in sustainability was noted with a 2007 Governor's Award for Pollution Prevention and Sustainable Practices, recognizing the first public higher education building in Washington to receive LEED Gold certification, courses in sustainability, and applying sustainable practices on campus.

See Also: Experiential Education; Green Community-Based Learning; Whole-School Approaches to Sustainability.

Further Readings

Evergreen State College. "Sustainability at Evergreen." http://www.evergreen.edu/sustainability/home.htm (Accessed June 2010).

Rappaport, Ann and Sarah Hammond Creighton. *Degrees That Matter: Climate Change and the University.* Cambridge, MA: MIT Press, 2007.

Stibbe, A. *The Handbook of Sustainability Literacy.* Totnes, UK: Green Books, 2009.

Ron Steffens
Green Mountain College

EXPERIENTIAL EDUCATION

Experiential education is rooted in experiential learning, a term that is used conventionally to refer to the pedagogical practices that allow learners to gain knowledge from experience. However, the term *experience* is elusive and difficult to define. Experience is grounded in subjective perception, and individuals may vary in the way they define and

The fieldwork activities of these students, who were conducting water quality tests in Virginia in 2002, are one example of experiential education.

Source: U.S. Department of Agriculture, Agricultural Research Service

give significance to an experience. David Boud, Ruth Cohen, and David Walker talked about "layers of experience" to emphasize the continuous process of internal organization of perceptions and emotional states within the individual. Similarly, experience should not be confined to the inside of an individual's body and mind. Eleanor Gibson stressed that experience is a property of the interaction between the individual and the surrounding context of "human and things." An important implication of this is that there is no such thing as a "piece" of experience. Rather, experience is inextricably connected to other experiences in specific contexts, and learning occurs through personal reflection and awareness. For such features, experiential learning has been associated with pedagogical practices aimed at promoting environmental consciousness and green behaviors. However, as indicated below, history and contrasting philosophical traditions have accounted for much variety as well as debate on experiential practices converging toward this aim.

It was John Dewey in the early 20th century who first looked at the value of experience in education. In *Experience and Education,* Dewey takes a critical stance toward classroom-based teaching, identified with a place of both intellectual and physical confinement, alien to the organic and physiological characteristics of the human body as a sentient and learning being. His critique started from a linguistic observation of key terms of the educational lexicon. For example, he criticized simple understandings of "growth" as accumulation of knowledge. Rather, a change to the present participle—growing—would bring education into the sphere of action and agency. In this way, experiential education, also known as progressive education, would involve physical and intellectual as well as moral and ethical development. As Dewey himself put it, education as growth is not the domain of the single individual. Starting from the premise of supporting physical development, Dewey brought experiential education in line with experiences of learning in the outdoors, including activities promoting development of practical skills, for example, in the transformation of natural resources and materials. In the industrial context of his time, Dewey was concerned with reinstating the value of human dignity and personal agency that had been overshadowed by the factory environment. From this perspective, action and reflection sit at the heart of a democratic society. Philosophically, there are connections with the writings of Mohandas Gandhi and Paulo Freire, concerned with people's life conditions in the fast cycle of industrial and environmental transformation. For all these authors, education cannot be disjointed from experience. Rather, education is an integral part of people's material, social, and spiritual emancipation.

More recently, the ideas of Dewey have found new interpretations within the fields of environmental education, green education, and some areas of curative education. In these contexts, emphasis is placed on giving children the opportunity to develop as individuals and members of a community. A plethora of different approaches may be found. Examples of experiential education activities in formal context may include fieldwork activities, such as monitoring of water quality and biodiversity integrated within broader areas of civic education, that is, citizens' science, encouraging students to gather data about their community and local environment to share at citizens' juries and democratic consultation. In informal contexts, we find activities aimed at developing a sense of personal reconnection with nature by means of outdoor experiences enhancing psychological and physical development and a sense of belonging (biophilia and development of an ecological identity).

In sum, subsumed within the umbrella of experiential education are a number of approaches promoting holistic learning experiences, bringing together practical and conceptual learning processes, or at least retracting from establishing power-based hierarchies between different types of knowledge.

However, concerns have been raised by commentators about experiential education being too narrowly understood as "learning by doing." This notion is rooted in a mind-body split that separates theory from practice and considers experience as a fixed entity, separated from knowledge-making processes. In a similar way, different conceptions of learning have increasingly grown apart. Following the classification produced by Brent Davis, on the one hand, there is learning as acquisition of knowledge from outside (the teacher, the books), a notion that has dominated traditional school teaching and that is still currently emphasized by assessment-driven educational systems. On the other hand, there is learning as growth and transformation, involving all spheres of human cognition: cognitive, emotional, linguistic, and sensorial. The latter view is associated with experiential education in a context of sustainability. It also characterizes play-based and outdoor learning experiences such as those offered in early childhood education and environmental education.

Further contributions to understanding experiential education in the field of green and sustainability education have been provided by some applications of complexity theory in education. Learners are defined as open systems who are continuously exchanging flows of matter, energy, and information between the internal and the external environment. The process of learning is thus a process of self-organization and biological evolution taking place within a complex choreography of nested levels, adapting to and affecting one another. The levels are hierarchically organized as bodily subsystems; the person or body biologic; communities; society or the body politic; and the planetary body. Learning is embodied within such systems. This view challenges binary, epistemological assumptions about knowledge in all disciplinary areas: mind/body; observer/observed; singular/plural; me/others, looking instead at experience as a property of interdependency. This resonates with current approaches in the field of sustainability education that are aimed at developing consciousness of humanity as dependent on Earth.

Activities of experiential learning based on reflection encourage the learners to consider the implications of everyday actions for both ecological and social equity (i.e., looking at local and global implications of buying and consuming goods, considering the qualitative and quantitative aspects of energy uses, and so on) or to reflect on commonly held perceptions of the environment (a container of resources, a stage for human performance, a sentient body). Experiential learning activities are thus used not only to acquire new information but to reorganize prior knowledge and learning according to different sets of values and relationships, with the learner being an active participant. As such, experiential education can be an instrument for cultural reform. It challenges ideas about knowledge as a static product by putting forward a view of knowledge as understanding of the network of relationships that sustain and make possible our existence in the world.

See Also: Early Childhood Education; Educating for Environmental Justice; Environmental Education Debate; Outdoor Education.

Further Readings

Boud, David, Ruth Cohen, and David Walker. *Using Experience for Learning.* Buckingham, UK: Open University Press, 1993.

Davis, Brent. *Inventions of Teaching: A Genealogy.* Mahwah, NJ: Erlbaum, 2003.

Dewey, John. *Experience and Education. The 60th Anniversary Edition.* Indianapolis, IN: Kappa Delta Pi Editions, 1998.

Fenwick, Tara. "Reclaiming and Re-Embodying Experiential Learning Through Complexity Science." *Studies in the Education of Adults,* 35/2 (2003).

Gibson, Eleanor. *The Ecological Approach to Visual Perception.* Boston: Houghton Mifflin, 1979.

Stibbe, A. *The Handbook of Sustainability Literacy.* Totnes, UK: Green Books, 2009.

Laura Colucci-Gray
University of Aberdeen

GEDDES, SIR PATRICK

Long before the environment was a social cause or interdisciplinary studies became a buzzword or sustainability an international movement, Sir Patrick Geddes (1854–1932) promoted all three. His passion for nature began in his father's garden and rambles in the woods and fields of his Scottish homeland, and it was formalized as a student and protégé of Thomas Huxley. Geddes's formal education and early career were in natural sciences, but even as a young man, he did not focus narrowly on science. He was interested in the organism and its environment and realized that the two should not be studied separately, a belief that led him toward what we would today call *ecology*, though the science of ecology was not fully developed until the 1930s. As early as the 1880s, he began to study humans and their environment in the same way he had studied other organisms. Geddes saw a continuity from the human organism through social organization to the built environment, the foundation for his "biosocial synthesis," which was a precursor to the study of human ecology and environmental sociology. His biographers suggest that it was his interest in human evolution and the human environment that led him to combine his education in the natural sciences with the emerging fields in social sciences—sociology, anthropology, social psychology, and geography.

Like many thinkers of his generation, Geddes developed a critique of the Industrial Revolution. By the late 19th century, Britain was displaying many of the environmental problems that would plague the modern world such as air and water pollution and toxic waste. Geddes witnessed the degrading conditions industrial workers lived in and sought reforms to improve their social and physical well-being. Geddes, along with his contemporary John Ruskin, wanted to create a new way of thinking that centered on the production of people rather than goods. He felt that the Industrial Revolution had turned the values of society upside down. People had been divorced from meaningful work and from nature, destroying the natural order of human ecology. His work in biology shaped this thinking about human society; he believed that humans, like all animals, depended on a healthy environment, but industrialization had compromised that relationship. The survival of the human species required a new equilibrium between the natural and the man-made environments. When he talked about the importance of environment for human life, he wanted people to understand nature, but his primary concern was with the built environment of cities.

Geddes's interest in humans and their environment led him, with his wife, Anna Morton, to work with Octavia Hill in London in the early settlement house movement in the 1880s. This work reinforced his belief in the important role of the environment, whether natural or built, to human society. The couple later moved back to Edinburgh, where they lived in one of the city's slums and took on the ambitious task of revitalizing it. There, Geddes had the opportunity to put his emerging ideas about urban planning and reform into practice. The Geddeses created an urban laboratory that combined the social and environmental sciences and trained students who would become the planners of the 20th century. The project successfully transformed the neighborhood for the workers who lived there. This was just the beginning of Geddes's planning and urban reform work, which became his lifelong concern. One of his greatest planning efforts took place in India, where he worked first in Madras and then in Bombay and other cities from 1914 to 1924. He wanted to help India avoid the problems of the European industrial city. His work in India and later Palestine has led to his recognition as one of the founding fathers of urban planning.

As a holistic thinker, much of Geddes's work anticipated the sustainability movement. In fact, he might be considered one its foundational thinkers. He anticipated sustainability's triple bottom line—people, planet, and prosperity—with his "integrated dynamics" of Place-Work-Folk. Like those working in sustainability, Geddes talked about maintaining equilibrium between the natural environment and the man-made world. He did not focus on resource depletion because it was not recognized as a problem in his day; rather, the environmental challenges he dealt with were poor cities and their effect on human lives. He also emphasized the importance of the region, believing that the region should be the major unit of analysis in understanding the human-environment relationship.

Though he held a chair in botany at the University of Dundee from 1888 to 1918 and a chair in sociology at the University of Bombay from 1919 to 1924, Geddes's most lasting work was as an urban planner. In addition to his contribution to many fields, two of his intellectual disciples, Lewis Mumford and Benton MacKaye, played significant roles in the development of U.S. urban and regional planning. Mumford's classic *Technics and Civilization* is built around Geddes's concepts of paleotechnic and neotechnic, according to the type of industrial technology they relied on.

Patrick Geddes could not have imagined the environmental problems that exist in the 21st century, but his writings about ecology, environmental sociology, geography, urban planning, regionalism, and human-environment equilibrium are as relevant to modern problems as they were to those of his own time.

See Also: Leadership in Sustainability Education; Social Learning; Sustainability Topics for K–12.

Further Readings

Boardman, Philip. *Patrick Geddes, Maker of the Future.* Chapel Hill: University of North Carolina Press, 1944.

Bud, Robert. "Biotechnology in the Twentieth Century." *Social Studies of Science,* 21/3 (1991).

Meller, Helen. *Patrick Geddes: Social Evolutionist and City Planner.* London: Routledge, 1990.

Novak, Frank G. *Lewis Mumford and Patrick Geddes—The Correspondence*. London: Routledge, 1995.

Odum, Howard. "Patrick Geddes' Heritage to 'The Making of the Future.'" *Social Forces*, 22/3 (1944).

Lindy Biggs
Auburn University

GLOBAL GREEN DAY

Landmarks such as Mount Rushmore and Rome's Colosseum, pictured here in 2008 with its lights turned down, have been part of Earth Hour observances, which have spread to more than 128 countries.

Source: Wikipedia

Climate change is an issue of increasing concern due to the rise of carbon emissions and other pollutants in the air from a variety of man-made sources, including vehicles and energy production. Energy consumption contributes greatly to this problem and, in an effort to stem the tide of energy consumption, many seek to teach the young to consume less energy, lessen the effects of climate change, and help keep Earth greener. "Going green" became increasingly popular after 2000. Suddenly, discussions about environmental problems became common: food was being chemically treated and genetically modified; water was being contaminated with toxic chemicals; resources were running out; wasteful habits were filling landfills; the city of New Orleans, Louisiana, was devastated by a hurricane; and gasoline prices were soaring. More people began recycling, composting, purchasing organic produce, growing their own food, and understanding the connection between saving money, improving health, and caring for the environment. This interest in environmentally appropriate practices spurred an interest in a global green day during which the public could take action to contribute to a greener planet.

Promoting Consumer Actions That Support a Greener Planet

Schools of all levels of education are taking steps to reduce their carbon footprint. Low Carbon Day allows all interested primary and secondary schools to participate by having students, parents, and teachers pledge to be more environmentally friendly. On Low Carbon Day, children are taught sustainable topics such as what constitutes renewable energy (e.g., wind, solar, and hydroelectric power), the impact of climate change on the environment, and the greenhouse effect. On Low Carbon Day, children also positively impact the environment by planting trees, reducing food waste, recycling, riding bikes, turning off unused lights and appliances, and many more activities.

Communities may also be involved with the green movement through Earth Hour and Earth Day. These events encouraging positive action are popular in the United States, but also involve communities from around the globe. In conjunction with the Peace Corps, Earth Day Network worked with local volunteers to implement environmental and civic education programs, tree plantings, village cleanups, and recycling seminars in rural areas such as Ukraine, the Philippines, Georgia, Albania, and Paraguay. Earth Day Network also sponsored city and village cleanups, environmental rallies, and educational programs for underprivileged children in India. In Afghanistan, Earth Day Network worked with more than 40 government and village leaders across the country to promote environmental sustainability practices, such as implementing recycling programs and promoting alternative energy.

Local Initiatives

Approximately 26 percent of the solid waste in the United States could be recycled instead of being buried in landfills. Using the slogan "Reduce, Reuse, Recycle," recycling efforts turn materials that would otherwise become waste into valuable resources. Collecting used bottles, cans, and newspapers and taking them to the curb for pickup or to a collection facility is the first in a series of steps that generate a host of financial, environmental, and social returns. Recycling not only keeps the Earth green but benefits people in individual communities by helping to expand and protect jobs in manufacturing. Recycling reduces the amount of landfill space needed; reduces energy consumption; prevents pollution caused by the manufacturing process; decreases greenhouse gases created during the extraction of raw materials from Earth; and conserves natural resources, such as trees, water, and minerals. Complementing recycling is the process of composting, which involves using yard trimmings and food scraps to create natural fertilizer.

Even though recycling has increased, many continue to purchase water bottled in plastic, which presents a different set of problems. Recycling water bottles is better than turning them into solid waste, but a great deal of energy is used in the transport of bottled water, putting a strain on the environment. In response to this, many are buying reusable water bottles, such as those made by Nalgene. This sort of environmentally friendly action is promoted by local Green Days and helps spread awareness about keeping the Earth green.

Global Green Day

Reducing the energy consumption of light bulbs and electronics is one of the more impactful steps an individual can take to affect climate change. Over 128 countries participate in what is called Earth Hour, where lights and all electronic devices in homes and businesses are to be shut off for an hour. Earth Hour contributes to the awareness of how much impact humans are having on the Earth through their consumption, and a time to focus on what they have contributed to global warming and climate change. Earth Hour is organized by the World Wildlife Foundation (WWF), and has resulted in lasting change. In Chicago, for example, the Building Owners and Management Association (BOMA) developed lighting guidelines to reduce light pollution and to reduce downtown buildings' carbon footprints. As of 2010, Mount Rushmore in South Dakota was powering down its lights each night at 9 p.m. instead of 11 p.m. Not only has Earth Hour involved many countries worldwide, it has also included major companies, such as Coca-Cola, Wells Fargo, IKEA Systems, Nokia, and Wyndham Worldwide, all of which sponsor green activities. This corporate buy-in helps increase awareness of the harm of diminishing the Earth's resources and encourages others to become greener.

For example, Wyndham Worldwide sponsored an event known as Global Green Day on October 12, 2008. This event, held by one of the world's largest hospitality companies, was intended to launch Wyndham Green, an environmental and sustainability program. Celebrating the company's global commitment to corporate social responsibility and the environment, Global Green Day was celebrated in more than 35 offices around the globe. Wyndham committed to a standard linen reuse program, the use of sustainable uniforms and cleaning products, and having new and existing properties pursue Leadership in Energy and Environmental Design (LEED) certification. Additionally, Wyndham enhanced recycling and water-reduction programs, initiated allergy-sensitive CleanAir rooms, and began to implement sustainable Blue Harmony Spa and Fitness Centers.

See Also: Early Childhood Education; Earth Day Network; Education, Local Green Initiatives; Green Week, National; Leadership in Sustainability Education.

Further Readings

Earth Day Network. http://www.earthday.org (Accessed July 2010).

Edwards, A. R. and D. W. Orr. *The Sustainability Revolution: Portrait of a Paradigm Shift.* Gabriola Island, British Columbia, Canada: New Society Publishers, 2005.

Global Green U.S.A. http://www.globalgreen.org/events/113 (Accessed July 2010).

Wyndham Green by Wyndham Worldwide. http://www.wyndhamgreen.com (Accessed July 2010).

Stephen T. Schroth
Jason A. Helfer
Lisa R. Blagg
Knox College

Grand Valley State University

Grand Valley State University (GVSU) is located in the town of Allendale on the outskirts of Grand Rapids, Michigan. Many of GVSU's interdisciplinary education courses, research projects, campus operations, and community-based initiatives are centered on its commitment to environmental, social, and economic sustainability. GVSU has established a sustainability plan, office, and director to implement its triple-bottom-line goals of environmental stewardship, economic prosperity, and social equity. Student demand has driven many of the school's sustainability programs. Grand Valley State University is at the forefront of sustainable practices and continued education.

Campus Sustainability Initiatives

Grand Valley State University's Strategic Plan includes sustainability goals and practices as a key part of its educational, research, and operational mission. GVSU adopted its campus-wide Guiding Principles of Sustainability in 2004, establishing the triple-bottom-line goals of environmental stewardship, economic prosperity, and social equity. Its Campus Sustainability Reports, instituted in 2005, focus on the triple bottom line. GVSU has also signed both the American College and University Presidents Climate Commitment and the Talloires Declaration. Campus sustainability governance groups include the Climate Action Committee, the Annual Campus Sustainability Week Planning Committee, the Campus Sustainability Report Steering Committee, the Energy Conservation Committee, and the Parking, Traffic, and Transit Committee.

Grand Valley State University established its Sustainability Initiative within the College of Interdisciplinary Studies in 2006, which later developed into the Sustainable Community Development Initiative (SCDI). The Sustainability Initiative Advisory Council oversees the implementation of the SCDI and development of the Sustainable Community Development Initiative Plan. GVSU maintains a sustainability initiative office and a full-time director of sustainability. Administrators, faculty, staff, and students work together with Grand Rapids city officials and community leaders to achieve campus and community sustainability goals.

GVSU conducted a greenhouse gas inventory in 2009 as a first step toward achieving climate neutrality. All new construction and major renovation projects are designed and built to U.S. Green Building Council's LEED certification standards. The university has eight LEED-certified buildings, including two with Gold certification and three with Silver certification, and is seeking LEED certification for its entire Allendale Campus. Many campus buildings include green features such as rain gardens, green roofs, improved storm water and irrigation management systems, and water-conserving bathroom fixtures.

GVSU's Mark A. Murray Living Center has received the EPA's Energy Star designation, the first university student housing in Michigan to be so recognized. Residential students receive energy usage data in order to promote energy use awareness and conservation. Sustainable energy projects save GVSU over $1 million annually. GVSU subsidizes free public transportation for students, faculty, and staff, and its vehicle fleet contains hybrid, electric, and fuel-efficient vehicles. GVSU participates in sustainable energy programs such as the Michigan Alternative Renewable Energy Center (MAREC), which explores such options as wind, solar photovoltaic, battery storage, microturbine, and biomass energies.

GVSU is a founding member of the West Michigan Purchasing Consortium, established to support local sustainable businesses. GVSU purchases Energy Star appliances and green cleaning products. It is slowly replacing its low-efficiency campus computers, recycling electronic waste, and has installed dual-sided printers. Campus Dining has increased the purchasing of local, organic, hormone- and antibiotic-free, and Fair Trade foods and purchases from the campus Community Garden. They also use recycled and compostable utensils, napkins, and food containers; recycle cooking oil for conversion to biodiesel fuel; and compost food waste. GVSU's master plan designates specific campus acreage for the preservation of green spaces and local natural habitats.

Student and Community Outreach Initiatives

The GVSU freshman orientation planner and handbook introduces new students to campus and community sustainability initiatives through the inclusion of a student sustainability guide. Campus student clubs and organizations with an environmental or sustainability focus include the Student Senate Sustainability Subcommittee, the Student Sustainability Roundtable, the Student Environmental Coalition, and the Garden Crew. Students also help plan and participate in annual sustainability events and competitions such as RecycleMania, the resident hall energy competition, and the Annual Campus Sustainability Week.

Students also play a key role in many campus and community sustainability projects. GVSU students are involved in over 200 sustainability projects in the Grand Rapids area. Students participating in Project Donation collected items for impoverished inner-city Grand Rapids residents. Students work alongside community members to maintain Grand Valley State Community Garden plots, one of the campus's largest sustainability projects. The organic food grown in the community garden is sold to Campus Dining services or through the university-sponsored campus farmers market. Student Food Pantry vouchers are accepted at the farmers market through a joint effort of the Women and Gender Studies Department, the Women's Center's Student Food Pantry, Health and Wellness, and the Brooks College of Interdisciplinary Studies. A number of undergraduate and graduate sustainability internship opportunities are also available.

Academic and Research Programs

GVSU's curriculum features over 200 courses with sustainability-related themes and over 60 courses specifically devoted to the university's triple bottom line of environmental stewardship, economic prosperity, and social equity. These courses represent approximately 13 percent of available student credit hours. Colleges involved in applied sustainability education include Liberal Arts and Sciences, Interdisciplinary Studies, Business, Engineering and Computing, Public Policy and Administration, Nursing, Health Professions, and Education. General education sustainability themes include Earth and Environment, Ethics, Cities, Global Change, and Health, Wellness, and Healing. Course projects and internships supplement the sustainability curriculum.

Students interested in environmental education may choose the flexible Liberal Studies major to create their own interdisciplinary program of study with a sustainability emphasis. Participating disciplines include anthropology, business, biology, sociology, engineering, health science, and chemistry. Key courses include Introduction to Sustainability and Environmental Studies and Environmental Problem Solving. All Liberal Studies majors must complete the capstone 400-level practicum course Sustainability in Practice, in which they perform a minimum of 125 hours of community service at a partner sustainable community organization of their choice in Allendale or Grand Rapids. Students also maintain journals analyzing their experiences. Students may also pursue the interdisciplinary Environmental Studies minor.

Sustainability certificates available through the Liberal Studies major and Environmental Studies minor include Sustainability, Sustainable Business, Sustainable Product Design, Renewable and Sustainable Energy, Sustainable Urban Environments, and Green Science. GVSU's Green Chemistry curriculum features courses such as Chemistry and Society, Pollution Prevention, Industrial Processes, Environmental Chemistry, and Environmental Chemistry Analysis. GVSU has also introduced several MBA courses in sustainable business management practices.

A variety of campus colleges and departments are also actively involved in applied research in sustainability with an emphasis on the triple-bottom-line goals. GVSU works with the Annis Water Research Institute in Muskegon, Michigan, on research dedicated to water conservation and the protection of the Great Lakes from invasive species and other threats.

GVSU also works with the Michigan Alternative Renewable Energy Center (MAREC) in Muskegon and the Michigan Higher Education Partnership for Sustainability (MiHEPS).

Recognition

Grand Valley State University has received a number of awards and recognitions for its sustainability leadership, including the Sustainable Endowments Institute Sustainability Innovator Award for Comprehensive Sustainability Strategy, the AIA GV Sustainable Design Merit Award, the APPA Award for Effective and Innovative Practices, the Sustainable Building Industry Council Exemplary Sustainable Building Award Honorable Mention (for MAREC), LEED certification for three campus buildings, and the U.S. Environmental Protection Agency Energy Star Award for efficient use of energy. The 2009 Kaplan Guide to Colleges and Universities named GVSU one of the top 25 Environmentally Responsible Colleges and Universities in the United States, and ARAMARK commended GVSU for its sustainable food service practices.

See Also: Leadership in Sustainability Education; Sustainability Officers; Sustainability Websites; Whole-School Approaches to Sustainability.

Further Readings

Barlett, Peggy F. and Geoffrey W. Chase. *Sustainability on Campus: Strategies for Change.* Cambridge, MA: MIT Press, 2004.

Grand Valley State University. "Sustainable Community Development Initiative." http://www .gvsu.edu/sustainability (Accessed June 2010).

M'Gonigle, R. Michael and Justine Starke. *Planet U: Sustaining the World, Reinventing the University.* Gabriola Island, British Columbia, Canada: New Society Publishers, 2006.

Moltz, David. "The New Sustainability U." http://www.insidehighered.com/news/2010/01/25/ gvsu (Accessed June 2010).

Rappaport, Ann and Sarah Hammond Creighton. *Degrees That Matter: Climate Change and the University.* Cambridge, MA: MIT Press, 2007.

Sustainable Endowments Institute. "College Sustainability (Green) Report Card." http:// www.greenreportcard.org/report-card-2010/schools/grand-valley-state-university (Accessed June 2010).

Marcella Bush Trevino
Barry University

GREEN BUSINESS EDUCATION

Attention to sustainability in management education in North America is a fairly recent phenomenon, with business schools beginning to offer environmentally focused courses around 1990. In the past decade, course and program offerings have expanded dramatically but are not yet commonplace in business schools. This article reviews major events and organizations facilitating green business education expansion in North America, its current status as of 2010, and suggests some questions regarding its future.

Development of Green Business Education in North America

Business schools paid little attention to the environment in the 1970s and 1980s. Environmental issues, the subject of a small amount of business faculty research, received some attention in business ethics/business and society classes, but in few others. This compartmentalization may have suggested to students that environmental issues had legal and ethical implications, but had little to do with economic performance, strategic decisions, or day-to-day business operations, other than as increasing costs.

Following several environmental crises in the late 1980s, the 20th anniversary of Earth Day in 1990 and the 1992 Rio de Janeiro Earth Summit increased and changed businesses' attention to the environment. Numerous books emerged aimed at managers but suitable for use in business school courses, in addition to increased attention to green business education by various organizations.

In 1987, the National Wildlife Federation's (NWF) Corporate Conservation Council (CCC) became interested in environment in management education. Designed to build bridges between the NWF and industry, the CCC consisted of a few NWF staff members and representatives from the environmental affairs units of about a half dozen corporations. Two issues were of concern: the understanding of business issues by environmental affairs staff and the understanding of environmental issues by managers and business school graduates. A survey of business faculty indicated environmental coverage occurred only in courses in business and society/business ethics. Additionally, faculty members believed few adequate teaching materials existed. The CCC funded a Curriculum Development Project (1988–1991) to increase curricular treatment of environmental topics in business schools. Several professors and graduate students in the business and society area identified existing cases and other curriculum materials, developed new cases, designed pilot sustainability courses, and offered these at the undergraduate and graduate levels in 1989, 1990, and 1991. Publications included project reports, a casebook, and a textbook.

In 1990, the Management Institute for Environment and Business (MEB) was founded. Over the next few years, MEB sponsored an international conference on greening business education, developed discipline-specific bibliographies of teaching materials, worked with faculty to generate case studies, and established a pilot project with five business schools to increase attention to the environment within the curriculum. MEB's Business Environment Learning and Leadership (BELL) program pursued projects, including an electronic newsletter, annual conferences, publishing cases, providing access to curriculum resources, providing technical assistance to participating schools, enabling student teams to consult with environmental entrepreneurs in developing countries, and conducting surveys to determine the state of business school environmental education. MEB merged with World Resources Institute (WRI) in 1996 as WRI's Sustainable Enterprise Program, and in the mid-2000s, BELL moved its focus to China, Brazil, and Mexico.

The Greening of Industry Network (GIN) was created in Europe and North America as a forum connecting researchers and practitioners from business, government, and nongovernmental organization (NGO) communities. GIN's conferences alternated between Europe and North America, and Asia joined the rotation in 2001. Conferences include presentations on business school environmental education and have resulted in books and annual issues of *Business Strategy and the Environment* devoted to conference papers.

The Organizations and the Natural Environment (ONE) Interest Group of the Academy of Management (AOM) was established in 1994, providing a forum for presentation of scholarly work on environmental management, symposia, and workshop sessions on environmental issues and education. BELL, GIN, and ONE/AOM meetings have provided substantial opportunities for interaction and exchange of ideas about materials and strategies for incorporating environmental topics into business school education. Of these organizations, BELL was the most education focused, GIN the most practice focused, and ONE the most research oriented.

In the late 1990s, the Aspen Institute assumed two of BELL's roles: providing assistance with course materials and ranking sustainability-oriented business school programs Caseplace.org and Beyond Grey Pinstripes. Caseplace.org is a free online library of syllabi, cases, articles, and other teaching resources related to business sustainability, particularly helpful to faculty interested in greening their courses.

Beyond Grey Pinstripes (BGP) assesses business school graduate curricula and faculty research on social/environmental business issues. It invites accredited business school programs to submit syllabi and research citations on sustainability topics to qualify for recognition in its "Global 100" listing. BGP seeks to advance innovation, inform prospective MBA students of green business education opportunities, inform business recruiters of sources of sustainability-literate business graduates, and disseminate best practices in business school scholarship and activity. An element of competition now exists, with schools commonly citing their BGP rankings.

The Association to Advance Collegiate Schools of Business (AACSB), the international accrediting body for business schools, has recently begun to attend to green business education. In 2008, AACSB transformed its annual business ethics conference into a sustainability conference both highlighting green education initiatives at various business schools as well as business sustainability initiatives by corporations and NGOs and increasing interaction across disciplines. The first three conferences were each attended by over 100 business school faculty and administrators.

Current Status of Green Business Education

In recent years, sustainability centers and institutes focusing on development and diffusion of sustainable practices and technologies have emerged. The Aspen Institute recently identified 240 different academic applied sustainability centers around the world, an increasing number of which focus on sustainability issues directly relating to business.

Teaching resources have expanded dramatically over the past two decades. Lists of resources, particularly syllabi and cases, are available from several organizations, including ONE, BELL, and the Aspen Institute, as well as casebooks and books emerging from conferences. In the early 1990s, only two environmental management texts were readily available in North America. This area has expanded significantly, but is still not a major one for U.S. publishers. Many European texts are available as well as trade books oriented toward various disciplines such as accounting, environmental marketing, purchasing, operations management, supply chain management, product design, and managing organizational environmental change, particularly from British publisher Greenleaf Press. While sustainable management readings options have increased dramatically, there is still a need for instructor- and student-friendly materials that exist for core subjects such as organizational behavior, principles of management, and strategy—particularly for materials dealing with human resources management, operations management, project management, and supply chain management. Attention to sustainability in other business disciplines such as marketing, finance, accounting, and information systems is growing but lags behind the management field.

Extended coverage of these issues by many traditional business publications, including the *Wall Street Journal, Economist, Financial Times,* and *Fortune,* has increased dramatically.

These and mainstream publications such as the *Washington Post* and *New York Times* frequently provide interactive Web-based content offering resources for business educators. Specialty organizations such as Environmental News Network, *Grist,* and Edie offer daily environmental news digests and links. Specialized business sustainability newsletters are offered by www.greenbiz.com and www.greentechmedia.com.

Green Business Education Programs

Courses on sustainability are now offered at hundreds of business schools, with a smaller subset offering sustainability-focused programs. This is at least partially due to the BGP reports issued since the late 1990s. Beginning with eight U.S. business schools in 1998, the most recent reports have highlighted 100 schools around the world that incorporate sustainability in education and research.

As of mid-2010, the Association for the Advancement of Sustainability in Higher Education (AASHE) listed 60 institutions in the United States and Canada offering either sustainability-focused bachelor's or master's degree programs (four bachelor's and 18 master's); concentrations, majors, emphases, specializations, or minors (8 and 19); or certificates in sustainable business (8 and 15). Additionally, several joint and dual degree programs exist.

Certificate programs typically consist of a relatively small number of courses and can be taken as part of or separately from a business degree program, making them particularly appealing both to working professionals and nonbusiness students.

Concentrations, majors, emphases, and specializations require more courses and must be taken within a broader business degree program. At the master's level, these generally require three or more courses, and at the undergraduate level, typically six or more, with minors requiring fewer.

Dual and joint degree programs were one of the earliest green business education structures, but their greater length has limited their growth. They usually partner a business school with a natural resources, engineering, or law school on its campus. A few interuniversity dual-degree programs exist.

Programs specifically focusing on sustainable business have shown some of the greatest growth. Originally pioneered by specialized institutions such as the Bainbridge Graduate Institute, the Presidio School of Management, and the Institute for Environmental Leadership, these programs are now emerging at traditional colleges and universities such as AACSB-accredited institutions Duquesne University and the University of Maine. Many are found at smaller schools, particularly those utilizing hybrid educational delivery involving a combination of online and intensive on-campus courses, such as Antioch of New England, Marlboro State, and Green Mountain State. The expansion of such programs represents one of the most exciting developments in green business education.

Questions Facing Green Business Education

While green business education has grown rapidly in the past few years, the need to dramatically increase its scope and effectiveness requires exploring questions such as the following:

- What are appropriate goals for sustainability courses and programs?
- What kinds of knowledge, attitudes, and skills are required to create environmentally sustainable organizations?
- What kinds of backgrounds facilitate developing these attributes?

- How well do various delivery mechanisms accomplish program goals?
- What are the relative merits of certificate, concentration, and fully integrated approaches?
- How long must a program be to provide adequate sustainability management education?
- Are interdisciplinary programs more effective than those in which courses are taught by business faculty with expertise in sustainability?
- How many environmentally focused courses are necessary?
- Which business sustainability topics are most important?
- Which teaching materials and methods are most effective?
- How do these various dimensions interact? Are there more and less effective program configurations that emerge from these interactions?

These are challenging questions and are the sort usually explored by education rather than business faculty. However, developing and deploying effective sustainability education efforts requires attention to these and similar questions if green business school education is to become integrated, commonplace, and effective.

See Also: Green Mountain College; National Wildlife Federation; Sustainability Officers.

Further Readings

Aspen Institute. "A Closer Look at Applied Sustainability Centers" (2008). http://www .aspencbe.org/documents/Applied%20Sustainability%20Centers%20Final.pdf (Accessed July 2010).

Association for the Advancement of Sustainability in Higher Education. "Sustainability in Business Education." http://www.aashe.org/resources/grad_business.php#masters (Accessed July 2010).

Rands, G. P. and M. Starik. "The Short and Glorious History of Sustainability Education in Management." In *Management Education for Global Sustainability*. J. A. F. Stoner and C. Wankel, eds. Charlotte, NC: Information Age Publishing, 2009.

Gordon Rands
Pamela Rands
Western Illinois University

Green Community-Based Learning

Green community-based learning is not yet a formally named movement. It is an outgrowth of community-based learning, also known as service learning, a widespread movement to improve student learning by making it clearly relevant to students' lives and communities. Most prevalent in institutions of higher education in the United States, but practiced worldwide in diverse educational contexts, service learning proposes to make faculty, students, and educational institutions agents of positive social change within their communities. Activities in which students engage range from direct action to community-based research that can form a basis for social change. In community-based learning, students engage in classroom learning relevant to one or more problems faced by the community, collaborate with community members and organizations outside the classroom to address aspects of these problems, and engage in structured reflection on their actions that enriches their classroom learning.

Community-based learning is a response to the abstract and fragmented nature of institutionalized education in modern industrial societies. Its problem-based foundations make it interdisciplinary in nature in that, in order to solve community problems, students must draw upon knowledge and methods associated with multiple disciplines, such as sociology, biology, mathematics, and history, to name a few. In interdisciplinary fashion, students must integrate knowledge and investigative methods from relevant disciplines in order to comprehend and address social problems so that their work recognizes and appropriately addresses real-world socioecological complexity. Community-based learning draws upon the educational philosophy of John Dewey, who promoted education of the whole person—the intellect and the affective domain—in service to social justice and grassroots forms of democratic engagement and community building.

Green community-based learning, as a form of service learning aimed at enhancing community sustainability, focuses attention on the relationships between people and the environments in which they live and construct community. It uses the social processes of community-based learning to address environmental problems, thereby explicitly recognizing environmental health as a foundation for community health.

The Process of Green Community-Based Learning

Community-based learning includes classroom study of issues or projects relevant to community improvement and health, student engagement with community members in work that is related to the issues or projects studied in the classroom, and structured reflection activities that call upon students to integrate their classroom learning with the insights they gain through community work. Reflection can be oral or written and can be done individually or in groups. In green community-based learning in particular, roles, uses, and/or conceptions of the environment are addressed explicitly as important aspects of community problems and problem solving. Community-based learning also often includes instructor collaboration with community members in developing learning experiences and projects in which students will engage. This process places the educational institution in a reciprocal relationship with the community so that the two become partners in the educational process. Partnership is widely recognized as a goal toward which educators, students, and community members should strive when engaging in community-based learning. In a true partnership, all parties guide the process, and all ideally benefit from the relationship. The partnership approach to community-based learning embodies community building within a grassroots democracy of equals in which education is part of the form and process of community itself. For faculty members, this approach entails moving away from lectures and other forms of information transfer as the central means for teaching and moving into roles as facilitators and coaches. For community members, community-based learning can mean assuming instructional roles formerly reserved only for faculty members.

Regarding the inclusion of extracurricular community service activities under the definition of community-based learning, faculty members and student services professionals are at times divided. Some faculty members are hesitant about extending the definition of community-based learning to include activities organized and facilitated entirely outside the classroom context. For traditional classroom learning, the emphasis has long been on cognitive development rather than on personal development in a broader sense that would include aspects of learning such as identity exploration and formation and development of social and civic responsibility. Still, growing numbers of faculty members and student services professionals alike are organizing student service projects with the goals of creating rich and valuable learning experiences and connecting students and the institution with the broader community to the benefit of all.

Principles Underlying Green Community-Based Learning

Green community-based learning is an expression of the inextricable intertwining of humans and the environment. It expands the notion of community to include more than people, the economy, social institutions, and other human-constructed organizations and structures. In its conception of the health of local and/or global environments as important to the health of human communities, it recognizes that all community is indeed ecological—that humans are reliant upon the environment as the source of all life and all material used in the building of societies.

The extension of the definition of community to include its environment and other species is a logical extension of the basic principles underlying community-based learning. This extension is highly relevant in current times in which environmental damage is posing ever-increasing risks to communities worldwide in the form of climate change, resource depletion, loss of biodiversity necessary for ecological health and resilience, contamination of food and water sources, soil erosion and salinization, depletion of ocean fisheries, and more. Increasingly, these phenomena not only represent risks, but directly impact people and communities. These phenomena, therefore, embody social justice concerns and pose challenges for grassroots democracy as a problem-solving strategy.

Because the interweaving of social and environmental concerns serves as a foundation for green community-based learning, it can be considered an active form of teaching and learning cultural ecology and/or political ecology. Cultural ecology is the study of relationships between particular cultural forms and practices and the environmental context in which they developed. Political ecology is the study of relationships between political, economic, and social phenomena on the one hand and environmental issues and problems on the other. Political ecology recognizes political aspects of environmental problems, including power relationships among involved and affected social actors and groups. Due to its socioecological foundations, green community-based learning is also related to place-based learning and indigenous forms of community education practiced for millennia in diverse societies. Green community-based learning can, however, engage students in service that addresses issues such as climate change that extend beyond the confines of the local community.

Study of localized food systems combined with local food projects or activism provides an excellent example of green community-based learning. Localizing food systems has many socioecological benefits that can be studied in classes. These benefits include reduction in fossil fuel use for food transportation and resulting positive impacts on the environment and the health of people; increased capacity of the community to take care of citizens' basic needs in times of economic instability and climate change; creation of new employment opportunities in urban and/or rural agriculture and horticulture; increased localized spending that stimulates the local economy; increased opportunities for students and people of all ages to learn about natural systems and how these systems relate to the production of food; increased self-reliance and economic resiliency among individual and community gardeners; food production among low-income people and/or development of affordable outlets for low-income residents to purchase healthy local foods; and increased nutritional value in foods available to community members.

Green community-based learning that studies and engages students in some aspect of local food production, distribution, and/or marketing represents learning that integrates knowledge and processes of knowledge making from diverse disciplines. It can also represent civic engagement among students, citizens, educational institutions, and citizen organizations. This form of teaching and learning frames social and ecological problems as community problems that can be addressed through citizen involvement. It, therefore, represents a basic form of democracy in action.

The educational philosophy of John Dewey serves as a foundation for community-based learning—and by extension, for green community-based learning—that ranges in its goals from social reform to the remaking of society itself. Dewey focused on what he saw as troubling dualisms that infuse modern education, particularly the distinctions drawn between knowing and doing and between emotions and intellect. He saw these distinctions, and the class and other social divisions they promoted, as deriving from the exercise of social power in society, and he advocated elimination of these divisions in the interest of authentic democracy. These social justice foundations of his educational philosophy continue to inform service learning today.

Dewey advocated active learning through which processes of knowing and doing could be holistically integrated in ways that would mirror engagement in community life and work. According to Dewey, the ultimate aim of education should be the engagement of students as citizens and the building of forms of democracy in which everyday citizens and communities learn about and effectively address issues of social justice and community health. Dewey proposed that educational institutions should function as facilitators of democratic engagement and problem solving and that students should participate intellectually, emotionally, and morally in active learning processes by assuming roles as citizens embedded in communities. The school, according to Dewey, should both represent a microcosm of democratic society and provide a platform for student engagement in society itself. Teachers should structure learning opportunities that would allow students to create their own meaning through community engagement. Anchored as it is in the educational philosophy of Dewey, community-based learning is not only interdisciplinary but transdisciplinary, in that it calls upon educators and students to contextualize and synthesize disciplinary knowledge and investigative methods and also to engage directly with community members and organizations in the process of solving community problems.

In addition to Dewey, the educational philosophies of critical pedagogues Paulo Freire and Peter McLaren, among others, may serve as foundations for educators who conceptualize their and their students' work as ultimately contributing to a complete reordering of society. Critical educators seek societal reordering resulting in social justice. By extension, with regard to an explicitly green approach, it is socioecological justice that is called for—a ceasing of the combined exploitation and destruction of both people and nature that is widespread in modern industrial societies. Green community-based learning's affinity to critical pedagogy derives from its tendency to highlight social contradictions, such as the gap between social and economic status quo that is supported by powerful economic and political interests on the one hand and the widespread changes necessary for realizing sustainability on the other.

Practitioners of community-based learning claim that it enhances institutionalized education in important ways. Various research studies confirm community-based learning's ability to increase academic engagement and achievement, develop civic attitudes and civic engagement, and develop social and personal skills. Those who promote service learning also claim that it can contribute to cognitive development. A series of books published by the American Association of Higher Education, titled *Service-Learning in the Disciplines* and edited by Edward Zlotkowski, supports this view.

Historical Development of Green Community-Based Learning

Community service has a long history on college campuses. Fraternities, sororities, and faith-based organizations, for example, have lengthy histories of community service. The social changes and awareness that developed during the civil rights movement of the 1960s and that also grew out of the war on poverty in the same era influenced the development of community-based learning as a means of working toward social justice and inclusive democracy. The formation of the Peace Corps in 1961 and Volunteers in Service to America in 1965 also added momentum to activist-oriented education by engaging the energy and knowledge of young people to make a difference in the world.

The expansion of the community-based learning movement since the 1980s can be viewed as a response to the idea that colleges and universities have actively participated in reproducing patterns of work (and their associated class divisions) that characterize modern consumer societies. In the 1980s, educational researcher and philosopher Ernest Boyer lamented what he saw as the disengagement of colleges and universities from community life. In his 1987 report "College: The Undergraduate Experience in America," Boyer decried the lack of attention devoted to undergraduates by college and university faculty members whose work had become increasingly focused on research, disconnected from the local community, and increasingly unrelated to deliberate action in the interest of the public good. Among other things, Boyer promoted community service programs as a means to reconnect faculty members with their students and with the surrounding community, thereby increasing the social relevance of higher education within communities and among the increasingly diverse students entering colleges and universities.

In the mid-1980s, community-based learning spread rapidly among institutions of higher education, and several national organizations were founded to support the movement. These organizations include the Campus Outreach Opportunity League, the National Association of Service and Conservation Corps, the National Youth Leadership Council, and Youth Service America. In 1985, Campus Compact was formed by the presidents of Brown, Georgetown, and Stanford Universities along with the Education Commission of the United States. Campus Compact is a national nonprofit organization representing a coalition of over 1,100 college and university presidents at institutions that comprise more than one-third of the higher education institutions in the country. Campus Compact promotes community service, civic engagement, and service learning among students and faculty at member institutions. The organization publishes material on service learning theory and on pedagogical practices that include service components. It maintains a roster of participating consultants available to visit member institutions to support development of community-based learning programs and publishes annual statistics on service learning in an effort to determine the state of community-based learning and identify emerging trends.

In 1993, the Corporation for National and Community Service was created as an independent agency of the federal government. The Corporation is the nation's largest grant-making organization supporting community-based learning and other forms of service. The Corporation's Learn and Serve America program specifically supports planning and implementation of service learning programs in schools, colleges and universities, community organizations, and Native American tribes. Campus Compact and the Corporation for National and Community Service have played important roles in stimulating interest and involvement in community-based learning nationwide. Work supported by these organizations has included green community-based learning.

Beginning in the 1990s, the community-based learning movement shifted its focus toward civic engagement of entire college and university campuses. This movement represents a step beyond involving individual classes in civically oriented action and toward transforming colleges and universities into civically engaged institutions that work in partnership with communities to improve community life and health. As the concept of sustainability has gained currency on college and university campuses, these forms of whole-institution engagement increasingly include a green focus. Green community-based learning is also prevalent among academic major programs in environmental studies or related fields offered by colleges and universities. Students in these programs are often required to participate in green community-based learning in introductory and/or capstone courses. These students may also be called upon to participate in internship programs or green research or consulting for community organizations.

Given that social justice and environmental health are closely related, it is difficult to separate the historical development of explicitly green community-based learning from the broader history of the service learning movement. Damage to environments often results in harm to humans and vice versa. Community-based learning and green community-based learning, therefore, form a continuum with regard to the degree of explicitness with which the environment is recognized as relevant to learning and social action.

The degree of explicitness with which educators and students have addressed environmentally relevant aspects of community problems has grown over time as recognition of widespread ecological damage has increased. Growing public awareness of the seriousness of issues such as climate change, groundwater depletion, and soil degradation prompt growing numbers of students and faculty members to search for and develop curricula relevant to addressing socioecological problems. A 2008 report by Campus Compact in which environment and sustainability were ranked sixth among the top 20 issues addressed through service learning emphasizes the growing prominence of green community-based learning.

Challenges to Green Community-Based Learning

Challenges to green community-based learning include challenges faced by service learning as a whole. In colleges and universities, faculty participation in research and publication is typically more highly valued than teaching, and service is most often ranked at the bottom of the typical tripartite structure of faculty work that includes scholarship, teaching, and service to the institution, the professional, and/or the local community. Community-based learning can be difficult to categorize within this tripartite structure because it typically overlaps the three categories, making its valuation complicated as well.

Research and pedagogical innovation by faculty members who engage their students in service learning also tends to fall into subject areas outside the knowledge-making domain of traditional disciplines. For example, a political scientist who calls upon her students to engage with the League of Women Voters to educate the public about local ballot issues and to encourage voting has moved into new pedagogical terrain, but this work would be unlikely to be seen by her colleagues as a contribution to the field of political science. Such a faculty member is likely to find her efforts undervalued by her colleagues and, at worst, even detrimental to her career. As colleges and universities increasingly embrace new roles as civically engaged institutions, the definition of scholarship is broadening, and these kinds of efforts are being rewarded more frequently, but long-held ideas about what constitutes worthy faculty work die hard, and innovative educators can find themselves discouraged or prevented from pursuing community-based learning pedagogies.

The social justice orientation of community-based learning is also seen as suspect by some faculty members and administrators, particularly those rooted in disciplines or subdisciplines that maintain a positivist orientation. The positivist tradition focuses on the scientific method as the only source of valid knowledge, and objectivity is considered an operational ideal. Working "for" social justice, for positivist educators, appears to be a subjective and questionable goal that threatens to politicize education.

Other challenges faced by faculty members who engage their students in communitybased learning are logistical in nature. Increasingly, institutions of higher education in the United States are facing budgetary constraints that impinge on the abilities of faculty members, academic departments, and institutional administrations to support communitybased learning. These pressures are likely to continue given the long-lasting impacts of the economic crash of 2008 on the national and global economies. Grant funds to support community-based learning are also likely to contract, and faculty members and other institutional professionals will likely be asked to do more with less in all areas of their work. Unless community-based learning becomes widely valued and rewarded, faculty members may face pressure to return to simpler, more time-efficient teaching strategies such as lectures that require less coordination among participants and reduced support in terms of logistics.

The Future of Community-Based Learning

Given the many converging socioecological crises of current and recent decades, the relevance of green community-based learning is increasing. The rise of interdisciplinary programs in environmental studies and related fields and strong student interest in these programs indicate a growing potential among institutions of higher education to foster student, faculty, and community participation in green community-based learning. The many environmentally related challenges faced by communities will increase over time due to the compounding of ecological disturbances and destruction. The socioecological challenges faced by human communities are many, and informed participation by large numbers of diverse people living in communities throughout the world will be required to end the self-reinforcing cycle of consumption and destruction in which the world's communities have become embedded and complicit.

Green community-based learning has been discussed here primarily as a higher education phenomenon in the United States, but it need not be limited to the institutional contexts of colleges and universities, nor even to institutionalized education. Neither is it limited to one national context. In the United States, even though budgetary support for institutionalized education for school-age children and for college and university students is rapidly falling, the potential for green community-based learning remains high. Whether and where it will be avidly practiced will depend on many social, economic, cultural, and environmental factors and their specific interplay in diverse social contexts.

See Also: Agenda 21: "Promoting Education, Public Awareness, and Training" (Chapter 36); Education, Local Green Initiatives; Place-Based Education.

Further Readings

Campus Compact. *Introduction to Service-Learning Toolkit: Readings and Resources for Faculty,* 2nd ed. Providence, RI: Campus Compact, 2003.

M'Gonigle, Michael and Justine Starke. *Planet U: Sustaining the World, Reinventing the University.* Gabriola Island, British Columbia, Canada: New Society, 2006.

Russo, Rosemarie. *Jumping From the Ivory Tower: Weaving Environmental Leadership and Sustainable Communities.* Lanham, MD: University Press of America, 2010.

Sterling, Stephen. *Sustainable Education: Re-Visioning Learning and Change. Schumacher Briefings,* No. 6. Totnes, UK: Green Books, 2001.

Ward, Harold, ed. *Acting Locally: Concepts and Models for Service-Learning in Environmental Studies.* Series on Service Learning in the Disciplines, Edward Zlotkowski, ed. Washington, DC: American Association for Higher Education, 1999.

Tina Evans
Fort Lewis College

GREENING OF THE CAMPUS CONFERENCE

Beginning in 1996, Ball State University (BSU) in Muncie, Indiana, has sponsored a series of Greening of the Campus (GOC) conferences. The impact of these conferences on the greening of higher education has been profound. They have facilitated personal networking among those interested in campus sustainability, widely disseminated various campuses' sustainability experiences, fostered research—and thereby the growth of journals—on campus sustainability, facilitated student involvement and presentations, and fostered growth of the emerging role of the sustainability coordinator. The successes of GOC contributed to the eventual formation of the Association for the Advancement of Sustainability in Higher Education (AASHE). Without the biennial GOC conferences, it seems unlikely that the campus sustainability movement would have developed nearly as rapidly as it did between 1996 and 2005.

In 1990, BSU's provost appointed the Green Committee, seeking recommendations to increase BSU students' environmental literacy. One of the committee's recommendations was to convene a national meeting of universities to address environmental issues in order to learn from them. A call for papers was issued, and conference proceedings were planned. This conference was very broad—addressing curriculum, operations, and other topics—and was intended to attract students, administrators, and faculty. At a final-day wrap-up session, attendees focused on the next steps and asked if a conference was planned for the following year. Although this was not their original intention, the organizers recognized a need and held a second conference in 1997. Conferences have since been held every other year.

This responsiveness has continued. The National Wildlife Federation's Campus Ecology program asked GOC organizers for the chance to offer a workshop at an early conference. Subsequently, organizers have sought and responded to collaborations with other organizations and groups, including sustainability coordinators. Early sustainability coordinators' main means of interaction was the Green Schools Listserv. GOC conferences hence became a key means of personal interaction, with sessions set aside for meeting and networking. As GOC sessions highlighted the contributions of coordinators, such a position was created at more institutions.

While the members of the GOC steering committees varied from year to year, the two faculty members consistently involved have been Robert Koester (architecture) and Stan Kiel (economics). The steering committees fostered highly interdisciplinary conferences typically featuring between four and seven keynote speakers from a wide variety of backgrounds (including business, nongovernmental organizations, and academia), representing a wide range of disciplines and perspectives, including international ones. Academic disciplines represented have included history, political science, geography, economics, English, and behavioral sciences. David Orr, director of Oberlin University's environmental studies program and a pioneer in campus sustainability, keynoted at eight successive conferences. Presentation sessions have featured students, staff, and faculty, frequently copresenting their school's initiatives. Conferences have typically included tours of notable sustainability initiatives both on and off campus.

Desiring a closer regional meeting, West Coast campus sustainability advocates created Education for Sustainability West (EFS West), with BSU's Robert Koester as an advisor for its conference held in Portland, Oregon (2004). In 2005, EFS West organizers and others, deciding that the time had come for a national organization, founded AASHE, which had over 650 attendees at its first conference in 2006 in Tempe, Arizona. Its second (2008) conference in Raleigh, North Carolina, attracted over 1,700 participants. AASHE's third conference was held in October 2010 in Denver, Colorado.

GOC organizers anticipated that AASHE meetings would be driven by its various program initiatives, while GOC conferences—smaller, mainly attracting a Midwest audience—would be a bit more philosophical in tone. Future GOC conferences will likely reflect BSU's emphasis on "immersive learning"—putting students in real-world contexts focusing on real-world problems to create opportunities for genuine innovation—and the impact of the "built pedagogy," reemphasizing David Orr's concept of student involvement in campus sustainability issues as a key to developing ecological literacy and change skills.

The reputation and legacy of GOC, its national and international standing, its continuation, and its important impacts—particularly, networking—are satisfying to its organizers. Additionally, many BSU staff and administrators not involved in GOC have been congratulated and applauded by their peers at other institutions for GOC's impacts, giving them a greater interest in and appreciation for greening at BSU. Recently, BSU has undertaken significant energy and green building initiatives, placing it among the leaders in sustainable campus operations.

See Also: Association for the Advancement of Sustainability in Higher Education; Leadership in Sustainability Education; Sustainability Officers.

Further Readings

American Association for the Advancement of Sustainability in Higher Education. http://www.aashe.org (Accessed August 2010).

Ball State University. "Greening of the Campus." http://cms.bsu.edu/Academics/Centersand Institutes/GOC.aspx (Accessed August 2010).

Gordon Rands
Pamela Rands
Western Illinois University

GREEN MATH

Real-world information or environmental data in numerical form assist in investigating many areas of our lives. These real-world data help us to study mathematical concepts while increasing environmental awareness in people. Green math is about the mathematical know-how that relates to the environmental awareness and preservation of resources. According to B. Schiller (2003), 21st-century textbooks use real-world data and their applications to solve real-world problems for different topics, including mathematics, but often cannot incorporate environmental issues. Schiller also suggested that even textbooks feature examples of environmental mathematics, which do not do much to encourage interest in either environmental or mathematical analysis. However, there is a greater scope of putting environmental-math or green-math topics into math textbooks that will enhance interest in understanding the environment and encourage solving environmental issues.

Math topics that do not deal with solving the environmental problems that we face are known as "simple math," but when mathematicians use math to promote environmental sustainability, that math is known as "green math." Green math can relate specifically to recycling, smog control, watershed study, fertilizer and pesticide application rate calculation, irrigation scheduling, soil nutrient analysis, and much more. The contextual numerical assessment of energy issues is a prime example of green math. The book *Sustainable Energy—Without the Hot Air*, written by David J. C. MacKay, is perhaps the best example of an environmental or green math application. In his book, MacKay numerically estimated both energy generation and energy usage in the United Kingdom. He used simple mathematics to explain the balance sheet of the energy generation and usage by scaling things down to the per-individual case. The main objective of the usage of green math in the entire book is to explain in very simple terms to the general public the necessity of sustainable or renewable energy. Many people remain skeptical about environmental degradation as they do not fully understand the extent of the damage to the environment done by the excessive use of nonrenewable energy sources. In a review to the book, Tony Juniper, former executive director of Friends of the Earth, wrote, "David MacKay sets out to find the answer through a forensic numerical analysis of what we use and what we can produce."

Though not particularly advanced, this type of math is only taught at a college level and as part of specialized course material. However, this math could be incorporated into middle and high school curricula as a means of providing children with a better understanding of the practical applications of such math, and raise awareness on how they could, at an early age, help preserve the environment.

An example of a green-math application can be seen in calculating the replenishment of groundwater. In most of the United States and other parts of the world, we are experiencing depleted groundwater levels. This is because of the unsustainable use of groundwater for irrigation, drinking, and other residential/commercial uses before the water table can be restored. It has very serious consequences, including water shortages, reduction in crop production, development of sinkholes, and so on. Green-math applications can empower the public. If green math is developed in this topic, as MacKay explained, for energy sustainability, then the general public would easily understand the grim scenario of groundwater depletion and its environmental consequences. Green math can be prudently used to explain the scenarios of groundwater contamination, stream and other water body impairment, and air pollution due to excessive fertilizer, pesticide, and insecticide application to crops, and toxic gases released into the atmosphere by burning nonrenewable energy sources. Simple numerical deliberation could help people understand the reality, and begin working toward environmental preservation. It is appropriately told that numbers and pictures say thousands of words in a simple, understandable way.

Green math is not simply the mathematical deliberation of environmental issues. It is about developing simple models by using statistics to show environmental inputs in discrete or numerical forms and their relationships toward the consequential environmental outputs. From these data, graphs can be created to provide helpful information in the form of a visual aid.

An example of a green-math application is shown in the equation below:

$$([\text{Waste Disposed or Generated—Recycled}] + \text{Imports} + \text{Exports}) \div \text{Population} = \text{Performance}$$

Green math is perhaps the best way to explain complex environmental issues to the general public or to skeptical environmentalists. Green math should to be taught at the appropriate level during educational development to make an impact on future decision making, and should be taught to a more general population and made available to those who knowingly or unknowingly partner in the environmental degradation. This would encourage informed decision making, leading to greater environmental preservation. Green math should be part of the K–12 curriculum as an additional math topic. It will enhance the math skills of young students and simultaneously help make our Earth greener.

The department of mathematics from Harvey Mudd College, Claremont, California, has hosted the first-ever green-math conference, the 2010 HMC Mathematics Conference on the Mathematics of Environmental Sustainability and Green Technology, and paved the way for encouraging green math.

See Also: Early Childhood Education; Sustainability Topics Correlated to State Standards for K–12; Sustainability Topics for K–12.

Further Readings

Ashman, M. R. and G. Puri. *Essentials of Soil Science.* Malden, MA: Blackwell, 2002.

Fetta, I. B. "Using Real-World Data to Understand Environmental Challenges." In *Environmental Mathematics,* B. A. Fusaro and P. C. Kenschaff, eds. New York: Mathematics Association of America, 2003.

Schiller, B. "Environmental News Teaches Mathematics." In *Environmental Mathematics,* B. A. Fusaro and P. C. Kenschaff, eds. New York: Mathematics Association of America, 2003.

Zero Waste. "Waste and Recycling: Data, Maps, and Graphs." http://www.zerowasteamerica.org/Statistics.htm (Accessed February 2010).

Sudhanshu Sekhar Panda
Gainesville State College

GREEN MOUNTAIN COLLEGE

Green Mountain College developed its Environmental Liberal Arts curriculum in the 1990s; today, the most popular major among its 800 students is Environmental Studies.

Source: Flickr/Doug Kerr

A liberal arts college in Poultney, Vermont, Green Mountain College (GMC) is known for its curricular focus on environmental literacy. Originally founded by the United Methodist Church as the Troy Conference Academy in 1834, the college went through several changes before adopting its current name in 1974. The Environmental Liberal Arts curriculum, an extension of the traditional liberal arts curriculum into environmental awareness and sciences, was formulated in the 1990s. The school remains small, with an enrollment of about 800 students. In addition to the standard majors, GMC offers degrees in Environmental Studies (the most popular major), Environmental Management, Youth Development & Camp Management, Sustainable Agriculture & Food Production, Natural Resources Management, and Adventure Education. GMC's small graduate school offers a Master of Business Administration in Sustainability and a Master of Science in Environmental Studies. As at other liberal arts institutions, all students must complete a core curriculum in addition to the requirements of their major (one of 27) and optional minor (one of 22). At GMC, that core curriculum is the Environmental Liberal Arts General Education Program (ELA), consisting of 37 credits. The four core courses of the ELA cover composition, literature, the history of scientific thought, and citizenship. Further ELA courses are selected from seven areas: Quantitative Analysis, Natural Systems, Human Systems, Aesthetic Appreciation, Moral Reasoning, Historical Context, and The Examined Life. In essence, the ELA covers the same ground as most liberal arts general education curricula, with an environmental slant. Like many progressive institutions (Hampshire College in Massachusetts, Evergreen in Washington), GMC has a greater-than-average expectation of scientific literacy from its nonscience majors and offers strong support for its self-designed major options.

The Sustainable Agriculture major was introduced in 2010, funded with a $110,000 grant from the Jensen/Hinman Family Fund. The program revolves around the maintenance, by students, of a 22-acre sustainable farm, Cerridwen Farm, and is designed to produce research in areas of sustainable agriculture such as thermal root-zone heating, while offering classes on science, business, art, philosophy, anthropology, and history. The farm grows dozens of varieties of fruits and vegetables, as well as raising heritage breeds of livestock (sheep, pigs, and cows) and poultry, and harvesting honey from the farm's beehives. The hay for the oxen and cows is grown on-site. As of spring 2010, Cerridwen Farm began accepting members from the local community into its Community Supported Agriculture (CSA) program.

The Bachelor of Science in Natural Resource Management (NRM) program trains students to work as park rangers, forest managers, or policy makers in an interdisciplinary program that emphasizes both conscientious leadership and a strong grounding in the collection, measurement, and analysis of natural resource data, including geographical information systems. The NRM program includes a mandatory 12-credit internship of at least 10 weeks and 400 hours; NRM majors have interned with the U.S. Forest Service, regional natural resource groups like the Squam Lake Association in New Hampshire, and technical internships working with geographical information systems. Some graduates of the program have gone on to study environmental law in law school.

The MBA in Sustainability is offered as a distance-learning program mediated by the Internet, aimed at students who work full-time and do their coursework in the evenings and on weekends. Core courses cover management, communications, information assets and technology, marketing and customer relationships, quantitative methods and statistics, economics, accounting, ethics and legal issues, and finance. Each student selects one of two concentrations, which determines his or her remaining coursework: the traditional General Business Administration or the Nonprofit Organization Management. Though the two-year program is conducted primarily online, it begins with a residency on campus and includes a shorter residency between the two years. Both residency periods include intensive coursework, workshops, and seminars, as well as social time during which students and faculty are encouraged to get to know one another.

Since 2009, GMC has also offered the Renewable Energy and EcoDesign (REED) Certificate Program, a certificate open to all majors. The program is field-oriented, interdisciplinary, and encourages participation by interested students regardless of whether they intend to pursue a career in green building or renewable energy. Partner institutions include the Yestermorrow Design/Build School in Warren, Vermont, and Solar Energy International, a Colorado school offering programs studying renewable energy resources. The REED certificate is also recommended to students interested in applying to engineering or design graduate programs.

The REED program consists of 22 credits, drawn from two electives and five core courses: The Nature of Design, Energy & Society, Renewable Energy Technology &

Applications, Ecological Design, and an external practicum conducted with one of the partner institutions or other appropriate organizations. Elective courses cover nonprofit management, geographical information systems, special topics courses on energy usage and the environment, advanced business courses, and other areas. An early independent study conducted through the REED program was completed by Ruth Larkin (class of 2010), who did a hands-on study of water purification and wastewater processing in homebuilt water systems, rebuilding the off-grid water system in her Vermont home in the process.

GMC's commitment to environmental awareness has extended beyond its curriculum design to its sustainability-minded campus. Light fixtures across the campus have been replaced with energy-efficient options, earning the school its designation as an Environmental Protection Agency (EPA) Energy Star Showcase campus. The campus greenhouse receives some of its power from a wind turbine; the student center, from solar panels on the roof. Both were installed by students. Thirty dollars from each student's activities fee goes toward the Student Campus Greening Fund, which is used to fund sustainability proposals, submitted by students and chosen by vote—modest but pragmatic projects like the installation of bicycle racks and recycling bins and the use of biodiesel in college-owned vehicles. On Earth Day 2010, a new heat and power biomass plant was opened, with the expectation of achieving carbon neutrality in the following year (in conjunction with the purchase of carbon offsets). Burning wood chips instead of oil fuel, the plant provides 85 percent of the school's heat and 20 percent of its electricity and is part of the school's ongoing efforts to reduce its carbon emissions.

Sustainability permeates the college experience of every student at GMC, regardless of course selection. Much of the food served in GMC's dining hall in the fall and spring is grown on the farm used by the Sustainable Agriculture program, and a recent project has built a root cellar using local, recycled, and repurposed materials, including locally grown hay and slate from local mines. The root cellar is used to store produce from the farm without the need for refrigeration or canning. The farm also uses solar panels to heat the water used by its dairy operations, and uses oxen instead of tractors for tasks like plowing.

See Also: Green Business Education; Leadership in Sustainability Education; Whole-School Approaches to Sustainability.

Further Readings

Filho, Walter Leal. *Sustainability and University Life*. Ann Arbor: University of Michigan Press, 1999.

Gough, Stephen and William Scott. *Higher Education and Sustainable Development*. New York: Routledge, 2009.

Hargreaves, Andy and Dean Fink. *Sustainable Leadership*. Hoboken, NJ: Jossey-Bass, 2005.

Pinsoneault, Eric. *The Environmental Impact of Green Mountain College: A Study of Resource Use and Waste Creation*. Poultney, VT: Green Mountain College, 2003.

Sandell, Klas, Johan Ohman, and Leif Ostman. *Education for Sustainable Development*. New York: Studentlitteratur AB, 2005.

Silka, Linda and Robert Forrant, eds. *Inside and Out: Universities and Education for Sustainable Development*. Amityville, NY: Baywood Publishing, 2006.

Wiland, Harry and Dale Bell. *Going to Green: A Standards-Based Environmental Education Curriculum for Schools, Colleges, and Communities.* White River Junction, VT: Chelsea Green Publishing, 2009.

Bill Kte'pi
Independent Scholar

Greens, The (PBS Website)

As sustainability issues and environmental awareness become more prevalent in our society, schools, publishers, broadcasters, and others have attempted to include children in efforts to build a greener planet. In conjunction with the Public Broadcasting Service (PBS), WGBH, a noncommercial television broadcast station in Boston, Massachusetts, has produced many children's television programs such as *Arthur, Curious George,* and *ZOOM.* To allow children an interactive environment to learn more about sustainability, WGBH and PBS also created a Website titled The Greens. Unlike other Websites supported by PBS, The Greens does not have an accompanying television program. The Website does include animated videos, however, as well as games, a blog, and other features. The Greens encourages children to explore the world of sustainability and prompts them to take positive action whenever possible.

Educational Children's Programming

Although children's programming has existed since the beginning of commercial broadcasting in the 1920s, shows with an educational concentration began to be produced after 1952, when a grant from the Ford Foundation enabled the founding of the Educational Television and Radio Center. Soon renamed as National Education Television (NET), it sponsored a number of well-known children's television programs, including *Mister Roger's Neighborhood* and *Sesame Street.* In 1970, NET was superseded by PBS, which continued to pioneer educational programming for children, including *The Electric Company, Liberty's Kids,* and *Reading Rainbow.* Children's programming on PBS has been awarded numerous Emmy and Peabody Awards. Beginning in 1993, a cable television channel, PBS Kids, began transmitting children's programming to selected audiences. In 2005, PBS Kids was transformed into PBS Kids Sprout as part of a commercial deal with cable giant Comcast Corporation, PBS, and several other partners.

PBS Kids is a Website that details programming offered on PBS that is intended for children. ("PBS Kids" also can refer to programming blocks that are intended for children offered on PBS television stations.) The PBS Kids Website includes games, videos, and other interactive resources, mostly tied into children's programs that are run on the network. This Website also offers resources for parents and teachers who want to better support their children in the classroom and at home.

Quality Websites for Children

Countless Websites are available on the Internet, many purporting to contain content that is available for children with an interest in a given subject area. Those familiar with best pedagogical practices have developed criteria that assist consumers in making determinations regarding the quality of a Website. These criteria include scope, content, graphic and multimedia design, purpose, and workability. Using these criteria to analyze The Greens Website assists in determining its relative value for children, parents, and teachers.

The scope of The Greens is fairly broad in that it explores practices and behaviors that will help make a more sustainable planet but is buttressed with a great deal of depth that allows interested children and their families or caregivers to put some of the theory into practice. Under the Reuse category, for example, is an episode titled *Landfill Blues*. The video shows the issues that can result when materials are put into landfills and suggests ways to "keep the planet green" by reusing items that might otherwise be discarded. Additionally, links are provided to games within the Website that allow children to work at projects that allow them to simulate recycling and energy-conscious behaviors, such as a game titled Thrifty Threads that encourages children to dye, restyle, and reuse old clothing rather than discard it.

With regard to content, The Greens is based almost exclusively on factual information, as opposed to opinion. Certainly The Greens advocates for children and others to make informed choices and meaningful changes in the world in which they live. All content contained on the Website, however, is grounded in factual knowledge about how best to engage in a sustainable lifestyle. The Website explores best practices that are simple to begin such as turning off unused electric lights and recycling unwanted items rather than disposing of them in a manner that harms the Earth. The Greens is composed almost exclusively of original content, with a few helpful links to other information included to assist students, parents, or teachers who may also access the Website. The content is integral to the many activities provided for users and meets the criteria of accuracy, authority, currency, and uniqueness that define a great Website for children.

Content on The Greens Website is accurate, in keeping with general understandings of the scientific community at large. Political or ideological biases are avoided, which is especially significant on a Website targeting children. The Greens Website also benefits greatly from the authority it generates from the organizations and experts who helped to create it. In addition to PBS and WGBH, both of which have a long-standing commitment to and experience with excellent children's programming, The Greens involves partners such as the National Geographic Educational Foundation, Project 3650, TVO Kids, and Zerofootprint. The project adviser is Alan Fortescue, the director of education for Earthwatch Institute, who has extensive experience assisting the United Nations Educational, Scientific and Cultural Organization (UNESCO) to develop programming regarding sustainable development. Updates of The Greens Website are common, and information and content continue to evolve. The content of The Greens makes for a unique experience. Unlike other PBS Websites, The Greens does not accompany a television program, which means that the videos posted there can be seen nowhere else. Games, blogs, and other features complement the content of the videos and provide multiple paths to access similar content.

Graphic and Multimedia Design

The Greens features an animated family, the Greens, that centers on two middle school students, Izz and Dex. Izz and Dex have green complexions and are cousins. Izz is female and lives with her father and grandmother. Izz's father is a software engineer who works from home, while her grandmother, known as "Granny," tends to her cat. Dex's mother is Izz's father's sister, so Granny is grandmother to both of them. Dex's mother is described as not green at all and involved in a retail business. She is involved in a relationship with her boyfriend, Hoyt, whom Dex does not like. Other characters include Izz's babysitter Jolie and her friend Hector; C. J., a researcher and teacher at the local aquarium; and Mrs. Greener,

Izz's next-door neighbor. The Greens is set in a suburban coastal setting, and the characters are all solidly middle class.

Since the purpose of The Greens is to encourage children to learn to live sustainable lives, the Website is very interactive and allows children to engage in a variety of activities that are intended to build their understanding of green issues. The Website presents an integrated environment, where familiar characters explore some of the tenets of environmental stewardship. In education, many advocate for a constructivist approach to learning, especially for children. Constructivism suggests that humans generate knowledge and meaning from their experiences. The Greens takes a constructivist approach to learning, one that encourages children to engage in hands-on activities to build their understanding of sustainable topics. Since content is delivered through the Internet, The Greens obviously cannot provide immediate hands-on experiences, but the site provides a variety of suggested activities related to recycling, reducing trash, and minimizing our carbon footprint that permit students to participate in the green revolution.

Visually, The Greens is stimulating, attractive, and enticing, as one might expect from the level of expertise of the organizations that have created and sponsored it. The animated episodes that are the focus of the Website are simple and contemporary, using animation to demonstrate a variety of perspectives. The visual effects enhance the content, which is thoughtful and relevant to the goals of the Website. The audio, video, and virtual reality modeling that are used are attractive and work without fail. The navigational design is especially useful. The Greens Website is convenient and effective to use. Rather than merely providing links to originating sites, The Greens has provided a set of materials that can either be viewed online or downloaded in PDF format. Any links that leave the Website are preceded by a warning page, which should provide assurance to parents or caregivers wary of some of the content on the Internet. The Website can be accessed with standard equipment and software and works well on either PC or Macintosh platforms. The Greens is available without charge and does not require a password or other form of login to use. In sum, The Greens provides an opportunity for children to ponder the world, their place in it, and ways that humans can engage in sustainable behavior.

See Also: Constructivism; Integrating Sustainability Education Concepts Into K–12 Curriculum; Social Learning; Sustainability Topics Correlated to State Standards for K–12; Sustainability Topics for K–12.

Further Readings

American Library Association. "Great Web Sites for Kids Selection Criteria" (2010). http://www.ala.org/ala/mgrps/divs/alsc/greatwebsites/greatwebsitesforkids/greatwebsites.cfm (Accessed June 2010).

Burden, P. *A Subject Guide to Quality Websites*. Lanham, MD: Scarecrow Press, 2010.

Public Broadcasting Service. "The Greens: A Site for Kids Looking After the Planet." http://meetthegreens.pbskids.org (Accessed June 2010).

Stephen T. Schroth
Emily K. Berkson
Levi W. Flair
Knox College

Green Week, National

National Green Week, officially the first week in February, was created by the Green Education Foundation (GEF) to bring environmental education to pre-K–12 students across the United States, emphasizing the importance of green education curriculum, waste reduction, recycling, and healthy living. The GEF, a nonprofit organization, was founded in 2008 with the goal of promoting and enhancing environmental education and sustainable habits in children, families, and communities. National Green Week is an indication of the environmental movement's growth and diversification into the field of education, starting at the elementary school level. Furthermore, this week reflects the role of nongovernmental organizations in shaping education for sustainable development. It might be said that the overall goal of the creators of National Green Week is to integrate the three Rs of the environment—reduce, reuse, and recycle—within the three Rs of a basic skills-oriented education—reading, 'riting, and 'rithmetic.

National Green Week is based on the Fisher Elementary Green School Experiment, implemented at the Fisher Elementary School in Walpole, Massachusetts, in 2008. During this two-week pilot study, the school reduced trash waste by 70 percent and showed a positive financial impact. The Green Education Foundation's efforts with the Fisher Elementary Green School Experiment to further environmental learning was rewarded in 2009 with the Secretary's Award for Excellence in Energy and Environmental Education by the Massachusetts Executive Office of Energy and Environmental Affairs. Subsequently, the Fisher Experiment was replicated nationwide during the first National Green Week in 2009. The results from the 2009 National Green Week show that nearly 500,000 students from 44 to 48 states participated, and roughly 250,000 participants eliminated 100,000 pounds of trash in five days by using reusable containers for snacks and lunches.

The Green Education Foundation uses National Green Week to foster environmental awareness and to promote a sustainable world by reshaping educational curriculum and implementing hands-on green programs. It attempts to increase the role of environmental education in pre-K–12 curricula across the United States by designing formally organized, standards-based curricula and participatory models that provide students with hands-on activities and knowledge related to the environment that can be integrated into the lessons of standard curriculum subjects such as math, science, social studies, language arts, and creative arts. The goals are to empower children to become environmental stewards and to promote well-being. The objective of the program is to provide environmental education to students and schools, to reduce or eliminate waste in schools, to promote recycling, and other "green" activities.

The Green Education Foundation provides the educational content and curriculum and environmental educational activities, projects, and games for National Green Week. Teachers also have the option to build their own unit plan using the Green Education Foundation's lessons organized by grade, subject, and topic. Subjects covered in the educational content includes climate change, energy, biodiversity, air and water quality, waste, pollution, recycling, school gardening, and air and water quality. Ongoing environmental educational opportunities that reinforce the goal of National Green Week are provided through after-school, summer, and home-based projects. The GEF also claims that it hosts the largest online global community dedicated to environmental education and the world's leading social networking site for K–12 educators.

Programs and activities promoted during National Green Week include waste reduction in snacks and lunches by using reusable containers; energy-saving measures such as reducing energy consumption via finding and fixing energy leaks at school and home, turning lights out where sunlight is a viable option, walking/biking/carpooling, and turning off car engines when the wait time is longer than 20 minutes; and a "Green Thumb Challenge" for schools to plant indoor and outdoor gardens during spring. The gardening promotes improved habits in health and nutrition.

Critics of green education in the classroom and National Green Week suggest that activists and educators overstep their bounds by trying to shape how families live. They believe that it should be left up to parents to teach their children about these values and lifestyles, especially since people still debate the magnitude of global environmental problems. Proponents of National Green Week, however, argue that many schools in the United States have little environmental education in the classroom and lunch programs that are unhealthy and have high ecological footprints. Among this group there is the belief that much more should be done than just a one-week program that is often only adopted by the more progressive teachers. Notwithstanding different points of view, the second National Green Week took place in February 2010, with the goal of mobilizing 2 million children in schools and organizations to participate in the environmental education programs of the week. The goal for 2010 was to eliminate 500,000 pounds of trash from schools and landfills. Nearly every governor in the United States made National Green Week 2010 an official event in his or her state.

See Also: Early Childhood Education; Integrating Sustainability Education Concepts Into K–12 Curriculum; Sustainability Topics Correlated to State Standards for K–12; Sustainability Topics for K–12.

Further Readings

Feinstein, N. "Education for Sustainable Development in the United States of America. A Report Submitted to the International Alliance of Leading Education Institutions" (2009). http://www.uspartnership.org/resources/0000/ 0074/ESD_in_the_USA_8–19–09.pdf (Accessed January 2010).

Green Education Foundation. "Fostering the Next Generation of Environmental Stewards." http://www.greeneducationfoundation.org (Accessed January 2010).

MacDonald, G. J. "Schools Cultivate Green Living." *USA Today* (January 27, 2009). http://www.usatoday.com/news/education/2009–01–26-green-schools_N.htm (Accessed January 2010).

Vanessa Slinger-Friedman
Kennesaw State University

GRIFFITH UNIVERSITY

Griffith University, Brisbane, Queensland, was the first Australian university to offer Environmental Studies in the 1970s and also offers the longest running, since 1991, master's-level environmental education professional development program in Australia. The university was established to provide an alternative university experience and was thus designed around a number of theme-oriented interdisciplinary faculties. The environmental education master's program, taught on Nathan campus—a campus located within Toohey Forest, Brisbane's largest patch of remnant bushland encompassing some 260 hectares—provides an innovative approach to teaching and learning through its curriculum.

The Master of Environment (Education for Sustainability) program was originally devised by Professor John Fien, an internationally renowned environmental education expert. The one-year full-time or part-time equivalent program, taught in face-to-face and distance education mode, was originally designed as a mid-career professional development opportunity for teachers with the broad aim of developing critical and reflective environmental education practitioners. Since then, an increase in students from the nonformal education sector has seen the content of the program change to include both formal and nonformal education sectors. The program is undertaken mainly through coursework, although options exist for a significant research component should students consider higher research degree study.

The program focuses on the foundations of environmental education, trends and issues in environmental education, school and community environmental education, and the role of environmental education in achieving sustainability and social change. Students complete four core environmental education courses along with four electives from a pool of ecological, social, or health-related courses. Students thus build specialist environmental education knowledge and also build their environmental knowledge.

A key feature of the program is its commitment to high-quality teaching and learning principles that are underpinned by environmental education and education for sustainability theory. These principles are derived from a sociocritical perspective that seeks to empower students to become critical thinkers and critically reflective practitioners. Inquiry-based teaching and learning strategies along with participatory, student-focused learning opportunities ensure that students experience and develop questioning and analytical skills and a range of strategies for problem solving and action taking, and rethink their personal and professional roles and responsibilities as environmental educators. The program seeks to enable students to locate the present within the past by, for example, identifying and examining assumptions underlying the environmental crisis and the role of environmental education in addressing the crisis.

This is achieved through a variety of interactive teaching and learning approaches, including workshops, mini-lectures, attending guest lectures by visiting national and international academics, hands-on activities, simulation games, student presentations, and real-life, relevant applications of education for sustainability theory. These learning activities provide students with opportunities to critically analyze issues, and clarify attitudes and values through discussion and debate in a supportive learning environment. Teaching staff in the program also regularly capitalize on opportunities for students to contribute to real-world projects such as creating published teaching resources and undertaking practical environmental education consultancies, which also provide a stimulus and real context for student assessment. A high degree of collegiality and community is also fostered in the program through a range of extracurricular field and social activities, which provide valuable opportunities for informal networking and induction into the community of practice that is environmental education.

Student evaluations indicate that students find the program motivating and inspiring, and many report that the program has had a significant positive impact on their own professional practice. Many graduates have found work in formal and nonformal education sectors as well as in government and industry. In these ways, the Griffith University program provides a successful example of environmental education theory in practice.

See Also: Leadership in Sustainability Education; Sustainability Officers; Sustainability Websites; Whole-School Approaches to Sustainability.

Further Readings

Griffith University. "Master of Environmental Education–Nathan." http://www17.griffith .edu.au/cis/p_cat/admission.asp?ProgCaode=5064&type=overview (Accessed June 2010).
Griffith University, Environment, Planning, and Architecture. "Educational Programs: Toohey Forest Environment Education Centre." http://www.griffith.edu.au/environment-planning -architecture/ecicentre/educational-programs (Accessed June 2010).

Elizabeth Ryan
University of the Sunshine Coast

Halifax Declaration

The Conference on University Action for Sustainable Development was held in Halifax, Nova Scotia, in 1991. At this conference, 16 Canadian universities adopted the Halifax Declaration, which recognized the leadership role universities could play in addressing environmental damage. Specifically, the Halifax Declaration stressed that universities must structure their environmental policies and practices so that they would contribute to environmental sustainability on local, national, and international levels. The Halifax Declaration resulted in many environmental initiatives and the restructuring of environmental practices and policies at the signatory universities. Although initially limited to 16 universities, the consequences of the actions spurred by the Halifax Declaration have been far-reaching and influential.

The Conference on University Action for Sustainable Development

The Conference on University Action for Sustainable Development was held December 9–11, 1991, in Halifax, Nova Scotia, Canada. The conference was sponsored by Dalhousie University, the Association of Universities and Colleges of Canada, the International Association of Universities, and the United Nations University. The 89 participants at the conference included university students, faculty, administrators, and presidents. Representatives of various businesses, nongovernmental organizations, and municipal, provincial, and federal governmental officials were also in attendance. Participants with a high level of interest in sustainability issues, previous experience in environmental education, or other green activism were invited to attend.

The conference was called for several primary purposes. These purposes included addressing universities' abilities to address environmental and development issues, determining their capacity in shaping public opinion, and discussing the consequences that the Talloires Declaration (the first written commitment made by university administrators to sustainable practices) had on Canadian institutions of higher learning. Specifically, the conference sought to discuss the role of universities in promoting environmental sustainability as an ongoing part of their missions. The conference was organized into sessions, which examined the role of university leadership in sustainability, activism that would lead to green development, and how environmental principles affected the curriculum. Throughout the conference, a group met to prepare a statement reflecting the ideas talked about at the sessions. This group's work resulted in the Halifax Declaration, which was presented to participants on the final day of the conference. The Halifax Declaration avowed that the following must occur:

- Universities must take a leading role in affecting positive environmental change.
- Institutional environmental policies must be reconceptualized, revised, and restructured.
- Administrators, faculty, and students must contribute to environmental sustainability on local, provincial, national, and international levels.

At the conclusion of the conference, the Halifax Declaration was endorsed by 16 Canadian institutions, including the universities of Calgary, Carleton, Dalhousie, Manitoba, McMaster, McGill, Memorial, Moncton, Montreal, Mount Saint Vincent, New Brunswick, Queens, Saint Mary's, Trent, Western Ontario, and York.

Action Plan

After the conclusion of the conference, an action plan was also created to assist with the implementation of the initiatives supported by the Halifax Declaration. This action plan was disseminated to all the signatory universities. The action plan outlined frameworks that would help guide actions taken by institutions, as well as short-term and long-term objectives for the signatory universities. The action plan, and indeed the Halifax Declaration itself, never provided a concise definition of either "sustainability" or "sustainable development." To remedy this, the action plan was intended to suggest key activities that each institution could engage in to meet the goals of the Halifax Declaration. Sustainable initiatives suggested cultural, economic, environmental, political, and social concerns that might be addressed. The action plan also emphasized the necessity of attending to these issues in their entirety rather than dealing with each one in a piecemeal fashion.

The action plan focused upon several aspects of sustainability, including public outreach initiatives; interuniversity cooperation; the cultivation of partnerships with private industry, government, and nongovernmental organizations (NGOs); and a scheme to increase green consciousness among members of the university communities. Specific activities suggested in the action plan included the following:

- Establishing a national university network focused upon green initiatives
- Developing local, provincial, national, and international sustainability education programming
- Sponsoring awards and prizes for environmental development for university students, faculty, administrators, and staff
- Creating advisory papers that would assist students and faculty in aligning their research with sustainability goals
- Integrating the national media with university initiatives to promote green development

Although the Halifax Declaration and the subsequent action plan were the result of coordination and cooperation among the various signatory institutions, no mechanisms were formally in place to enforce participation or even to monitor involvement.

In addition to the action plan, recommendations were also made regarding activities that a university could engage in to support sustainability. The recommendations included means of assisting green initiatives at the local, regional, national, and international level, and covered a wide range of suggestions, from simple concepts such as calling a meeting to explain the Halifax Declaration to university constituencies, to more difficult ideas like establishing a national university network to promote sustainability development.

Implementation of Action Plans

A variety of initiatives were begun as a result of the Halifax Declaration action plan. All 16 universities that were signatories to that document implemented at least some of the initiatives, although none of the institutions fully implemented the action plan. Some activities were more popular with universities than others. Those activities that were implemented at a majority of the 16 included public forums to explore and exchange ideas about sustainability, programs that related to sustainability education and other environmental literacy training, and collaborative research projects that examined green development. Other activities, however, were less common. Few universities fully designed and implemented an environmental curriculum, provided scholarships for work in sustainable fields, encouraged innovative educational technologies for disseminating sustainability to the general public, or established chairs in sustainable development. Although some found the level of implementation disappointing, others believed that the Halifax Declaration began an important process that, while imperfect, raised the profile of sustainability development for the universities that adopted it as well as for others.

Part of the problems with implementation of action plans may have stemmed from the manner in which the Halifax Declaration was disseminated. At the conclusion of the conference, representatives from each university were asked to become Halifax Declaration "ambassadors" within their institutions. Ambassadors were drawn from many levels within their universities, including administrators, faculty, and staff. Since the change literature suggests that successful initiatives need clear and consistent support from university leadership, the inconsistent authority held by ambassadors may have hindered the implementation of action plans at certain universities. However, because sustainability initiatives often are led at least in part by students, this lack of central support may not have been as significant as in other situations.

Since 2000, many of the 16 universities that endorsed the Halifax Declaration have added sustainability coordinators to help build a culture of environmental development on their respective campuses. Sustainability coordinators are frequently responsible for the coordination and implementation of campus activities related to green initiatives, including seminars, conferences, workshops, public events, and other programming. Sustainability coordinators are also charged with encouraging and facilitating environmental programs and research initiated by students, faculty, and administrators, and to identify materials or resources that would assist these initiatives. The presence of sustainability coordinators appears to have channeled campus support for sustainable development and the Halifax Declaration's goals and objectives. This in turn has increased both the visibility and progress of environmental development initiatives, building support for projects supportive of a sustainable future.

Influence of the Halifax Declaration

The Halifax Declaration has proved to be influential in other attempts to begin sustainability initiatives that build bonds between universities, environmental development, and other sectors. The International Association of Universities (IAU), for example, in 1993 adopted a declaration in support of sustainable development similar to the Halifax Declaration. Indeed, the IAU declaration specifically embraced both the language and substance of the Halifax Declaration. The IAU declaration supported collaborative teaching and research about sustainability, combined with green practices, to move universities from what it deemed a traditionally passive role into an energetic and central leadership promoting environmental issues. Working with partners such as the United Nations Educational, Scientific and Cultural Organization (UNESCO), University Leaders for a Sustainable Future (ULSF), and others, the IAU has used its declaration to expand conversations about and actions supporting sustainable development in ways the Halifax Declaration envisioned.

See Also: Association for the Advancement of Sustainability in Higher Education; Association of University Leaders for a Sustainable Future; Sustainability Officers; Whole-School Approaches to Sustainability.

Further Readings

Halifax Declaration. http://www.iisd.org/educate/declarat/halifax.htm (Accessed August 2010).

Wright, T. S. A. "A Tenth Year Anniversary Retrospect: The Effect of the Halifax Declaration on Canadian Signatory Universities." *Canadian Journal of Environmental Education,* 8/1 (2003).

Stephen T. Schroth
Jason A. Helfer
Daniel T. Stafford
Knox College

HARVARD UNIVERSITY

Harvard University, located in Cambridge, Massachusetts, and a member of the Ivy League, is a private university founded in 1636, and is the oldest institution of higher education in the United States. Ten academic units serve over 21,000 undergraduate and graduate students. In addition to its long history of academic excellence, Harvard has been regularly recognized for its commitment to campus sustainability. While currently under the leadership of the Harvard Office for Sustainability, prior to 2008, Harvard's sustainability initiatives were organized under the Harvard Green Campus Initiative (HGCI). In 1999 a group of faculty, staff, and students came together to develop ideas on creating a greener campus environment. After seeing a presentation from Leith Sharp and the success at the University of South Wales, the task force secured internal funding to bring Sharp to campus to oversee sustainability efforts at Harvard, creating the HGCI. Sharp and the HGCI were able to create and receive funding for a five-year plan at $150,000 per year for administration, along with the establishment of the $3 million Green Campus Loan Fund to encourage sustainability innovation on campus. The university's efforts in sustainable

Harvard University's sustainable campus operations include an organic landscaping program on 25 acres of campus and the composting of 100 percent of landscaping waste.

Source: Wikipedia

campus operations begin at the administrative level, which Harvard refers to as "Sustainability Principles." Introduced in 2005, these guidelines provide leadership for green building, energy, emissions, and transportation, among other areas. The master plan provides district-level guidance. A number of committees help coordinate these efforts, including a Greenhouse Gas (GHG) Reduction Goal Executive Committee, a GHG Reduction Goal Student Advisory Group, a planned Faculty Sustainability Advisory Committee, and the University Construction Management Committee. In total, 32 full-time staff members are employed to assist in sustainability efforts in the Office of Sustainability, which was created in 2008 at the school and unit level. The director of sustainability at Harvard reports to the executive vice president.

Purchasing, Energy, Facilities, and Planning

While Harvard does not have a formal green purchasing policy, environmental language is included in many vendor contracts. In addition, all computers, printers, and copiers through university vendors are Energy Star rated. Much of the paper purchased is from recycled content, and the institution is preparing to have all fine printing paper purchased from a merchant meeting Forest Stewardship Council certifications. Approximately 90 percent of the university's cleaning products, hand towels, and bath tissue are Green Seal certified as well. Harvard Facilities Maintenance Operations (FMO) has an Organic Landscaping program across 25 acres of campus, with a goal of having 50 percent of all landscaping being treated organically.

Harvard Green Building Requirements detail how their Sustainability Principles apply to building projects, which includes a commitment of a 30 percent reduction in greenhouse gases by 2016. Greenhouse gas emissions are inventoried annually among all 652 buildings, and showed a modest decrease between 2006 and 2008 from 297,000 metric tons of carbon dioxide to 290,000 metric tons, a reduction of 2.4 percent. University policy requires that all capital projects over $5 million must at least meet LEED Silver certification. Those under that amount, between $100,000 and $5 million, still have energy efficiency targets, but are not required to be LEED certified. A total of 20 buildings are LEED certified, and 40 buildings are LEED registered, and an additional building meets criteria but is not certified. The Harvard School of Public Health's International House is a certified Energy Star building, with several more meeting the required criteria.

In an effort to address energy efficiency, all new buildings and retrofits have an in-house energy auditing and commissioning team. The Office for Sustainability also works to address behaviors on campus through emails, posters, signage, and a Green Office and Green Lab Certification Program, as well as through student engagement efforts. The office also regularly distributes Smartpower strips, compact fluorescent lamp (CFL) bulbs, and plug timers to students and staff.

In addition, a $12 million Green Campus Loan Fund provides funding to facilities departments for conservation efforts. This fund addresses energy efficiency through behavioral change, HVAC upgrades, insulation, lighting, solar, and sub-metering, saving the institution over $4.6 million. The university has a number of small-scale renewable energy projects under way. Photovoltaic solar cells are located on three buildings, and create 58 kW of electricity combined. Six wind turbines have been installed on the Holyoke Center, with plans to add additional units on Harvard Stadium. Harvard also utilizes solar hot water systems on campus, including the Sacramento Street dorm and two residential properties owned by the university. Most campus buildings utilize occupancy sensors and energy-efficient exterior and interior lighting, including light-emitting diodes (LEDs), super T-8s, and ceramic metal halide. Buildings that are new or retrofitted have insulation and windows that exceed code, as well as some nonhistoric buildings that use a white (Energy Star) roof membrane. The university also utilizes a variety of heating, ventilating, and air conditioning (HVAC) technologies to achieve energy efficiency, including ground source heat pumps (eight projects), energy recovery systems (10 buildings), and combined heat and power (three areas). New buildings on campus have low-flow showerheads and aerators and high-efficiency condensing boilers, as well as dual-flush toilets and low-flow urinals.

Harvard purchases renewable energy credits to help offset carbon emissions. These include a landfill gas project and hydro facilities (10,000,000 kWh), totaling 8.7 percent of electricity usage. In addition, Green-e certified voluntary credits were purchased equivalent to 3.8 percent of the electricity used.

Composting and Recycling

The institution recycles a variety of materials, including aluminum, cardboard, glass, paper, and all plastics, with a diversion rate in 2009 of 55 percent. An electronic waste program also recycles materials, including batteries, cell phones, computers, light bulbs, printer cartridges, and compact discs. Harvard Recycling operates a surplus center that collects used furniture that is available for free to the local community. In addition, the organization has assisted in collecting shoes and cosmetics, and in the reformatting and selling of surplus computers to a local nonprofit organization. Composting also has a strong presence on campus, with 100 percent of landscaping waste being composted. Twenty-five academic buildings offer composting receptacles, as well as the entire campuses of the School of Public Health and Divinity School. On LEED building projects, 92.6 percent of waste from construction and demolition was diverted from landfills.

Transportation

Harvard's motor fleet of 232 vehicles includes three hybrids and 78 shuttles that run on B20 biodiesel. Employees who carpool to work receive reduced rate charges, as well as a preferred program for those who drive low-emitting and fuel-efficient vehicles. The university's

Transportation Services department created the CommuterChoice program to provide a variety of sustainable transportation options and planning services.

For employees who choose public transportation, Harvard provides a 50 percent subsidy for passes, as well as pre-tax savings on private transit passes. This subsidy totals $3.1 million annually. In 2000, Harvard began a Zipcar program, which offers 10 hybrid vehicles to affiliates for a $25 membership fee. A departmental bicycle program was created in 2004 that provides the usage of 33 bicycles in 13 departments for free to employees. In addition, the program offers repair and winter storage, as well as monitoring usage. The program is overseen by the division of transportation and university operations services. In all, at least 81 percent of students, faculty, and staff utilize environmentally friendly transportation methods, including walking, biking, carpooling, and public transit.

See Also: Greening of the Campus Conference; Leadership in Sustainability Education; Whole-School Approaches to Sustainability.

Further Readings

Harvard University. "Transportation Services: Commuter Choice Program." http://www .commuterchoice.harvard.edu (Accessed July 2010).
Keller, Morton and Phyllis Keller. *Making Harvard Modern: The Rise of America's University.* New York: Oxford University Press, 2007.

Justin Miller
Ball State University

HESA: Higher Education Sustainability Act

The hallmark of HESA is the University Sustainability Program (USP) administered by the U.S. Department of Education. In addition, the Department of Education is required to host a summit on sustainability in higher education, bringing together higher education faculty and staff, federal staff, and leaders from both the private and public sectors to collaborate on sustainability projects and to share best practices between campuses. If appropriated at the full approved level of $50 million included in the legislation, the USP will support up to 200 sustainability projects per year for both individual institutions and consortia.

Funding for individual institutions will be available for a number of projects, including best practices at the administrative and facilities level; programs that assist in research and education of sustainability in the social, economic, and environmental disciplines; and initiatives for new methods in facilities such as energy, building practices, waste, and transportation. In addition, those faculty, students, and staff working on sustainability projects at institutions could be eligible for subsidized compensation. In the classroom, USP will provide funding for developing programs in sustainability literacy as well as creating sustainability curricula that can be applied across multiple disciplines, such as the sciences, engineering, and business.

Consortia that work with sustainability in higher education would be involved in funding for projects that affect multiple institutions, such as training programs for faculty and staff; disseminating best practices in an effort to create sustainability standards for higher education; developing partnerships with the private sector for research; and partnering with accrediting agencies so that standards can be easily reported nationwide, creating benchmarks for institutions, and making analysis more standardized and easy to compare across campuses.

As a program, USP aims to financially support the work of those engaged on campus in sustainability programs and to encourage faculty and students to create new practices and research projects in sustainability. USP also intends for higher education institutions to develop and grow partnerships with their local communities, industry, government agencies, and nonprofit organizations with the idea that practices developed in higher education can be implemented with partner agencies.

Goals of the USP are numerous and include graduating 3 million "sustainable-literature" college students; introducing future professionals to sustainable methods in areas such as business, the sciences, and planning; ensuring that sustainability becomes common practice in higher education administration; and producing methods that have applications off-campus, including the development of products and services for the public.

Appropriations

Funding for the USP has not been officially appropriated by Congress, and, as of the end of 2009, the U.S. Department of Education had not announced any specific funding programs in sustainability. There is an expectation that the upcoming round of FIPSE (Fund for the Improvement of Secondary Education) grants for fiscal year 2011 will include sustainability as one of its areas of focus. If this is confirmed, applications could be accepted in late spring 2011, with projects beginning the following fall or spring. There has been no announcement made by the Department of Education about a summit on sustainability in higher education.

See Also: AAAS Forum on Science, Technology, and Sustainability; Sustainability Officers; Sustainability Teacher Training.

Further Readings

Campaign for Environmental Literacy. "The Higher Education Sustainability Act (HESA)." http://www.fundee.org/campaigns/hesa (Accessed November 2009).

Campaign for Environmental Literacy. "The University Sustainability Program." http://www.fundee.org/campaigns/usp (Accessed January 2010).

GovTrack.us. "H.R. 3637—110th Congress: Higher Education Sustainability Act of 2007." http://www.govtrack.us/congress/bill.xpd?bill=h110-3637 (Accessed January 2010).

GovTrack.us. "H.R. 4137—110th Congress: Higher Education Opportunity Act." http://www.govtrack.us/congress/bill.xpd?bill=h110-4137 (Accessed December 2009).

Justin Miller
Ball State University

HIGHER EDUCATION ASSOCIATIONS SUSTAINABILITY CONSORTIUM

As campus sustainability has become a more popular goal of colleges and universities a growing awareness has developed that cooperation with and collaboration between different higher education associations can assist in obtaining mutually held objectives. The Higher Education Associations Sustainability Consortium (HEASC) is an informal network comprised of a variety of higher education associations that seek to advance sustainability amongst their members as well as advocating for environmentally friendly policies within the broader world of higher education. Through collaboration and partnerships, HEASC members strive to develop capacity to better address sustainability through their home associations. HEASC attempts to be as inclusive as possible in its mission, and as a result welcomes all higher education associations in an effort to receive the broadest perspectives and to create the strongest possible mandate for sustainable development. HEASC offers a variety of services to its members, including access to green resources, advocacy for sustainable development, and forums for the discussion of environmental issues.

Background

In 2005, the leaders of a variety of higher education associations decided to form HEASC to advance their ability to fulfill the critical role of colleges and universities to promote an educated citizenry, especially with regard to sustainability issues. It was felt that through HEASC more opportunities could be created for a broader, more systemic, and collaborative approach to research and operations related to sustainable development. To that end, HEASC encourages its members to collaborate on joint projects, share expertise and perspectives, and to maintain a collective and constant focus on advancing education for a sustainable future. Current members of HEASC include the following:

- American Association of Community Colleges
- American Association of State Colleges and Universities
- Association for the Advancement of Sustainability in Higher Education
- Association of Collegiate Conference & Events Directors International
- American Council on Education
- ACPA-College Student Educators International
- Association of Higher Education Facilities Officers
- Association of College and University Housing Officers International
- Association of Governing Boards of Universities and Colleges
- Council for Christian Colleges & Universities
- National Association for Campus Activities
- National Association of College and University Business Officers
- National Association of Educational Procurement
- National Association of Independent Colleges and Universities
- National Intramural-Recreational Sports Association
- Society for College and University Planning

Unlike most other groups promoting sustainability, HEASC promotes collaboration between higher education associations rather than amongst the members of these groups themselves. To that end, most of HEASC's work focuses on connecting the leadership and staff of higher education associations with professional development opportunities, funding opportunities, and joint projects that these groups can best work on together.

Collective Efforts

HEASC works to further support for campus sustainability through a variety of means. These efforts include advocacy and collaboration on a number of fronts. For example, HEASC was an official endorser of the American College and University Presidents' Climate Commitment (ACUPCC). The ACUPCC seeks to build support for sustainability by obtaining institutional commitments to concrete actions that will reduce and ultimately eliminate net greenhouse gas emissions on campus. To support the ACUPCC, HEASC members have promoted it to their constituents, provided toolkits and guidance documents to assist member institutions support implementation, and developed Webcasts to share best practices regarding sustainability. HEASC also has supported Campus Sustainability Day, which is celebrated annually on various campuses across the United States. HEASC has helped provide programming for Campus Sustainability Day via a Webcast panel discussion that was broadcast to participating campuses to highlight recent research developments related to climate disruption and ways to address climate change.

HEASC has also sponsored the Higher Education Sustainability Fellows (Sustainability Fellows) program. The Sustainability Fellows program, launched in 2008, is intended to support campus sustainability projects by providing them with experts who can provide guidance and know-how related to issues at hand. HEASC assists each year in the selection of a group of individuals with the knowledge, skills, and training to help make campus-based projects successful. Sustainability Fellows work five hours each week with a higher education association to support and accelerate the drive toward making the transition to a more sustainable society. An integral part of the Sustainability Fellows program is to create lasting tools that can benefit higher education associations that desire assistance in the future. To date, Sustainability Fellows have assisted with the following projects:

- Scaled renewable energy financing strategies
- Curricular and co-curricular leadership starter kits
- Community partnerships fostered out of college and university best practices
- Social media networking strategies to communicate sustainability to the general public
- Sustainability conference and events database
- Sustainability professional development resources, marketing strategies, and lobbying materials

The Sustainability Fellows serve for two-year terms and make their expertise available to interested higher education associations.

Additional Resources

HEASC provides additional resources related to sustainability to its member higher education associations through its Website. The resources HEASC provides are varied, and include news digests, professional development materials, notice of upcoming events, and links to other online resources. HEASC also provides a series of Webinars where members are able to learn additional information about pertinent topics, such as media strategies to promote sustainability or ways to improve campus practices with regard to purchasing, student life, operations, and other areas of administration. HEASC maintains a number of project committees, each with designated areas of responsibility. The project committees include those for Sustainable Practices in Conferences, Operations, Purchasing & Publications; Member Education; Communicating Sustainability to the General Public and Media; Communicating Sustainability to Association Members; Financing Sustainability; and Sustainability Awards. Each of these project committees coordinates with HEASC members to provide services pertinent to group interests and needs.

As more higher education associations commit energy and resources to advancing sustainability, HEASC provides a valuable network that allows collaboration and permits the sharing of resources. HRASC promotes sustainability amongst higher education associations through supporting each others environmental programming, allowing for a free and open exchange of ideas between members, engaging in joint projects that buttress members' interests, and coordinates resources that the members possess. Although fairly new, HEASC has provided members with benefits in their efforts to promote sustainability. These benefits include reducing redundancies between member programs, increasing the quality of sustainability projects through synergistic cooperation, building a community of support for higher education associations, and providing sustainability expertise to members. As more higher education associations, and their member colleges and universities, decide to pursue opportunities to collaborate regarding programming related to sustainability, HEASC's services may become more popular. Certainly the financial difficulties facing many higher education associations make sharing the costs of programming attractive. The joint development and maintenance of resources that promote and move forward higher education associations' commitment to sustainable practices will help to expedite the green movement on campuses across the United States.

See Also: American College and University Presidents Climate Commitment; Association for the Advancement of Sustainability in Higher Education; Association of University Leaders for a Sustainable Future; Halifax Declaration; International Alliance of Research Universities: Sustainability Partnership.

Further Readings

American College and University Presidents' Climate Commitment. http://www.presidents climatecommitment.org (Accessed March 2010).

Association for the Advancement of Sustainability in Higher Education. http://www.aashe.org (Accessed March 2010).

Association of University Leaders for a Sustainable Future. http://www.ulsf.org (Accessed March 2010).

Disciplinary Association Network for Sustainability. http://www2.aashe.org/dans (Accessed March 2010).

Higher Education Association Sustainability Consortium. http://www2.aashe.org/heasc (Accessed March 2010).

Stephen T. Schroth
Jason A. Helfer
Leslie Y. Kang
Angelina A. Rosa
Knox College

I

INTEGRATING SUSTAINABILITY EDUCATION CONCEPTS INTO K–12 CURRICULUM

As awareness of environmental issues has grown in recent years, so has an interest in integrating sustainability education concepts into the K–12 curriculum in schools. Some states, school districts, and schools have developed thorough and rigorous standards and curricula that address sustainability education. Other administrators, teachers, and parents, however, desire to have students examine environmental issues in school but are provided with little material with which to do so. Fortunately, a variety of resources exist that allow sustainability education concepts to be integrated into the K–12 curriculum. A variety of approaches, including differentiated instruction and guided investigations, can be used to integrate sustainability into elementary, middle, and high school instruction.

Sustainability education is best when tailored to each grade level and integrated into many parts of the curriculum. For example, students in grades 5–8 may learn about natural resources, biodiversity, ecosystems, and ecological footprints, as well as resource scarcity and energy economics.

Source: iStockphoto.com

Sustainability Standards

The U.S. Partnership for Education for Sustainable Development (U.S. Partnership) is an advocacy group comprising educators, organizations, and schools. The United Nations Decade of Education for Sustainable Development inspired the partnership to foster such education within schools in the United States. To that end, the partnership has devised National Education for Sustainability K–12 Student Learning

Standards, the third version of which was unveiled in October 2009. The standards covered a variety of issues, including the following:

- Intergenerational responsibility
- Ecological systems
- Economic systems
- Social and cultural systems
- Personal action
- Collective action

These standards are intended to be used by students in kindergarten through 12th grade. The U.S. Partnership has devised a summary chart that details how the standards can be used with students at different grade levels.

The focus of each sustainability standard shifts depending upon the grade level taught. The U.S. Partnership discusses the concepts aligned with each standard in terms of three bands, with each band representing a different level of schooling. The three bands represent students enrolled in grades K–4, 5–8, and 9–12; the bands designed for the upper grades build on topics and areas of study covered in the lower grades.

For "intergenerational responsibility," students enrolled in K–4 study concepts such as family and generations (grandparents, parents, children), while those in 5–8 explore the responsibility to future generations, and those in 9–12 grapple with intergenerational equity. When examining "interconnectedness," K–4 students explore relationships, historical connections, and sense of place; those enrolled in 5–8 look at systems and interdependency; and those in 9–12 consider systems thinking and cradle-to-cradle design. Areas of study that focus upon "ecological systems" begin in grades K–4 by looking at connections to nature and plants, animals, and habitats; while those in 5–8 move on to natural resources, biodiversity, ecosystems, ecological footprints, carrying capacity, environmental stewardship, and nature as model and teacher; and students in grades 9–12 explore respect for limits, respect for nature, the tragedy of the commons, environmental justice, biomimicry, urban design/land management, and natural capital. Students examining "economic systems" in grades K–4 look at human needs and wants; those in grades 5–8 study equity, resource scarcity, energy economics, ecological economics, and food systems; and those in grades 9–12 look at poverty, ecosystem services, alternative indicators and indexes of progress, globalization, true cost accounting, triple bottom line, and micro credit. Students examining "social and cultural systems" in grades K–4 focus upon family and friends, personal identity, happiness, fairness, and collaborative learning; those in grades 5–8 examine cultural diversity, multiple perspectives, citizenship, resource distribution, population growth, quality of life indicators, and education; and those in grades 9–12 look at human rights, social justice, peace and conflict, multilateral organizations, international summits, conferences, conventions and treaties, global health, appropriate technology, and governance. Explorations of "personal action" begin with setting goals, communicating ideas, and making a difference in grades K–4; move on to personal responsibility, personal footprint calculation, critical thinking, problem solving, and project planning and action in grades 5–8; and examine accountability, lifelong learning and action, and personal change skills and strategies in grades 9–12. Students in grades K–4 examining "collective action" concentrate on stating goals and working together, while those in grades 5–8 begin exploring concepts such as designing a sustainable system, structural versus personal solutions, democracy, society footprint calculation, and local, state, and national sustainability plans; leaving 9–12 grade students to concentrate on local to global responsibility, community-based and societal-level decision making, public discourse and policy, and organizational and societal change skills and strategies.

These concepts allow sustainability to be integrated into almost any subject at any grade level. Teachers who wish to integrate sustainability topics into their curriculum may use a variety of instructional strategies to do so. Allowing students to pursue an investigation that is personally meaningful allows them to have more invested in the subject studied because they are studying something they care about. Teachers can craft guided investigations and differentiate those investigations so that they will support student learning and the core concepts of sustainability.

Guided Investigations

Effective teaching considers which curricular goals, inclusiveness, and instructional methods will best reach and support students at a given grade level. When planning, teachers devise the most compelling means to relay information to students. In class, mental models are introduced that allow students to organize and segregate the information learned. Material must be reviewed over time to circumvent memory loss that otherwise occurs. Problems investigated must touch upon the child's world so that learning is meaningful and relevant. Teachers must assess and address student misconceptions about material learned, both to provide a safe environment in which mistakes can be made and also to prevent misinformation from interfering with the learning process. The interaction between children and society is crucial because things studied do not come within their experience unless they touch children's own well-being or that of their families and friends. When children investigate real-life problems that touch upon and affect their world, instruction is much more likely to take root. Such problems, with open-ended solutions, intrigue and interest children, spurring their best efforts and the most complete engagement possible.

Guided investigations (GI) rely on the teacher playing an active role in student learning. Simply put, teachers "guide" children's progress. GIS assist children in investigating matters that affect them, their families, and their communities. Central to any guided investigation is the classroom teacher. These teachers are best able to guide the investigations of students because they are able to balance development and disciplines to assure maximum learning. The GI model is supported by classic concepts of learning theory, such as Lev Vygotsky's zone of proximal development (ZPD). The ZPD is the distance between a child's independent problem-solving level and that same child's level of potential development at problem solving when working under an adult's guidance. Using the GI model, the teacher ascertains what a student's independent problem-solving level is and then provides that student with the supports and structures necessary for him or her to work at the next level. For example, a native English speaker working on a problem regarding the volume required to fill a bottle must expect to do most of the work, but her classmate, because of his English language development (ELD) level, may need assistance that a native English speaker would not. Within GI, the teacher is not expected to help a great deal initially, but to stand back, allowing the student to manage as much as possible on his or her own. When a student's attempts go askew, however, an expert teacher raises questions, rather than helping the student directly. The expert teacher will ask the student to explain how he or she progressed through a particular step of a problem, how he or she might best describe what happened, how an answer was arrived at, or how one answer deviates from another attempt. The truly exemplary practitioner even manages to use this situation to transform the student from an extrinsic to an intrinsic motivational source. Rather than praise the student for getting the correct solution after it is solved, the expert teacher discusses how difficult the problem is before it is tackled. GIS assist teachers in meeting student learning needs through curriculum that both challenges and supports them directly.

Differentiating Guided Investigations

Students learn best when they are provided with a moderate challenge. When tasks are far too difficult for a learner, that learner feels threatened and will not persist with thinking or problem solving as a self-protection mechanism. Conversely, tasks that are too simple also suppress a learner's thinking and problem solving; rather than learning, such a learner drifts through school unchallenged by and indifferent to the learning process. Either situation is problematic, especially for those students for whom school represents the sole connection with learning. Schools that are interested in decreasing the time students spend with inappropriate tasks seek to have teachers differentiate instruction so that the needs of all learners are met. Differentiation involves adjusting the complexity of content, the processes used for instruction, and the products students produce in order to afford each student an appropriate challenge.

Differentiation presents such a compelling model for classroom modification of instruction because it is deep, profound, and multifaceted. It provides a configuration that novice teachers can use to set up their practice, yet also provides a challenge for the competent veteran teacher. Differentiation asks teachers to determine the readiness and needs of each student and then to provide that student with instruction and activities that are appropriate, cogent, and beneficial. Classroom teachers are, of course, the experts regarding the children in their care. Teachers who examine any set curriculum will understand that parts of it may need modification to best meet individual or group needs. Some suggestions for how to ensure a successful change in the provided curriculum are detailed below.

Elements of Differentiation

No single formula produces a differentiated classroom. Instead, a few key ideas guide the practitioner who seeks to differentiate instruction. These principles include the following:

- Teachers focus on the essentials
- Teachers attend to student differences
- Teachers use ongoing and diagnostic assessment to guide instruction
- Teachers modify content, process, and products
- All students participate in respectful work
- Teachers and students collaborate in learning
- Teachers balance group and individual norms
- Teachers and students work together flexibly

These ideas are designed to assist students who learn in different ways and at different rates and who bring various talents and interests to school to feel comfortable and valued. In their planning and instruction, teachers must address the student traits of readiness, interest, learning profile, and affect. Readiness refers to students' knowledge, understanding, and skills related to a particular sequence of learning. Interest concerns events and subjects that spur learners' curiosity and evoke their passions. Learning profiles relate to learning style, intelligence preference, culture, and gender. Affect concerns students' social and emotional response to themselves, their work, and the classroom as a whole. Finally, classroom elements such as process, product, and learning environment are also concerns of differentiation. These refer, respectively, to how a student makes sense of information, ideas, and skills; the assessments or demonstrations of what a student knows, understands, or is able to do; and the operation and tone of the classroom. In the successful differentiated classroom, student traits and classroom elements are linked to achieve the optimal learning situation for each student.

Elementary School

Studying sustainability assists children in seeing the connection between their lives, community, and society in which they live and the curriculum with which they interact in school. Elementary school teachers work to assist children in seeing themselves as connected to the world. As part of their role as citizens, students must grapple with concepts such as sustainability, environmentalism, and civic responsibility. Including these concepts expands the standard curriculum and provides students with a greater understanding regarding how human actions affect the Earth.

Primary Grades (K–2)

Students in the primary grades, traditionally kindergarten through second grade, focus on exploring a variety of subjects related to each child. An array of information is presented, allowing children to see the breadth of information in their world while exploring few topics in depth. Incorporating sustainability and knowledge of environmental issues can be done so that students realize their impact on the classroom and school environment. In discussing classroom procedures, for example, instructing children about the process of recycling can begin an age-appropriate discussion of how resources are used in the classroom and what this means for the school and society as a whole.

With primary grade students, incorporating literature into subject matter assists in driving home complicated points. This is important when teachers seek to expose children to concepts such as sustainability because it allows the perspective to be moved away from actions taken within a specific classroom and shows how children's actions affect the world. Doing so allows students to gain an understanding of sustainable behavior and to begin to understand how their actions result in consequences.

Classroom projects that can build a sense of how personal action can make a difference might include using old cans as pencil holders in the classroom. Engaging in activities such as this begins to instill in students an understanding that reusing materials has benefits for them and the world in which they live.

Intermediate Grades (3–5)

After initial exposure to a variety of sustainability topics in the primary grades, students in the intermediate grades (historically third through fifth) begin to explore sustainability concepts in greater depth. These concepts can be explored throughout the curriculum, including during social studies, mathematics, reading/language arts, science, music, and art. Drawing connections to sustainability in all content areas is vital as the increasingly abstract nature of assignments demands that children be able to draw clear links between the material that they are learning and its applicability in their lives.

Using literature to introduce topics being explored enables these connections. Dr. Seuss's *The Lorax* is a book that provides a variety of lenses through which to examine sustainability. This story touches upon social studies with regard to geography and economics but also has a science base that explores human involvement and its effect on animals and natural resources. Incorporating children's literature into the classroom allows teachers and students to investigate the environment and human behaviors. Well-chosen books allow children to shift their perspectives on environmentalism from merely a scientific issue to one that investigates all facets of daily life. Other books that may assist teachers in crafting sustainability lessons include Donald Hall's *The Oxcart Man*, Vera Williams's *A Chair for My Mother*, Marcia Brown's *Stone Soup*, and Virginia Burton's *The Little House*.

The Sustainability Standards allow incorporation of many instructional strategies to focus upon environmental issues. Schools that desire to explore civic responsibility and raise awareness regarding environmental issues can easily do so. Simple ways to build sustainability into the curriculum include scheduling field trips to nature reserves, inviting guest speakers with expertise regarding recycling efforts, and allowing students to interact with their environment in ways that effect a positive change.

Children of color, English language learners, or students from low-SES backgrounds can prosper when working with sustainability issues. Understanding the environment in which they live and the positive steps they can take to affect it often increases their interest in and enthusiasm for school. Although an appropriate project for any student, engaging diverse learners in investigations that make better informed and more engaged citizens is especially helpful. Gaining an understanding of the local nature of many environmental factors is often empowering, especially since diverse students frequently live in the most environmentally unfriendly environments. Allowing them to identify risk factors and to take actions to improve their community is empowering. Possible activities might include writing letters to elected officials, making presentations regarding sustainable activities within their community, or initiating a recycling program.

Middle School

Elementary education focuses on a rudimentary understanding of and engagement with sustainability. As students progress to middle school (grades 6–8), they should be able to think about these issues in more abstract ways. Integrating sustainability into middle school education requires dealing with abstract ideas. This "fuzziness" enables students to increase their critical-thinking skills while also engaging in instructional activities that build their content knowledge. When integrating sustainability education into the middle school curriculum, it is important to link elements of different concepts together in a way that forces children to think critically about choices they and society make. At this level, instructional strategies should emphasize higher-order thinking skills through less-structured investigations and more project-based learning.

At the middle school level, students should take a greater role in defining the problems they wish to explore. As knowledge constantly changes, the curriculum should provide students with the basic tools needed to investigate sustainability issues: identifying a problem, observing and gathering resources, formulating a possible solution, and taking action that will help bring about the solution. Integrating sustainability into the curriculum allows students to encounter, accept, and embrace challenges in learning. The middle school curriculum allows students to move from their own understandings regarding sustainability to grappling with how environmental issues challenge society as a whole.

Students can explore a variety of topics that build on those examined in elementary school. Potential topics that might be explored include examinations of ecological systems with regard to biodiversity, ecosystems, and carrying capacity. For example, students living in a given area might explore how housing developments and other construction affect wetlands in their community and how the depletion of these wetlands changes the environment in that area. Once students conclude that human actions affect the environment in which we live, they might investigate how their own choices shape their personal carbon footprint, learn how to calculate that footprint, and then determine what steps, if any, they might take to reduce their negative influence on their surroundings. Such an investigation will result in discussions regarding key concepts that shape sustainability, such as equity, resource scarcity, citizenship, and quality of life.

High School

High school students (grades 9–12) who have been exposed to sustainability topics in earlier grades understand that the environment, broadly speaking, is made up of their surroundings. More specifically, the environment includes all the living organisms around them, the landscapes humans inhabit, the things people construct, the oceans and other elements on Earth, and practically all with which they interact. It is vital that students strive to preserve the environment in its healthiest state manageable. Sustainability is a concept integrated within environmental studies. Its aim is to find the means with which to continue life on Earth indefinitely. By high school, students can explore ways to either develop new approaches to satisfy consumers or conserve and limit the amount of resources that are being consumed. Students can also understand that if steps are not taken, then global climate change, overpopulation, consumption crises, species extinction, and other problems caused by human interaction with the environment may continue.

Although it is important to educate students about these issues, many children never have the opportunity to take an environmental science class because not all high schools offer them. In those schools where environmental science is offered, students are able to learn in depth about these topics. If such classes are not offered, then it is incumbent upon teachers to incorporate such topics throughout the high school curriculum, as students will have few other options to explore these issues. Fortunately, student interest in sustainability issues is high, and their enthusiasm makes integrating these topics into classes fairly easy. The breadth of sustainability topics also makes their connection with other content areas relevant and germane. Teachers and administrators working in schools without an environmental science course might consider incorporating sustainability topics into other areas of the curriculum.

Certainly, students who have been made aware of these issues from an early age are often more likely to fully comprehend sustainability issues and to make an effort toward reducing or resolving them. If, however, adolescents enter high school lacking the environmental knowledge needed to live a sustainable lifestyle, there is still hope. By integrating sustainability education into other areas of study through interdisciplinary lessons, teachers can provide students with important environmental information without the benefit of an environmental science course. Students can learn what they need to know about being an environmentally responsible citizen even without being enrolled in any science course.

Thinking beyond the realm of science, sustainability instruction can be integrated within English, social studies, math, or even art classes. In an English class, an easy writing assignment can entail having students write an essay or a letter to their mayor, senator, or any political representative, choosing a sustainability topic that is of interest to students. Another idea is to have students imagine they are writing an article for a sustainable living magazine. Teachers can always assign a research paper on an environmental topic, which then requires students to research and learn more about a particular sustainability issue.

There are many moments in history and social studies classes that allow integration of sustainability topics. The Industrial Revolution provides an excellent point at which to integrate such topics. One can have students discuss or write about inventions that proved to be most harmful to the environment. A twist on this might be for students write a certain inventor's perspective regarding whether he or she would have created the item if he or she had known how harmful it would be to the Earth beforehand. A teacher can also have students recall how poor agricultural practices affected the environment and in turn helped lead to the Dust Bowl.

Mathematics offers many opportunities to teach sustainability as well. Teachers can develop math problems dealing with fuel efficiency in automobiles. Students can, for example, discover the miles per gallon at which a certain model of car performs in city or highway driving, or the total miles of a driving trip, by doing so algebraically, leaving out certain pieces of information. Students can also be asked to determine the volume of water that a solar water heater with specific dimensions and measurements holds, or the total area of a deposit of bauxite ore found in a measured tract of land.

Within the art curriculum, there are numerous possibilities for incorporating environmental and sustainability education. A simple way is to have students go around the school grounds and pick up trash. The students can then use this trash to create recycled art sculptures. Students benefit from raised environmental consciousness and are also able to create and enjoy their works. Another assignment that students can work on is to paint a portrait of the same popular landscape twice, once from the perspective of 100 years ago and once from the present day (Glacier National Park is ideal for this). Students will be able to better comprehend just how much environmental damage humans are doing to our landscape.

No matter what subject one teaches, it is important to attempt to include sustainability education within the curriculum. Obviously, sustainability can be incorporated into branches of science other than environmental science. For example, an earth science lesson observing erosion and deposition, a biology lesson exploring endangered species, a physics lesson regarding the velocity of a rotating wind turbine, a chemistry lesson dealing with the reactions in a hydrogen fuel cell, and a physical science lesson on plate tectonics and its relationship to specific natural disasters all support sustainability education while remaining true to the subject at hand.

Conclusion

Students who are aware of the importance of sustainability are more likely to act upon such understandings. Sustainability issues can, of course, be addressed in a high school environmental science course, but many other opportunities exist to integrate this subject matter into the K–12 curriculum. Students at any grade level can benefit from sustainability education, and a wealth of resources exists to aid teachers in weaving this into their curricula.

See Also: Constructivism; Early Childhood Education; Experiential Education; Sustainability Topics Correlated to State Standards for K–12; Sustainability Topics for K–12; United Nations Decade of Education for Sustainable Development 2005–2014.

Further Readings

Michaels, S., A. W. Shouse, and H. A. Schweingruber. *Ready, Set, Science! Putting Research to Work in K–8 Science Classrooms.* Washington, DC: National Academies Press, 2008.

National Research Council. *Inquiry and the National Science Education Standards: A Guide for Teaching and Learning.* Washington, DC: National Academies Press, 2000.

Schroth, S. T. "Gifted English Language Learners: Developing Talent While Supporting English Language Acquisition." *Gifted Education Press Quarterly,* 20/2 (2007).

Tomlinson, C. A. *Fulfilling the Promise of the Differentiated Classroom: Strategies and Tools for Responsive Teaching.* Alexandria, VA: Association for Supervision and Curriculum Development, 2003.

Stephen T. Schroth
Jason A. Helfer
Derek M. LaRosa
Jordan K. Lanfair
Christian D. Mahone
Knox College

International Alliance of Research Universities: Sustainability Partnership

United around their shared values, global perspective, and commitment to educating future world leaders, 10 prestigious institutions formed the International Alliance of Research Universities (IARU) in 2006. The IARU supports a wide variety of initiatives among its member institutions including summer courses, conferences, student exchange programs, development of joint and dual degree programs, and faculty exchanges. The 10 members of the IARU include Eidgenossische Technische Hochschule (ETH) Zurich; National University of Singapore; Peking University; Australian National University; University of Tokyo; University of California, Berkeley; University of Cambridge; University of Copenhagen; University of Oxford; and Yale University.

Given the breadth of intensive research underway at member institutions, and the collaborative environment fostered among them, the IARU is uniquely positioned to advance efforts to address global societal challenges. Institutional sustainability, although not originally a focus of the IARU, has been championed and advanced by this prominent group. The IARU has provided an opportunity for presidents to build on their regional leadership and establish a global platform for sharing best practices for institutional sustainability.

Beginning with the establishment of Sustainability Principles in 2006, and evolving to the 2009 the IARU Presidents' Statement on Campus Sustainability, the IARU sustainability efforts have grown tremendously. The Presidents' Statement highlighted the role these institutions plan to play in advancing global sustainability. They state that they will seek to provide regional leadership and help to encourage engagement on a global level. Efforts to date focus on campus attributes and challenges that all institutions share. Methods for measuring and addressing the global impacts of a variety of campus activities, including housing, local and international transportation, facility renovation and construction, and grounds maintenance are some of the areas being assessed.

The IARU has also attempted to recognize the impact of the variation between institutional settings when considering campus sustainability efforts. Factors including age of the institution, the length of time they have had dedicated sustainability programs, availability of historical data, the pressures of regional environmental priorities and climate conditions, projections for growth, and the regionally available energy and natural resources all play important roles when considering the most sustainable approach to any campus issue. These factors are of particular concern when establishing greenhouse gas reduction targets. As of 2009, all 10 IARU institutions had established targets appropriate for their universities, taking into account the conditions by which each institution is bound. The IARU projects that by 2020, member institutions will be reducing greenhouse gas emissions by 340,000 tons per year under business-as-usual emission levels.

IARU Efforts

Each IARU institution has a number of exciting and innovative efforts underway, lessons that are shared among the group and beyond. At Australian National University, an organic composting machine is in place that can quickly transform food wastes and animal bedding to usable compost material. It is expected that the machine can compost up to 230 tons per year for the campus with very little cost, odor, or related pest concerns, all common complaints associated with urban compost projects. At the University of California, Berkeley, the dining system has been recognized for its efforts in improving overall sustainability. The system consists of green building certified facilities, and has 100 percent organic salad bars in each of the residential dining halls. Food wastes are donated to nonprofits or composted, and visitors receive a discount for using reusable mugs and bags. The University of Oxford has a sustainable transportation program that actively supports cycling and walking as transportation options, while reducing parking around campus. At Yale University, a staff leadership program engages with staff throughout the institution to empower them with information and support to institute basic sustainability improvements in their spaces. These examples offer a glimpse into the wide array of efforts underway at IARU campuses.

Sustainability Toolkit

Building from all of these examples, IARU has developed a Sustainability Toolkit to guide other institutions in tackling the complex challenges faced when implementing similar sustainability programs. The toolkit highlights the steps that institutions must take, and provides a wealth of informational sources for further reading. First, institutions must map their current situation and develop a governance structure. Next, institutions must clearly understand where their systems currently are by measuring their environmental impacts. With this understanding, they can then begin to work to integrate campus activities, and determine goals and strategies for the process. Next, institutions must establish strategies to create a sustainable campus, while continually placing emphasis on education and awareness. This toolkit is not meant to be exhaustive or prescriptive, but instead offers guidance to institutions embarking on the journey toward becoming sustainable.

The Future of IARU

Moving forward, the IARU has pledged to work collaboratively to expand institutional research on climate change and the challenges of sustainability. They have also committed to enhancing their student's ecological literacy, enabling the leaders of tomorrow to be prepared to handle the difficult work of building a more sustainable society. With these goals in place, IARU plans to offer a forum for better sharing of information and further action improving campus sustainability on a global basis.

See Also: Agenda 21: "Promoting Education, Public Awareness, and Training" (Chapter 36); Association of University Leaders for a Sustainable Future; Lüneburg Declaration; Millennium Development Goals; United Nations Decade of Education for Sustainable Development 2005–2014.

Further Readings

Association for the Advancement of Sustainability in Higher Education. http://www.aashe .org/index.php (Accessed March 2010).

Association of University Leaders for a Sustainable Future. http://www.ulsf.org (Accessed March 2010).

International Alliance of Research Universities (IARU). http://www.iaruni.org (Accessed June 2010).

International Alliance of Research Universities (IARU). "IARU Presidents' Statement on Campus Sustainability." http://www.iaruni.org/sustainability/statement (Accessed June 2010).

International Alliance of Research Universities (IARU). "Prospectus." http://www.iaruni.org/ about/IARU_prospectus.pdf (Accessed June 2010).

International Association of Universities. "GHESP: Global Alliance to Promote Higher Education for Sustainable Development." http://www.iau-aiu.net/sd/sd_ghesp.html (Accessed March 2010).

Sara E. Smiley Smith
Yale University

INTERNATIONAL INSTITUTE FOR SUSTAINABLE DEVELOPMENT

The International Institute for Sustainable Development (IISD) is a nonprofit research and policy organization based in Canada. It was founded in 1990 at the recommendation of Canada's National Task Force on Environment and Economy, which was modeled after the Brundtland Commission at the United Nations. The Brundtland Commission (named after its chair, the Norwegian prime minister Gro Harlem Brundtland, and formally known as the World Commission on Environment and Development) published an influential report in 1987 titled *Our Common Future*. The report recognized that "poverty, environmental degradation, and population growth are inextricably related" and called for a new focus on sustainable development as "development that meets the needs of the present without compromising the ability of future generations to meet their own needs." IISD advocates for development that meets the needs of people and the environment by engaging over 100 staff members and consultants in more than 30 countries to conduct its work, receiving support for its $12 million budget from the Canadian government, United Nations agencies, foundations, and the private sector. It publishes papers and books on various topics related to sustainable development; provides news coverage of international negotiations and development topics; coordinates formal networks of organizations working to achieve similar goals; and places young people in relevant internships around the world.

IISD organizes its research into five topic areas: Climate Change and Energy, Measurement and Indicators, Trade and Investment, Natural Resources Management, and Economic Policy. The institute's work within these topic areas ranges from analyses of local resource management practices to studies of broader trade and governance issues. In order to further its mission of informing policy makers, practitioners, and citizens about pressing environmental concerns, all of its publications are available for free on its Website, www.iisd.org. IISD's news reporting services are also particularly successful as communication tools. Since the first installment of its online newsletter *Earth News Bulletin,* which detailed the proceedings of the 1992 Earth Summit in Rio de Janeiro, Brazil, IISD's reporting has grown to cover 370 large negotiating conferences, 140 smaller conferences and workshops, and other news related to its research areas. It maintains several Internet mailing lists as well as an online research library with a sustainable development focus.

Sustainable commodities, subsidies, and water issues receive particular attention as areas of research as part of the institute's three major initiatives: the Sustainable Commodity Initiative (SCI), the Global Subsidy Initiative (GSI), and the Water Innovation Centre (WIC). These projects are collaborative (see below for information about IISD's partnerships) and have their own newsletters and policy briefs, with the exception of the newly established Water Innovation Centre. The SCI is a joint project of IISD and the United Nations Conference on Trade and Development (UNCTAD) that seeks to measure and share information about the environmental and market impacts of different commodities. The GSI, a joint project with the Earth Council, tracks the effects of government subsidies on sustainable development outcomes, emphasizing the need for transparency and accountability to the taxpayers who pay for them. Among others, the initiative is supported by the governments of the Netherlands, New Zealand, Denmark, and Sweden, although its analysis is not limited to these nations. IISD's work is more regionally specific at its Water Innovation Centre, which will build on insights from its 2005 Prairie Water Policy Symposium to inform its recommendations for the management of water resources in Canada's Lake Winnipeg Basin.

In its initiatives and other projects, the International Institute for Sustainable Development has a history of collaboration with other organizations, having partnered with over 200 other groups in 20 years and launched several cooperative networks. It currently manages five of these communities, including the Canadian Sustainability Indicators Network, which has nearly 1,000 members that work to measure and advance best practices in sustainability; the Ookpik Network, which identifies opportunities for youth from Arctic nations; SDplanNet (Sustainable Development Planning Network), which shares best practices in sustainable development for the Asian and Pacific region; the Sustainable Coffee Partnership, which engages stakeholders to improve the economic and environmental sustainability of the coffee industry; and the Trade Knowledge Network, which brings research institutions together to share information about the impact of trade and investment policies on development.

Finally, youth engagement in the field of sustainable development is a priority for IISD. It has conducted research on school sustainability programs and on existing training programs for young professionals interested in environmental and development work. It also manages its own youth programs, including the Leaders for a Sustainable Future program, which participates in the Canadian government's Youth Employment Strategy to arrange internships for 20 Canadian college graduates at sustainable development organizations, and the Circumpolar Young Leaders Program, which arranges work experiences for northern youth at organizations in Arctic countries. IISD leverages its network of relationships with environmental groups around the globe to train its young interns each year.

With its international relationships and its strength in research and policy work, the International Institute for Sustainable Development fills the role of the respected research institution as originally envisioned by the National Task Force on Environment and Economy. It continues to make the case for combining environmental and economic strategy into a vision for sustainable development and to provide resources for decision makers in the field, and its broad support from multiple governments is a sign that sustainable development will continue to be an important aspect of policy dialogue in the future.

See Also: Brundtland Report; United Nations Conference on Environment and Development 1992 (Earth Summit); United Nations Decade of Education for Sustainable Development 2005–2014.

Further Readings

International Institute for Sustainable Development. http://www.iisd.org (Accessed July 2010).
International Institute for Sustainable Development Library. http://ic.iisd.org/?page_id=255 (Accessed July 2010).
World Commission on Environment and Development. *Our Common Future*. Oxford, UK: Oxford University Press, 1987.

Sophie Turrell
Independent Scholar

INTERNATIONAL SUSTAINABLE CAMPUS NETWORK

The International Sustainable Campus Network (ISCN) is devoted to promoting sustainability on campuses, both educational and corporate, around the world. The ISCN is based on the principle that universities can lead society to adopt sustainability as a cornerstone for individual and organizational behavior, by example, because of their prominence as centers of learning. Campuses, by acting as demonstration sites, can play a catalytic role in disseminating the concepts and practices associated with sustainability. ISCN is partnering

A brushtail possum photographed on the campus of the Australian National University, which features nearly 10,000 trees in addition to its many other sustainability initiatives.

Source: Wikipedia

with institutions from Singapore, Japan, and Taiwan, to Switzerland, Denmark, and England, to the United States.

The ISCN efforts are comprehensive and incorporate multiple aspects of life on campuses. In an effort to lay a foundation for collaboration between institutions, the ISCN is finding common ground with a focus on energy use and efficiency, building design and transport infrastructure. The underlying focus is that these structural aspects of campuses involve significant investments of capital, knowledge, time and effort, or high sunk costs. Once put in place, these infrastructural projects are difficult and costly to change. Therefore, careful planning and implementation should guide the choice of these projects.

The ISCN places emphasis on both the ends and means of achieving sustainability on campuses. Setting targets, goals, and benchmarks for each institution is important, as is the process by which to achieve these objectives. A successful process tends to include multiple stakeholder engagement and one that seeks input and feedback. Community involvement in campus projects is important for both ethical and practical reasons.

ISCN and Higher Education

The ISCN, along with the Global Universities Leaders Forum (GULF)—a component of the Davos-based World Economic Forum—has created a charter and guidelines for international educational institutions. The charter is designed to help universities set their sustainability targets and to report their accomplishments regularly and with transparency. In 2009, the ISCN-GULF organized a conference in Lausanne, Switzerland. Over 50 participants attended the conference, representing a variety of international educational institutions and corporations. The three major objectives of the conference were to (1) facilitate exchange of information on campus sustainability issues and foster convergence among different networks engaged in a variety of different but related activities; (2) incorporate representation from key industries involved in providing sustainability services and expertise; and (3) promote a common approach not just toward sustainability objectives but also the pathways and processes to achieve the desired goals.

The ISCN considers university campuses to be laboratories where cutting-edge technology and concepts can be put into practice. Buildings can incorporate the latest energy conservation features while ensuring that an active learning environment is fostered. Since students encounter the buildings on a daily basis, novel opportunities for learning are created. The ISCN seeks to encourage exchange of knowledge between campuses, so that mutual learning can take place. To this end, it has created a number of Working Groups, consisting of members drawn from universities and corporations worldwide. The activities of these Working Groups include creating an international award for excellence in sustainable campus development, harmonizing differing sustainability standards in use internationally, creating a library of best practices, developing curricula, and instituting student exchange programs.

Building a 2,000-Watt Society

Novatlantis, the parent body of the ISCN, chose 2,000 watts as the energy consumption target for a global individual (that is, regardless of nationality). Two thousand watts is the global average energy consumption per capita, per day, with the level of consumption in the developed countries being much higher, and developing countries having levels below the average. The choice of 2,000 watts represents an attempt to find a compromise solution to the problem of energy overconsumption—it would simultaneously involve lowering the energy expenditure in the developed countries and increasing it in the developing countries.

Novatlantis is the sustainability implementer in the ETH domain (ETH is a renowned Zurich-based technical university). Its main function is to connect sustainability research—mainly in the fields of energy, building design, and transportation—to societal needs through promoting appropriate practices and policies. Although the ISCN has collaboration and partnerships with universities and corporations around the world, one of its main showcases is the Science City campus of ETH, which is being developed as a model campus.

ETH Science City

The Science City initiative involves designing and creating a sustainable campus that will serve as an international benchmark that will inspire similar efforts elsewhere. The Science City incorporates a multipronged approach to sustainability, with improved efficiency and reduced waste in constructing and maintaining buildings, managing transport, and energy solutions. Efforts are under way to reduce the Science City's greenhouse gas emissions. Since about 95 percent of the CO_2 emissions are composed of the travel undertaken by the faculty and students, alternatives to transport are being explored. In the area of transport, the Science City has made major strides, with over 90 percent of its staff and students using public transport. Paper consumption in offices has been lowered, and a target of 50 percent reduction was set for 2010.

Insights From Comparative Studies

A specific aspect of the ISCN program is the development of campuses in the developing countries. Due to globalization, the economic, scientific, and educational partnerships between countries in Europe and North America, the potential for transfer of technology and creating model campuses in the developing countries has increased greatly. Some companies in these countries have voluntarily started adhering to LEED standards, probably due to fierce competition, in which an environmental certification may help distinguish them from their competitors.

Comparative studies that examine different approaches to sustainability in campus projects in different contexts are also valuable. Campus sustainability projects may differ in terms of how general or specific their thrust is. For instance, a project may involve incremental changes in multiple areas, such as energy, building construction and design, and waste management, whereas another project may involve a prototype that deals with installing an alternate energy source. A campus project may include a broad array of stakeholders and involve the off-campus community, or stakeholder participation may be restricted to on-campus groups of teachers, students, and administrators. The level of integration of sustainability objectives with the broader institutional mission and goals will vary across campus projects.

See Also: Australian National University; Harvard University; Leadership in Sustainability Education.

Further Readings

Elder, J. L. "Higher Education and the Clean Energy, Green Economy." *Educause Review,* 44/6 (2009).
International Sustainable Campus Network. http://www.international-sustainable-campus -network.org (Accessed June 2010).
Myers, N. J. and C. Roffensperger. *Precautionary Tools for Reshaping Environmental Policy.* Cambridge, MA: MIT Press, 2006.

Neeraj Vedwan
Montclair State University

KYOTO DECLARATION ON SUSTAINABLE DEVELOPMENT

The Kyoto Declaration on Sustainable Development is a formal agreement made by the International Association of Universities (IAU) at its Ninth Round Table, held in Tokyo, Japan, on November 19, 1993. The goal of the commitment was to stress and identify a clear baseline for universities to pursue concrete actions in matters of sustainable development policy. Based on the precedent dictates of the association on the matter of natural conservation, this declaration originates from the so-called Talloires meeting, but embodies the language and the substance of both the Halifax Declaration and the Swansea Declaration, which tried to pursue a common vision on such a pivotal and urgent topic.

The foundation of this body of commitments came from the evidence, clear to many university leaders, of a complete absence of significant contribution by part of the academy to the United Nations Commission on Environmental and Development Conference in Rio de Janeiro. Conscious of the precious role that the university should have to interpret through its unique contribution in keeping with the mission of advancing learning, research, and teaching, the representatives of the IAU decided to put a major effort into understanding and promoting sustainable development. Four key elements have guided the representatives in this decision: the serious and pressing nature of the problems linked to environmental overexploitation, the interdisciplinary scope of potential solutions, the international scale of their impact for the civic and intellectual communities, and finally, the ethical imperatives of self-knowledge, self-discipline, moderation, fairness, and justice for all.

More precisely, the Kyoto Declaration aims to urge universities worldwide to seek, to establish, and to disseminate a clearer understanding of what sustainable development is. According to this agreement, sustainable development is intended to be a development that meets the needs of the present without compromising development principles and practices at the local, national, and global level in the future. Starting from this definition, the IAU established precise interventions to implement a sustainable development strategy of collaborative teaching and research, as well as good practice, which should have propelled universities out of their passive role into an energetic and influential central role for the elaboration of the major social orientations. A special task force was also formed to provide working elements for strategic solutions.

The Kyoto Declaration encouraged utilization of the resources of the university to foster a better understanding on the part of governments and the public at large of the inter-related physical, biological, and social dangers facing the planet Earth as well as recognition of the significant interdependence and international dimensions of the sustainable developments themselves. Further, the declaration aimed to emphasize the ethical obligation of the present generations to overcome those practices of resource utilization and those widespread disparities of income and labor possibilities that lie at the root of social degradation, intended as direct causes for the environmental unsustainability communities are forced to cope with. For this reason, the agreement strove to enhance the capacity of the university network to teach and undertake research and actions that could be easily activated in the social reality of turning sustainable development principles into practices, besides having the objectives of increasing environmental literacy and enhancing the understanding of environmental ethics among experts and the public at large.

For this reason, the International Association of Universities committed itself to making scientific institutions and academies of sciences cooperate with one another and with all segments of civil society in the pursuit of practical and policy measures to achieve sustainability of natural and social spaces, thereby safeguarding the needs and the interests of future generations.

The Kyoto Declaration wanted to encourage single universities to review their own choices and practical operations to reflect actions more coherently oriented to sustainable developments practices. Further, a specific purpose of the 1993 agreement was to request that the IAU board consider and implement ways to give life to its actuation in the mission of each of its members through the common enterprise of the IAU itself.

Guidelines of the Declaration

More precisely, in adopting the Kyoto Declaration, representatives of different universities underlined the following points:

- Sustainable development must not be interpreted in a manner that could jeopardize the natural path of growth for certain communities, thus blocking their legitimate aspiration to raise their standard of living.
- Sustainable development must take into account existing diversities and disparities in consumption and distribution patterns, with unsustainable overconsumption in some parts of the world contrasting with dramatic states of deprivation in others.
- Global sustainable development implies changes of existing value systems, a task in which the educational system (and in particular, the university level) has an essential role, due to the inner mission of the academic institution. To create the international consciousness and the global sense of responsibility and solidarity that sustainable development requires, a solid cognitive and normative base has to be built within local and global communities. For this reason, the delegate subscriber of the Kyoto Declaration underlined that university cooperation is aimed to further the process of making aware public opinion on the matter of natural conservation, further assuring that universities from countries with insufficient proper resources may play an active role in this process.

Practical Recommendations for Each University

Drawing from these general objectives, it was recommended that each university turn into practice the consciousness requested by the Kyoto Declaration by the following:

- Making an institutional commitment to the principle and practice of sustainable development within the academic milieu and to communicate that commitment to students, employees, and to the broader civic society

- Promoting sustainable consumption practices in its own operations and daily organization
- Developing the capacities of its academic staff to teach environmental literacy
- Encouraging among both staff and students the sharing of an environmental perspective, regardless of particular field of study and carrier
- Utilizing its own intellectual resources to build strong environmental educational programs
- Encouraging interdisciplinary and collaborative research programs related to sustainable development, intended as being the central mission
- Overcoming traditional barriers between disciplines and departments in the name of natural resources preservation and conservation
- Emphasizing every ethical obligation of the immediate university community (current students, faculty, and staff), to understand and defeat the forces that lead to environmental degradation, north–south disparities, and the intergenerational inequities
- Working on social identities and governmental incentives that could support the single university in this ambitious project
- Promoting interdisciplinary networks of environmental experts at the local, national, and international level in order to disseminate knowledge and consciousness and to collaborate on joint projects in both research aspects and educational matters
- Promoting the mobility of staff and students as an essential stage for allowing free trade of knowledge and the development of free networks of data exchange
- Forging partnerships with other sectors of the civic society in transferring innovative and appropriate technologies that can benefit and enhance sustainable development practices

The Kyoto Declaration on Sustainable Development was a fundamental step toward an effective policy fostering awareness and urgency in matters of natural conservation. In particular, with its goal of voicing the need for global action, it remains a key body of rules to help regulate efforts and forces within each university and provides broad and coordinated courses of action for scientific programs, research committees, and the ethical obligations of the educational system

See Also: Environmental Education Debate; Halifax Declaration; Swansea Declaration; United Nations Conference on Environment and Development 1992 (Earth Summit).

Further Readings

Corcoran, Peter Blaze and Arjen E. J. Wals. *Higher Education and the Challenge of Sustainability. Problematics, Promise and Practice.* Dordrecht, Netherlands: Springer, 2004.
International Association of Universities. http://www.iau-aiu.net (Accessed March 2010).
Wright, Tarah S. A. "Definitions and Frameworks for Environmental Sustainability in Higher Education." *International Journal of Sustainability in Higher Education,* 3 (2002).

Beatrice Marelli
University of Milan

L

LEADERSHIP IN SUSTAINABILITY EDUCATION

The primatologist Jane Goodall, pictured here, has had a great impact on sustainability education through the Jane Goodall Institute for Wildlife Research, Education and Conservation, and through the educational program Roots & Shoots, which has reached as many as 100 countries.

Source: Wikipedia

Sustainability education takes many forms, reaching pre-K–12 and university students, as well as life-long learners, through formal and informal educational networks. The basic ideas of sustainability education can be traced to a variety of movements. For example, the Environmental Education movement that arose in the 1960s and 1970s followed outdoor education, nature study, and conservation education. Also in the 1960s, international conservation and development organizations, such as the United Nations Educational, Scientific and Cultural Organization (UNESCO) and then-new World Conservation Union (IUCN), began developing curricula that informed or at least complemented rising public education emphases on ecosystem analyses and species preservation. The 1970s saw three major declarations regarding environmental education: the Stockholm Declaration, the Belgrade Charter, and the Tbilisi Declaration. More recently, in December 2002, the UN General Assembly passed Resolution 59/237, which declared the years 2005–2014 as the UN Decade of Education for Sustainable Development (DESD), with UNESCO as the lead coordinating agency. Through the resolution, UN member states committed to the International Implementation Scheme for the DESD and its four objectives: "to facilitate networking and exchange among stakeholders in ESD; to foster increased quality of teaching and learning in ESD; to help countries attain the Millennium Development Goals through ESD efforts; and to provide countries with new opportunities to incorporate ESD into education reform efforts." In sum, major international bodies have repeatedly deemed sustainability education key to individual and social well-being.

Whether intentionally aligned with sustainability education, education for sustainable development, environmental education, or another related title or movement, there are many talented people and productive organizations engaged in sustainability and environmental education, and they practice certain key principles in their leadership. These principles, as summarized by Janet Moore, include the following:

- Incorporating sustainability in all decision-making
- Collaboration
- Transforming education to become transdisciplinary
- Focusing on both personal and social sustainability
- Integrating planning, decision making, and evaluation
- Integrating research, service, and teaching
- Allowing for, and encouraging, pedagogical transformation

The following elaborates on these principles and the many ways in which they overlap.

Sustainability in Decision Making

Environmental leaders integrate sustainability into every decision across the institution, including curriculum, finances, facilities, and student life. Sustainability becomes the vision of the institution, with the campus serving as a learning lab.

One way to regularly accomplish this integration is through the promotion of ecological literacy by studying ideas about environmental sustainability and then helping to transform them, to concrete examples on campus, such as composting bins in dining or hybrid buses on campus. Sustainability leaders actively seek to learn best practices and view it as their responsibility to model those, in and out of the classroom. Decision makers have a responsibility to be aware of campus energy, water, and materials consumption. One way this occurs is through campus building projects, which can demonstrate commitment to ecological literacy and ecological design. The Adam Joseph Lewis Center (AJLC) for Environmental Studies is one such integrated building–landscape system. It houses the Environmental Studies Program while serving as a core part of the curriculum at Oberlin College. Environmental literacy projects also often integrate many strands of sustainability, including urban revitalization, sustainable agriculture and forestry, green development and jobs, advanced energy technologies, and, of course, education.

Sustainability leaders also share similar characteristics and duties as managers, whether overseeing a campus-wide environmental management program to engage campus stakeholders or creating an office dedicated to greening the campus. As change agents, sustainability professionals often work across financial, human resource, and organizational lines to help all campus stakeholders more fully embrace sustainability on campus. On any given campus, sustainability professionals are asked to manage a diverse range of projects, including LEED-certified building projects, local and organic food programs, on-site renewable energy projects, all with the goal of reducing greenhouse gas emissions.

Collaboration

Those considered leaders in environmental education excel at bringing together diverse groups to confront environmental issues, in the classroom and beyond. This does not just take place within the university, but between institutions, within the community, and in the sustainability community at large.

As champions of collaboration, sustainability professionals must bring together faculty, staff, and students to advance sustainability in the development of curriculum, institutional policies, and community outreach—all focused on institutional sustainability initiatives such as biodiversity, climate, culture, and food. Off-campus, they regularly collaborate with a number of local, regional, and national organizations and associations related to sustainability, often in leadership roles.

Transdisciplinary Education

Environmental education has often been confined to science courses or one-off classes within specific disciplines. Transdisciplinary education aims to create a holistic approach to environmental education and provide students with a full understanding of sustainability across, between, and beyond disciplinary boundaries. This requires greater program flexibility for undergraduate students, often including rethinking and redesigning disciplinary programs.

Leaders in sustainability education emphasize sustainability across the curriculum—a mission that directs coursework, research, and outreach toward solving today's most important problems. Institutions of higher education can accomplish this through the creation of research centers under a central administration to concentrate on research and teaching to better understand the Earth, and to increase sustainability efforts worldwide. These research centers bring together multidisciplinary teams of faculty, staff, students, and research fellows to focus on sustainable development in areas such as water, climate, energy, urban planning, health, nutrition, and poverty. In addition, sustainability research centers maintain partnerships with academia, corporations, nonprofits, and individuals, as well as governmental, multilateral and private institutions to develop and implement effective and sustainable solutions for the world's challenges.

One standout example, the School of Sustainability at Arizona State University, places special emphasis on urban environments in research and education, particularly focusing on the greater Phoenix area as an urban laboratory where solutions to water, energy, transportation, and livability are largely applicable to other rapidly urbanizing areas around the world. As an interdisciplinary academic unit, the School of Sustainability also offers all levels of degrees concentrating in sustainability, including the nation's first Ph.D. in sustainability, as well as assists other disciplines in creating concentrations in sustainability (e.g., business, engineering, and law). This example of collaborative research brings faculty, researchers, and students together to concentrate on finding real-world solutions to environmental concerns and works closely with other institutions of higher education, schools, communities, nonprofits, and corporations.

Social and Personal Sustainability

Understanding one's environmental footprint—how lifestyle impacts environment—is a vital concept in environmental education. Leaders have to help move organizational missions and/or curricula beyond personal understanding of the environment to encompass how society both influences and can address "the environment." This is often demonstrated not only in the classroom, but in providing opportunities within the local community for green building, cleaner transportation, and more sustainable dining choices.

For leadership in sustainability education, these ideals can be linked to the term "natural capitalism," first coined by Paul Hawken, Amory Lovins, and L. Hunter Lovins. Natural capital refers to natural resources and ecosystem services that make any economic activity—and indeed all life—possible. As described in their book, *Natural Capitalism: Creating the Next Industrial Revolution* (1999), natural capitalism promotes four central strategies for social and personal sustainability: "radically increasing resource productivity; redesigning industry on biological models with closed loops and zero waste; shifting from the sale of goods (for example, light bulbs) to the provision of services (illumination); and reinvesting in the natural capital that is the basis of future prosperity."

Leaders in sustainability education are often asked to connect the private sector to the educational world. This can occur in a variety of ways, such as through conducting research on socially responsible investing, social entrepreneurship and environmental funding, and natural resource management. These opportunities can become excellent learning experiences for students as well. Often sustainability professionals and their collaborators are asked to write on or present their accomplishments to educate senior decision makers in business, government, and civil society.

Integrating Planning, Decision Making, and Evaluation

As mentioned earlier, decision making is an important aspect of being a leader in environmental education. Since these leaders often influence institution-wide decisions, they often integrate sustainability in the planning process and in evaluation opportunities. For example, many colleges and universities have master building plans and strategic plans that specifically address sustainability within each goal. In addition, most institutions have regular schedules for evaluating their various units and processes; these become opportunities to address sustainability within each unit.

It should be noted that sustainability leaders may come from all institutional levels. These leaders must practice such integration and work to strengthen their institution in the local community, nationally, and abroad. On campus, this is often focused within offices of sustainability, which oversee campus initiatives, such as developing strategies for meeting greenhouse gas reduction targets. Through these offices, institutions can begin to address sustainability in on-campus buildings and construction projects through the protection of land, water, and biodiversity, and through energy initiatives.

Integrating Research, Service, and Teaching

The core duties of educators typically fall under the categories of research, service, and teaching. Often these categories are treated separately by educators, especially when evaluating performance for promotion and tenure. The broad concept and applicability of sustainability allows educators to integrate their duties while addressing environmental concerns, and encouraging this is a major responsibility of a leader in environmental education.

One organization that can assist institutions is Second Nature, whose mission is to work with college and university leaders in putting sustainability at the center of all aspects of higher education. Since 1993, Second Nature has brought this mission to over 500 institutions, reaching over 4,000 faculty members. Leadership of Second Nature has also helped to launch the American College and University Presidents Climate Commitment, with 676 signatories by late 2010. Second Nature also supports the Association for the Advancement of Sustainability in Higher Education (AASHE), an association of colleges and universities working to create a sustainable future. AASHE provides resources, professional development, and a network of support for higher education institutions to model and advance sustainability in governance, operations, education and research.

Pedagogical Transformation

Leaders in environmental education work to develop models of instruction that promote, enhance, and reward experiential learning, community-service learning, participatory group learning, critical and self-reflective thinking, and problem-based learning—all while considering the concept of sustainability. UNESCO has hosted Summits on Sustainability to pursue pedagogical transformation. An important result of these summits is Chapter 36 of Agenda 21, the Rio Earth Summit Action Plan on Education, Public Awareness and Training. It lays out four major thrusts: (1) improving the quality of and access to basic education, (2) reorienting existing education to address sustainable development, (3) developing public understanding and awareness, and (4) training. UNESCO leadership recognizes that there is not adequate funding to retrain all existing 60 million teachers worldwide to meet these standards of ESD. Thus, the UNESCO committee for ESD works directly with institutions and ministries that direct teacher education to develop curricula for new teachers and those who undergo recertification processes to achieve competencies in ESD. Their broad, international network links to many of the other kinds of organizations and even some of the specific institutions noted above.

For leaders in sustainability education to succeed, they must make sustainability education relevant to learners, teachers, and their social and institutional settings. With the variety, depth, and breadth of resources now available, there are ample opportunities to do so.

See Also: Agenda 21: "Promoting Education, Public Awareness, and Training" (Chapter 36); Arizona State University; Green Community-Based Learning; Oberlin College.

Further Readings

Arizona State University, "Global Institute of Sustainability." http://sustainability.asu.edu (Accessed December 2009).

Moore, J. "Barriers and Pathways to Creating Sustainability Education Programs: Policy, Rhetoric and Reality." *Environmental Education Research*, 11 (2005).

Moore, J. "Seven Recommendations for Creating Sustainability Education at the University Level: A Guide for Change Agents." *International Journal of Higher Education*, 6 (2005).

University of New Hampshire. "University Office of Sustainability." http://www.sustainable unh.unh.edu (Accessed July 2010).

Justin Miller
Ball State University

Jennifer Coffman
James Madison University

LEEDS UNIVERSITY

Leeds University's School of Earth and Environment, founded in 2004, has 80 faculty members and includes the Institute for Climate and Atmospheric Science, the Earth Surface Science Institute, the Institute of Geophysics and Tectonics, and the Sustainability Research Institute.

Source: Wikipedia

The University of Leeds, in West Yorkshire, England, is one of the largest universities in the United Kingdom (UK), with over 30,000 full-time students. A highly rated research institution, it is ranked in the top 10 British universities by research funding and is one of the original six "red brick universities" established at the turn of the 20th century for the study of engineering and science. In particular, the university is considered among the top research institutions in the world in the areas of electronic and electrical engineering, mechanical engineering, food science, and transportation planning, and is the leading British research university in chemical engineering. The university offers 700 different undergraduate programs and 470 graduate programs, including specialty areas like fire science, color chemistry, nanotechnology, and aviation technology (a program that includes flight training). A great amount of money flows through Leeds's research programs, with the school investing considerable capital in exploring the commercial potential of its research developments; more private sector capital is spent on joint ventures with Leeds than at any other British university. Alumni of Leeds include Nobel Prize–winning chemist George Porter, who devoted much of his work to the advocacy of hydrogen energy.

The School of Earth and Environment at Leeds (SEE) was formed in 2004 when the School of Earth Sciences merged with the School of Environment. The faculty of 80—with an additional 50 support staff members and 100 postgraduate students—comprises one of the largest schools of any British university. Since its inception, SEE has emphasized multidisciplinary approaches to the study of the environment "from core of the Earth to its atmosphere," including the social and economic dimensions of sustainability. SEE has strong links with industry—particularly in the fields of civil engineering and hydrocarbon exploration—and access to considerable laboratory resources, but is known for the informal sense of community among its faculty and students.

SEE's research is divided among four institutes operated by the school: the Institute for Climate and Atmospheric Science, the Earth Surface Science Institute, the Institute of Geophysics and Tectonics, and the Sustainability Research Institute (SRI). The SRI is home to about 30 researchers, working in sustainable development and environmental change, environmental policy and governance, ecological economics, the environment and corporate responsibility, and sustainable production and consumption. SRI research is conducted across one of two themes. The Adapting to Environmental Change Research

Theme studies climate change, the factors determining its impacts, and the differences in climate change vulnerability across different groups; the interactions between natural and socioeconomic systems, and the development of integrated models thereof within ecological economics; and the role of environmental management in responding to environmental change. Organizations collaborating with this research theme at the SRI include the Intergovernmental Panel on Climate Change; the UN Human Development Program; the UN Environment Program; the UN Development Program; the World Meteorological Organization; the Food and Agriculture Organization; the UK Department for International Development; the UK Department for Environment, Food, and Rural Affairs; and the UK Climate Impacts Program. Research projects have included Adapting to Environmental Change, Participatory Rangeland Monitoring and Management in the Kalahari, Transforming Ghana's Land Policy for Sustainable Development, Determining the Socio-Economic Implications of Different Land Management Policies in Yorkshire Water's Catchments, Desertification Mitigation and Remediation of Land, Sustainable Uplands, and Integrating Economic and Land Use Models to Anticipate Rural Vulnerability to Climate Change.

The Governance and Sustainability Research Theme studies environmental governance and the nature and influence of the various institutions that impact the sustainability of social and economic behaviors, the planning and formulation of environmental policy and private and voluntary environmental governance, and the environmental impacts of new technologies. Institutions collaborating with this research theme include the UN Environment Program; the UK Office of the Deputy Prime Minister; the UK House of Lords Science and Technology Select Committee; the UK Department for Innovation, Universities, and Skills; the UK Department of International Development; the UK Department for Environment, Food, and Rural Affairs; the UK Academy for Sustainable Communities; the International Standards Organization; the British Standards Institute; and the Friends of the Earth. Research projects have included The Environmental Effect of Car-Free Housing, Multi-Level Governance of Natural Resources, Evaluating Variations in Corporate Environmental Performance, Sustainability as a Vehicle for Competitive Advantage, The Governance Implications of Private Standards Initiatives, Trade-Offs in Decision-Making for Sustainable Technologies, Transitions to a Low-Carbon Economy, and Exploring the Conditions for Innovation and Uptake of New Energy Technologies.

A good deal of sustainability research is conducted outside the SRI. The multidisciplinary "water@leeds" center is the largest university-based water research center in the UK, connecting water research students and faculty from disparate disciplines, while offering peer review of work, the benefits of interdisciplinary networking, and assistance in establishing collaborations. Recent research conducted by water@leeds members includes Methane Production From Peatlands, Optimizing Carbon Storage in Yorkshire Water Peat Catchments, Analysis of Networks and Institutions of Acts in Wetland Management: A Study of Niger Delta Wetland, and Sustainable Flood Risk Assessment and Management. Yorkshire Water is a major private sector sponsor of water@leeds research; public sector funding comes from both British and international sources.

Other significant contributors to sustainability and green research at Leeds include the Center for Integrated Petroleum Engineering and Geoscience, which devotes much of its research to carbon capture and storage and sustainability and is involved with the UK Carbon Capture and Storage Consortium. The Center for Climate Change Economics and Policy, funded by the UK Economic and Social Research Council and private company Munich Re, includes researchers from both Leeds and the London School of Economics and Political Science. Research at the center falls under five interconnected research programs: developing climate science and economics; climate change governance; adaptation to climate change and human development; governments, markets, and climate change mitigation; and the Munich Re Program, which evaluates the economics of climate risks and opportunities in the insurance sector.

The Leeds campus has become increasingly greener in recent years. Recycling doubled at the university when recycling bins were installed alongside all trash bins, and in 2008, students voted to ban bottled water from all on-campus vending sources.

See Also: AASHE: STARS (Sustainability Tracking, Assessment & Rating System); Sustainability Officers; United Nations Decade of Education for Sustainable Development 2005–2014.

Further Readings

Filho, Walter Leal. *Sustainability and University Life*. Ann Arbor: University of Michigan Press, 1999.

Gough, Stephen and William Scott. *Higher Education and Sustainable Development*. New York: Routledge, 2009.

Hargreaves, Andy and Dean Fink. *Sustainable Leadership*. Hoboken, NJ: Jossey-Bass, 2005.

Sandell, Klas, Johan Ohman, and Leif Ostman. *Education for Sustainable Development*. New York: Studentlitteratur AB, 2005.

Silka, Linda and Robert Forrant, eds. *Inside and Out: Universities and Education for Sustainable Development*. Amityville, NY: Baywood Publishing, 2006.

Wiland, Harry and Dale Bell. *Going to Green: A Standards-Based Environmental Education Curriculum for Schools, Colleges, and Communities*. White River Junction, VT: Chelsea Green Publishing, 2009.

Bill Kte'pi
Independent Scholar

LUND UNIVERSITY (SWEDEN)

Lund University is located in Lund, a small city of approximately 100,000 residents in southern Sweden. Lund was founded in 1666 and has become an international center for research and education, including in the field of sustainability science. Lund enrolls approximately 41,000 students annually, many of them international, making it one of Sweden's largest universities. Lund's key academic program is the International Master's Program in Environmental Studies and Sustainability Science. Lund also offers a master's degree in Environmental Sciences, Policy, and Management. In 2008, Lund offered an interdisciplinary Ph.D. program in Sustainability Science through the Lund University Centre of Excellence for Integration of Social and Natural Dimensions of Sustainability. The Lund University Centre for Sustainability Studies is an internationally recognized education and research center.

The LUMES Program

Lund University is a leader in international sustainability education for its International Master's Program in Environmental Studies and Sustainability Science (LUMES). The LUMES program has hosted over 300 students from 75 countries around the world since its inception in 1997. Admission to the LUMES program requires a bachelor's or higher degree in a relevant discipline such as natural and social sciences, law, economics, engineering, agronomy, or medicine. Students must also demonstrate proficiency in English because the program is conducted entirely in that language. Undergraduate programs in environmental science are available at Lund, but are not taught entirely in English. Approximately 36 students are admitted into the two-year program annually.

LUMES courses are measured according to the Swedish credit system, and graduates receive the Swedish equivalent of a master's degree, known as a Magisterexamen. The overall goal of the LUMES program is to prepare students to contribute to the long-term sustainable development of local and global natural, societal, and human resources through a critical- and system-thinking approach. The two main themes integrated throughout the program are system thinking and effective academic communication.

Faculty from a variety of disciplines work together in teams to create interdisciplinary coursework centered on the program's two main themes. The program's first term consists entirely of mandatory coursework that introduces students to issues such as current environmental problems, societal structures, sustainable development, causal relationships in nature-human systems, and cultural differences and varied worldviews. Required core courses include environmental problem awareness; ideas behind economy, environment, and society; environmental governance; and sustainability science.

During the program's second and third terms, students learn the methodology of systems analysis, the use of software tools, and different research methods and research design as well as how to apply this knowledge to real-life situations. Required core courses center on methodology, methods, and tools; environment and health; and development and sustainability. Optional courses include energy and sustainability, transport and sustainability, industry and sustainability, water and sustainability, urban systems and sustainability, and rural systems and sustainability. All students must complete the final core course, Making Change Happen. In the fourth and final term, students produce an interdisciplinary thesis.

Other Academic Degree Programs

Lund University offers a master's degree in Environmental Sciences, Policy, and Management (MESPOM) as part of the Erasmus Mundus program Geo-Information Science and Earth Observation for Environmental Modeling and Management (GEM), approved by the European Commission. Lund University is a leading teaching partner in the GEM program, working with a consortium of international universities to provide coursework and thesis research opportunities. Other key participating institutions include the Universities of Manchester and Southampton in the United Kingdom, the Central European University in Hungary, the University of the Aegean in Greece, and the International Institute for Geo-Information Science and Earth Observation ITC in the Netherlands. MESPOM graduates are awarded a master of science (M.Sc.) degree.

The Lund University Centre for Sustainability Studies (LUCSUS) hosts the Erasmus Mundus GEM program. The interdisciplinary environmental studies curriculum trains students to identify, develop, and implement integrated solutions to environmental challenges in an international context. Academic coursework is supplemented by training in research and communication skills needed to work effectively in a professional environment. Scholarships are available for students and scholars outside the European Union (EU) in order to broaden the school's international mission.

Lund University offers an innovative interdisciplinary Ph.D. program in Sustainability Science. LUCSUS hosts the sustainability science research school, known as the Lund University Centre of Excellence for Integration of Social and Natural Dimensions of Sustainability (LUCID). LUCID is a Linnaeus Grant program sponsored by the Swedish Research Council for 2008–2018.

Ph.D. candidates as well as postdoctoral students work in interdisciplinary groups under professional scientific leadership. Ph.D. students examine various research methods, techniques, and tools through their coursework, which they then apply to their research. Scientific theory and methodology courses explore and attempt to bridge the gap between the different theories and methodologies that exist across the natural and social science disciplines. Studied theories include ecological economics, resilience theory, cultural theory, transition theory, and world systems theory. Ph.D.-level courses have included Methods in Sustainability Science, Introduction to Sustainability Science, Modeling Human-Induced Change in Aquatic Ecosystems, and System Analysis, Risk, Uncertainty, and Precaution. Academic skills such as writing, publishing, and rhetoric are also integrated throughout the program.

Education and Research Centers

LUCSUS, formerly known as the Centre for Environmental Studies, is an internationally recognized education and research center. LUCSUS was established in 2000 to coordinate Lund's sustainability education, research, and cooperative activities both within and outside the campus. This interdisciplinary center hosts academic programs such as LUMES and LUCID as well as sustainability research programs and workshops. LUCSUS's educational and research agendas include local, regional, and global issues, events, and relationships related to sustainability.

LUCSUS initiates and coordinates all university courses related to environmental studies and sustainable development. The overarching methodological approach is system analysis of human-nature system interactions in the context of sustainability. The LUCSUS research program centers on the role of science in the transition toward sustainability and the necessity of new approaches that can bridge the divides between social and natural sciences. Key sustainability issues under study include climate change, global health, loss of biodiversity, the global water crisis, and land use change. Both quantitative and qualitative perspectives are studied.

LUCSUS serves as the European Campus in the Right Livelihood College (RLC) network. The Swedish Parliament annually presents the Right Livelihood Award Foundation award for personal courage and social transformation to people who offer workable solutions to the most urgent challenges of our time. RLC was founded to establish a global network of organizations and institutions that provide Right Livelihood Award winners a forum for sharing their ideas. LUCSUS also serves as an Earth System Governance (ESG) Project research center to support the implementation of the ESG science plan. The ESG science plan is based on five analytical problems—Architecture, Agency, Adaptiveness, Accountability, and Allocation and Access—with the goal of preventing, mitigating, and adapting to global and local environmental change and Earth system transformation within the context of sustainable development. Lund University is also a member of the Öresund University, a regional confederation of universities in Denmark and southern Sweden.

See Also: Leadership in Sustainability Education; Sustainability Officers; Whole-School Approaches to Sustainability.

Further Readings

Barlett, Peggy F. and Geoffrey W. Chase. *Sustainability on Campus: Stories and Strategies for Change*. Cambridge, MA: MIT Press, 2004.

Lund University International Master's Programme in Environmental Studies and Sustainability Science (LUMES). http://www.lumes.lu.se (Accessed June 2010).

M'Gonigle, R. Michael and Justine Starke. *Planet U: Sustaining the World, Reinventing the University*. Gabriola Island, British Columbia, Canada: New Society Publishers, 2006.

Rappaport, Ann and Sarah Hammond Creighton. *Degrees That Matter: Climate Change and the University*. Cambridge, MA: MIT Press, 2007.

Marcella Bush Trevino
Barry University

LÜNEBURG DECLARATION

The Lüneburg Declaration on Higher Education for Sustainable Development (the Lüneburg Declaration) is a policy initiative that sought to increase the role of institutions of higher education in the move to more sustainable development. Specifically, the Lüneburg Declaration strove to mobilize institutions of higher education, nongovernmental organizations (NGOs), and other stakeholders to obtain certain goals. These goals included revising curricular materials to reflect current scientific understandings, reorienting teacher education to support sustainable development, providing continuing education to teachers, having colleges and universities embrace sustainability as a goal, advocating for more environmentally friendly practices and policies, and networking with other stakeholders to ensure that these changes took place. In addition to these goals, the Lüneburg Declaration sought to provide opportunities for institutions of higher education to collaborate toward sustainable initiatives and to establish a system that could serve as a clearinghouse for tools, initiatives, and other innovations developed by signatories.

Background of the Lüneburg Declaration

The Lüneburg Declaration was adopted on October 10, 2001, at the University of Lüneburg, located in Lüneburg, Germany. The Lüneburg Declaration resulted from the International Copernicus conference *Higher Education for Sustainability—Towards the World Summit on Sustainable Development (Rio+10)*. The conference was jointly organized by the University of Lüneburg and the Copernicus Programme of the European University Association (EUA), an inter-university cooperative project that was endorsed by more than 200 European universities. The Lüneburg Declaration was instigated by the Global Higher Education for Sustainability Partnership (GHESP) formed by Copernicus, the International Association of Universities (IAU), the Association of University Leaders for a Sustainable Future (ULSF) and the United Nations Educational, Scientific and Cultural Organization (UNESCO). The conference sought to coordinate efforts to increase collaboration for and encourage research into sustainable practices and environmental research.

Experts in the field in attendance at the conference, as well as GHESP affiliate agencies, proposed that the propositions and findings of numerous forums, conferences, and commissions on sustainable development be taken into consideration as the basis for any declaration issued. A variety of other resolutions were considered, including the following:

- United Nations Conference on Environment and Development (UNCED): Chapter 36 of Agenda 21 (1992)
- The International Work Programme on Education, Public Awareness and Training for Sustainability adopted by the United Nations Commission on Sustainable Development (1996)
- International Conference on Environment and Society (Thessaloniki, 1997)
- World Conference on Higher Education (Paris, 1998)
- World Conference on Science (Budapest, 1999)
- World Education Forum (Education for All) (Dakar, 2002)

The goal of the conference attendees was to craft from these a declaration that would provide guidance and direction to institutions of higher education.

Goals of the Lüneburg Declaration

The Lüneburg Declaration maintained that education plays an essential role in addressing the challenges of implementing sustainable development and seeks to increase the involvement of institutions of higher education in shaping and nurturing this implementation. Through collaboration and cooperation, the Lüneburg Declaration sought to build a global culture of learning, through which the interrelated issues of globalization, poverty relief, social justice, democracy, human rights, peace, and environmental protection could be confronted with comprehensive consociation, where groups with differing perspectives could put aside ideological disputes in pursuit of the common goal of increased sustainability. The principal objective of education for sustainable development, as envisioned by the Lüneburg Declaration, was the promulgation of the knowledge, values, dispositions, and proficiency requisite to enable individuals and organizations to facilitate necessary, sustainable social change. Excellence in sustainability education necessitates access to current technological and scientific knowledge. As the base of technological and scientific knowledge is constantly in flux, institutions of higher education must engage in a perpetual analysis and revision of learning modules and materials. Only through this process can colleges and universities ensure that information presented in their classes is accurate and pertinent to the conversation regarding sustainability.

The Lüneburg Declaration also called for greater consolidation of and cooperation between the social networks of the three aggregate academic institutions that make up the GHESP. The GHESP includes more than 1,000 colleges and universities that pledged, as signatories of various charters and declarations including the Lüneburg Declaration, to strengthen the promotion and advancement of sustainable development. Increased attention to and advocacy for sustainable development and environmentally friendly practices and policies was hoped to increase attainment of these goals. The Lüneburg Declaration advocates for institutions of higher education, NGOs, and others with an interest in the advancement of sustainable development to take actions that ensure that the curriculum presented to students is as green as possible, as are management practices that promote social and economic stability. Furthermore, the Lüneburg Declaration encouraged all institutions of education to include in their activities a strong component of investigation into the values and norms underlying the promotion of sustainable development. Real change is required across a given university to ensure that students, faculty, and staff have the requisite attitudes and skills to allow sustainability strategies to lead to real outcomes.

To encourage real change to take place, the Lüneburg Declaration sought to heighten awareness and understanding of the importance and relevance of technological and risk-based assessment in the advancement of sustainable development. Also, as a means of obtaining this goal, the declaration sought to promote creative, all-encompassing sustainability projects in higher education, and to extend these projects to all levels of education. Also proposed was increasing the attention paid to the international nature of sustainable development, as well as expanding its focus to provide the greater provision for global intercultural exchange in the learning environment. In order to enable this, the declaration alleged that intensified institutional networking, as well as stronger integration of training and research, is necessary within educational institutions at all levels.

Next Steps

The Lüneburg Declaration called upon governments and their agencies to work to ensure that the World Summit on Sustainable Development recognize the importance of education in general, and specifically higher education, in its future affairs. To that end, the United Nations was asked to highlight in the Secretary-General's main policy report the indispensable role that education, and specifically higher education, plays in achieving sustainable development. Also, it was the recommendation of the Lüneburg Declaration that education be made a discussion topic during the multi-stakeholder dialogue sessions held during the preparatory committee meetings for the Johannesburg Summit and during that summit itself. To this end, the Lüneburg Declaration called upon UNESCO, as the appropriate United Nations agency, to support promotional efforts regarding the Johannesburg Summit.

To further the Lüneburg Declaration's objectives, EUA-Copernicus, the IAU, and the ULSF committed themselves to a five-year plan focusing on several concrete steps that would assist in implementation of these aims. These steps included establishing a global learning environment; promoting expanded endorsement of the Talloires, Kyoto, and Copernicus declarations; and the development of a proactive and utilitarian body of instructional materials, including strategies for implementation and reform, a comprehensive listing of available resources and case studies, and an inventory of best practices. Finally, the Lüneburg Declaration recommended that an effective network of regional centers of excellence operating in both developed and developing countries be enhanced, and that the objectives of implemented sustainable development within the realm of education be realized.

Conclusion

The Lüneburg Declaration further emphasized a desire for sustainability in institutions of higher education. Its emphasis upon both theoretical and practical ways to support sustainability provided a needed outlet to those who desired more active roles for colleges and universities in advocating for environmentally friendly practices. Its stress on the value of continuously reviewing and updating the scientific body of knowledge upon which sustainable practices are based satisfied those who were suspicious of the green movement. The Lüneburg Declaration's emphasis upon cooperation and collaboration has ensured its continued significance and consequence.

See Also: AASHE: STARS (Sustainability Tracking, Assessment & Rating System); American College and University Presidents Climate Commitment; Halifax Declaration; Higher Education Associations Sustainability Consortium.

Further Readings

Blewitt, J. and C. Cullingford, eds. *The Sustainability Curriculum: The Challenge for Higher Education*. London: Earthscan Publications, 2004.

Jones, P., D. Selby, and S. Sterling. *Sustainability Education: Perspectives and Practice Across Higher Education*. London: Earthscan Publications, 2010.

United Nations Educational, Scientific and Cultural Organization (UNESCO). "Lüneburg Declaration." http://portal.unesco.org/education/en/ev.php-URL_ID=37585&URL_DO=DO_TOPIC&URL_SECTION=201.html (Accessed September 2010).

Stephen T. Schroth
Jason A. Helfer
Daniel O. Gonshorek
Knox College

M

MAASTRICHT UNIVERSITY (NETHERLANDS)

Maastricht University is a public university in Maastricht, the Netherlands, and was founded in 1976 as Rijksuniversiteit Limburg. Since 2008, the school's official name has been Maastricht University—not *Universiteit,* the Dutch word for "university"—in order to signal the school's international character. About two-fifths of the school's 13,000 students come from outside the Netherlands, and half of the undergraduate programs and most of the graduate programs can be undertaken entirely in English. A thoroughly modern university, but distinctly different from the American progressive institutions founded in the same era, Maastricht is among the leading European institutions in green curricula.

Originally founded as a medical school relying on problem-based learning as its core teaching method, Maastricht expanded to include law and economics in the 1980s, followed by psychology and arts in the 1990s and a full liberal arts college within the university (University College Maastricht) in the 21st century. Each decade has thus seen the university expand or alter its character dramatically, with internationalism as the school's guiding ambition in the second decade of the 21st century. This international focus at today's Maastricht goes hand in hand with the school's curricular focus on sustainable development, which is particularly clear at the school's International Centre for Integrated Assessment and Sustainable Development (ICIS).

ICIS was founded in 1998 as an international research center for integrated studies of complex issues, studies that incorporate analyses of causes, effects, and the complex relationships between the economic, environmental, institutional, and social and cultural processes involved in or impacting a particular issue. The center's mission statement focuses on the use of science and education and "acting on different social levels" in the promotion of "sustainability of life on Earth." ICIS's work is inherently interdisciplinary—ideally, not just incorporating multiple disciplines but transcending the boundaries between them. The director of ICIS as of 2010 is Professor Pim Martens, the university's chair of Sustainable Development, who directs a number of projects on sustainable development, environmental change and society, and the implications of globalization. The staff of 22 people includes research fellows, professors, and advanced doctoral students. ICIS's sustainability focus carries through to its "commitments for internal sustainability," the current version of which was agreed to in an October 2, 2006, session. Those commitments call for conserving energy in a number of specific ways, including switching off devices not in use, printing double-sided documents, using low-emission travel and only traveling for conferences when video conferencing is not an option, and using sustainable supplies in the center to the best ability allowed by the center's resources.

Maastricht's Sustainable Development work group (with a staff of 10) keeps sustainability issues on the university agenda and in the spotlight and identifies concrete measures the university community can implement to further its commitment to sustainability. Sustainability-minded regulations are in effect campus-wide, governing in detail the treatment of waste from confidential memos to medical waste to data storage media. Furthermore, they are built into the campus's laboratory work procedures.

UNU-MERIT is a joint venture between Maastricht University's Economic Research Institute on Innovation and Technology and United Nations University's Institute for New Technologies. UNU-MERIT researches the factors that impact technological change across social, political, and economic spheres, from a policy context that seeks to address national and international governance questions related to science, technology, and access to knowledge. The venture is another example of Maastricht's recent strong focus on internationalism and sustainability, and one of the goals of the research is to explore the inequalities of the global playing field and the hope of reducing those inequalities while continuing to drive science and technological innovation forward with low environmental costs. UNU-MERIT research is organized across five research themes: micro-based evidence research on innovation and technological change; the role of technology in growth and development; knowledge and industrial dynamics; innovation, global business strategies, and host country development; and the governance of science technology and innovation. Recent publications have dealt with the growth of businesses in Côte d'Ivoire, nanoscience and the social sciences, and the development of a global science and technology policy agenda for sustainable development.

UNU-MERIT offers a Ph.D. program in Economics and Policy Studies of Technical Change, as well as a master of science in Infonomics and a Master of Economics and Business Administration. Courses offered by UNU-MERIT, all at the advanced graduate level, include Innovation Dynamics and Industrial Change; Social Construction of Technology in Development; Environment and Sustainable Development; and Technology, Industrial Development, and Economic Growth in Developing Countries.

The School of Management at Maastricht has become internationally known for its sustainable development courses such as the environmental management course that examines policy making and environmental technology. The Sustainable Development Center (SDC) at the school is devoted to research on sustainable economic development in the private and public sectors, in Europe and elsewhere, at the local, regional, and national levels. Specifically, the SDC explores the premise that private sector companies must play a key role in sustainable development. Like so much of Maastricht's research and curriculum, the SDC is multidisciplinary; staff members include Ph.D.s, MBAs, and Masters of Science. The SDC's current focus is on introducing the Round Table concept of sustainable economic governance to more countries; the Round Table concept involves the collaboration between policy makers, international businesses, and local companies approaching economic and social development from a business perspective. In 2010, the SDC operated Round Table Africa (in conjunction with the Eastern and Southern African Management Institute and the Expert Center for Sustainable Business and Development Cooperation) and Round Table Indonesia (with Bogor Agricultural University).

Other SDC projects draw on the center's expertise in value chains and the structure of business partnerships. The Partnerships & Value Chains ongoing research project examines sustainability in international commodity chains, using the Fair Trade mango value chain in Burkina Faso and Mali, the sorghum value chain in Ghana, and the organic cotton value chain in Burkina Faso as case studies.

See Also: Green Business Education; Leadership in Sustainability Education; Sustainability Officers.

Further Readings

Filho, Walter Leal. *Sustainability and University Life.* Ann Arbor: University of Michigan Press, 1999.
Gough, Stephen and William Scott. *Higher Education and Sustainable Development.* New York: Routledge, 2009.
Hargreaves, Andy and Dean Fink. *Sustainable Leadership.* Hoboken, NJ: Jossey-Bass, 2005.
Sandell, Klas, Johan Ohman, and Leif Ostman. *Education for Sustainable Development.* New York: Studentlitteratur AB, 2005.
Silka, Linda and Robert Forrant, eds. *Inside and Out: Universities and Education for Sustainable Development.* Amityville, NY: Baywood Publishing, 2006.
Wiland, Harry and Dale Bell. *Going to Green: A Standards-Based Environmental Education Curriculum for Schools, Colleges, and Communities.* White River Junction, VT: Chelsea Green Publishing, 2009.

Bill Kte'pi
Independent Scholar

Maine Green Campus Initiative

Bowdoin College, like other Maine colleges in the Green Campus Consortium, promotes sustainability efforts on campus through such actions as buying renewable energy credits to offset carbon dioxide emissions from campus energy use.

Source: iStockphoto.com

American college and university campuses have been at the forefront of efforts to promote sustainability in response to a plethora of scientific studies dealing with the impact of global warming and climate change. In Maine, statewide commitment is evident through the Maine Green Campus Initiative, which resulted in the formation of the Green Campus Consortium, made up of sustainability coordinators and faculty from campuses across the state. The consortium's focus is concentrated on using education and communication to promote sustainability. The group meets three times each semester to discuss issues of sustainability, establish goals, address problems, and engage in proactive sustainability planning. As early as 2002, the Green Campus Consortium of Maine was involved in bringing together groups from various cities in Maine to address ways to create a greater statewide commitment to sustainability. The following year, Colby, Unity, and the College of the Atlantic founded the Maine Green Power Connection. By 2004, Unity had become fully "green powered," purchasing only clean, renewable energy that had been generated within the state.

Consortium Member Activities

The University of Maine (UMA) has provided leadership to other schools in the state on ways to use education to further the goals of sustainability. In October 2006, the institution became a signatory to the newly created American College and University Presidents Climate Commitment. The following summer, UMA adopted a comprehensive Climate Action Plan in conjunction with the 650 other American institutions of higher learning that had become extremely concerned about the impact of global warming on the environment. The UMA plan established new guidelines for campus operations, transportation, administrative policies, financing strategies, and sustainability education. At the campus in Augusta, the state capital, the student body is chiefly made up of commuters who emit large amounts of carbon dioxide into the atmosphere. Throughout the state of Maine, various campuses have joined UMA in expressing their commitment to sustainability through participation in the Green Campus Coalition of Maine (GCCM). UMA defines that commitment in its mission statement, which pledges "to develop a broad-based, student-led effort toward environmental awareness and stewardship by promoting sustainable relationships between the ecological, economic, and social systems" that make up the UMA campus and the surrounding community.

Although sustainability actions at the various institutions have much in common, each institution has contributed its own unique efforts to advancing sustainability within education. At Unity College, a LEED Platinum residence was built with a five kW photovoltaic system, solar hot water, and a cold-climate heat pump with the result that the building produces more electricity than it uses. Bowdoin College in Brunswick buys renewable energy credits to offset carbon dioxide emissions. At Colby, the decision to devote a fifth of the food budget to local suppliers, many of whom grew only organic foods, not only proved environmentally sound, it also improved the finances of 72 of Waterville's food suppliers. Bates College in Lewiston regularly holds an annual Clean Sweep when students vacating residences at the end of the year swap or sell items rather than trashing them. All proceeds from sales are donated to charity. At the University of Maine, the Green Campus Initiative sponsors monthly free rentals of donated Blue Bikes from fall until winter snows make cycling impractical. The University of New England in Biddeford has instituted a Free Bike program that goes even further, donating bikes, helmets, and locks to entering freshmen for permanent use. When the Farmington campus of the University of Maine was established in 2000, the Green Campus Coalition was the first group formed.

Outside Assessments

At the College of the Atlantic (COA), which was established in Bar Harbor in 1969, much of the curriculum is devoted to sustainability. Core areas of study include Environmental Science, Ecological Planning and Policy, Field Ecology and Conservation Biology, Sustainable Business, and Sustainable Food Systems.

While the school received excellent ratings for its efforts in creating a sustainability council, adopting a master plan to improve campus sustainability, meeting green building standards, and making a commitment to achieving carbon neutrality by the year 2040, it was determined that more needed to be done in the areas of food and recycling, student involvement, and investment priorities. The institution achieved a low rating in the field of endowment transparency because of its failure to develop policies dealing with the disclosure of endowment holdings and shareholder voting records.

Membership in the Green Campus Consortium has led the colleges of Maine to become actively green, working both together and individually to address all issues of sustainability and to produce cleaner environments for themselves and future generations.

See Also: American College and University Presidents Climate Commitment; Leadership in Sustainability Education; Sustainability Officers; Whole-School Approaches to Sustainability.

Further Readings

"Campuses Making the Connection." *Maine Green Power Newsletter,* 1 (Winter 2004).

"The Campus Green." http://www.uma.edu/campusgreen.html (Accessed June 2010).

"The College Sustainability Report Card: The University of Maine." http://www .greenreportcard.org/report-card-2010/schools/university-of-maine (Accessed June 2010).

Dernbach, John C. *Agenda for a Sustainable America.* Washington, DC: ELI Press, Environmental Law Institute, 2009.

"Forest Service Renews Partnership with College of Menominee Nation in Wisconsin." *Indian Country Today* (December 2, 2009).

Franson, Melissa. *The Impact of Classroom Exposure to Sustainability, Course Content, and Ecological Footprint Analysis of Student Attitudes and Projected Behaviors.* Unpublished Thesis. Auburn University, AL, 2008.

"Green Campus Initiative." http://www.umaine.edu/property/gci.htm (Accessed June 2010).

Hignite, Karla, et al. *The Educational Facilities Professional's Practical Guide to Reducing the Campus Carbon Footprint.* Alexandria, VA: APPA, 2009.

"The *Princeton Review*'s Guide to 286 Green Colleges." http://www.princetonreview.com/ green-guide.aspx (Accessed June 2010).

Pyles, Jesse. "Advancing Education for Sustainability." http://secondnaturebos.wordpress .com/2010/02/03/nine-maine-elements-of-a-sustainable-campus (Accessed June 2010).

Rockwood, Larry, et al., eds. *Foundations of Environmental Sustainability: The Coevolution of Science and Policy.* New York: Oxford University Press, 2008.

Elizabeth Rholetter Purdy
Independent Scholar

MICHIGAN STATE UNIVERSITY

Michigan State University (MSU), founded in 1855 as the Agricultural College of the State of Michigan, was the first U.S. institution of higher learning to teach scientific agriculture and served as the prototype for land-grant institutions established under the Morrill Act of 1862. MSU is located in East Lansing, Michigan, and has a 5,200-acre campus as well as 15,000 acres located throughout Michigan that are used for agricultural, animal, and forestry research. In fall 2009, the enrollment was 47,278, including 36,489 undergraduate and 10,789 graduate and professional students. In 2008–2009, funding for sponsored research at MSU totaled nearly $405 million, with major funders including the National Science Foundation, U.S. Department of Health and Human Services, U.S. Department of Energy, and U.S. Department of Agriculture. MSU is a member of the Association of American Universities, a group of 60 research-intensive institutions in the United States and Canada, and has achieved high national and international rankings, including being ranked 29th among U.S. public universities by *US News and World Report* and in the top 100 world universities in the Shanghai Jiao Tong University Academic Ranking of World Universities. MSU has a strong international focus and has been the national leader in study abroad participation for the last five years (2,969 MSU students studied abroad in 2007–2008). MSU was also the first major U.S. university to appoint a dean of international programs, is the fourth-largest producer of Peace Corps volunteers among universities, and is sixth nationally in terms of international student enrollment.

Office of Campus Sustainability

The University Committee for a Sustainable Campus (UCSC), a grassroots effort of faculty, staff, and students, began in 1999 with the goal of making the university more environmentally aware and proactive. In 2000, the Office of Campus Sustainability (OCS) was created with a $250,000 Environmental Protection Agency (EPA) Sustainable Challenge Grant, and Terry Link was chosen as the first director. The office has four main teams and committees: the University Committee for a Sustainable Campus, the Environmental Stewardship Systems Team, the Behavior Team, and the Communications Team.

The OCS produces an annual report on facilities and infrastructure that includes consideration of MSU's environmental stewardship. Key indicators in the 2009 report included MSU's purchasing of copy paper, recycling and landfill waste, greenhouse gas emissions, electrical consumption, coal and gas usage, and steam and electric demand. For instance, in the area of copy paper, overall usage decreased by 10 percent in 2008–2009 (meaning that 28,616 fewer reams of paper were purchased), use of virgin pulp paper was reduced by 28.5 percent, use of 30-percent-recycled paper was increased by 25.2 percent, and use of 100-percent-recycled paper was increased by 19.9 percent.

Since 2004 (when the EPA grant expired), the OCS has been funded by the offices of the vice president of finance and operations, the provost, and the associate deans of research. In 2005, Environmental Stewardship was included in the university's Boldness by Design initiative, which was implemented to make the university a world leader in teaching, education, and outreach. A systems team, now known as the Environmental Stewardship Systems Team (ESST), was formed by students, faculty, and staff to study ways to make systemic changes in MSU's environmental stewardship, considering technical, cost, and behavioral issues, and looking at campus systems, processes, inputs, and outputs.

The ESST makes annual recommendations to the university: all 50 recommendations made in 2007 and 2008 were implemented. The 24 recommendations made in 2009 include the areas of energy reduction and offsets; waste reduction; transportation; behavior, communication, and education; and compliance and technology. Recommendations in the area of energy reduction and offsets include a more aggressive plan for switching to alternative fuels, development of a decision-support tool for financial and environmental performance for the T.B. Simon Power Plant (a coal-fired power station on the MSU campus that provides electricity and steam), improving campus databases concerning carbon offset planning and monitoring, and expanding MSU forest properties as well as planting more trees on the main campus. In the area of waste reduction, ESST recommendations included studying the feasibility and best practices for composting campus food waste, extending recycling and reuse of electronic equipment, expanding student and faculty knowledge of alternative textbook sources (e.g., rental, electronic downloads) that minimize paper and transport waste, establishing a campus-wide information hub to encourage use of electronic rather than paper communications and reports, and expanding the bin sensor system (which monitors fullness) and evaluating the labor savings available by using this system.

Recommendations in the area of transportation included a study of bicycle use on campus and assessment of bicycle safety issues, establishing a pilot car-share program (e.g., Zipcar, Enterprise WeCar) on campus, and surveying student transportation behavior to complement a recently completed survey of faculty and staff. In the area of behavior, communication, and education, recommendations included differentiating communications intended for students, faculty, and staff; increasing education on the relationship between personal energy use and climate change; improving visibility of campus environmental and sustainability projects to increase their visibility and perceived benefits; and creating a streamlined process for reporting and publicly cataloging campus sustainability projects. Recommendations in the area of compliance and technology included beginning a strategic planning process for emissions compliance and integrating Web conferencing technology into campus life.

Environmental Science & Policy Program

The Environmental Science & Policy Program (ESPP) at MSU was established in 2003 as an umbrella organization for research and graduate education. Goals of the ESPP include creating innovative, interdisciplinary graduate education programs, facilitating interdisciplinary environmental research at MSU, linking MSU research with national and global research priorities, and increasing the visibility of MSU's areas of excellence. MSU currently offers a doctoral specialization in Environmental Science and Policy and will soon offer a master's-level Environmental Policy Graduate Specialization, and offers information about environmental emphases within MSU undergraduate programs as well. The ESPP established the Environmental Research Initiative to strengthen interdisciplinary collaboration: 12 teams have been funded as of 2010 in emerging and critical areas of environmental research. The ESPP also maintains a searchable database of faculty members with expertise in environmental subjects at MSU.

Recognition

MSU was named one of the top five environmental campuses for sustainability in 2008 by the U.S. National Wildlife Federation, and was ranked high in several different areas of environmental sustainability on the National Wildlife Federation 2008 National Report Card on Sustainability. This report card, based on information from 1,068 campuses (27 percent of all U.S. colleges and universities), listed MSU as an exemplary school for environmental or sustainability goal-setting, noting that MSU had a written commitment to environmental sustainability, a written commitment to educating students about sustainability as part of the university's academic mission, and was a leader in setting and reviewing goals related to conservation and sustainability. MSU was also recognized for having recruiting programs and offering interdisciplinary degrees in environmental or sustainability studies, demonstrating commitment to energy efficiency, conservation and use of renewable energy, commitment to making its transportation program more sustainable, increasing recycling, materials exchange, and sustainable purchasing programs.

The American Association of Sustainability in Higher Education (AASHE) awarded MSU a Campus Sustainability Leadership Award in 2007. MSU participates in the AASHE STARS (Sustainability Tracking, Assessment & Rating System) program, which was developed after three years of an open development process. STARS ratings are notable for their transparent methodology, comprehensive approach, and use of high-quality and comparable data, and are based on an absolute scale of progress (i.e., the ratings are not relative to the performance of other institutions). The first STARS ratings will be available in 2011.

See Also: AASHE: STARS (Sustainability Tracking, Assessment & Rating System); Association for the Advancement of Sustainability in Higher Education; Leadership in Sustainability Education.

Further Readings

Association for the Advancement of Sustainability in Higher Education. http://www.aashe.org (Accessed May 2010).

Michigan State University, Office of the Vice President for Finance and Operations. "Facilities and Infrastructure Report" (February 2010). http://www.sustainability.msu.edu/us/fac_reports.htm (Accessed May 2010).

Michigan State University Environmental Stewardship Systems Team. "Phase III Environmental Stewardship Recommendations" (December 2009). http://www.sustainability.msu.edu/us/phases.htm (Accessed May 2010).

National Wildlife Federation. "Campus Environment 2008: A National Report Card on Sustainability in Higher Education." http://www.nwf.org/Global-Warming/Campus-Solutions/Resources/Reports/~/media/PDFs/Global%20Warming/CampusReportFinal.ashx (Accessed May 2010).

Sarah Boslaugh
Washington University in St. Louis

MIDDLEBURY COLLEGE

Middlebury College is a private liberal arts college founded in 1800 and located in Middlebury, Vermont. The college serves over 2,400 students through 44 majors, offering both undergraduate and graduate degrees. Middlebury College's high dedication to environmental sustainability is a natural outgrowth for an institution that awarded the first bachelor's degree to an African American and was one of first all-male institutions to accept women.

Middlebury College's environmental studies program is based in this LEED Platinum certified building, whose solar panels may help the college achieve its goal of carbon neutrality by 2016.

Source: Wikipedia

Middlebury's sustainability efforts in campus operations are an integral part of the college's Environmental and Sustainability Policy, which includes a goal of carbon neutrality by 2016. Middlebury is a signatory to both the American College and University Presidents Climate Commitment (ACUPCC) and the Talloires Declaration, and sustainability is the focus of the institution's Master Plan, including sustainable development recommendations for each chapter of the plan. Within the strategic plan are recommendations for energy efficiency, renewable energy investments, sustainable agriculture, and green building, and sustainability principles are included in the college's mission statement. An Environmental Council of administrators, faculty, staff, and students helps organize and spearhead these initiatives and reports directly to the president. To provide central coordination for multiple groups on campus, Middlebury employs a number of sustainability-related staff, including a dean of Environmental Affairs, a Sustainability Integration Office, and a Sustainability Communication and Outreach director.

Purchasing

Middlebury has a formal policy regarding green purchasing and utilizes environmentally friendly vendors as well as stating that only 100 percent postconsumer recycled and chlorine-free paper can be purchased for printers and copiers. The college uses the Betco Green Earth product line for cleaning, including campus bathrooms, windows, and general cleaning needs. In addition, a central vacuum system is used instead of traditional vacuum cleaners so that waste is collected in central locations. When possible, Energy Star options are utilized when making purchases, as are Green Seal–certified general cleaning products. Most computers are purchased in accordance with the Electronic Product Environmental Assessment Tool (EPEAT).

Energy, Facilities, and Planning

Middlebury College has performed an annual greenhouse gas inventory since 2005 with the goal of becoming carbon neutral by 2016. A biomass facility was opened in 2009 that burns wood chips (a carbon-neutral fuel), and the college anticipates a 40 percent reduction by the end of 2010. Working closely with its state's energy-efficiency utility, upgrades have been made to HVAC equipment in addition to re-insulating many campus buildings. The cogeneration of campus electricity (20 percent of total demand) at the school's thermal plant was recognized by the U.S. Environmental Protection Agency.

In addition to the wood chip biomass gasification plant, the Farrel House uses a 3 kW solar array, and an 8 kW array is a part of the Franklin Environmental Center. Wind powers 25 percent of the electricity demand at the campus recycling facility by means of a 10 kW turbine. Solar hot water systems are being tested in one dorm, while "cow power," which comes from farm manure digestion electricity, helps power the President's House and the Franklin Environmental Center.

Middlebury's Green Building Policy requires LEED Silver certification for all new construction and major renovations. Currently, 13,707 square feet are LEED Platinum certified, with an additional 401,448 square feet that meet LEED Silver criteria but are not certified. In total, Middlebury's construction projects have diverted 85 percent of construction and demolition waste.

A Reporting and Measurement Committee works to inventory data related to addressing carbon neutrality on campus, working closely with college facilities. Specifically, this partnership is assisting in the creation of a thermostat policy that maintains temperatures at 75 degrees in the summer and 68 degrees in the winter.

Compost and Recycling

Recycling is an integral part of the sustainability efforts at Middlebury, including aluminum, cardboard, glass, paper, and all plastics as well as tires, paints, motor oil, and scrap metal, for a diversion rate of 62 percent. The college also has an electronic waste recycling program, which includes batteries, computers, light bulbs, and printer cartridges.

Middlebury's Dining Services acquires products from 47 local (state) food producers as well as a student-run organic garden. In total, 25 percent of food on campus is local, and 75 percent of the waste is diverted to the campus composting program. The dining halls also address sustainability, with one featuring a green roof covered in shrubs and grasses, and the other with variable-speed fans in the hood exhausts, which conserves electricity.

Transportation

Of the 96 vehicles in Middlebury's transportation fleet, four are hybrid and one is electric, in addition to 20 electric golf carts. Beginning in the 2009 school year, students were charged to park their vehicles on campus. The $50 fee goes to finance sustainability initiatives on campus. For those who choose to take public transportation, free transit bus passes are available. The college covers weekday fares for faculty and staff and offers an "on demand" service during nights and weekends to get around campus. For those who choose to car-share, Middlebury has a partnership with Zipcar, which includes two hybrid vehicles for a $35 membership fee. Finally, the Nordic ski team utilizes a converted truck that runs on waste vegetable oil from the college's dining halls.

The Yellow Bike Cooperative was started in 2001 and serves as a bike repair shop and sells abandoned bikes. In addition to repairing bikes, the cooperative also offers free maintenance lessons. The bikes sold to faculty, staff, and students, which number 100, help fund the shop. To better serve bike traffic on campus, the Master Plan specifically addresses bike and pedestrian needs. Student green fees and an alumni green fund help to finance alternative transportation and support the school's carbon neutrality goals.

See Also: American College and University Presidents Climate Commitment; Sustainability Officers; Whole-School Approaches to Sustainability.

Further Readings

Middlebury College. "Sustainability." http://www.middlebury.edu/sustainability (Accessed June 2010).

Neapolitan, Jane E. and Terry R. Berkeley, eds. *Where Do We Go From Here? Issues in the Sustainability of Professional Development School Partnerships.* New York: Peter Lang, 2006.

Sustainable Endowments Institute. "Green Report Card 2010." http://www.greenreportcard .org/report-card-2010/schools/middlebury-college (Accessed June 2010).

Justin Miller
Ball State University

MILLENNIUM DEVELOPMENT GOALS

The intergovernmental Millennium Development Goals (MDGs) aim to tackle the extreme poverty experienced by over a billion people globally. Education is identified as one of the methods through which widespread sustainable change may be promoted, something recognized in the MDGs and through Education for Sustainable Development (ESD). Although there is evidence of MDG progression, some barriers may be seen to provide implementation challenges.

Introduction

The Millennium Development Goals represent eight time-bound targets, committed to by 189 states as a consequence of the United Nations Millennium Declaration and adopted in September 2000 by world leaders at the United Nations (UN) headquarters in New York. There are eight Millennium Development Goals:

- Eradicate extreme poverty and hunger
- Achieve universal primary education
- Promote gender equality and empower women
- Reduce child mortality
- Improve maternal health
- Combat HIV/AIDS, malaria, and other diseases
- Ensure environmental sustainability
- Develop a global partnership for development

Each goal has a headline target and a number of indicators that measure progress toward attainment. Progress is measured by the UN on an annual basis. The overall aims of the MDGs are to support the reduction of poverty and to improve quality of life. In the introduction to the 2009 UN MDG progress report, Secretary-General of the UN Ban Ki-moon summarized the aims of the goals as "to free a major proportion of humanity from the shackles of extreme poverty, hunger, illiteracy, and disease." Also, the MDGs establish targets to achieve gender equality and the empowerment of women, environmental sustainability, and a global partnership for development. There has been some concern that the economic downturn of the late 2000s may affect progress toward the goals, the majority of which are to be achieved by 2015. However, despite this, there have been clear successes, especially in the reduction of deaths from avoidable diseases and in the number of people catching incurable diseases. There have also been improvements in the number of children attending primary education.

While the MDGs are broad and far-reaching in nature, this article focuses on their relationship with education and sustainable development.

Education and Sustainable Development

Education plays a pivotal role in achieving sustainable development, over and above its intrinsic societal benefits. The most commonly used definition of *sustainable development* comes from the Brundtland Report of 1987: "development that meets the needs of the present without compromising the ability of future generations to meet their own needs." Agenda 21 (the principal output of the Rio Earth Summit in 1992, which is seen as an attempt to put the principles from the Brundtland Report into practice) highlights the pivotal role that education plays within sustainable development. First, it is said that society's poorest communities are likely to destroy their immediate environment in order to survive. Higher educational attainment may provide households with more employment options, making them less dependent on such natural resources. More directly, higher levels of educational attainment are said to foster more positive environmental attitudes and behaviors. In addition to this, the profile of education and sustainable development has been raised through the United Nations Educational, Scientific and Cultural Organization (UNESCO) Decade of Education for Sustainable Development (DESD) 2005–2014, with a goal of giving students the skills to address global environmental issues through integrating sustainable development into all areas of education in four key ways:

- The reorientation of all existing education toward climate change and its associated issues
- The introduction of new climate change material into education
- The development of critical thinking skills
- An emphasis on participation and finding solutions

The Millennium Development Goals and Education

The main target associated with Goal Two: Achieve Universal Primary Education is to ensure that all children, regardless of gender, will be able to complete a full course of primary schooling. As of 2009, progress has been made toward this goal; however, this is viewed as too slow to meet the 2015 deadline, with a coverage of 88 percent being achieved in 2007 in developing countries. As explored above, this goal is inextricably linked to the success of the other seven goals and to sustainable development more broadly. Indeed, the UN suggests that higher levels of education (even at the primary level) are linked to reduced child malnutrition rates and mortality, a reduction in human immunodeficiency virus (HIV) infection, and improved employment prospects.

However, enrollment varies greatly between developing countries. While the greatest improvements are in sub-Saharan Africa and southern Asia, levels in these areas are still below average; for example, in sub-Saharan Africa, rates have improved from 58 to 74 percent between 2000 and 2007.

There are large variations in educational enrollment between developing countries, something measured through categories: enrolled but dropped out, expected to enter late, expected never to enroll. For example, 65 and 63 percent of children in west Asia and in sub-Saharan Africa, respectively, were expected never to enroll at school in 2007. On the other hand, only 32 percent of children were in this category in south Asia, but 63 percent had enrolled but later dropped out (compared to only 8 percent in sub-Saharan Africa). These figures are the most recent available (2007) at the time of writing this article and do not reflect the possible impact of the economic downturn, which may threaten funding in education in developing countries, funding that is thought to have had a directs positive impact on improved educational enrollment and retention.

Barriers

One of the key barriers to achieving the Millennium Development Goals is a global inequality in educational opportunities. These inequalities are generally said to stem from gender, ethnicity, income, language, or disability-related obstacles. Once again, these vary markedly by country and by development status. The most problematic barriers to equality are thought to be income and gender, with children in the poorest 20 percent of the population being three times less likely to be enrolled in primary school than those in the richest 20 percent. The 2009 MDG progress report found that educational attendance and retention could be improved upon by abolishing school fees, constructing new schools in underserved areas, recruiting teachers, and actively encouraging educational participation among girls worldwide. In 2008, the UN suggested a number of factors that needed addressing in order to further progress toward this goal. First, it suggests that education itself must be sustainable, with sufficient staff who have received adequate training and who remain in their posts. Second, it stresses the importance of universal education, regardless of the demographic and geographical factors described above, especially as it has such a positive effect on health. Third, it highlights national capacity to deliver education and to ensure the effectiveness of aid programs; closely related to this, it advocates increased national spending on education. Fourth, it suggests the importance of a range of financial incentives to encourage school attendance, including the removal of school fees, cash transfers to poor families conditional on school attendance (especially among girls), and the provision of school meals and transport. Fifth, it stresses the importance of improved school resources such as textbooks and other learning materials. Sixth, it encourages the availability of preschool facilities.

Limitations of the MDGs

As with any performance indicator, the MDGs are criticized for being reductionist and limited in scope. More specifically, some argue that the time-bound nature of the goals makes policy responses short term in nature, and something of a "quick fix," and will encourage countries to focus resources on the areas of measurement, sidestepping those that are not measured such as freedom of speech or the rights of minority groups. Equally, there are questions about the reliability of data that is collected, especially in the least developed countries. It is also argued that the goals are very broad and are not relevant or appropriate to the range of national settings to which they are applied.

See Also: Agenda 21: "Promoting Education, Public Awareness, and Training" (Chapter 36); Early Childhood Education; ESD to ESF (Education for Sustainable Development to Education for a Sustainable Future); Second Nature: Education for Sustainability.

Further Readings

Franklin, Thomas. "Reaching the Millennium Development Goals: Equality and Justice as Well as Results." *Development in Practice,* 18/3 (2008).

United Nations. "End Poverty 2015: Millennium Development Goals Fact Sheet, Goal 2." New York: United Nations, 2008.

United Nations. "The Millennium Development Goals Report: 2008." New York: United Nations, 2008.

United Nations. "The Millennium Development Goals Report: 2009." New York: United Nations, 2009.

United Nations Educational, Scientific and Cultural Organization (UNESCO). *Education for Sustainable Development and Climate Change.* Paris: UNESCO, 2009.

Carolyn Snell
Sarah Wilson
University of York

NAEP Sustainability Institute

The National Association of Educational Procurement (NAEP) is an organization made up of college and university purchasing officers. The NAEP's history dates to about 1920, when a group of college- and university-affiliated purchasing officers met and decided to form a group. This group, which evolved into the NAEP, allowed them to share information with each other and to engage in cooperative or "pool" purchasing in order to receive bulk discounts and to achieve better economy of scale.

As these efforts developed and were refined over the years, cooperation extended to other areas as well. As sustainability became a topic of increasing interest to colleges and universities, the NAEP began to explore avenues through which purchasing decisions could be made in a more environmentally sustainable manner. To that end, the NAEP in 2007 began to offer a Sustainability Institute for its members and other interested parties. The first NAEP Sustainability Institute was held in Tempe, Arizona, near the Arizona State University campus. The NAEP Sustainability Institute assists colleges and universities to engage in environmentally preferable purchasing decisions. By providing participating colleges and universities with information and methods that will help purchasing agents to make more environmentally friendly decisions, the NAEP Sustainability Institute provides both tools and support to encourage such change.

Environmentally Preferable Purchasing

The NAEP Sustainability Institute is designed to assist colleges and universities to better maintain campuses' green environmental systems to continue to renew themselves and prosper. A major goal is building the understanding of all stakeholders, from administration to faculty to students, that the current generation has an obligation to future generations. One key step that the NAEP encourages to assist in this transformation is an awareness of environmentally preferable purchasing. Environmentally preferable purchasing is predicated on the desire to obtain goods and services that have a reduced negative effect on human health, the environment, or both when compared to alternative goods and services. The NAEP Sustainability Institute suggests that environmentally preferable purchasing decisions consider three factors: the environment, price, and performance. When all three factors are considered holistically, purchasing decisions can be made that are green and sustainable.

Environmentally preferable purchasing has several advantages for the colleges and universities that embrace it. These advantages include an improved ability to meet the institution's goals and requirements regarding sustainability; potential cost savings; improvements for worker health and safety; lower operating and maintenance costs, including hazardous material handling and disposal costs; creation and sustainment of markets for environmentally responsible and recycled content products; and preservation of natural resources. While sustainable practices are advocated by many, the NAEP Sustainability Institute teaches participants to understand how such procedures can be implemented and put into practice.

Guiding Principles of Environmentally Preferable Purchasing

To obtain these advantages, stakeholders must understand, appreciate, and embrace the guiding principles of environmentally preferable purchasing. These principles include the following:

- Including environmental considerations as part of the normal purchasing process
- Emphasizing pollution prevention early in the purchasing process
- Examining multiple environmental attributes throughout a product's or service's life cycle
- Comparing relevant environmental impacts when selecting products and services
- Collecting accurate and meaningful information about environmental performance and using it to make purchasing decisions

These principles can be enacted into purchasing practices at a college or university rather simply, but in ways that will greatly affect the institution's realization of sustainability goals.

Achieving sustainability goals requires certain modifications of practice. To include environmental considerations as part of the normal purchasing process, for example, the NAEP Sustainability Institute suggests including a blanket sustainability statement in Requests for Proposals (RFPs) and Requests for Information (RFIs). Doing so reminds all involved in the purchasing process from the beginning that making sustainable decisions is one part of the process. To emphasize pollution prevention early in the purchasing process, an emphasis must be placed upon procuring products that eliminate or minimize use of toxics through the substitution of less harmful substances. Similarly, to increase energy and water efficiency on campus, procurement agents must seek out products that reduce demand on these resources. To examine multiple environmental attributes throughout a product's or service's life cycle, college or university purchasing agents must engage in a systems analysis across the life cycle of that product or service. The main production sequence to be considered might take into account the energy used for material production, manufacture, use, and disposal of a product.

To compare relevant environmental impacts when selecting products and services, the NAEP Sustainability Institute suggests examining certain impacts. Key factors to consider involving these environmental impacts include their potential reversibility, their geographic scale, the relative increase in environmental performance provided, and the importance of protecting human health. Finally, to collect meaningful data on environmental performance, purchasing agents must keep accurate records. Such practices are aided by the use of creating purchasing comparison charts and comparing rating systems, in part by using third-party certification systems, such as the U.S. Environmental Protection Agency's Energy Star program, to identify appropriately green computers and appliances.

Sustainable College and University Procurement

To assist with environmentally preferable purchasing, the NAEP provides a "sustainability microsite," which is a Website dedicated to sustainability issues. This Website provides a host of resources for over 1,200 college and university procurement agents who are members of the NAEP. Areas of the Website are devoted to topics that assist in environmentally preferable practices, including renewable energy, green building, green purchasing, and climate change. The Website also contains links to selected energy efficiency studies, sustainability events, and recommended reading lists. All of these resources are useful for procurement agents who might have to build the case for sustainable practices on their campuses. Especially useful are those links to other Websites, such as the Responsible Purchasing Network, that allow colleges and universities to join forces and work together to obtain more sustainable solutions regarding procurement and building decisions.

Campus purchasing agents often take positive actions as a result of the NAEP Sustainability Institute that make the impact of their decisions more powerful. By including a sustainability statement in RFIs and RFPs, for example, colleges and universities alert suppliers that green practices are valued. Incorporating environmentally preferable purchasing into campus policies and procedures makes clear the commitment to sustainable practices and provides guidance for administrators and faculty about how best to proceed. Many colleges and universities survey their top suppliers with regard to their sustainable activities, gaining valuable insights and information in the process. Working together on sustainability planning with other colleges and universities as well as comparing progress toward attaining goals allows purchasing agents to benefit from the experiences and successes of others. The NAEP Sustainability Institute assists greatly with all of these efforts as it enables good intentions with regard to sustainability to coalesce into actions that support greener campuses through shared information, useful resources, and partnerships with like-minded colleges and universities.

See Also: AAAS Forum on Science, Technology, and Sustainability; Association for the Advancement of Sustainability in Higher Education; Greening of the Campus Conference; Halifax Declaration; Higher Education Associations Sustainability Consortium; Sustainability Websites.

Further Readings

Gough, Stephen and William Scott. *Higher Education and Sustainable Development.* London: Routledge, 2009.
Hargreaves, Andy and Dean Fink. *Sustainable Leadership.* Hoboken, NJ: Jossey-Bass, 2005.
National Association of Educational Procurement (NAEP). *A Green Purchasing Policy Roadmap.* Baltimore, MD: NAEP, 2007.

Stephen T. Schroth
Jason A. Helfer
DeAndre A. Henderson
Knox College

NATIONAL ASSOCIATION OF INDEPENDENT SCHOOLS (SUSTAINABILITY TEACHER TRAINING)

The membership of the National Association of Independent Schools (NAIS) consists of approximately 1,400 American schools and associations. While schools outside the United States cannot become NAIS members, many have affiliated themselves with the organization in various capacities. NAIS exists chiefly to provide strength and unity for independent schools and to train teachers in those schools to practice sustainability and diversity in education. NAIS was established in Boston, Massachusetts, in 1964 through the merger of the Independent Schools Education Board, a group made up of heads of schools concerned about curriculum development, testing, and teaching, and the National Council of Independent Schools, a loosely knit group of state and regional associations that banded together to monitor accreditation and government involvement in independent schools and to discover methods of improving the public image of independent schools. NAIS moved to Washington, D.C., in 1993, and being located in the nation's capital strengthened its voice with both the government and the public and renewed cohesiveness within the organization. Over time, the most significant contribution of NAIS became its focus on providing teacher training and support. By the 21st century, the entire world was becoming more focused on the impact of global warming and climate change. In response to this new awareness, NAIS took action to advance the integration of sustainability education into all independent American schools. The organization also responded to the changing world by developing a new focus on diversity and technology in member schools and on forming global partnerships to address all of these issues.

The National Association of Independent Schools identifies five dimensions that it considers integral to achieving its goals for promoting sustainability: financial sustainability, environmental sustainability, global sustainability, programmatic sustainability, and demographic sustainability.

- The concept of financial sustainability focuses on increasing efficiency and lowering costs through actions that are nonthreatening to the environment.
- Environmental sustainability, on the other hand, calls for action on the part of teachers and the campus community by motivating them to incorporate sustainability into the classroom and throughout each campus.
- By encouraging members to focus on global sustainability, NAIS acknowledges that solving the dilemmas caused by irresponsible use of the planet requires international cooperation.
- NAIS has issued a call for programmatic sustainability by promoting innovation and seeking out interdisciplinary and nontraditional ways of teaching and advancing the goals of sustainability within the realm of education.
- The notion of demographic sustainability encompasses the belief that involving students of all ages in sustainability efforts is essential to the success of the sustainability agenda.

Principles of Good Practice

All members of NAIS are expected to be responsive to the Principles of Good Practice, which demand that individual members exhibit ethical behavior at all times. These principles, familiarly known as the PGPs, were first developed by a task force charged with developing methods for including environmental sustainability into independent school systems. These principles are constantly being examined and revised, and specific sets of principles are generated to target appropriate issues and/or groups. In 2010 at its annual meeting in San Francisco, NAIS issued the board-approved "Principles of Good Practice for Environmental Sustainability," which established guidelines for member schools to incorporate sustainability into all elements of school life by (1) demonstrating the willingness to engage in sustainability education; (2) making sustainability an integral part of the curriculum, teacher training, student life, operations, and finances; (3) involving parents in sustainability efforts; (4) forming community partnerships to promote sustainability; and (5) measuring sustainability success through regular assessments and disclosures. Within that framework of guiding principles, NAIS has also issued the "New Guidelines of Professional Practice," which target particular positions within independent school systems. The guidelines for librarians, for example, call for increased use of information technology in schools. They also establish specific requirements for granting equal access to information.

Professional Opportunities

Teaching educators how to teach and promote sustainability is considered key to creating sustainable schools, and NAIS has played a significant role in providing independent school professionals with the tools necessary for those purposes while offering them a myriad of opportunities for interacting with their counterparts and with professionals in the field. In the summer of 2010, for instance, NAIS sponsored the three-day seminar "Schooling for Sustainability—Strategies That Make Learning Come Alive." Topics discussed included the need for sustainability in education, food services, operations, and sustainability outreach.

Other seminars have included the Introduction to Education for Sustainability Workshop, held at the Cloud Institute in New York City, and the CELF Summer Institutes for Sustainability Education, held in Westchester County, New York, and Boston. The Project Seasons Summer Workshop in 2010 emphasized the interrelationship between agriculture and sustainability. Education in a Changing Climate in August 2010 addressed issues arising from global warming and its impact on climate change. The Independent School Experiential Education Network Institute (ISEEN) is scheduled for winter 2011, and will encourage outdoor sustainability and conduct training for developing experiential curricula.

Attendance at national conferences also enhances teacher training among NAIS member schools, and interactions with teachers from public schools sometimes add new dimensions to those meetings. The Green Schools National Conference, for instance, held in Minneapolis in 2010, gathered public and private school teachers and administrators together with politicians and business and community leaders to address the issue of environmental literacy and the development of a "green culture" at the K–12 level. Believing that students play a significant role in sustainability training, NAIS encourages student participation at many events it promotes.

Global Sustainability Education

Through the Global Initiative Objective, the National Association of Independent Schools has expanded sustainability in education beyond the borders of the United States. Two of the most important objectives of this program are fostering partnerships among NAIS members and affiliates and expanding channels of communication and support. In order to accomplish those goals, NAIS believes that teachers must be adequately trained to promote ecological literacy and encourage students to become active participants in green education. Elements involved in achieving the goal of adequate training include the promotion of diversity and respect for all races, nationalities, and creeds; designing curricula that recognize the right to diversity; offering resources and support for all teachers engaged in teaching sustainability from a global perspective; engaging in global partnerships; encouraging parents to take an active role in sustainability education; and recruiting teachers and administrations who represent a range of cultural, national, and ethnic backgrounds. One important part of NAIS's global sustainability agenda is the China Connection Program, which was first implemented in 2006–2007. Working with HANBAN, a government-funded Chinese organization, heads of schools from the United States spend time in China observing the educational system, experiencing Chinese culture, and recruiting Mandarin Chinese teachers to work in the United States. Program goals are twofold: adding diversity to the curricula of American independent schools and providing teachers and schools in both China and the United States with more extensive tools for promoting sustainability in education.

NAIS encourages nonmembers as well as members to engage in efforts to seek out solutions to global problems. One way they do this is through the Internet-based Challenge 20/20, which pairs K–12 classes from American schools with classes in other countries that are interested in solving the same problems. While the issues are of global concern, implementation is expected only at local levels. The issue addressed by each team consists of one of the global problems posed by Jean Francois Rischard in *High Noon: 20 Global Problems, 20 Years to Solve Them.* Each 20/20 challenge takes place over the course of the fall semester. In 2006, students representing 32 countries participated in the challenge, which had been launched the previous year when more than 51 schools from 16 U.S. states and 31 countries participated. In that competition, one lower-school team consisted of students from the Children's School in Georgia and those from the William Hulme Grammar School of the United Kingdom. They chose to address problems of peacekeeping and conflict resolution issues. One of their solutions was to create a "peace table" to allow students in conflicted areas to meet in a safe place to discuss ways of resolving points of dissension. Other students addressed issues such as deforestation and global infectious diseases.

Resources for Incorporating Sustainability Into Education

Fulfilling its function as a major resource for teacher training, the National Association for Independent Schools has issued a number of publications dealing with various aspects of global sustainability in education. In *101 Ways to Go Global and Green,* NAIS promotes the practice of teaching foreign language at all educational levels, including global education in all mission statements, employing the Internet to establish connections and partnerships with counterparts in other countries, participating in the Challenge 20/20 Program, choosing parent/child teams to participate in short-term exchange programs with similar teams in other countries, reading relevant literature on global sustainability education, establishing groups for international parents in American communities, funding summer travel grants for teachers to travel to other countries, inviting speakers with international experience to American schools, teaching anthropology, educating parents as well as students in global sustainability issues, and holding World Fairs and Global Summits.

NAIS also educates teachers in ways to improve their schools' green credentials. The organization suggests that teachers be involved in greening projects such as starting a school garden; establishing outdoor adventure programs; planting trees; promoting energy-saving practices; recycling; ride sharing; spreading sustainability cooperation by recruiting trustees, parents, and alumni to the sustainability cause; composting; and establishing a biodiesel plant that can also be used to teach sustainability to chemistry students. Other suggestions include constructing LEED-certified buildings, teaching courses on sustainable development, incorporating sustainability across the curriculum, and conducting workshops to keep all teachers informed about sustainability issues and trends.

The National Association of Independent Schools also offers support to its members through the Demographic Center. In 2008, NAIS announced that the center had increased its support capabilities by generating Metropolitan Area Reports designed to track demographic trends in metropolitan areas as a means of helping independent schools address the needs of their communities. The first reports covered details and analyses of school-age populations, income, race, and ethnicity in Boston, New York City, Dallas/Fort Worth, Los Angeles, Seattle, and Washington, D.C. Another new service was offered by Multiple Area Reports, which provided demographic information on neighboring counties, multiple cities, or by ZIP code or census tract.

Use of Technology

The National Association of Independent Schools often uses technology as a channel for promoting sustainability and encouraging teacher interaction. In January 2009, under the auspices of Teachers of the Future, the association launched an Internet-based discussion forum run by 24 educators who had established reputations for excellence in sustainability education. They were all recognized leaders in the field, and all had demonstrated a commitment to innovation, environmentalism, equity, justice, globalism, and technology. Initial discussion topics included mathematics, science, globalism, journalism, art, and music.

In 2007, NAIS established the Schools of the Future initiative as a means of addressing the interrelationship between technology and sustainability in education. The initiative was implemented by the 21st Century Curriculum/Technology Task Force, which examined submissions from 20 schools known for incorporating technology into classrooms. The results were published in *Stories of Excellence: Case Studies of Exemplary Teaching and Learning With Technology*. The project was deemed an excellent way to demonstrate the effectiveness of teaching with technology and of engaging in creativity and innovation.

In 2007, NAIS surveyed member schools about their use of technology, paying particular attention to curriculum, professional development of faculty, school policies, staffing, systems, and future goals. Their overall conclusion was that technology was rapidly changing the way that educators transmit information and the means by which students learn. The survey revealed that schools were incorporating technology into curricula at all grade levels. Methods of doing so ranged from requiring students to engage in Internet-based research to using technologically savvy presentations in the classroom. As might be expected, technological skills of students advanced with age. As early as first and second grade, some students were already learning how to use educational software, do word processing and desktop publishing, and use the Internet. By the third grade, teachers were educating their students in how to create multimedia presentations. The ability to use spreadsheets and e-mail was introduced to fifth and sixth graders. By the time they reached the secondary level, many students were being taught computer programming, blogging, and Web development.

Through the use of Internet tools such as Web 2.0, both teachers and students were able to interact and share resources with their counterparts around the world. Respondents reported that they were using Wikis, YouTube, Moodle, podcasts, and blogs to facilitate these interactions. They were also making use of teacher development tools offered through Teacher Tube, Ning, SecondLife, and Flickr. Faculty development was generally strongly supported by individual schools. More than 90 percent of schools provided laptops for teachers, but only 61 percent offered this access outside the normal teacher workday. Regular training sessions were conducted at more than 60 percent of responding schools.

Listservs are also a popular NAIS tool, and the association operates the NAIS Sustainable Schools listserv. Other listservs target groups as diverse as those for teachers of the learning disabled and one for those interested in solving peer-related problems at their individual schools.

The National Association of Independent Schools has generally been successful in leveling the playing field when independent schools compete with public schools for recognition. The organization also continues to provide valuable resources for teacher training and for promoting sustainability, diversity, and globalization within the curricula.

See Also: Cloud Institute for Sustainability Education; Early Childhood Education; Integrating Sustainability Education Concepts Into K–12 Curriculum; Sustainability Teacher Training; Unity College.

Further Readings

Booth, Susan. "Technology Use in Independent Schools." *Independent School,* 68/2 (2009).

Dernbach, John C. *Agenda for a Sustainable America.* Washington, DC: ELI Press, Environmental Law Institute, 2009.

Franson, Melissa. *The Impact of Classroom Exposure to Sustainability, Course Content, and Ecological Footprint Analysis of Student Attitudes and Projected Behaviors.* Unpublished Thesis. Auburn University, AL, 2008.

Hignite, Karla, et al. *The Educational Facilities Professional's Practical Guide to Reducing the Campus Carbon Footprint.* Alexandria, VA: APPA, 2009.

National Association of Independent Schools. http://www.nais.org (Accessed June 2010).

National Association of Independent Schools. "Principles of Good Practice." http://www .nais.org/about/index.cfm?ItemNumber=146811&sn.ItemNumber=146810 (Accessed June 2010).

Neapolitan, Jane E. and Terry R. Berkeley, eds. *Where Do We Go From Here? Issues in the Sustainability of Professional Development School Partnerships.* New York: Peter Lang, 2006.

Rockwood, Larry, et al., eds. *Foundations of Environmental Sustainability: The Coevolution of Science and Policy.* New York: Oxford University Press, 2008.

"Stories of Excellence Profile Best Technology Teaching." *Independent School,* 68/2 (2009).

"Students From 31 Nations Took on the Challenge 20/20." *Independent School,* 66/1 (2006).

Thuermer, Kitty. "New Principles of Good Practice for Environmental Sustainability." *Independent School,* 69/4 (2010).

Elizabeth Rholetter Purdy
Independent Scholar

NATIONAL WILDLIFE FEDERATION

Thousands of caribou north of the Brooks Range mountains in the Arctic National Wildlife Refuge, an area of particular concern for the National Wildlife Federation.

Source: U.S. Fish & Wildlife Service

As the largest and one of the oldest conservation organizations in the United States, the National Wildlife Federation (NWF) has grown from an eclectic mix of hunters, anglers, gardeners, and bird watchers into a renowned association of outdoor enthusiasts and environmental activists. Founded in 1936 and boasting over 4 million members today, NWF seeks to provide inspiration for protecting and restoring wildlife and its habitat. The tax-exempt NWF is also unique in its federation structure, with 46 state affiliates across the country as well as, since a 1961 bylaw change, individual members who receive associate, nonvoting member status. NWF's mission addresses three main areas: global warming (particularly since 2004), safeguarding wildlife and restoring wildlife habitat, and connecting people with nature. Along these lines, NWF maintains a notable legislative and political agenda, highlighted by efforts such as raising fuel economy through its lobbying on the 2007 Energy Independence and Security Act as well as winning landmark Endangered Species Act rulings in 1984 and 2005 on protection of the Florida key deer by prohibiting federally subsidized development within its habitat.

Goals and Activities

Despite such successes, educational initiatives clearly remain at the heart of NWF's focus. Indeed, including its wealth of curriculum materials for primary and secondary school teachers, NWF's educational publications have evolved into a strategic niche in the crowded field of environmental interest groups within the United States. Three notable publications here include *Ranger Rick,* with a target audience of children ages 7 to 12; *Your Big Backyard,* with a target audience of ages 3 to 7; and *Animal Baby,* for toddlers and preschool children under age 4. Such an evolution is more than a bit ironic considering initial fears within NWF leadership that its flagship publication, *National Wildlife,* would become just another environmental magazine. Since that first issue in December 1962/ January 1963, nothing could be further from the truth. And while e-mail updates continue an activist tradition at NWF and new endeavors over the past decade include cooperative television projects with Animal Planet and Discovery Communications, NWF continues to stand out for its traditional children's literature.

Seeking to build an environmental constituency from an early age, NWF also targets experiential learning through such initiatives as its Backyard Wildlife Habitat program and the Great American Backyard Campout. The Backyard Wildlife Habitat program, for example, began in 1973, and teaches children while developing a sense of environmental responsibility. Since the 1970s, some 120,000 backyards and school plots have received certification by meeting fundamental criteria for food, water, and shelter as well as converting gardens to native species, emphasizing low-water-usage landscaping, and minimizing chemical fertilizer and pesticide dependency.

The Great American Backyard Campout follows a similar line of thinking, encouraging kids across the country to sleep under the stars at least one night each June. Indeed, both of these programs illustrate what became a more cohesive effort to entice children outside with NWF's "Be Out There" campaign, now highlighted by the push for a daily "green hour" as key to overcoming childhood obesity and statistics showing television, Internet, and video games dominated 44 hours of children's play time a week.

NWF has also been active since the late 1980s in the field of higher education, expressly targeting college audiences, with a goal of improving sustainability by linking campus activists and lobbying administration through its Campus Ecology and Campus Report Card programs, respectively. Since 1989, a series of NWF workshops, conferences, webinars, and email correspondence has enabled networks to develop and reinforce sustainability projects on campuses across the country. Over 100 colleges and universities have enrolled at no cost, and receive membership benefits such as the biweekly e-newsletter, ClimateEdu, which includes a mix of book reviews, editorials, event listings, and other resources. NWF further shapes campus sustainability through its Campus Report Card, which recognizes postsecondary institutions for performance on energy, landscaping, transportation, waste reduction, and water usage by awarding a collective letter grade (A to D) for national performance. This comprehensive study, conducted most recently in 2008 in partnership with Princeton Survey Research Associates, highlights trends and new developments in sustainability at 1,068 institutions.

With its headquarters in Reston, Virginia, plus a national Advocacy Center in Washington, D.C., as well as nine regional centers across the country, NWF has built a national presence over the past 75 years. But it is with the aforementioned affiliates such as the Tennessee Wildlife Federation and the Florida Wildlife Federation that NWF continues to emphasize its local foundation in the environmental community. This continued connection between a local and national presence did not come without conflict, however, and arguably continues to define the group today.

NWF Beginnings

NWF traces its origins to two-time Pulitzer Prize–winning cartoonist Jay Norwood "Ding" Darling and his efforts to organize conservationists in the 1930s. More specifically, Darling's sketches caught the attention of President Franklin Delano Roosevelt, and in 1934, he was appointed head of the U.S. Biological Survey, the forerunner to the U.S. Fish & Wildlife Service. From this position, Darling created two major environmental initiatives still with us today. One was the conception of the famous duck stamps that are required as a permit to hunt in a refuge, but also serve as a mechanism for generating revenue to acquire wetlands for the national wildlife system. Darling even drew the first design in 1934. The other major initiative, of course, was NWF. Darling garnered presidential support to convene the North American Wildlife Conference at the Mayflower Hotel in Washington, D.C., in February 1936. There, over 2,000 hunter, angler, and conservationist conference participants formed the General Wildlife Federation, renamed the National Wildlife Federation in 1938.

An avid waterfowl hunter himself, Darling was elected the organization's first president and argued vociferously that conservationists and hunters needed to unite if their interests were to be taken seriously in Washington. He famously asserted at the 1936 conference, for instance, that despite the presence of 36,000 different organizations across the country involved in wildlife habitat conservation, not a single election result, even for that of a dogcatcher, had ever been determined by their groups. Over the next several years, Darling set about to rectify that dismal record, establishing a creative federation of sportsmen and garden clubs where each state had one and only one affiliate with voting privileges. These affiliates, in turn, represented still additional organizations back in their home states.

Early NWF programs under Darling brought much-needed national attention to conservation issues, but failed to develop into the fundraising entities that were initially envisioned. National Wildlife Restoration Week, subsequently shortened in name to National Wildlife Week, was one such example. Begun in 1938, this annual event corresponds to a week around the first day of spring each year. Focal topics over the years ranged from water to pesticides to soil, and spokespersons have included Bing Crosby, Walt Disney, Robert Redford, and Kermit the Frog, among others. Similarly, the NWF wildlife conservation stamp program experienced early financial frustrations as a fundraising device. Here, Darling tried to mimic the success he created with the duck stamp program for the federal Migratory Bird Hunting Act, but due to a number of complications with affiliates the popular stamps failed to raise sufficient funds. In short, fear of bankruptcy dominated the early years of NWF, and the organization remained in a fiscal deficit until 1941. A true financial turning point came in the 1950s as the group became self-sustaining and reenergized itself with the aforementioned 1961 bylaw change, offering associate membership to individuals across the country, and with that, a dependable source of income.

Over time, NWF has proven adept at both evolving to address new challenges such as climate change and keeping core philosophical approaches like its emphasis on public education as a mainstay. Despite criticism about financial partnerships with fossil-fuel giants like BP/Amoco—arrangements that critics labeled blatant corporate branding by big oil— NWF remains a leader in setting the agenda for climate change politics within the environmental community. Removal of its previous affiliate from Pennsylvania, the Pennsylvania Federation of Sportsmen's Clubs, in 2008 is one example of this concerted effort—the Pennsylvania group refused to accept human activities as a cause of climate change. Since then, Penn Future: Citizens for Pennsylvania assumed the Pennsylvania-affiliate role and actively pursues a climate change agenda.

See Also: Agenda 21: "Promoting Education, Public Awareness, and Training" (Chapter 36); Experiential Education; Outdoor Education.

Further Readings

Allen, Thomas B. *Guardian of the Wild: The Story of the National Wildlife Federation, 1936–1986*. Bloomington: Indiana University Press, 1987.

Better Business Bureau. "Charity Review of National Wildlife Federation." Wise Giving Report. http://www.bbb.org/charity-reviews/national/environment/national-wildlife-federation-in-reston-va-1199 (Accessed February 2010).

National Wildlife Federation. "Our Mission." http://www.nwf.org/About/Our-Mission.aspx (Accessed March 2010).

PBS. "Nature: Secret Garden Introduction." http://www.pbs.org/wnet/nature/episodes/secret -garden/introduction/3043 (Accessed March 2010).

Senser, Kelly L. "Plant a Garden, Help a Child Grow." *National Wildlife* (August/ September 2004).

Stauber, John. "Endangered Wildlife Friends Are Here!" *PR Watch,* 8/3 (2001).

Michael M. Gunter, Jr.
Rollins College

NEW JERSEY HIGHER EDUCATION PARTNERSHIP FOR SUSTAINABILITY

The New Jersey Higher Education Partnership for Sustainability (NJHEPS), located at New Jersey Institute of Technology, was founded in 1999 and consists of 42 member institutions representing a broad array of universities and colleges in the state. NJHEPS is a body that connects institutions of higher education with private corporations and other state and business entities. The corporate sponsors of NJHEPS include utilities and energy and engineering companies. The chief objective of NJHEPS is to promote sustainability practices across higher education, and, to this end, use the considerable intellectual and cultural resources located at universities and colleges to establish and implement an active agenda in teaching, research, outreach, and engagement with communities.

Vision

The central vision guiding NJHEPS is that institutions of higher education are in a unique position to lead on matters related to sustainability by setting examples. The choices that they make about sustainability policies and practices have ripple effects far beyond academia. The universities are also ideal for sustainability initiatives because of the opportunities they provide for influencing numerous courses of studies in diverse disciplines. Since sustainability is an inherently interdisciplinary concept, universities are well positioned to produce advances in the relevant science and knowledge.

Goals

The major goals of NJHEPS fall in two broad areas: first, communicating and disseminating the knowledge of the theory and practice of sustainability to institutions of higher education and the broader public and, second, forging partnerships between higher education institutions and providing technical expertise related to conceptualizing and executing sustainability measures. Communication of sustainability takes place through conferences and seminars that provide forums for in-depth discussion and dissemination of the various issues involved in sustainability. The technical know-how is provided through a number of specialized teams that possess the technical knowledge in topics such as energy, green design, waste management, and reducing greenhouse gas emissions.

Activities and Accomplishments

NJHEPS organizes a wide range of informational activities aimed at the institutions of higher education and the general public. A recent initiative of NJHEPS consisted of organizing Communities of Practice (COP). These are groups of people who have a common interest in one or more components of sustainability. The groups meet on a regular basis, share information and experience of various practices, strive to learn from each other, and improve their policies and actions. Issues that are discussed include sustainability education across colleges and universities and beyond, healthy and sustainable food systems, sustainability in business education, and promotion and marketing of sustainable products.

NJHEPS created a regional climate change alliance that consists of all of its member institutions. The climate change alliance committed to reducing greenhouse gas emissions as a response to the problem of climate change. The presidents of New Jersey colleges and universities signed a Sustainability Covenant that committed them to reducing, by 2015, their greenhouse gas emissions by 3.5 percent from the 1990 level.

Among the topics of recently organized workshops by NJHEPS was "Winter Energy and Design." Various aspects leading to energy efficiency, building design, and power purchase in higher education were discussed. A similar event titled "Spring Energy and Design" is being planned for April 2010.

An energy efficiency improvement project was undertaken by NJHEPS at the College of New Jersey (TCNJ). The project grew out of the realization at TCNJ that energy expenditure comprised a large part of its operating budget, and even marginal improvements in energy use and conservation could translate into large savings. A number of changes were made to the measurement and use of energy on campus in order to determine opportunities for reducing waste and improving efficiency. Automatic energy-monitoring systems were installed in buildings across the campus to constantly record usage. Based on the data, energy use was controlled, and substantial reductions were made. The project was so successful that the cost savings achieved during the first year of operation paid for the installation of the new equipment. A similar project was completed in Richard Stockton College, which succeeded in achieving substantial cost savings and lowering of operating expenses. It also enabled the college to become a regional showcase for geothermal technology. This, in turn, has helped to draw federal and state funds for upgrading the college's energy infrastructure and for offering greater research and teaching opportunities to its faculty and students.

NJHEPS also provides support to universities and colleges in making environmentally responsible purchases that are also economical. For example, they arranged for the joint purchase of fuel cells by several universities in the state. This enabled participating institutions to lower the purchase price of each unit, while lowering their energy expenditure. Similarly, joint purchasing of recycled paper and recycled toner cartridges has been made available to a consortium of community colleges.

Among the publications of NJHEPS are an Energy Toolkit and two volumes of a High Performance Campus Design Handbook. Other publications include monthly newsletters of NJHEPS activities and events. NJHEPS has also contributed to the redesign and changes in the curriculum of several universities and colleges, to ensure that sustainability is being integrated into course offerings across many departments. Topics that have been introduced into curriculum at some of the participating universities include climate change, studies of physical plants on campuses, and the use of engineering in energy audits.

Future Directions

NJHEPS plans to continue increasing its membership and to broaden its appeal to higher education and beyond. Among the activities planned are compiling best practices related to sustainability in higher education and disseminating it widely. It will also seek to increase participation of volunteers and others in its activities.

On the curricular front, NJHEPS will help develop courses of study and degree programs in sustainability. Faculty will be provided opportunities to upgrade their skills through summer institutes. NJHEPS will also work with existing programs to help place students in field learning experiences (such as internships and service learning courses).

In the future, NJHEPS plans to develop relationships with similar regional and national organizations in an effort to reach out to a larger audience as well as foster an exchange of best practices. Another major goal is to enhance the capacity of university and college campuses to effectively implement sustainability in curriculum, research, and practice. This will involve changes in staff, faculty, and student attitudes while fostering a culture that values research and teaching in sustainability. NJHEPS will continue to pursue a strategy of convincing the upper echelons of higher education in the state to create a conducive and rewarding environment for promoting sustainability in every aspect of life on campuses.

See Also: Leadership in Sustainability Education; Sustainability Officers; Whole-School Approaches to Sustainability.

Further Readings

Dernbach, John C. *Agenda for a Sustainable America*. Washington, DC: ELI Press, Environmental Law Institute, 2009.

Franson, Melissa. *The Impact of Classroom Exposure to Sustainability, Course Content, and Ecological Footprint Analysis of Student Attitudes and Projected Behaviors*. Unpublished Thesis. Auburn University, AL, 2008.

New Jersey Higher Education Partnership for Sustainability. http://www.njheps.org (Accessed June 2010).

Neeraj Vedwan
Montclair State University

North American Association for Environmental Education

The North American Association for Environmental Education (NAAEE) is a professional association that supports a diverse membership of professionals, students, volunteers, and organizations based primarily throughout the United States, Canada, and Mexico. An important network for the exchange and discussion of ideas among individuals and entities engaged in research, teaching, and service in environmental education, the NAAEE plays a significant role in enhancing the practice and efficacy of environmental education through activities such as its annual conferences, publications, advocacy, and work in program and resources development.

History and Mission of the NAAEE

The NAAEE was founded in 1971 in the United States by a group of community college educators as the National Association for Environmental Education (NAEE). The establishment of the group coincided with historic national and international events in environmentalism such as the passing of the National Environmental Policy Act in the United States, the first Earth Day observance, and the United Nations Conference on the Human Environment. Furthermore, while other organizations at the time accorded consideration to issues in environmental education, the NAEE's incorporation of "environmental education" in its name was a clear indication of its focus on this area, underscoring its importance.

Initially, the geographic focus of the NAEE was the United States, as reflected in its name, and the group was concerned with developing and disseminating instructional materials pertaining to the environment for use by community college educators. The association's focus soon expanded to encompass various educational levels and formal and nonformal education, and its name was subsequently changed to that of the NAAEE to recognize increased participation from members in Canada and Mexico, as well as to encourage geographical involvement beyond North America. While the majority of its membership is from North America, the network now has members from over 50 countries. This broad-based membership is composed of individuals, including university faculty, schoolteachers, conservationists, and environmental scientists, as well as entities such as universities, nongovernmental organizations, museums, and government agencies.

As outlined in its vision and mission statements, the NAAEE aspires to be the leading organization in North America supporting environmental educators working in all spheres and encouraging excellence in environmental education to ensure an environmentally literate populace and, by extension, equitable access to the use and benefits of resources among generations.

NAAEE Activities

Since its inception, principal activities of the network have been its conferences and publications. The NAAEE organizes an annual conference in various locations throughout North America, offering a forum for the sharing of ideas, research, and work in environmental education alongside opportunities for networking, the exhibition of resources, the recognition of accomplishments in the field, and other activities. The NAAEE also promotes the field through its publications, which include a bimonthly newsletter (the *Environmental Communicator*), a monograph series, conference proceedings, occasional publications, and other resources.

During the 1990s, the NAAEE initiated the National Project for Excellence in Environmental Education to develop standards for excellence. Also known as the "standards project," outputs include guidelines for excellence for environmental education materials, student learning, professional development and nonformal programs, and resources for educators. These guidelines can be used by educators, organizations, and the public as benchmarks by which to develop meaningful environmental education programs and resources and to assess their effectiveness. The National Wildlife Federation, for instance, has adapted the NAAEE guidelines specifically for climate change education.

The NAAEE has also collaborated with the Environmental Education and Training Partnership (EETAP), first as its manager and subsequently as a partner. A training program of the U.S. Environmental Protection Agency's Environmental Education Division, EETAP is engaged in various activities, including professional development and training, the development of guidelines, and facilitating accessibility to environmental education resources.

NAAEE supports other initiatives such as the VINE (Volunteer-led Investigations of Neighborhood Ecology) Network, an urban environmental education program, and the Environmental Issues Forums (EIF), through which individuals discuss divisive environmental issues in an effort to identify common ground for collaboration. Through its support of projects such as these, often in partnership with other entities, the network has actively engaged with communities and the public to advance environmental education.

Additionally, the NAAEE is engaged in advocacy, involving themselves, for instance, in efforts to institutionalize environmental education in the U.S. education system by calling for federal legislation such as the National Environmental Education Act and the No Child Left Inside Act, which are concerned with lifelong environmental learning and the inclusion of environmental education in the curricula of formal education, respectively.

Significance of the NAAEE

The NAAEE supports its members through the development of programs and resources for various levels and stakeholders, its conferences, publications, advocacy, and professional development opportunities. Moreover, through initiatives such as the National Project for Excellence in Environmental Education, the VINE Network, and the EIF, the NAAEE has worked to enhance the effectiveness, accessibility, and relevance of environmental education practice.

See Also: American College and University Presidents Climate Commitment; Education, Federal Green Initiatives; National Wildlife Federation.

Further Readings

Disinger, John F. "Tensions in Environmental Education: Yesterday, Today, and Tomorrow." In *Essential Readings in Environmental Education*, H. Hungerford, W. Bluhm, T. Volk, and J. Ramsey, eds. Champaign, IL: Stipes Publishing, 1998.

Simmons, Deborah. "Developing Guidelines for Environmental Education in the United States: The National Project for Excellence in Environmental Education." In *Environmental Education and Advocacy: Changing Perspectives of Ecology and Education*, Edward Johnson and Michael Mappin, eds. Cambridge, UK: Cambridge University Press, 2005.

Therese Ferguson
Independent Scholar

NORTHEAST CAMPUS SUSTAINABILITY CONSORTIUM

The Northeast Campus Sustainability Consortium (NECSC) is a vibrant network of institutions of higher learning from the northeastern United States and the Canadian maritime region with a shared interest in advancing sustainability on their respective campuses and as a region. The consortium was established in October 2004 to meet the challenges set forth by the United Nations Decade of Education for Sustainable Development (UNDESD). The NECSC is one of the first regional sustainability consortiums established in the United States.

The model of collaboration and information exchange developed by the NECSC network is indicative of the growing collaborative nature of the campus sustainability profession. Participating institutions have recognized the value in sharing best practices and strategic plans in an effort to enhance the work on their own campuses. The NECSC helps to provide a community of support for instituting operational changes, integrating sustainability into curriculum, and advancing research and community outreach.

The NECSC is led by a steering committee of 20 professionals from colleges and universities, as well as the by the U.S. Environmental Protection Agency (EPA). Participating universities demonstrate the diverse nature of institutions of higher learning committed to campus sustainability—from small liberal arts colleges to large research universities and Ivy League institutions. These pioneering institutions have opted to establish a grassroots network of professionals that communicate via a monthly conference call, in addition to an annual meeting. There are neither dues required nor a formalized membership mechanism.

The NECSC participants committed to 10 years of action as a region, from 2005 to 2014. A series of annual meetings was proposed within the UNDESD time frame that would take place in each of the participating northeastern states and in the Canadian maritime region. NECSC member institutions maintain an ongoing dialogue throughout the year, and have committed to hosting an annual meeting held at a different member campus throughout the 10-year time period. The annual conference provides an opportunity for networking, professional development, and a chance to learn about and share best practices. In addition, steering committee members also attend biannual meetings and participate in monthly conference calls. These invaluable opportunities to build expertise and strengthen toolkits are central components of the NECSC's work.

See Also: Leadership in Sustainability Education; Sustainability Officers; United Nations Decade of Education for Sustainable Development 2005–2014.

Further Readings

Barlett, Peggy F. and Geoffrey W. Chase. *Sustainability on Campus: Stories and Strategies for Change.* Cambridge, MA: MIT Press, 2004.

Meadows, Donella. "The Global Citizen, 1996–2001." http://www.pcdf.org/meadows (Accessed November 2009).

Yale Office of Sustainability. "Northeast Campus Sustainability Consortium" (2010). http://sustainability.yale.edu/necsc (Accessed June 2010).

Sara E. Smiley Smith
Yale University

NORTHLAND COLLEGE

The Congregational Church (now the United Church of Christ) founded Northland College in 1906 as an outgrowth of the North Wisconsin Academy founded in 1892. Northland is located in Ashland, Wisconsin, and utilizes its proximity to Lake Superior and the Chequamegon National Forest as an important component of its environmental education mission. Northland has integrated environmental and sustainability issues throughout its liberal arts curriculum since 1971, making it a national leader in sustainability education. Northland also houses the Sigurd Olson Environmental Institute to supplement its educational, research, and outreach programs.

Academic Programs and Student Groups

Northland College has a distinctive environmental mission. Environmental and sustainability issues are woven throughout the Northland curriculum, including general education classes. Classes offered include Sustainable Business, Introduction to Environmental Studies, Sustainable Agriculture, Renewable Energy, and Natural Resources. Two environmental student scholarships are available: the Eco-Visionary Leadership & Service Scholarship and the Sigurd Olson Environmental Scholarship.

The Foundations in Nature block integrates natural science, social science, and humanities perspectives in a single set of courses. There is also a similar block for Sustainable Agriculture. Superior Connections students utilize the proximity of the Lake Superior watershed in their geology, biology, religion, writing, and other liberal arts courses. Students also contribute to Northland's environmental mission through class projects and activities. Capstone projects have included a regional conference on organic farming and the development of an action plan to increase use of local foods in the cafeteria.

Northland offers a variety of majors with an environmental emphasis. Northland has offered an Environmental Studies program since 1970. An outgrowth of this program was the Peace Studies program implemented in the early 1980s. Other ongoing sustainability-related majors include Natural Resources, Native American Studies, and Outdoor Education. Majors introduced in 2009 as part of the redesigned and updated curriculum include Humanity and Nature Studies, with emphases in Environmental Humanities, Native Cultures, and Ecopsychology; a Social Justice emphasis in the Sociology and Social Justice major; and a Sustainable Community Development major.

Northland students also gain environmental leadership experience through participation in a variety of student organizations and campus sustainability initiatives, many of which begin as class projects. New students receive a sustainability brochure and participate in an Outdoor Orientation program led by upper-class students. Students may live in the McLean Environmental Living and Learning Center (MELLC), an environmentally conscious residence hall, or other campus houses devoted to aspects of sustainability. Residence halls hold energy reduction competitions.

The Northland College Student Association (NCSA) annually provides $20,000, funded by student fees, for a campus sustainability project. Past projects include the implementation of renewable energy systems, the purchase of hybrid cars, expansion of the local bus service and free rides for students, and a Web-based monitoring system for the campus's photovoltaic arrays and wind towers. Students gain environmental experience through internships, work-study, and volunteer positions. Students are creating the first

North American chapter of OIKOS International, dedicated to everyday sustainability and social responsibility. Graduation ceremonies have been carbon-neutral since 2008.

Research and Outreach

Northland College held its first environmental conference in 1971 and sponsored the Midwest Regional Collaborative for Sustainability Education 2010 Summer Conference. The college also hosts numerous guest lecturers and commencement speakers on sustainability topics. Northland participates in the Lake Superior Binational Forum, a model partnership featuring members from various sectors—including businesses, environmental organizations, Native Americans, and educational institutions—that provide information to governments about critical environmental issues related to Lake Superior. The college's Pathfinders program provides leadership training to community middle and high school students.

Northland College is in the Leadership Circle of Signatories of the American College and University Presidents Climate Commitment and has signed the Talloires Declaration. Northland's memberships and affiliations include the Association for the Advancement of Sustainability in Higher Education (AASHE), the Upper Midwest Association for Campus Sustainability, the National Wildlife Federation's Campus Ecology Program, the Campus Consortium for Environmental Excellence, the Chequamegon Bay Green Team and Natural Step Framework of Sustainability, the Ecoleague, and the Dark Sky program. Awards include a 1993 Renew America award for environmental achievement, the awarding of Tree City USA and Tree Campus USA status from the Arbor Day Foundation, a 2004 Distinguished Growth Award, and a 2005 Project Partnership Award from the Wisconsin Urban Forestry Council.

Northland's chief community outreach organization is the Sigurd Olson Environmental Institute (SOEI), founded in 1972. SOEI coordinates regional environmental initiatives regarding Lake Superior and northern ecosystems. SOEI also presents the annual Sigurd F. Olson Nature Writing Awards and the Sigurd T. Olson Loon Research Award. Students can volunteer with SOEI through its renowned LoonWatch research and protection program or participate in the Apostle Island School environmental Mayterm program for area youth offered in cooperation with the Northland College Outdoor Education Department and the Apostle Islands National Lakeshore. SOEI also offers students, faculty, and staff courses and internships.

Campus Sustainability

Northland committed itself to a Sustainability Charter in 1997 and implemented a Sustainability Policy in its effort to serve as a sustainability model for students as well as for other organizations. The college maintains a full-time sustainability coordinator as well as an Environmental Council composed of faculty, staff, and students. Students can also participate in sustainability initiatives through work-study. Sustainability goals include the areas of building and landscaping, energy, food, transportation, waste, and composting. Students maintain the campus Environmental Commitment Website section, the electronic newsletter, and the Environmental Council MySpace page.

Northland is committed to attaining at least the U.S. Green Building Council's LEED Silver certification on all new buildings and major renovations. Dexter Library is Northland's first certified building. Alternative sources of campus energy include wind towers, hot-water arrays, photovoltaic arrays, and geothermal heating and cooling. Students helped install the photovoltaic array at the college president's house. Students built the Strawbale Demonstration Energy Lab as a model of alternative building techniques. Northland's food service provides students, faculty, and staff with organic, local, and Fair Trade foods as well as vegetarian and vegan options. Other sustainable food practices include composting, free reusable mugs, and trayless dining.

Campus sustainability initiatives provide students with firsthand opportunities to gain experience. Students participate in composting and campus gardens and manage storm water ponds and bioswales. Students also participate in recycling, reuse, and conservation programs such as the Reuse Center and RecycleMania competition. Students run a Sustainable Vending Initiative to gain small business experience and provide healthy beverage alternatives. Students participate in carbon footprint transportation surveys, the creation of an electronic ride-share board, and the implementation of free public transit passes and free access to bicycles. Students also participate in campus events, including the Sustainability Fair, Stewardship Week, and Energy Awareness Month.

See Also: Association for the Advancement of Sustainability in Higher Education; Outdoor Education; Sustainability Officers.

Further Readings

Association for the Advancement of Sustainability in Higher Education. "Northland College 2008 Campus Sustainability Leadership Award Application." http://www.aashe.org/resources/profiles/cat2_133.php (Accessed June 2010).

Barlett, Peggy F. and Geoffrey W. Chase. *Sustainability on Campus: Stories and Strategies for Change.* Cambridge, MA: MIT Press, 2004.

M'Gonigle, R. Michael and Justine Starke. *Planet U: Sustaining the World, Reinventing the University.* Gabriola Island, British Columbia, Canada: New Society Publishers, 2006.

Rappaport, Ann and Sarah Hammond Creighton. *Degrees That Matter: Climate Change and the University.* Cambridge, MA: MIT Press, 2007.

Marcella Bush Trevino
Barry University

OBERLIN COLLEGE

Oberlin College is a four-year, private, liberal arts college located in rural Oberlin, Ohio. Establish in 1833, Oberlin is home to over 2,800 students and 1,000 staff. The college is strongly committed to sustainability, and has served as a leader in the discipline, including in its Center for Environmental Studies.

Oberlin's advances in sustainability stem from a strong central administration, beginning with its Board of Trustees, which adopted an Environmental Policy Statement in 2004, as well as within the institution's strategic plan. The college is also a signatory of the American College and University Presidents Climate Commitment (ACUPCC) and the Talloires Declaration.

A Committee on Environmental Sustainability (CES) serves as the center point for implementation of sustainability initiatives on campus, as well as to promote sustainability practices on campus. CES consists of administrators, faculty, staff, students, and a member of the City Council, and reports to the special assistant to the president responsible for Sustainability and the Environment. In addition to the CES, a full-time sustainability coordinator reports to the vice president for finance, who oversees the Office of Environmental Sustainability, which was created in 2006.

Purchasing

To assist the college in its sustainability goals, a Green Purchasing Policy was enacted in 2007, which states that purchased products should meet environmental standards. All available Energy Star products are purchased, and environmentally preferable paper products when possible, including Forest Stewardship Council certified paper for official college publications. When possible, environmental responsibility cleaning products and electronics are purchased, while organic fertilizers and pesticides are being tested on campus.

Energy, Facilities, and Planning

Oberlin has tracked its greenhouse gas emissions annually since 2005. Emissions dropped in 2006, but were on the rise again, from 50,104 metric tons in 2005 to 51,049 in 2007.

Included in the campus facilities are 126 buildings encompassing 2,640,040 square feet. The central cooling system was recently replaced with new chillers that are twice as efficient as the former units, while the Adam Joseph Lewis Center for Environmental Studies, the Litoff Building, and Allen Art Building are either currently utilizing geothermal heat pumps, or are having systems installed. In addition, the Lewis Center features a 60 kW photovoltaic grid, with a 100 kW solar parking pavilion next to it.

Forty percent of Oberlin's electricity is purchased through renewable energy credits from its local municipal utility. To help reduce energy use, Oberlin uses sign placement, student competitions, peer outreach, and a Web-based monitoring system, among other methods, including 60 percent of building square footage being under the control of a building automation system. In addition, 98 percent of campus lighting comes from non-incandescent sources. Within residence buildings, 80 percent of laundry machines are Energy Star certified, with low-flow showerheads and faucets being placed in 75 percent of dorms. The Annex of the Lewis Center for Environmental Affairs contains a waterless urinal, a composting toilet, and a Living Machine that utilizes wastewater for toilets and landscaping needs.

A Green Building Policy helps guide the college's efforts, and states that all new construction and major renovations must meet LEED Silver standards. Currently, 140,000 square feet are certified at the LEED Gold level, with an additional 2,500 square feet meeting the LEED Silver level, and 14,000 square feet (the Center for Environmental Studies) meeting the LEED Platinum level, but not officially certified. Funding for sustainability-related projects comes from a number of resources, including a student green fee, a green fund for alumni, and a revolving loan funds for various campus projects.

Compost and Recycling

Recycling is encouraged at Oberlin, and the institution has a diversion rate of 45 percent, including aluminum, cardboard, glass, paper, most plastics, and carpeting. In addition, the college has an electronic recycling program that collects batteries, cell phones, computers, light bulbs, and printer cartridges. When the college undergoes construction projects, 75 percent of waste is diverted from landfills. Aside from dining facilities, 100 percent of landscaping waste is composted or mulched.

Transportation

Commuting is widely adapted at Oberlin, including 99 percent of students and 48 percent of faculty. Oberlin's transportation fleet of 80 vehicles includes three hybrids, as well as a tractor that has been converted to run on used vegetable oil for lawn mowing and snow plowing. In terms of public transportation, the college provides a free student shuttle service as well as an indirect subsidy for public transportation. A bike co-op, created in 1986 and run by students, provides repair, rental, and education services. On average, 45 bikes are available for rental for $15 per semester. Two vehicles are also available as part of a car-sharing program for students for a $50 annual membership fee. These transportation services, and goals for future plans, are included within Oberlin's Environmental Policy.

See Also: American College and University Presidents Climate Commitment; Leadership in Sustainability Education; Sustainable Endowments Institute.

Further Readings

Oberlin College. "Environmental Sustainability." http://www.oberlin.edu/sustainability (Accessed June 2010).

Sustainable Endowments Institute. "Green Report Card 2010." http://www.greenreportcard .org/report-card-2010/schools/oberlin-college (Accessed June 2010).

Justin Miller
Ball State University

Outdoor Education

Outdoor activities may help children learn respect for the environment and improve group dynamics, and have been found to be appropriate even for very young children.

Source: U.S. Fish & Wildlife Service

Outdoor education centers upon learning experiences that take place outside. Making the natural environment a focal point of education has been endorsed by many different cultures as a means to prepare their children for an active and fulfilling life. Outdoor education seeks to allow journey-based experiences that allow students to engage in activities such as hiking, climbing, canoeing, and other group games. Outdoor education combines outside activities and environmental learning into a cohesive whole that allows for experimentation and manipulation of the natural world, but also creates an appreciation for activity, cooperation, and the natural environment. Exposure to outdoor learning teaches the ability to overcome adversity, enhances personal and social development, and encourages a deeper relationship with the natural world. Outdoor learning encourages respect for the environment, is appropriate for a variety of age groups, and is grounded in accepted learning theory.

Background

Environmental education examines how the natural environment functions, the effects humans and their activities have on the environment, and how certain behaviors and practices can be changed to foster sustainability. Environmental education impacts and affects outdoor education, which allows for immersion into the natural world and exposure to the issues facing society. Outdoor education also has the ability to explore many other social norms and values. Outdoor education encourages activities that build interaction with the natural environment, allowing individuals to learn about and participate in the shaping of social norms, to undergo and surmount challenges in a supportive environment, and to gain appreciation of and knowledge about the process of learning and engaging in activities with others.

The theory supporting outdoor education is predicated upon the belief that an individual's true nature is revealed through working with nature, as he or she is removed from the distraction of modern comforts and conveniences. The stresses facing individuals during this experience create an appreciation for their own abilities to survive and interact, and thus also serve as a vehicle for self-analysis and as a catalyst for personal growth. Through these processes, it is believed that a greater awareness of society and the necessity of teamwork are gleaned.

As a result of outdoor education, teachers often observe differences in student behavior following trips outside of the school. Educators have also witnessed marked changes in the relationships between individuals comprising the group, further supporting the idea that outdoor education enhances group dynamics and an understanding of teamwork. This connection between outdoor education and group dynamics has its roots within the foundations of education, thanks to the influential work of educators and school reformers such as John Dewey, whose belief in experiential learning and views on the goals of education have shaped educational policy.

Benefits

Interest in outdoor education has grown as awareness of environmental issues has increased. With spreading urbanization, "nature deficit disorder" has become increasingly common. This disorder, caused by a lack of time spent outdoors, causes behavior problems linked to a lack of connection to nature. The situation is exacerbated by a concurrent increase in parental fears concerning germs, physical harm, and other issues. Although this disorder is not uniformly recognized by medical journals, it has become an issue that educators and other societal leaders have sought to address, especially for children from low socioeconomic status backgrounds. Against this background, outdoor education has become a popular way to remedy some of these perceived ills. Studies have been conducted on students engaging in outdoor activity for prolonged periods of time. Outdoor activity has been demonstrated to have a positive impact on children's physical development, improve their balance and agility, and enhance other fine motor skills. Studies of outdoor education suggest that students engaging in it experience a decrease in the number of injuries caused by accidents as they develop skills in assessing potential danger in situations and recognizing ways to avoid danger. Additional benefits observed include an increase in the operation of students' immune systems as a result of outdoor activity, a benefit to both students and their caregivers.

Age Appropriateness

The idea of outdoor education and utilizing the natural environment has been implemented with students of all grade levels. Typical activities for students involve canoeing, climbing, games, hiking, and rope courses. Some of these activities were originally believed to be too rigorous for younger students. As a result, outdoor education in some areas centered mainly upon older students, such as those in high schools. As educators have become more familiar with outdoor education, it has increasingly been used with younger students. One of the more interesting recent manifestations of outdoor education involves the formation of forest kindergartens. Forest kindergartens serve children between the ages of 3 and 6, who, despite their youth, spend the majority of their time outdoors. This emphasis on outdoor instruction extends to all curricular areas, including teaching lessons on dressing appropriately for the weather. Although originating in Europe, forest kindergartens have spread to the United States. These schools are common in areas that do not easily lead to outside activities. For example, the private Waldorf School in Saratoga Springs, New York, has insisted that children enrolled there spend three hours a day outdoors, using a 325-acre portion of Hemlock Trail, a New York state park, as its campus. The realization that inclement weather does not hinder outdoor education has spurred its growth.

Forest kindergartens refuse to allow children to play with commercial toys with scripted uses. This policy encourages children to form their own language of play and encourages the creation of a stronger community environment. Language creation assists in the development of a child's understanding of both the value of language and its use in assigning meaning. Better understanding of the conventions of language, and its creative nature, contributes to improvement in children's verbal aptitude and builds imagination.

Learning Theory

Models of instruction, such as outdoor education, are examined to see how well they are supported by learning theories. Howard Gardner's theory of multiple intelligences, influential since the time that he first identified different expressions of intelligence related to cognition, is one of the most popular learning theories. Outdoor education embraces Gardner's work insofar that it allows students to negotiate their own learning in ways that increase performance. Gardner's naturalistic intelligence, which centers on nurturing nature and relating information to natural surroundings, would seem a logical source of support for outdoor education. The work of other learning theorists, however, also buttresses the goals and objectives of outdoor education, such as Robert Sternberg's view of practical intelligence, Joseph Renzulli's emphasis on the need for task commitment, and David Perkins's stress on the necessity for active use of knowledge. As such, outdoor education aligns with goals and objectives of classical learning theory.

Conclusion

Outdoor education's benefits allow renewed relationships between groups, changes in the awareness of self, an increased ability to succeed, and a marked difference in how individuals utilize their personal abilities. Outdoor education also allows increased awareness of the environment and humankind's role in its stewardship.

See Also: Constructivism; Early Childhood Education; Integrating Sustainability Education Concepts Into K–12 Curriculum; Sustainability Topics Correlated to State Standards for K–12; Sustainability Topics for K–12.

Further Readings

Brooks, J. and M. Brooks. *In Search of Understanding: The Case for Constructivist Classrooms,* 2nd ed. New York: Prentice Hall, 2000.

Gardner, Howard. *Multiple Intelligences: New Horizons in Theory and Practice.* New York: Basic Books, 2006.

Gilbertson, K., T. Bates, T. McLaughlin, and A. Ewert. *Outdoor Education: Methods and Strategies.* Champaign, IL: Human Kinetics, 2006.

Knight, Sara. *Forest Schools and Outdoor Learning in the Early Years.* Thousand Oaks, CA: Sage, 2009.

Louv, Richard. *Last Child in the Woods: Saving Our Children From Nature-Deficit Disorder.* New York: Algonquin Books, 2008.

Perkins, David. *Making Learning Whole: How Seven Principles of Teaching Can Transform Education.* Hoboken, NJ: Jossey-Bass, 2010.

Renzulli, Joseph S. and Sally M. Reis. *Enriching Curriculum for All Students.* London: Corwin Press, 2007.

Sternberg, Robert J. *Wisdom, Intelligence, and Creativity Synthesized.* New York: Cambridge University Press, 2007.

Stephen T. Schroth
Jason A. Helfer
Jordan K. Lanfair
Knox College

P

PLACE-BASED EDUCATION

Place-based education can help students make connections between their local communities and regional, national, and global issues. For example, a trip to a local orchard might include activities calculating the food miles that apples in a grocery store travel.

Source: iStockphoto.com

Resulting from significant scholarship, the concept of place has come to the forefront of education in recent years. Although learning within traditional cultures has historically occurred within communities' natural surroundings, contemporary school systems are increasingly emphasizing curricula that either occur within, or at least draw on, local surroundings. Such education that places particular pedagogical value on the local is referred to by a variety of names: pedagogy of place, place-based learning, experiential education, community-based education, sustainable education, or most generally, environmental education. What unites both these terms and these forms of education together is an emphasis on the local context. Place-based education is distinguished from contemporary school settings and defined by its emphasis on the pedagogical value of students' local communities. This emphasis on "the local" can take innumerable forms. Depending on the community, ecology, and cultural context, place-based education can engage students in learning activities regarding the history, environment, culture, economy, literature, or art of a particular landscape. While this learning often focuses on students' own "place," such as their immediate schoolyard, neighborhood, town, park, or community, in doing so, it can also engage students in processes of reflection, making connections between local and larger regional, national, and global issues and processes. An example of this reflection is a school program that brings students to a local apple orchard and then engages them in activities calculating the food miles that grocery store produce travels. Place-based education is frequently applied and can take a project-based approach. The Foxfire program in Rabun, Georgia, is a famous example of place-based education; as part of a journalism class, rural Appalachian high school students interview family and community members about cultural traditions and publish a magazine based on their stories.

Place-based education can occur in various spatial contexts and with various types of populations. Although frequently classified as "education," it is noteworthy that these types of activities can also occur in the context of both formal and informal learning. Following trends in education scholarship, education, whether classified as formal or informal, is that which occurs within an organized and institutionalized setting. Formal education is separated from informal education by the context of the educational setting; normally, formal education is that which largely takes place in a contemporary school, whereas informal education could occur at a nature center. Formal examples of place-based education within contemporary schools can consist of a lesson that discusses the hydrologic cycle and involves students in tracing the flow of a local watershed. Informal place-based education might occur at a state park as part of a school field trip, where students can get a hands-on opportunity to visualize microorganisms living beneath rocks in a stream.

The pedagogical value of the local can also be imparted and gained through formal and informal learning. An example of formal learning might be a Boy Scout hiking trip, where scouts learn to identify local plants and their cultural usages. Informal place-based learning, distinguished by the lack of institutionalized organization, could occur on a family hiking excursion where parents relate stories and memories related to specific landmarks.

Place-based education became formalized and promulgated as a concept in the 1990s through work of the Orion Society, but it has existed as a practice for much longer. In fact, depending on one's definition, most traditional forms of education are place-based, and it is modern education that has become separated from place. Teachers now learn both pedagogies and methodologies of place-based education in graduate schools and engage students in this form of education both domestically and internationally.

There are two main currents in place-based education: uncritical and critical. These two forms are not mutually exclusive, but making a distinction is useful for understanding place-based education initiatives. Uncritical place-based education initiatives are characterized by unquestioning analyses of places, and the cultural, historical, and political constructions of these places. An example of an uncritical place-based education lesson might focus on students learning to identify local species of trees by their bark. Critical place-based education, conversely, includes elements of either individual or collaborative deconstruction. Such analyses might involve examining the economic, social, historical, and political forces that are responsible for the production of environments. An example of a critical place-based education approach might involve students looking at their local landscape, such as the Pacific Northwest, learning to identify trees by their bark, and then discussing how that particular assemblage of species is a product of both indigenous management practices and contemporary economic and political processes. Critical reflection is the process that distinguishes these two forms of place-based education. Whether or not critical reflection is involved in place-based education is not a function of the student demographic, or the environmental context; rather, it is a matter of teacher choice.

This distinction between uncritical and critical place-based education has its roots in academic approaches to the study of education. From a scholarly perspective, education researchers see the renewed interest in "the local" as a response to the traditional educational system that obscures peoples' relations with places they inhabit and the lives they live within these places. Place-based education, whether critical or uncritical, seeks to promote dynamic engagements with local settings, making teachings relevant to the daily experiences of educators and students in fundamentally practical ways.

The practical engagement of students with their local environments is one of the defining features of place-based education. This applied engagement can be either critical or uncritical. As the local community and its landscape are resources within place-based education, addressing community problems is a way to facilitate reflection in students through service-based learning. An example of an uncritical place-based education activity can take the form of a class adopting a watershed. A critical version of this activity might involve adopting a watershed, investigating sources of point pollution, and presenting information gained to community leaders.

Regardless of its formal or informal context, or the critical or uncritical aspect of its pedagogy and implementation, place-based education is distinguished by its emphasis on the pedagogical value of students' local communities.

See Also: Environmental Education Debate; Experiential Education; Outdoor Education.

Further Readings

Gruenewald, D. "Foundations of Place: A Multidisciplinary Framework for Place-Conscious Education." *American Educational Research Journal,* 40/3 (2003).

Hutchison, D. C. *A Natural History of Place in Education.* New York: Teachers College Press, 2004.

Sobel, D. *Place-Based Education: Connecting Classrooms and Communities.* Great Barrington, MA: Orion Society, 2004.

Thomashow, M. *Bringing the Biosphere Home.* Cambridge, MA: MIT Press, 2003.

David Meek
University of Georgia

PORTLAND UNIVERSITY

Among higher education institutions, Portland University is paving the way in sustainability research. Through innovation, ideas, and community partnerships, Portland University seeks to enable students, faculty, and campus leaders to meet contemporary environmental, economic, and social challenges. Portland University (or Portland State University, PSU) is located on a 49-acre campus within the city of Portland, Oregon, which is recognized as America's most sustainable city. One of the goals of the university is to enable all undergrads to become literate in the concepts, methodologies, and applications of sustainability. The campus itself is considered a living laboratory, and its motto, "Let Knowledge Serve the City," exemplifies the applied relationship between teaching, research, and service to an increasingly global community.

In 1946, the university was first established under the name Vanport Extension Center as a way to fulfill the need for higher education among Portland's returning World War II veterans. In 1969, the school gained university status through the Oregon State System of Higher Education, becoming Portland State University. In 1994, PSU made controversial headlines within the higher education community by canceling the traditional undergraduate distribution system and adopting a novel interdisciplinary general education program termed "university studies."

Portland State University has received numerous sizable financial contributions that have advanced its facilities and research in sustainability studies. In 2004, Dr. Fariborz Maseeh made a donation of $8 million, through the Massiah Foundation, to the College of Engineering and Computer Science. Following this grant, the college was renamed the Maseeh College of Engineering and Computer Science. At the time, it was the single largest donation to the university, and along with other financial gifts, led to the opening of a new LEED Gold-certified engineering building, the Northwest Center for Engineering, Science and Technology, which houses much of that college. Additional recognition of the university's commitment to sustainability came in 2006, when the nonprofit organization Salmon Safe declared Portland State to be the nation's first Salmon Safe University. This award recognized the campus's efforts at treating storm water runoff before it flowed into the local watershed. In 2008, Portland State University received a $25 million challenge grant from the James F. and Marion L. Miller Foundation. A condition of both the $25 million Miller grant and the funds raised to match it was that the monies be used solely to develop and support sustainability programs. Another milestone in the development of the university's sustainability focus was in 2008, when the Jimmy and Rosalynn Carter Partner Foundation presented the university with the Jimmy and Rosalynn Carter Partnership Award for Campus-Community Collaboration for their innovative Watershed Stewardship Program. That program recruited more than 27,000 community volunteers, donating a quarter of a million hours to install 80,000 plants and restore 50 acres of watershed along two miles of river.

Degrees and Departments

At Portland University, the study of sustainability is not a single class, major, or even department. Rather, like the subject itself, it is interconnected, and integrated on campus throughout the curricula, buildings and grounds, research initiatives, general education requirements, and university objectives. Sustainability research and education within the university is focused on four primary areas of inquiry: urban sustainability, the sustainable integration of human societies with the natural environment, methods for implementing sustainability, and metrics for measuring change and sustainability. The university's first focal subject explores the integration of natural and human communities, and asks its faculty and students to consider how we can build societies that are integrated with those natural systems that support our lives.

Given the ever-expanding population and increasing migration to urban centers, the university believes the answers to these questions lie in the study of urban communities. As the second thematic focus area is urban sustainability, the City of Portland offers an ideal research laboratory to analyze the dynamic ecological and socioeconomic conditions of a city that is already adopting various sustainable policies and principles. Specifically, the university is studying what it terms the triple bottom line (economic, ecological, and social needs) and how this mediates sustainable planning in urban centers. The third area of focus is working toward understanding transitions to achieve sustainability. Within this area, research seeks to understand why organizations, people, and societies transition to more sustainable policies and behavior, and what motivates these behaviors. Equally important, faculty and students at PSU study why sustainable practices are not adopted. The fourth focus area is concerned with developing and applying metrics for sustainability. By providing tested measurement tools, the university envisions enabling individuals, organizations, and governments to choose practices based on previous knowledge of what works, thereby increasing sustainable behavior.

As a result of these foci, faculty within the university are engaged in cutting-edge research and quality teaching on local issues of critical importance at various scales within the Pacific Northwest and abroad, such as how to create and support sustainable urban communities, understand climate change, measure ecosystem services, and develop renewable energy sources. More specifically, targeted research areas in sustainability within the university include community food systems, economic sustainability, environmental science and green technology development, environmental sustainability, humanities and sustainability, integrated water resource management, social sustainability, sustainable business processes and practices, sustainable urban development, and transportation systems.

PSU approaches sustainability as an integrated way of living and working, and not as a novel separate issue. As a result, sustainability is actively integrated into both general education requirements and specific disciplinary majors. With cross-campus integration in mind, the university has developed the following undergraduate programs that emphasize economic, environmental, and social sustainability: University Studies, Environmental Sciences and Management, Maseeh College of Engineering and Computer Science, and the School of Urban Studies and Planning. Graduate programs that emphasize sustainability include Environmental Sciences and Management, Graduate School of Education, Mark O. Hatfield School of Government, Maseeh College of Engineering and Computer Science, School of Business, and the School of Urban Studies and Planning. To explore a few examples of how these programs emphasize sustainability, students within the master's program in Environmental Management learn how to examine and manage natural environments to ensure both human benefit and ecosystem health. Another innovative application of a traditional graduate program is PSU's Master of Public Administration, which focuses on providing students with the training necessary for administrative positions in governmental and nonprofit organizations. For students interested in pursuing graduate degrees in sustainable technology, the Master of Science in Mechanical Engineering includes courses in Building Energy Modeling, and Solar Energy Engineering. The most focused graduate program in sustainability is the Graduate Certificate in Sustainability, which provides an integrated series of postgraduate courses that constitute an interdisciplinary study of the environmental, social, and economic dimensions of sustainability. The certificate comprises a two-part program. The first element of the program consists of four core courses that provide students with knowledge concerning the diversity of sustainability concepts. In these courses, students learn about the major theories and concepts related to the pivotal aspects of sustainability, as well as engage in case studies on private and public projects. During the second part of the program, students take two elective courses that provide the opportunity to focus on specific issues related to sustainability within the student's area of interest. The certificate program is administered by the Center for Sustainable Processes and Practices. To complete the certificate, students need to complete six courses, which must total at least 22 credits. The certificate program is open to students admitted solely to the certificate program, as well as those currently enrolled in a master's or doctoral program in the university. Portland State University also offers a professional certificate in green building. This program focuses on developing knowledge and strategies for building professionals interested in sustainable construction who are looking to expand their knowledge. They also offer LEED Green Associate and LEED AP Exam prep courses.

Campus Sustainability

Sustainability at Portland State University is coordinated by the Campus Sustainability Office (CSO). The CSO coordinates strategic planning related to resource conservation, and monitors resource use at PSU. The office facilitates communication between departments, faculty, students, and the administration, so that all individuals have the information necessary to make sustainable choices. The CSO also serves as a liaison between the university and students, fostering sustainable leadership opportunities. The CSO has piloted several novel initiatives, including the Sustainability Tracking, Assessment & Rating System (STARS), which monitors and quantifies all aspects of sustainability at PSU. The CSO is also actively involved in a partnership with the City of Portland in planning an EcoDistrict. Specifically, the CSO is working to identify new means of tracking and reducing resource flows within the city. The Green Team Program is another CSO initiative that seeks to help departments adopt sustainable practices.

Portland State University's campus is considered a living laboratory because it utilizes the campus's location as a place-based teaching resource. The fundamental principle underlying this living laboratory is that the campus is an ideal location to test, analyze, and review novel approaches to solving problems related to sustainability. As a laboratory, the campus also serves as a data repository for information regarding past projects, ensuring that lessons learned are dovetailed into new projects with the end goal of disseminating best practices to the larger academic, professional, and public communities.

Sustainability at PSU is not reserved solely for the classroom. Various student groups promote environmental, economic, and social sustainability. These groups include Engineers Without Borders, the Permaculture Guild, bicycle and cycling clubs, and the student-run *Ooligan Press*. The students also manage the Food For Thought Café, which has been committed to serving local, organic, and Fair Trade menu items since 2003.

Portland State University is a leader both in sustainability research and in integrating sustainability throughout campus buildings, programs, and projects. As a learning laboratory, the university is a center for place-based higher education. Its location in the city of Portland provides the university the ideal environment in which to study urban sustainability. With leading professors, substantial funding for sustainability research and programs, and a student body increasingly matriculating into sustainability-related programs, it is clear that Portland State University will continue to push the boundaries of sustainability research and the greening of higher education.

See Also: Green Community-Based Learning; Leadership in Sustainability Education; Outdoor Education; Place-Based Education.

Further Readings

Filho, Walter Leal. *Sustainability and University Life*. Ann Arbor, MI: University of Michigan Press, 1999.

Gough, Stephen and William Scott. *Higher Education and Sustainable Development*. London: Routledge, 2009.

Hargreaves, Andy and Dean Fink. *Sustainable Leadership*. Hoboken, NJ: Jossey-Bass, 2005.

Portland State University. http://www.pdx.edu (Accessed September 2010).

Sandell, Klas, Johan Ohman, and Leif Ostman. *Education for Sustainable Development*. New York: Studentlitteratur AB, 2005.

Silka, Linda and Robert Forrant, eds. *Inside and Out: Universities and Education for Sustainable Development*. Amityville, NY: Baywood Publishing, 2006.

Wiland, Harry and Dale Bell. *Going to Green: A Standards-Based Environmental Education Curriculum for Schools, Colleges, and Communities*. White River Junction, VT: Chelsea Green Publishing, 2009.

David Meek
University of Georgia

R

RENSSELAER POLYTECHNIC INSTITUTE

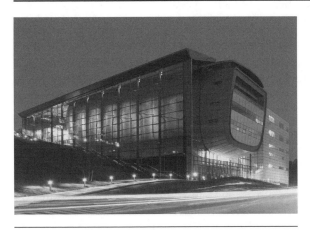

Rensselaer Polytechnic's new Experimental Media and Performing Arts Center, completed in 2008, features 36,000 square feet of sustainably grown Western Red Cedar from Canada and is in the process of becoming LEED certified.

Source: Wikipedia

Rensselaer Polytechnic Institute (RPI), in Troy, New York, is a private research university founded in 1824 by Stephen Van Rensselaer, the Dutch American politician whose inheritance of the 1,200-square-mile Rensselaerswyck quasi-feudal estate 40 years earlier had made him the 10th richest American of all time. Van Rensselaer appointed the school's first administrators and set forth its mandate—to research "the application of science to the common purposes of life"—making RPI the oldest technological university in the English-speaking world. It has remained one of the top technological universities in the world.

Over the years, RPI has repeatedly made headlines for its newsworthy initiatives and innovations, from paying students to participate in weekly course feedback meetings to becoming the first major American university to require that students have laptops. In 1999, the new Rensselaer Plan was introduced by new (and current as of 2010) university president Shirley Ann Jackson, who sought to preserve RPI's position on the cutting edge of research institutions with nanotechnology, biotechnology, interdisciplinary, and sustainability initiatives. Although Jackson has repeatedly come under criticism both on and off campus, by 2006, the school met her $1 billion research capital goal and subsequently reached its next goal nine months early. In the 21st century, the *Princeton Review* called RPI "simultaneously the oldest technological school in the country and the most modern . . . like George Jetson meets Archimedes."

RPI is in the early stages of integrating its curriculum to create interdisciplinary study programs covering the scientific, engineering, social, and architectural aspects of sustainability. The goal is for this sustainability focus to permeate all levels of the curriculum: coursework, minors, and degree programs.

The "2010 *Princeton Review*'s Guide to 286 Green Colleges" named RPI one of the most environmentally responsible American colleges. The *Princeton Review* found, specifically, that sustainability was a strong concern of students, that it was well represented in academic programs and research projects, and that it informed the development of infrastructure and operations on campus. Highlights of RPI's most recent commitments to sustainability include the creation of the Sustainability Studies minor and the opening of the Vasudha living and learning community for first-year students. The Sustainability Studies minor was created in 2009, and consists of four courses for students in any degree program: one of four 1000-level courses focusing on intersections of the environment and social issues, two 4000-level courses focusing on more specific topics like Environmental Philosophy or 21st Century Risks, and the 2000-level course Environment and Society. The Vasudha project, named for the Sanskrit word for "Earth," provides a living space for first-year students interested in sustainability issues. Vasudha residents are encouraged to take the Nature/Society course (part of the Sustainability Studies minor) together, blending the living and learning environment, and have the opportunity to take other courses in tandem.

In contrast, the Green Report Card awarded RPI only a B, with high marks for on-campus, curricular, student, and faculty initiatives, but lower marks for the involvement of the administration and shareholders. For instance, though the school is committed to sustainability, it has no dedicated sustainability office, nor has its president signed the American College and University Presidents Climate Commitment. It also has no official greenhouse gas emission reduction goals. All of these things resulted in a significantly lower grade for the administration than other areas of the university, despite having a formal sustainability policy. Further, because the equity holdings of the university's endowment are invested in mutual funds, it lacks the ability to communicate with and involve shareholders, tasks it must delegate to its investment managers.

The chemical-free Garden and Greenhouse Project produces lettuce, tomatoes, cucumbers, and herbs for use on campus. RPI's dining commons, operated by Sodexo Hospitality Services, emphasizes its sustainable options for a student population that is increasingly impatient with traditional cafeteria food and more likely to demand offerings from more adventurous cuisines, with local and organic ingredients. Fair Trade sustainable coffee is offered in all dining halls, and the campus has been converting to recycled napkins, sustainable packaging, and environmentally friendly dishwashing. The student-managed Terra Cafe, founded in 2007, serves organic meals made from ingredients from Troy-area farms every Wednesday afternoon. As of 2010, plans were under way to launch an on-campus farmers market and develop a biodigester that will use campus food waste to generate energy through anaerobic digestion.

In 2010, members of the RPI student group Engineers for a Sustainable World (ESW) traveled to Haiti to install solar panels, which they had designed and built, in order to provide power to impoverished schools; the panels are sufficient to run laptop computers donated by RPI (with cosponsorship from Troy's St. John's Episcopal Church and General Electric). The communities gifted with the solar panels had no reliable access to electricity and were taught how to maintain the solar panels to keep them running. The laptops allow digital books to be delivered to the school for free, in addition to their many other uses and applications, such as language-learning software and mathematics tutorials. An earlier project had installed solar-powered milk pasteurization systems in Peru in order to help dairy farmers produce pasteurized cheese that could be sold on the international market. ESW is a nationwide group, the RPI chapter of which launched in 2005 and grew to a roster of 40 students. Many ESW projects carry course credit as independent studies or part of the Introduction to Engineering Design course; the RPI ESW group also actively recruits nonengineers in order to build an interdisciplinary team to generate more effective and innovative projects.

See Also: Experiential Education; Sustainability Officers; Whole-School Approaches to Sustainability.

Further Readings

Filho, Walter Leal. *Sustainability and University Life*. Ann Arbor: University of Michigan Press, 1999.

Gough, Stephen and William Scott. *Higher Education and Sustainable Development*. New York: Routledge, 2009.

Hargreaves, Andy and Dean Fink. *Sustainable Leadership*. Hoboken, NJ: Jossey-Bass, 2005.

Sandell, Klas, Johan Ohman, and Leif Ostman. *Education for Sustainable Development*. New York: Studentlitteratur AB, 2005.

Silka, Linda and Robert Forrant, eds. *Inside and Out: Universities and Education for Sustainable Development*. Amityville, NY: Baywood Publishing, 2006.

Wiland, Harry and Dale Bell. *Going to Green: A Standards-Based Environmental Education Curriculum for Schools, Colleges, and Communities*. White River Junction, VT: Chelsea Green Publishing, 2009.

Bill Kte'pi
Independent Scholar

S

SANTA FE COMMUNITY COLLEGE

Santa Fe Community College (SFCC) is a two-year public community college established in 1983 and located in Santa Fe, New Mexico. SFCC offers more than 72 degrees and certificate programs, including several in sustainable technologies, and enrolls over 14,500 students annually. SFCC and the nearby Oshara Village community serve as models of sustainability. The new Sustainable Technologies Center will integrate sustainability across the SFCC curriculum and campus operations, and train current and future workers in emerging green technologies.

Campus and Community Sustainability Initiatives

Santa Fe Community College introduced a strategic plan including a commitment to sustainability and the adoption of new emerging technologies in 2006. Santa Fe is preparing a separate Campus Sustainability Plan to set core values and goals, inform and involve the campus community, gain funding for green initiatives, and integrate sustainable education throughout the curriculum. Santa Fe also has a Green Task Force composed of faculty, students, staff, and community members. Santa Fe seeks to be a model of sustainable operations, energy generation, and conservation.

New campus buildings will aim for LEED Silver level or higher through the U.S. Green Building Council's certification system. The Health and Sciences Building and the Trades and Advanced Technologies Building seek LEED Gold certifications or higher. Other campus sustainability initiatives now under way include a recycling program and use of recycled paper, recycling wastewater for campus landscaping irrigation, the purchase or lease of more flex-fuel, hybrid, and fuel-efficient vehicles, and the availability of free filtered water to reduce demand for water in plastic bottles.

College President Sheila Ortego is an original signatory to the American College and University Presidents Climate Commitment. Part of this commitment includes using a carbon calculator system to determine campus energy usage and greenhouse gas emissions as a first step to reducing them. Accounting students aided in the carbon calculator survey. Greenhouse gas and energy reduction initiatives include the installation of CO_2 monitoring devices, an automated biomass-heating system, solar thermal collectors to heat the swimming pool, a grid-tied solar photovoltaic system, and campus-wide low-consumption lighting.

SFCC works with neighboring Oshara Village, a model of sustainable and green technologies where many technologies taught at the college are tested. Local builders and building inspectors as well as students have observed the construction of homes in Oshara to gain firsthand experience in green building. Green features of Oshara homes include walkable designs, super-efficient structures, on-site energy generation, 100 percent water reclamation, and photovoltaic panels. Oshara also features a Sustainable Enterprise Zone, which houses an SFCC algae research laboratory and classroom to study the production of ethanol and biodiesel from plants. The first truly zero-energy home has been completed and is generating more electricity than it consumes.

Academic Programs and Sustainability Initiatives

Santa Fe Community College seeks to integrate sustainability throughout its curriculum. SFCC offers green general education courses such as Introduction to Sustainability and seeks acceptance of these courses by the New Mexico Higher Education Department as part of the statewide transferable-credit general education course offerings through the Social Sciences component. SFCC hosted a Sustainability Forum in 2007. SFCC offers several undergraduate and workforce environmental and sustainability degree and certification programs in renewable energy, green building, environmental technology, and sustainability.

The Environmental Technologies Associate in Applied Science (AAS) and Certificate program focuses on issues regarding sustainability, renewable energy, and water. The Biofuels Certificate program focuses on the process of making biodiesel, ethanol, and algae oil. The Green Building Systems Certificate focuses on building methods and materials that relate specifically to green building systems and the environmental impact of built structures. The Green Building Construction Skills Certificate focuses on skills needed for the green building construction industry, including computer-aided drafting and modeling, traditional construction skills, and green practices. The Solar Energy Certificate program focuses on the design and installation of photovoltaic solar electric energy systems and related skills. These programs also require students to gain real-world working experience while in the program.

Santa Fe Community College operates the $11.4 million Sustainable Technologies Center (STC). The building seeks LEED Gold or higher certification and utilizes alternative energy sources and green design features such as rainwater catchment. The STC integrates SFCC's sustainability curriculum and campus operations. The STC partners with community, state, and national organizations to expand the Santa Fe region's renewable energy cluster. The STC advisory board ensures the center's industry focus is maintained by seeking the input of task forces from the different trades and curriculum areas.

The STC will offer credit and noncredit courses and industry certification programs to an estimated 250 to 300 students annually along with customized workforce training, seminars, symposiums, and special events. The center's emphases include science, technology, engineering, and mathematics (STEM) education and workforce training in the fields of green construction, home weatherization, plumbing and HVAC, electrical trades, and renewable energy fields.

Santa Fe Community College has received several awards and recognition for its sustainability efforts. Representatives from Tucson, Arizona, toured the campus in 2009 as part of their research into sustainability and green jobs training. New Mexico Governor Bill Richardson praised the college's leadership role in the use of green technology in the same year. President Ortego was selected to present the college's campus sustainability model at the American Association of Colleges' Future Leaders Institute in Chicago, Illinois. Santa Fe was one of only two community college representatives.

See Also: Green Business Education; Leadership in Sustainability Education; Sustainability Officers.

Further Readings

Barlett, Peggy F. and Geoffrey W. Chase. *Sustainability on Campus: Stories and Strategies for Change.* Cambridge, MA: MIT Press, 2004.
M'Gonigle, R. Michael and Justine Starke. *Planet U: Sustaining the World, Reinventing the University.* Gabriola Island, British Columbia, Canada: New Society Publishers, 2006.
Rappaport, Ann and Sarah Hammond Creighton. *Degrees That Matter: Climate Change and the University.* Cambridge, MA: MIT Press, 2007.

Marcella Bush Trevino
Barry University

SECOND NATURE: EDUCATION FOR SUSTAINABILITY

Second Nature is a nonprofit, nongovernmental organization that has trained thousands of faculty members, students, and administrators in more than 250 institutions of higher education across the United States to make sustainability a core concern of their college or university. Founded in 1993 in Boston, Massachusetts, by Anthony Cortese and Senator John Kerry, among others, Second Nature has helped advance the concept of education for sustainability in the United States. According to Second Nature's Website at www.second nature.org, it aims to serve and support leadership in higher education to adopt "sustainable, just and healthy living [as] the foundation of all learning and practice in higher education."

Philosophical Underpinnings

The philosophy of Second Nature emphasizes the interconnectedness of the social, economic, and security dimensions of the contemporary environmental crisis. Further, it considers that environmental problems arose from faulty policies and practices in the social, economic, or other arenas. Thus, in this view, environmental problems do not comprise a separate phenomenon by themselves, but are offshoots of deeper systemic problems and distorted policy priorities. Second Nature recognizes the limitations of disciplinary approaches to understanding environmental issues, and emphasizes transdisciplinary contributions in understanding sustainability. One of the corollaries of considering environmental problems as systemic is that technical solutions alone would not bring about a transition to sustainability. What is needed to create a sustainable society is a change in mindset, values, aspirations, and knowledge—in other words, in culture itself.

Capacity Building for Sustainability

The core function of Second Nature is to promote capacity building in colleges and universities around the nation in order to incorporate sustainability as a primary value in all of their sectors. Second Nature undertakes networking to mobilize the public opinion in support of sustainability measures. Specifically, it seeks to identify high-leverage players and institutions whose participation in sustainability will impart credibility and momentum to the efforts to create a sustainable society. It was upon the establishment of the Association for Advancement of Sustainability in Higher Education (AASHE) that Second Nature decided to focus specifically on the leadership in higher education. As part of its outreach efforts, Second Nature provides consultancy services, organizes workshops, and provides speaking engagements. In the future, it plans to use online communication tools to reach out to a broader audience in a cost-effective way.

Strategic Alliance Building

Climate change has emerged as the defining environmental problem of our times. Second Nature promotes alliance building with a wide array of organizations to promote learning and collective action aimed at addressing climate change and supporting the transition to a sustainable society. The Higher Education Associations Sustainability Consortium, which includes 15 higher education organizations, is also coordinated by Second Nature. The goal of the consortium is to share information and promote co-learning in order to advance sustainability in member organizations. On the legislative front, Second Nature has partnered with Campaign for Environmental Literacy, and supported the passage of the Higher Education Sustainability Act and funding for the Energy Independence and Security Act of 2007. Tompkins County Climate Protection Initiative also involves Second Nature—in partnership with a multisectoral team of stakeholders, including private businesses, government agencies, and nonprofits—to work toward substantial reductions in the energy expenditure in Tompkins County, New York.

Influencing National Policies and Policy Priorities

Second Nature recognizes the critical influence that national policies have in affecting the leadership in higher education, especially in the areas of education and the environment. It works with organizations like the Campaign for Environment Literacy to secure more federal funding for environmental education. Second Nature's policy priorities for 2010 include securing $25 million for university sustainability programs, promoting a climate-sensitive national transportation policy, and increasing environmental awareness among the public, especially on the strong interconnections between the environment, economy, and energy.

American College and University Presidents Climate Commitment: Second Nature and Climate Change

Climate change, a global environmental crisis that can inflict catastrophic damage worldwide, has become a focal concern of Second Nature. The complexity and the enormity of the problem require unprecedented collaboration among a broad range of social players and institutions to produce the needed innovative solutions. Second Nature, along with

AASHE and ecoAmerica, both nonprofit organizations, supports initiatives to combat climate change on campuses nationwide. Specifically, Second Nature supports the core activities of American College and University Presidents Climate Commitment (ACUPCC), founded in 2006. These include recruitment, training, and educational activities.

ACUPCC aims to achieve a state of climate neutrality. This is defined as zero net emission of greenhouse gases, which can be achieved by a reduction in emissions as well as by using carbon offsets. Signatories to ACUPCC are required to keep an inventory of their greenhouse gas emissions and to set a target date to achieve climate neutrality within two years of signing the commitment. The member colleges and universities are required to take short-term steps to lower their carbon emissions. They also commit to incorporate sustainability into their curriculum. The climate action plans, inventories, and progress reports are also made public.

The approach of ACUPCC allows for drawing up appropriate local plans driven by the needs and priorities of the institution. Although the commitment is not legally binding, the release of an annual compliance report and the naming of institutions that fail to implement their climate action plans will create public pressure to streamline their efforts. The ACUPCC recognizes the plurality of motivations and circumstances that have driven the members to sign on to the commitment to transition to climate neutrality. There is no attempt to straitjacket sustainability in a top-down fashion, and it is acknowledged that among the objectives leading to the adoption of climate change abatement strategies are anticipated present and future benefits. These include increasing the institution's share of federal and private funding, improving the profile of the institution with potential students and faculty, gaining a competitive edge over peers, reducing operational costs, and potentially benefiting from the development of carbon trading.

In 2000, Second Nature organized a conference of U.S. college and university presidents at Oberlin College in Ohio. The joint declaration made at the end of the conference clearly carried the imprint of Second Nature's philosophy and thinking. The university presidents recognized the importance of raising awareness of the critical role that higher education will play in achieving the ambitious goal of a sustainable society. The declaration also noted the importance of social networking, sharing information, establishing a framework for assessing sustainability, and creating a community of higher education leaders who will take the lead in finding and strengthening synergies between the ongoing reforms and sustainability initiatives.

Green Building in Higher Education

In a bid to translate sustainability ideas into action, Second Nature established the initiative Advancing Green Building in Higher Education. Its purpose is to provide technical knowledge to campuses, especially those that cater to underserved populations and tend to have inadequate funding, to help facilitate their transition to greener infrastructure. This will enable cost savings, improve public health, promote environmental education, and increase student enrollment. A grant of $1.2 million from the Kresge Foundation helps to fund the initiative.

Second Nature has also established a Web portal called Campus Green Builder. The portal is the first one of its kind in higher education and will provide an extensive list of Web resources, lists of experts, and information on funding and training for colleges that lack the resources to make the transition to a greener infrastructure. A part of the grant will be used to provide fellowships to university executives for technical training in the field of green building. A portion of the grant will also be used to subsidize the institutional membership fee for AASHE.

Approach to Sustainability: Current Situation and Vision for the Future

Anthony Cortese, the founder and president of Second Nature, has provided a comprehensive view of sustainability that encompasses economic, social, and institutional dimensions and their interrelatedness. According to Cortese, the transition to sustainability will require changes in actions as well as in thinking, perception, and imagination. In this view, a sustainable society can only be created if methods of analyzing contemporary problems and their interconnectivity are fundamentally transformed. It is suggested that higher education, due to its influential social position, has a pivotal role to play in catalyzing the sweeping changes that are required. Additionally, since most societal and business leaders are trained in institutions of higher education, and acquire not only professional skills, but perhaps more importantly, values that will guide their decision making and thinking broadly, the culture of colleges and universities holds the key to the success of efforts to create a sustainable society. The impact of higher education goes beyond teaching and research to serving as a model for practices related to areas ranging from energy purchase and use to waste disposal and building design.

In a paper titled "Visions of Sustainability in 2050," Cortese lays out a bold and detailed vision of what a sustainable world might look like in 2050. In this scenario, Earth's population would have stabilized at a level where it is within the short-term and long-term carrying capacity; there would be broad acceptance of the dependence of health and socioeconomic progress on the integrity of the biosphere; fossil fuels would have been largely replaced by renewable resources; substantial reductions would been achieved in throughput by making manufacturing more efficient; production would have been redesigned by mirroring natural cycles to reduce wastage and pollution; socioeconomic inequality would be considerably reduced; and universal health and literacy would have been achieved. Cortese also describes reconfigured institutions and global governance structures that would be facilitative of a transition to a sustainable society. What is significant about this vision is its sweeping scope. The integrity of the ecosystems figures as prominently as the well-being, in the broadest sense, of the human population. The local and the global, the present and the future, the First World and the Third World, and the economy and the environment are envisioned in a seamless manner.

The philosophical basis of Second Nature is rooted in a belief in the possibilities that knowledge and an increased level of awareness lead to positive action. The constraints involved in achieving a sustainable society, however, go beyond knowledge. For instance, there are entrenched economic and political interests that may be committed to upholding the status quo, and to whom a transition away from the current technologies and socioeconomic systems may seem deeply destabilizing. It can be argued that the type of large-scale changes required to produce a sustainable way of life will produce distinct winners and losers. The transition to sustainability may therefore require more than knowledge and leadership. More specifically, political action and alliance building to bring together diverse constituencies will be necessary. Legislative action and legal changes that alter the mix of incentives and disincentives to promote sustainable actions, at both individual and institutional levels, will be indispensable.

In conclusion, Second Nature has helped create a powerful movement in support of the idea of education for sustainability. In doing so, it has helped reconceptualize education as a source of the knowledge, beliefs, values, and skills needed to be responsible global citizens in today's highly interconnected world. Higher education is therefore not a mere instrument for solving problems, but is seen as being able to shape minds and hearts to undertake the monumental task of reimagining and creating a new society. Second Nature thus redefines the emancipatory potential of education for today's world and the unique challenges presented by it.

See Also: American College and University Presidents Climate Commitment; Brundtland Report; Earth Day Network; National Wildlife Federation; United Nations Decade of Education for Sustainable Development 2005–2014.

Further Readings

Cortese, Anthony. "Visions of Sustainability in 2050." http://www.secondnature.org/pdf/snwritings/articles/AccTheTrans.pdf (Accessed June 14, 2010).

Cortese, Anthony and William McDonough, "Education for Sustainability: Accelerating the Transition to Sustainability Through Higher Education." http://www.secondnature.org/pdf/snwritings/articles/AccTheTrans.pdf (Accessed June 2010).

Second Nature: Education for Sustainability. "Catalyzing Sustainable Strategies for Higher Education." http://www.secondnature.org (Accessed March 2010).

"U.S. Presidents' Meeting at Oberlin College, November 2000." *International Journal of Sustainability in Higher Education*, 2 (2000).

Neeraj Vedwan
Montclair State University

SOCIAL LEARNING

Social learning is an approach to environmental problem solving and policy making that emphasizes the decision-making process over achieving predefined objectives. Equity, democratic participation, and fairness are considered as important as efficiency in managing natural resources. Typically, social learning involves bringing together a broad range of stakeholders with diverse interests for sustained interaction to discuss an issue. The stakeholder interaction can range from relatively informal and open-ended to fairly formal, where preferences and perceptions are elicited and mapped using modeling tools. The stakeholders engage in a sustained process of deliberation and are encouraged to define the problem in a holistic and inclusive fashion wherein all or most views are represented. The assumption underlying social learning is that stakeholders possess the capacity and willingness to learn from each other and, where necessary, to modify their positions to arrive at a shared definition of public good as it relates to the problem at hand. It is postulated that the decision-making process will necessarily involve give and take, but the main advantage for stakeholders lies in reducing the level of acrimony and conflict that often characterizes natural resource management.

The impetus for social learning has come from sources within and outside the field of natural resource management. The increasing recognition of uncertainty in the knowledge of natural systems, with the concomitant awareness of gaps in understanding, has increased the importance of the information derived from a multiplicity of sources. Also, over time, an exclusive reliance on "end of pipe" solutions, as part of the top-down approach to natural resource management, has revealed itself to be ineffective, even counterproductive, and expensive.

In light of the poor track record of conventional approaches to managing natural resources, the attention of policy makers and managers has shifted to determining and influencing the fundamental causes of environmental problems. This has resulted in a broader problem definition to include the role played by factors such as individual choices, land-use changes, local-global linkages, and patterns of consumption. The recognition that human perceptions, values, and behaviors are important drivers of biophysical processes has made social learning an indispensable part of management philosophy and practice. Additionally, the natural resource management paradigm of "command and control" that relied on top-down (usually state management of natural resources) approaches has lost its dominant status. As environmental degradation has escalated and legal and political challenges have mounted, a participatory and democratized approach to decision making has come to be practiced widely. The advantages include a lower likelihood of litigation, resulting in speedier and more cost-effective environmental decision making.

Social learning is also used in environmental education when the objective is not to simply convey conceptual knowledge but to engage multiple stakeholders in problem definition and evaluation of possible solutions. It usually incorporates educators, other professionals, and stakeholders, and the objective is to train the service providers, who, in turn, are expected to involve the students. Stakeholders are specific to a local environmental issue, with its characteristic social and political dimensions. The educational strategy is developed through a collaborative process and is regularly tested and revised. The teachers are therefore trained to take an analytic and experimental approach to their teaching. Social learning in education may involve learning communities, often a network of diverse stakeholders who interact on a continuous basis and, using dialogue and structured communication, arrive at a consensus definition of the issues and systematically consider the possible solutions and the trade-offs involved in various choices.

Social learning is predicated on the notion of a practice-based approach to learning. Participants start with a basic recognition of their interdependency, thus underscoring a shared interest. Also important is the acknowledgment of the diversity of perceptions and mental models among stakeholders. Stakeholders learn through their membership in social networks, and not simply from receiving information. In becoming members of communities centered on specific goals, norms, and values, participants gain understanding of natural resources that is grounded in the history and culture of a specific region. The iterative nature of interaction helps new social practices and routines to emerge, which are considered as important as the content that is exchanged. Over time, trust is built, which helps in consensus building.

A common assumption underlying social learning is a belief in the ability of iterative dialogue to alter the framing of issues. Different stakeholders, by virtue of their differing interests and experiences, frame the problem in divergent ways. These frames are representations of reality that are partial and are compelling to those who hold them because of their exclusion of certain elements. Entrenched frames are often vested with emotional salience and become impediments to achieving consensus. Social learning helps render implicit assumptions explicit. Instead of decisions by default, conscious choice with an awareness of the trade-offs becomes the norm. Interaction and brainstorming over a period of time makes stakeholders aware of the existence of a plurality of views, thereby increasing the likelihood of modification of frames.

Tools utilized for social learning include participatory modeling and group envisioning exercises. Instead of focusing on models derived from formal decision theory, emphasis is placed on understanding decision making and individual mental mode. Collective models are used that seek to utilize actor perceptions and values. Reliance on approaches devoted to finding a single optimal solution is discarded. In order for social learning to take place, effective facilitation is extremely important. A skilled facilitator can create scenarios and deploy formalistic modeling using specialized software programs. Stakeholders who lack verbal, technical, and other expertise may require additional assistance.

Natural resource problems have been referred to as "wicked," meaning that they are complex, involving many trade-offs, and therefore require out-of-box thinking. In part, the high level of uncertainty that characterizes natural resources makes the application of conventional modeling approaches problematic. To complicate matters further, the uncertainty is not limited to the biophysical parts of the natural resources but also to the linkages between the social and ecological components of the system. The cumulative effect of uncertainty is therefore amplified with the increasing scale of the system. Social learning focuses on building capacity in the socioecological system to deal with surprise. No assumption of system stability is made; on the contrary, change and unpredictability are considered inherent aspects of the system. Thus, using the social learning approach, mechanisms for dialogue, sharing of information, and awareness of multiple perspectives are fostered. Adaptive management, which involves hypothesis testing—that is, evaluating decision making in terms of the outcomes produced—also utilizes social learning.

Social learning in natural resources management can potentially induce transformative outcomes. This is because of the effect on the wider context and the scope for shift in values and inducing political realignment. Instead of merely causing incremental changes, social learning can shift the stakeholders' worldviews, producing increased congruence in problem definition and the search for solutions. Natural resource management regimes are examined for their underlying assumptions and biases. The process of unpacking these assumptions makes the regimes a target for improvement and change. Social learning is therefore particularly compatible with promoting a systems approach to managing natural resources. Similarly, social learning is a good fit with integrated views of the socioecological system.

See Also: Agenda 21: "Promoting Education, Public Awareness, and Training" (Chapter 36); Educating for Environmental Justice; Vocational Education and Training.

Further Readings

Berkes, Fikret and Carl Folke, eds. *Linking Social and Ecological Systems: Management Practices and Social Mechanisms for Building Resilience.* New York: Cambridge University Press, 2000.

Keen, Meg, Valerie A. Brown, and Rob Dyball, eds. *Social Learning in Environmental Management: Building a Sustainable Future.* London: Earthscan Publications, 2005.

Mostert, E., M. Craps, and C. Pahl-Wostl. "Social Learning: The Key to Integrated Water Resources Management?" *Water International,* 33/3 (2008).

Neeraj Vedwan
Montclair State University

SOCIETY FOR COLLEGE AND UNIVERSITY PLANNING

The Society for College and University Planning (SCUP) supports the work of administrators and staff in higher education who contribute to their institutions' planning efforts. Established in 1965, SCUP offices are located in Ann Arbor, Michigan. There are five geographic member regions: Mid-Atlantic, North Atlantic, North Central, Pacific, and Southern. According to the organization's Website, SCUP's mission is to provide its members "with the knowledge and resources to establish and achieve institutional planning goals within the context of best practices and emerging trends."

SCUP promotes an integrated planning approach, focusing on physical as well as academic and operational planning issues. Through integrated or "smart" planning, the organization believes that colleges and universities will be able to engage broad constituencies, develop synergistic opportunities among different campus units, and gauge their overall performance more effectively. The American College and University Presidents Climate Commitment (ACUPCC) is one recent and well-known example of integrated planning in that it requires signatories to pay simultaneous attention to numerous facets of campus operations while pursuing that commitment's goal of climate neutrality.

SCUP provides several resources to assist its members in joining their various campus planning efforts together with their overall missions and academic plans. These resources include weekly updates and an annual report on trends in higher education. SCUP compiles these documents through an environmental scanning process that systematically gathers and analyzes current information about higher education from a variety of sources. The annual reports that SCUP prepares from this data contain sections describing trends in areas such as demographics, economics, environment, and technology. The environment section of the 2009 report, for example, observes that opportunities exist for higher education to reduce energy use through careful planning for information technology units in general and data centers in particular. This section of the report also observes that China has become the world's largest emitter of greenhouse gases and that significant local action to reduce pollution is not likely in the near future.

Among SCUP's other resources is the peer-reviewed journal *Planning for Higher Education,* which it publishes quarterly. The journal contains articles on timely planning issues in academia. For those interested in communicating with others on specific planning topics related to higher education, the organization also maintains several virtual "knowledge communities." Subscribers to these communities can engage in e-mail discussions with colleagues from other campuses on a wide range of topics that include budgeting and resource planning, campus open space, classroom design, disaster planning, sustainability, and town-gown relationships.

Events hosted by SCUP include both regional conferences and an annual international conference. In addition, SCUP sponsors a three-part Planning Institute at different locations every year. Each sequential part comprises a one- or two-day workshop on integrated planning and the tools to support it. Participants gain competencies in six areas: people, language, process, the plan, planning context, and resources.

In recent years, SCUP has developed a focus on sustainability issues as they pertain to higher education. This emphasis pertains both to the organization's own operations and to the resources it provides for its members. For example, SCUP has moved from print materials to online resources for many of its reports and other publications. It has adopted a number of green practices within its offices, including use of recycled paper and carbon offsetting. Perhaps most notably, SCUP founded the national Campus Sustainability Day in 2002. This annual event, which takes place in October, involves a streamed Webcast of a panel of campus sustainability experts; in addition, individual campuses are encouraged to develop their own events to recognize and promote campus sustainability. SCUP produces this event together with the Higher Education Associations Sustainability Consortium (HEASC).

See Also: American College and University Presidents Climate Commitment; Association for the Advancement of Sustainability in Higher Education; Higher Education Associations Sustainability Consortium.

Further Readings

Calhoun, Terry. "Smart Planning: Why Do Any Other Kind?" *University Business* (October 2008). http://www.universitybusiness.com/viewarticle.aspx?articleid=1139&p=2#0 (Accessed June 2010).

Holmes, Jeffrey. *20/20 Planning: A Short History of the First Twenty Years of the Society for College and University Planning.* Ann Arbor, MI: Society for College and University Planning, 1985.

Planning for Higher Education. http://www1.scup.org/PHE/FMPro?-db=PHE.fp5& -lay=Home&-format=home.htm&-FindAny (Accessed June 2010).

Society for College and University Planning. http://www.scup.org/page/index (Accessed June 2010).

Stacey Swearingen White
University of Kansas

Stanford University (Global Ecology Research Center)

Stanford University, founded in 1891 by California businessman and politician Leland Stanford, is one of the premier research universities in the United States. Located in Palo Alto, California, in 2008, Stanford enrolled 6,812 undergraduates, 8,328 graduate students, and had 11,910 tenure-line faculty, medical center tenure-line faculty, senior fellows, and center fellows at specified policy centers and institutes. Academics at Stanford are organized around seven schools: the Graduate School of Business, the School of Earth Sciences, the School of Education, the School of Engineering, the School of Humanities and Sciences, the Law School, and the School of Medicine. Sponsored research at Stanford was funded at $1.13 billion in 2009–2010 and included over 4,400 projects. In 2008–2009, Stanford received over $65 million in royalties from 517 licensed technologies, and the Office of Technology Licensing at Stanford concluded 77 new licenses. Stanford has received high rankings by many services that compare universities, including second overall on the Academic Ranking of World Universities produced by Shanghai Jiao Tong University and 17th on the Times Higher Education Supplement—QS World University Rankings (including a ranking of third in life sciences and biomedicine, and eighth in natural sciences).

In addition to the grasslands of the Jasper Ridge Biological Preserve, the Stanford Arboretum has preserved rare trees, many types of eucalyptus, and a 100-year-old Santa Lucia Fir.

Source: Wikipedia

Department of Global Ecology

The Department of Global Ecology (DGE), founded on the Stanford campus in 2002, includes faculty members, postdoctoral fellows, and graduate and undergraduate students as well as technical and administrative staff. It is one of six research departments of the Carnegie Institution for Science (previously known as the Carnegie Institution of Washington), a private organization founded in 1902 by Andrew Carnegie to advance scientific research and understanding. The DGE is by far the youngest of the departments: The next youngest, the Department of Embryology, was founded in 1913. The purpose of the DGE is to conduct basic research on the interactions among Earth's ecosystems, land, atmosphere, and oceans and to increase understanding of how these interactions affect the Earth system. The DGE receives funding from the Carnegie Institution, and federal and state agencies. The Global Ecology Research Center building, designed by the San Francisco architectural firm EHDD and completed in March 2004, was chosen as an American Institute of Architecture Top Ten Green Project in 2007. The building was designed to provide high-functioning research and lab space while also reflecting the DGE's concerns with reducing carbon impact and addressing biodiversity and water issues. Methods used in the building include natural ventilation, a "night sky" radiant cooling system, an evaporative katabatic cooling tower, orientation, daylighting, and sunshading. Much of the wood used in construction and to create furniture is salvaged, and about 20 percent of the site's concrete aggregate is recycled. Water use is reduced by no-irrigation landscaping, low-flow faucets, waterless urinals, and dual-flush toilets. A reduction of 72 percent in carbon emissions was achieved for building operations as well as a 50 percent reduction in embodied carbon for building materials.

Research Labs Within the Department of Global Ecology

There are four labs within the DGE. The Asner Lab, or Laboratory for Regional Ecological Studies, researches how ecosystems are changing at regional levels. Areas of expertise within the lab include remote sensing, biogeochemistry, land-surface modeling, ecosystem ecology, biodiversity, geographic information systems, terrain modeling, and cultural geology. Among the lab's projects, as of 2010, were studying remote sensing of the structure and chemistry of the biosphere, satellite analysis of disturbances in the Peruvian rainforests, selective logging and forest disturbance in Borneo, effects of invasive species on ecosystem structure and function in Hawaii, and regional carbon storage responses to woody encroachment in western pinyon–juniper systems.

The Berry Lab is headed by Joseph Berry, Ph.D., from the University of Denver. Berry specializes in remote sensing and is the principal of BASIS (Berry & Associates Spatial Information Systems), which consults and develops software for geographic information systems (GIS) technology. Berry's research and consulting emphasizes understanding spatial patterns and relationships through grid-based map analysis and GIS modeling. Topics of Berry's studies include the spatial characterization of timber supply, wildlife habitat, wildfire risk and impact, marine ecosystem populations, surface and groundwater hydrology, retail market analysis and in-store movement analysis, hazardous waste site location, and air pollution modeling.

The Caldeira Lab is headed by Ken Caldeira, Ph.D. The main focus of the Caldeira Lab is to improve the science base to protect the environment, while also allowing human civilization to develop. Specific areas of study within the Caldeira Lab include predicting changes in ocean chemistry and how those changes will affect marine life; geoengineering, that is, deliberate intervention in the climate; the biophysical and carbon cycle effects of changes in land cover; and improved understanding of ocean biogeochemistry and ocean carbon cycle change.

The Field Lab, headed by Chris Field, Ph.D., focuses on basic ecological research at all scales from the ecosystem to the global. Current areas of research include the impact of climate change, ecosystem responses to global changes, the terrestrial carbon cycle, the biogeochemical consequences of changes in species composition, and interactions between the biosphere and atmosphere.

Department of Global Ecology Projects

Since 1992, a number of studies led by Stanford University scientists and scientists from the Department of Global Ecology with the participation of scientists from other institutions have been conducted at the Jasper Ridge Biological Preserve. This grasslands area, owned and operated by Stanford University and located on the east side of the Outer Coast Range of central California, is located near the Stanford campus. The ongoing studies use grasslands to study how ecosystems (as opposed to individual plant species) respond to climate change. The Jasper Ridge area is ideal for this type of study because it includes many species and functional types of grasses, is frequently disturbed by gophers (so there is less concern that soil disturbance necessary for some scientific experiments is making a major change in the ecosystem), and is dominated by annual grasses whose complete life cycle can be studied each year. A research station was established in the Jasper Ridge Eddy Flux in 1998 and continues in operation today. The station records data, including fluxes of water vapor and carbon dioxide (CO_2) as well meteorological parameters such as wind, temperature, net solar radiation, and barometric pressure, and transmits them by radio every second to the DGE.

The Carnegie Airborne Observatory (CAO) allows the remote measurement of biospheric structure and function through a system of a 3D laser scanner and imaging spectrometer that can be flown on fixed-wing aircraft and can thus be used to study ecosystems anywhere in the world. The CAO mission is to understand how ecosystems change with changes in land use, climate, and natural disturbances, and how this affects people living in or near the ecosystem. As of 2010, the CAO was operating in South Africa, Madagascar, and the Amazon Basin, and received funding from multiple sources.

The DGE serves as the host for content produced by SCOPE (the Scientific Committee on Problems of the Environment), an interdisciplinary body focused on global environmental issues. In 2010, 39 SCOPE books were available from the DGE Website, covering a wide range of topics, including the global cycles of nitrogen, phosphorous, and sulfur; principles of ecotoxicology; the biogeochemistry of small catchments; and the environmental consequences of nuclear war. The DGE Website also contains a wide range of press releases, scientific and technical publications, and a bibliography of materials on global ecology dating back to the 1980s.

Recent Department of Global Ecology Research and White Papers

In May 2010, a report coauthored by Ken Caldeira underlined the importance of including vegetation in models predicting global climate change. This study found that increased CO_2, already associated with global warming as a greenhouse gas, also makes vegetation less effective in counteracting the effects of global warming. The study attributed as much as 25 percent of warming due to CO_2 in some regions to the direct effect on plants. For instance, trees are generally believed to combat global warming by absorbing CO_2 through stomata (tiny holes or pores in the leaves) and releasing water through the same pores (evapotranspiration), which directly cools the air. High levels of CO_2 cause the leaf pores to shrink and make the process of evapotranspiration less efficient, thus increasing global warming. The model also predicted higher land runoff associated with increased CO_2 levels because of the interference with evapotranspiration.

Caldeira also coauthored a March 2010 study with Steven Davis, a postdoctoral researcher in DGE, that suggested that measures of CO_2 emissions should include not only those emissions produced within a country's geographical boundaries, but also those from goods and services produced elsewhere but used within the country. This study found that many industrialized countries were successfully "outsourcing" their CO_2 emissions to developing countries: for instance, about 2.5 tons of CO_2 per person is associated with products and services consumed in the United States but produced elsewhere. Switzerland actually outsources over half of its carbon emissions in this way, while China is on the other end of the spectrum, with about one-quarter of its CO_2 emissions associated with products that are ultimately exported.

In March 2010, DGE director Chris Field released a white paper on climate change, coauthored with Peter Darbee, CEO of Pacific Gas and Electric, and intended for policy makers and business leaders. Field and Darbee's paper summarized the best current scientific evidence for climate change, and was intended, by the combined authorship of a business leader and a respected scientist, to make the case to other business leaders and policy makers for why the United States needs to begin decreasing greenhouse gas emissions through policy changes and investments in clean-energy technologies and infrastructure.

In December 2009, a team of scientists from the DGE announced release of CLASlite, a desktop software package that allows the marriage of free satellite imagery and powerful analytic methods to quantify global greenhouse gas emissions due to destruction of tropical forests (believed to account for about 20 percent of emissions). About 70 governmental, academic, and nongovernmental organizations have adopted the software, which makes it easier to use satellite imagery to identify deforestation and forest degradation by converting raw satellite images into highly detailed maps. In 2010, the group planned to provide a Web-based version of the software to allow tropical forest mapping around the world, and will also provide training and technology to countries in the Amazon region to improve forest monitoring and management in that region.

See Also: Educating for Environmental Justice; Leadership in Sustainability Education; Sustainability Websites; Whole-School Approaches to Sustainability.

Further Readings

American Institute of Architects. "Carnegie Institution of Washington Global Ecology Center (Global Ecology Research Center)." http://www.aiatopten.org/hpb/overview.cfm?Project ID=809 (Accessed May 2010).

Carnegie Institution for Science. "CLASlite: User Friendly Forest Monitoring Policy." http://claslite.ciw.edu/en/index.html (Accessed May 2010).

Carnegie Institution for Science. "Department of Global Ecology." http://globalecology .stanford.edu (Accessed May 2010).

Darbee, Peter A. and Christopher B. Field. "Climate Change for Policymakers and Business Leaders" (March 3, 2010). http://www.pgecorp.com/corp_responsibility/pdf/climate paper_final.pdf (Accessed May 2010).

King, John. "Setting a Higher Green Standard: Stanford's Ecology Center Earns Kudos for Its Low Impact" (May 15, 2007). http://articles.sfgate.com/2007–05–15/bay-area/ 17243237_1_energy-and-environmental-design-climate-change-building (Accessed May 2010).

Tollefson, Jeff. "Climate: Counting Carbon in the Amazon." *Nature News* (October 21, 2009). http://www.nature.com/news/2009/091021/full/4611048a.html (Accessed May 2010).

Sarah Boslaugh
Washington University in St. Louis

STELLENBOSCH UNIVERSITY (SOUTH AFRICA)

Stellenbosch University, in Stellenbosch, South Africa, is a public research university established as Stellenbosch Gymnasium in 1866. It took its current name in 1918 when it acquired university status. Although Stellenbosch ranks low in terms of international publication output and citations of Stellenbosch scientific publications, the University of Stellenbosch Business School (USB) is the center of sustainable energy research in South Africa, one of the leading business schools in the world, and the only African business school regularly included in lists of top business schools. Stellenbosch itself is South Africa's second-oldest town and has become a university town of 90,000 year-round residents and a further 25,000 students (about 9,000 of them graduate students). Language use has been an ongoing issue at Stellenbosch; while most South African colleges conduct their courses in English, Stellenbosch is predominantly Afrikaans, though assignments and examinations in English are accepted, and a third of the student body are native English speakers.

Stellenbosch's Department of Mechanical Engineering has a long relationship with the South African power industry, including extensive research on efficiency improvements in power stations and renewable energy research focusing on solar power, wind power, and the use of biofuels. Stellenbosch researchers have been responsible for significant advances in waterless cooling of power stations, an important development in dry South Africa, where water is an especially limited resource in the northwest of the country. Bulk solar power generation has been a major focus of the department's research since the late 1990s, augmented by ongoing work on wind turbines. The Department of Electrical and Electronic Engineering has also worked with renewable energy, focusing on projects on wind power and hydroelectric power generation applications, as well as the energy storage systems key to economically feasible long-term use of renewable energy. A project group in the Department of Process Engineering built an ethanol extractor for producing ethanol for biodiesel from fermentable waste products, and a research group in the Microbiology Department developed new yeast strains for use in such fermentation.

In 2006, Stellenbosch was designated by the South African National Energy Research Institute as the hub of a postgraduate program in Renewable and Sustainable Energy Studies, associated with research chairs forming the "spokes" at other institutions. The goal of the resulting Center for Renewable and Sustainable Energy Studies is to enhance the nation's renewable energy capacity, in part because of recognized problems with the price volatility of fossil fuels. The program is tasked with enhancing human resource capacity through education, stimulating innovation through research, and encouraging enterprise in the renewable energy field as well as identifying potential applications of research conducted at Stellenbosch. A variety of degrees are offered through the center and the other South African universities associated with it, which in 2010 included North West University, the University of Pretoria, the University of Cape Town, Nelson Mandela Metropolitan University, and University Fort Hare.

Coursework at the center takes the form of brief intensive modules lasting about a week. There are three specific modules required of students at the center, the core courses of its program: Sustainable Development is considered the foundation course of the program and examines an overview of the social, economic, and environmental challenges facing humankind in the early 21st century. Climate change, biodiversity, waste and pollution, carrying capacity problems, pandemics, poverty, and global economic trends are all addressed over the course's week, along with an introduction to the history and rhetoric of sustainable development.

Renewable Energy Systems surveys the most significant renewable energy resources: hydroelectric power, geothermal energy, tidal/wave/ocean power, wind power, solar power, and photovoltaic systems. Students are presented with case studies of renewable energy systems and discuss the strengths and weaknesses with regard to efficiency, practicality, and climate change effects as well as the challenges of moving forward and adopting renewable energy on a wider basis.

Renewable Energy Policy examines policy initiatives in South Africa and introduces students to policy instruments, financial structuring tools, and alternative finance sources outside the traditional commercial financial institutions, like carbon finance. Significant discussion is devoted to the financial and political barriers faced by renewable energy proponents and entrepreneurs.

In addition to these core courses, students without technical undergraduate backgrounds are expected to take Conventional Energy Systems, a module on the five main conventional sources of energy: coal, oil, natural gas, hydroelectric power, and nuclear power.

Other modules offered by the center include the following:

- *Solar Energy*—studying thermodynamics, energy storage, bulk solar thermal power generation systems, life cycle costing of solar power systems, and large-scale solar power plant specifics
- *Energy-Efficient Cities*—examining energy in the operation of buildings, the construction and demolition of buildings, and the transportation to and from such buildings, as well as mitigation and adaptation strategies at macro- and microscales

- *Sustainable Biomass Production*—covering sustainable biomass production systems including agricultural, forestry, agroforestry, and aquatic systems, as well as greenhouse gas and energy balances, conflicts of interest among industries and land users, bioremediation, crop science, and land use planning
- *Wind and Hydro*—encompassing the study of wind power and hydroelectric power in the present and future
- *Bio-Energy*—considering the economic feasibility, sustainability, and technical issues of bio-energy production in Africa
- *Sustainable Land Use*—presenting an integrated approach to the use and management of land resources and introducing students to geo-informatics tools for use in land management and related decision making

Recent and current research projects at the center have encompassed a vast variety of sustainability studies, including linear wave energy converter systems, novel magnetic gears for renewable energy power generators, determination of air drying times for energy crops, solar thermal gas turbine systems, optimization of the mechanics of a water wave energy pump, adaptive thermal engineering, solar power plant performance evaluation, solar thermal energy storage, bio-energy production in the Western Cape, fermentation of municipal solid waste into bio-ethanol, and sustainable urban infrastructure. The center's biofuels research team is actively involved in developing advanced raw starch-fermenting yeasts to be used for bioethanol production in order to improve the cost-effectiveness of biofuel derived from corn and wheat. The solar research team is engaged with work on optimizing solar chimney power plants. The Ocean Energy Network research team studies the usability of the abundant wave- and ocean-current energy to South Africa, which enjoys close proximity to the South Atlantic Ocean and the Agulhas Ocean Current.

One of the partners in the center, the Sustainability Institute is an international living and learning center located outside the town of Stellenbosch and operated by Stellenbosch's School of Public Management and Planning. Established in 1999, the institute studies sustainable living in South Africa in the hopes of reducing the nation's level of poverty. More than just a sustainable living dormitory, as some schools offer, the institute as a living space includes a child-centered home base and over 50 participants in its Early Child Development Program, plus 40 young workers in its teen-centric Sustainable Construction training program. The institute's focus in the second decade of the 21st century is on community development, capacity building, sustainable cities, and the nonprofit sector.

The institute offers a master's degree in Sustainable Development Planning and Management in conjunction with Stellenbosch. A part-time degree program, it is constructed of a number of six-day intensive-seminar course modules, including Applied Economics, Biodiversity and Sustainable Agriculture, Sustainable Cities, Ecological Design, Leadership and Ethics, Complexity Theory, Development Planning Theory and Practice, and Development Planning and Environmental Analysis. The master's program in Sustainable Agriculture is designed to train commercial farmers to operate sustainable large-scale farms and includes coursework on sustainable agriculture, financial management, and marketing and information technology as well as a three-month internship on a currently operating sustainable farm. Furthermore, doctoral candidates working with the Sustainability Institute and overseen by its academic director attend biannual workshops in addition to their ongoing supervision as they pursue sustainability studies from what the institute calls a transdisciplinary approach. There is also a recent bachelor's degree program, offering majors in Sustainable Development, Development Planning, Managing Sustainable Agriculture for Development, and Renewable and Sustainable Energy.

One of the institute's initiatives is Sustainable Stellenbosch, which promotes sustainability throughout the Stellenbosch municipality, both on and off campus. The Science and Society group within Sustainable Stellenbosch, for instance, promotes renewable energy, biodiversity, and food security in Stellenbosch. The Reinventing Stellenbosch Partnership is a formal partnership between the university and the municipality, focusing on land reform to provide access to common land for emerging farmers, creating jobs, promoting food security, encouraging household and community gardens, and addressing challenges of energy, waste, transport, and housing in the municipality. The Stellenbosch Development Forum pursues similar goals within the local business community.

See Also: Experiential Education; Green Business Education; Sustainability Officers.

Further Readings

Filho, Walter Leal. *Sustainability and University Life*. Ann Arbor: University of Michigan Press, 1999.

Gough, Stephen and William Scott. *Higher Education and Sustainable Development*. London: Routledge, 2009.

Hargreaves, Andy and Dean Fink. *Sustainable Leadership*. Hoboken, NJ: Jossey-Bass, 2005.

Sandell, Klas, Johan Ohman, and Leif Ostman. *Education for Sustainable Development*. New York: Studentlitteratur AB, 2005.

Silka, Linda and Robert Forrant, eds. *Inside and Out: Universities and Education for Sustainable Development*. Amityville, NY: Baywood Publishing, 2006.

Bill Kte'pi
Independent Scholar

Sustainability Officers

The position of sustainability officer is quickly becoming a professional role embraced by universities, municipalities, large businesses, and industry alike. The emphasis of each role may vary from organization to organization; however, the intention shares common ground. In the United States and Canada, as many as 215 two- and four-year colleges and universities have hired full-time sustainability professionals as of July 2010. This increase from approximately 90 in 2007 represents a dramatic upsurge in the creation of these positions. Hiring a sustainability officer may be one of the most critical steps that a university can take to significantly advance sustainability on campus in that it establishes a central point of accountability. When the appropriate skill set is hired, this position has the potential to provide the leadership necessary to enable a university to be successful in its endeavor to become sustainable over the long term. Almost no two positions are quite the same from university to university, as the reporting lines, rank, and responsibilities tend to vary. The reporting lines range from positions that are situated within facilities and environmental, and health and safety departments to those that share joint appointments with a provosts office, an academic department, or the division of finance and administration, to name a few variations. Some positions report directly to an executive vice president or provost. Very few report directly to a university president. For the sake of this entry, the position is referred to as a "sustainability officer"; however, depending upon assumed rank within the college or university, the title of coordinator, manager, or director may apply.

The sustainability officer position assumes the role of change manager, and thus calls upon a unique skill set, which requires one to build horizontal and vertical relationships throughout the institution. The change management function manifests as an assigned role within the organization dedicated to fostering and sustaining a continuous process that drives organizational learning, innovation, systems thinking, and continuous improvement toward sustainability. The inherent challenge for sustainability officers is that they must exert influence without power. Such positions could be classified within higher education as a "hybrid" position. The hybrid nature of the position refers to the need to bridge the academic and operational functions of the university as they relate to the sustainability objectives of that institution. On the operational side, they are responsible for integrating sustainability principles into campus systems that do not directly report to them, for example, the power plant, the university fleet, design standards, purchasing systems, grounds maintenance, or custodial services. The academic response to sustainability can be multidimensional in that, if successful, the university develops a set of curriculum and research objectives. Due to the meritocratic nature of higher education, influencing curricular and research objectives can be quite challenging for the sustainability officer, depending upon reporting structure and academic standing.

The process of integrating sustainability into a college or university is a complex and nonlinear endeavor that requires patience, a clear and in-depth understanding of how the organization functions, and a systems framework. Understanding the fundamental structure of higher education can either impede or enhance the sustainability officers' best efforts to be successful. The successful sustainability professional will know how to build coalitions vertically and horizontally throughout the system in an effort to bring together people and departments that may not be accustomed to collaboration under current conditions. Sustainability officers typically have multidimensional responsibilities. A skilled officer will be able to foster the vision for a sustainable campus, facilitate a multistakeholder process, build and maintain working relationships across systems, and manage multiple projects and personalities. Management and leadership skills must be complemented by a set of quantitative skills that initiate the development and analysis of measureable goals as well as financial mechanisms.

Sustainability officers tend to have varied backgrounds and degree levels. The degree and background that is sought varies from school to school. The profession today is represented by individuals with a range of backgrounds that vary from environmental studies to natural resources management and policy to sustainable development, to environmental engineering and economics. Professionals in the field boast a range of degree levels from bachelor's to J.D. to Ph.D. The salary range of $35,000–$150,000 reflects an extreme variation of degree levels and experience.

Sustainability calls for organizational reform, shaped by new ways of doing business, identifying new priorities, revised goals, and targets. To be successful in this change management endeavor, the institution must be prepared and poised for this change, and the internal actors must be responsible for decisions throughout the institution. It is up to the emerging and multidimensional role of the sustainability officer to shape this emerging field and transform our universities in the short term for the long term.

See Also: Association for the Advancement of Sustainability in Higher Education; Greening of the Campus Conference; Leadership in Sustainability Education.

Further Readings

Association for the Advancement of Sustainability in Higher Education. "Campus Sustainability Officers." http://www.aashe.org/resources/sust_professionals.php (Accessed August 2010).

Rands, G., B. Ribbens, and D. Connelly. "The Sustainability Coordinator: A Structural Innovation for Managing Sustainability." In *Innovative Approaches to Global Sustainability,* C. Wankel and J. A. F. Stoner, eds. Basingstoke, UK: Palgrave, 2008.

Sharp, L. "Campus Sustainability Practitioners: Challenges for a New Profession." In *Conference Proceedings: Greening of the Campus VI: Extending Connections,* R. Koester, ed. Muncie, IN: Ball State University, 2005.

Julie Newman
Yale University

SUSTAINABILITY ON TELEVISION

This NASA satellite image shows the oil slick from the BP spill near New Orleans, Louisiana, on July 9, 2010. The widespread television coverage of the spill has been seen as an opportunity to renew efforts to promote climate change legislation.

Source: National Aeronautics and Space Administration

Television changed an audience's relationship with the mass media content by changing the location and consumption of its delivery. After decades of going out to see a movie and staying in to read or listen to radio, the advent of television changed the media ecosystem by elevating (and isolating) the living room as a place to watch the world. The delivery mechanism brought families indoors and placed them into the often-passive role of watcher. When coupled with advertisers whose marketing spawned a mass-consumer society, we find television in the 1960s as an environmental *Twilight Zone*: a decade of landscape-based Westerns, Lassie the rescue dog and Flipper the rescue dolphin, Yogi Bear in Jellystone Park stealing picnic baskets, and sitcoms like *Gilligan's Island* that conveyed nature as an isolating yet supportive paradise. At the same time, viewers explored nature with adventure-conservation shows like *Wild Kingdom* and Jacques-Yves Cousteau's undersea wonders. The decade ended with televised moon landings (and the image of "Earthrise" from the moon) and news footage of the Santa Barbara oil spill and the Cuyahoga River on fire in Ohio. In the 1970s, Earth Day was celebrated across the country, and a public service announcement showed an Indian chief crying after he discovered a polluted and trash-filled America.

The technology that marketed what some saw as a degraded and unsustainable relationship with the environment had in turn become a key vehicle in launching environmental initiatives. This topsy-turvy relationship continues today, with shows like *Lost* and *Survivor* sharing the same delivery mechanism and screen as news coverage of islands threatened by global warming and coastlines smeared with oil. The shopping channels are a remote control's click away from the Weather Channel and "green" channels that promote sustainable lifestyles (though some such programs focus on healthier and greener ways to consume). The medium is the message, as Marshall McLuhan observed, yet, when considering television as a medium for environmental messages, one must consider its inexplicable opposites: televised nature documentaries persuade us to save the world even as TV's entertainments and advertisements serve to disengage us from an endangered world (and to consume its products even faster).

Critics have observed that the medium itself separates us from nature. In his book *The Age of Missing Information,* Bill McKibben contrasted 24 hours spent observing a mountain with 24 hours of cable television—to note how our primary senses are overwhelmed by the sheer frenetic noise of television. This transition is detailed by science writer Carol Kaesuk Yoon in *Naming Nature: The Clash Between Instinct and Science;* Yoon claims that our human *umwelt* (the manner in which our sensory perception of the world shapes our mental images of that world) has changed from the ability to find and name distinctions of the species around us to the ability to differentiate product brands in a grocery store. Richard Louv in *Last Child in the Woods: Saving Our Children From Nature-Deficit Disorder* finds that much of modern culture—from our schools to our media—have served to disconnect our children from the basic tools of sustainable choices. "Today," Louv notes, "the life of the senses is, literally, electrified," by televisions and computers but also by the air conditioning that allows children to spend summers indoors, in front of a screen and divorced from primary forms of sensory stimulation and exploration.

Television: A Medium of Disconnect and Discontent

Television engages audiences in the sense of community, despite a commonsense realization that television separates us from place and community. And in its robust engagement of audience, its impact as an agenda-setting and framing medium serves to both support and undercut sustainability. Ronald Reagan, the actor who sold Borax on *Death Valley Days,* would become the president who removed solar panels from the White House, expanded energy consumption, and sought to isolate the environmental movement as extremists. He quipped that "trees cause more pollution than automobiles do," yet his effective use of television demonstrated the agenda-setting potential of charismatic leaders paired with a commonly distributed visual rhetoric. So it is little surprise that television in the Reagan era continued its historic role of promoting consumption, not the values of conservation and self-reliance that were posited by President Jimmy Carter.

Today, television continues to connect a consumer society (and an electorate) with both the market and the marketplace of ideas. But television programming and management are evolving as markets, issues, and audiences evolve; television executives and channels are engaging niche and general audiences who support Anne Leonard's claim (from *The Story of Stuff*) that our consumption is an unsustainable process, one that leads to global inequity and insecurity as well as global warming. (Notably, her influential mixed-media *Story of Stuff* was distributed not via television broadcast but as a viral Webcast.)

Television had risen as the most prominent media in history and with it there emerged a television culture that altered existing lifestyles and reengaged communities around common icons—the actors and the landscapes and stuff they acted with. The 1970s environmental movement introduced these sustainability concepts to television audiences, though the post-energy-crisis era of the 1980s and 1990s was typified by consumerism and deregulated business practices. Against the mass-media messages, advocates like Leonard raised concerns regarding the widespread impact of television culture. A common claim by media critics, social activists, and environmentalists observed that the ads that support the majority of television programming operate by developing feelings of insecurity—leveraging discontent with who we are and what we possess into profligate purchasing. Hence the claim that television contributed to overconsumption of resources and energy and an industrialized agriculture.

Environmentalists and educators also stated that simply the amount of time spent watching television (and later, video games and computers) has resulted in less time in contact with a nonmediated "real world," thus disconnecting the audience from nature, from our relationships with food production (gardening and farming), and from the solace and physiological processes that occur when a person connects with nature. With less time to form a physical relationship to the outside, we're more reliant on the dramatized worldview of televised nature—to see nature as something rare, dirty, or dangerous, filled with germs and sharks and bears. We champion the charismatic megafauna of beached whales and threatened polar bears, yet overlook the ecosystem—or, as writer Joy Williams titled her essay, we elect to "Screw the Shrimp and Save the Whales."

History of Sustainability in Television: From "Harvest of Shame" to *Silent Spring*

Edward R. Murrow, the celebrated CBS broadcast journalist, may be one of the first broadcast journalists to bring an audience's focus to sustainability issues in his examination of food production in his 1960 documentary. In "Harvest of Shame," broadcast on the series *CBS Reports,* Murrow investigated the lives and working conditions of the United States' migrant agricultural workers, previewing the farm workers and environmental justice movements and also highlighting the growing disconnection between food production and consumers—an issue that would eventually transform into the current "locavore" movement.

"Harvest of Shame" was Murrow's final documentary for the network, but it offered a stepping-stone into other environmental issues that *CBS Reports* covered during the 1960s. From this series, the key documentary focusing on sustainability issues aired in May 1963. It focused on Rachel Carson and her acclaimed yet controversial book *Silent Spring,* about the danger posed by widespread application of toxic chemicals and pesticides (primarily DDT) on farms, suburbs, and cities that threatened not only the target pests but all living creatures, from frogs and birds to humans. *CBS Reports* featured interviews with Carson and several opponents of her scientific claims. Carson's book, with the additional confirmation and broadcast distribution provided by *CBS Reports,* became one of the influential steps in the U.S. environmental movement; it attracted the attention of President John F. Kennedy, who ordered a presidential science advisory committee to investigate the issue. The U.S. Senate also conducted investigations into the use of pesticides, which eventually led to the foundation of the Environmental Protection Agency and a U.S. ban on DDT use in 1972.

In 1967, *CBS Reports* again energized support around an environmental concern with another award-winning documentary, "The Poisoned Air," which highlighted the threat of air pollution to public health and property as well as the long-term financial impacts of the damage caused by air pollution.

The documentary exposés of this era aired within a media ecology based on relatively few channels that reliably gathered a large demographic swath of U.S. consumers. Broadcast licensing, including entertainment value, was balanced to some degree with information due to the public service requirement associated with broadcast licenses. Yet the medium was expensive and supported by sponsors and advertisers; the search for marketable "nature" shows included documentary series profiling the work of charismatic explorer-naturalists such as Marlin Perkins, whose hosting of *Mutual of Omaha's Wild Kingdom* famously pointed out that the risks of the world—both to Marlin and his front-line assistants—might be mitigated by the insurance company that sponsored the adventures. The work of Jacques Cousteau blended charismatic open-seas adventure with never-before-filmed scenes of underwater life. Likewise, the promise of exotic or amazing wildlife supported the expanding role of *National Geographic* specials, the precursor of today's travel shows, as well as the Travel Channel, the Discovery Channel, Animal Planet, and the National Geographic Channel.

Environmental Television in the 1970s

The social turmoil of the late 1960s and early 1970s was reflected in the growth of an active environmental movement that was chronicled and supported by television. Along with antiwar movements, the issues of racial, gender, and environmental justice were well discussed and promoted throughout the 1970s in the United States. Television networks provided full coverage of mass demonstrations across the country calling for the government to implement sustainable policies. Analysts of the era have noted that the sight of average people appearing on television in the role of active protesters in the civil rights movement served to authorize civic engagement in a wide variety of issues, including the antiwar and environment movements. Millions of people, including a large cohort of college and university students, marched in demonstrations marked as the first Earth Day celebration on April 22, 1970, which garnered support in part because it was less confrontational than the antiwar movement. Another milestone from 1970 was National Educational Television's airing of the eight-part series *The Vanishing Wilderness*, noted as the first environmental TV series to air in the United States.

While television producers explored environmental issues in news reports, advertising agencies teamed the themes of peace and harmony in their commercial ads, such as the Christmas 1971 advertisement for Coca Cola, "I'd like to teach the world to sing." The ad featured singers on a bucolic hilltop voicing their support for the slogan "I'd Like to Buy the World a Coke." The campaign not only sold Coke, but also promoted world harmony that was based, in part, on the goodwill that might come from buying the world a soft drink.

Earlier that same year, on Earth Day 1971, one of the more memorable public service commercials aired—the "Crying Indian" public service announcement for the Keep America Beautiful campaign. The Indian chief canoed through apparently clean waters but soon was overwhelmed by polluted waters (reminiscent of the burning Cuyahoga River and the Santa Barbara oil spill, widely televised environmental events in 1969), then past smoke-belching smokestacks and even smokier cars as he walked along a crowded highway to become the target of tossed litter, then cut to a close-up on his tear. The commercial challenged Americans about consumption, pollution, and even racial issues as it showed the audience how they affected the environment and the Earth they shared. While its major objective was to support an antilittering campaign, its message was much broader.

From Environmental Education to Persuasion to Entertainment: Sustainability Television Today

A number of television shows have sought to engage children in sustainability issues, primarily through the educational nature of the medium. *Sesame Street* offered a range of story lines supporting environmental responsibility. Oddly, the catch phrase "It's not easy being green" came from a song by Kermit the Frog about accepting individual differences, yet the slogan transformed over time into an ironic yet engaging claim for environmental sacrifice. Maintaining a safe (and clean) neighborhood was core to the message of *Mr. Roger's Neighborhood,* a message expanded globally by the superhero activist Captain Planet.

Science and nature programs came to typify family-friendly and educational television, including shows such as *Bill Nye the Science Guy.* While such environmental themes also appeared on commercial television shows, an initiative to manage the commercialism of children's television was weakened when a key effort by the Federal Trade Commission was stopped in 1980. But the momentum to apply television as a medium for sustainability education has continued. For instance, the North American Association for Environmental Education (NAAEE) has collaborated with national educational broadcasting associations to produce television programs to target schoolchildren ages 10 to 15, with programs designed to both guide the study of their local environments and demonstrate the need to work together to support sustainability. The organization also highlights a comprehensive range of curriculum guides that utilize educational television programs.

Informational television saw declines during 2000–2010, as all traditional media (print and broadcast) responded to diversifying claims for audience time, from the proliferation of cable channels to digital gaming to online media. First, attention to sustainability topics was drawn away by the 9/11 terrorists attacks. As media outlets recovered, debates over climate change engaged a wide range of media outlets, with much of the coverage focused on the argument, not the substance or potential solutions. More recently, the 2008 recession continued to erode support for some aspects of sustainability coverage. Major television channels resized teams assigned to their environmental programs, and some even dropped their green television programs. For example, CNN, the Cable News Network, cut its entire unit of science, technology, and environment in December 2008, when the network integrated its environmental and science reporting into the general editorial structure. This was seen as a strategic response to the economic downturn and led to more dramatic coverage under the title "Planet in Peril" produced by the *Anderson Cooper 360* program. NBC Universal made similar changes at about this time, including cutting the entire staff of its *Forecast Earth* environmental program and reducing staff and on-camera meteorologists on the Weather Channel. These cuts occurred while NBC was focusing its on-air efforts on repeating Green Weeks themed around a corporate-wide campaign, "Green Is Universal."

Aside from education and information, sustainability content often finds its way on the air as elements in survival, travel, and food shows and as avenues to demonstrate celebrity activism with shows like *Living With Ed,* providing a green alternative to reality programming. Such niche programming came to exemplify the transition to the cable-television and convergent-media environment, with new channels that leverage cable and new media distribution channels. In fact, one might ask whether television as we know it—a mass medium distributed through a set number of channels—is itself a sustainable media model. Small-market initiatives, such as the niche-focused and action-oriented Green Energy TV, are competing against a mass-market approach like Planet Green, a part of the Discovery Channel. Other models feature a regular block of sustainability-focused programming, such as an evening of *The Green* show on the Sundance Channel. Programming includes a reprise of Bill Nye, whose show *Stuff Happens* builds from Anne Leonard's claims. And Nye guests on his neighbor Ed Begley's show *Living With Ed* in a good-natured competition to out-green each other.

This competition to live a green life is not limited to on-air story lines. A number of organizations within the industry support environmentally sound media production, led by the Producers Guild of America "Green Initiative" and the Environmental Media Association, which supports green production and also offers story-line leads and annual awards for integrating environmental messages into television and film productions.

Sustainability and Convergent Media

In the case of convergent media, both production and distribution become easier and cheaper. On aggregated online video-channel Websites such as You Tube, independent media and individuals use new technology tools to produce their own videos to educate viewers about the concept of sustainability. Popular social networking sites such as Facebook and Twitter allow video producers to share their own videos and advocate for sustainability. Environmental organizations such as Greenpeace and Earth Day Network develop their own television channels on the Internet and use video to educate viewers about how individuals have contributed to the destruction of Earth. The Earth Day Network's television channel Earthday Television has produced a wide range of documentaries in different categories—global warming, alternative energy, greener living—and also a series of videos about Earth Day celebrations tied to K–12 and college classrooms. Earthday Television also developed media and book review channels with a focus on writing about sustainability and environmental education.

As media choices expand, so does the challenge of sorting out sustainable choices amid the noise of information and entertainment. The fight for audience continues even as another telegenic environmental disaster plays out on the nightly news. In May and June 2010, amid increasingly somber and angry reports concerning the Gulf of Mexico oil spill on national news broadcasts, the chief executive officer of British Petroleum (BP) appeared in commercials, claiming "We will do right." At the same time, the corporation had purchased the search words "oil spill" in order to direct Google searches to their Website. Amid this debate, some saw an opportunity to translate the images of oily beaches into a renewed effort at climate change legislation, reminiscent of the advocacy that resulted from broadcast images of degradation that spawned the first wave of media-era environmentalism.

See Also: Greens, The (PBS Website); North American Association for Environmental Education; Social Learning.

Further Readings

Louv, Richard. *Last Child in the Woods: Saving Our Children From Nature-Deficit Disorder.* New York: Algonquin Books, 1992.

McKibben, Bill. *The Age of Missing Information.* New York: Random House, 1992.

"Our Vanishing Wilderness/THIRTEEN." http://www.thirteen.org/ourvanishingwilderness (Accessed June 2010).

"Planet Green: Sustainable Living, Energy Conservation, Earth Day." http://planetgreen .discovery.com (Accessed June 2010).

Public Broadcasting Service. "The Seeds of a Revolution: Earth Days." *American Experience.* http://www.pbs.org/wgbh/americanexperience/films/earthdays (Accessed June 2010).

Ron Steffens
Green Mountain College

SUSTAINABILITY TEACHER TRAINING

A U.S. Fish & Wildlife Service conservation training session in 2008. Sustainability teacher education programs worldwide include the World Wildlife Fund's Thinking Futures in the United Kingdom, the European Union's Environmental Education into Initial Teacher Education, and UNESCO's Teaching and Learning for a Sustainable Future.

Source: U.S. Fish & Wildlife Service

If there is to be any hope of a sustainable society, all citizens need to have the knowledge, attitudes, and skills to bring this to fruition. Education is a key strategy; however, its potential to help achieve a sustainable society will not be realized if there are not teachers able to teach sustainability education. The success of sustainability education in schools is in no small way dependent on the sorts of preparation for this task that teachers receive through their teacher education. Sustainability teacher education needs to prepare trainee teachers not only through the provision of new content on sustainability but also through experiencing and becoming skilled at using pedagogies that are interdisciplinary, holistic, inquiry based, experiential, and action oriented.

Green education has since the 1960s been seen as a means to bring about changes to the way people make decisions and act in relation to the natural environment. In the 1970s and 1980s, green education, or environmental education, focused on increasing individuals' awareness and knowledge and on changing attitudes toward the environment, believing that this would lead to individuals behaving in less environmentally damaging ways.

However, such behavior change did not materialize. Sustainability education emerged following the release of the Brundtland Report in the 1980s. This report led to an increased focus among environmental educators on the notion of sustainability. It provided a challenge to environmental educators to think more widely than the natural environment by introducing issues such as international development, economic development, cultural diversity, social and environmental equality, and human health and well-being as issues either impacting or affected by the health and well-being of the natural environment.

Sustainability education is different from environmental education in that it seeks to develop a "frame of mind" that requires educators and learners to be open to and engage with the complexity of environmental issues. Sustainability education seeks to address the systemic causes of environmental problems through holistic and integrated means. This means that issues are understood in their totality: not just as environmental issues but also as economic, social, and political issues. In addition, sustainability education sees people as agents of change who have the capacity and ability to bring about change themselves, rather than have it imposed on them. The question now becomes how to prepare the youth of today to live in a sustainable fashion.

It is teachers who hold the key to change, as they are the implementers of sustainability education in schools. Over the past 15 years, much has been written about the need to reorient teacher education toward sustainability. In all such writings, teacher education is identified as a key strategy that is yet to be effectively utilized to embed sustainability education in schools.

In fact, as far back as 1971, the International Union for Conservation of Nature (IUCN) saw the importance of teacher education in developing skills in trainee teachers if environmental education were to be included in school education. However, the education of teachers was accorded special status in 1990 when United Nations Educational, Scientific and Cultural Organization–United Nations Environment Programme (UNESCO-UNEP) stated that developing environmentally educated teachers was the "priority of educational and environmental priorities" and identified teacher education institutions and teacher educators as key change agents in reorienting education to address sustainability. Most recently, the University Twinning and Networking Programme (UNITWIN)/UNESCO chair on Reorienting Teacher Education to Address Sustainability confirmed the vital role of teacher education in the global community in transforming society for a sustainable future by acknowledging the vast and varied roles teacher education institutions can play.

The most significant reason to focus on the role of teacher education in achieving sustainability is, therefore, that the introduction of sustainability education into schools depends largely on the nature and extent of education received by trainee teachers. Many current teachers, while interested in the educational benefits for their students of sustainability education, feel they have neither the skills nor the training to be able to deliver sustainability education to their students. Teacher training in sustainability education content and approaches would not only prepare teachers for the complexities of teaching sustainability education but would also broaden the base of staff in schools that are committed to the educational benefits that sustainability education offers.

There has been much discussion over what sustainability teacher education might look like. Internationally, a number of innovative sustainability teacher education programs, such as World Wildlife Fund–UK's Thinking Futures, the European Union's Environmental Education into Initial Teacher Education, and the UNESCO Teaching and Learning for a Sustainable Future projects, laid the foundation for much current thinking about sustainability teacher education. Such projects provide a range of useful parameters for the development of a framework for sustainability teacher education. These include possible aims and objectives, course content, pedagogical approaches, and implementation strategies.

Possible Aims and Objectives

Trainee teachers, if they are to be prepared to teach sustainability education in schools, need not only new knowledge but also new ways of teaching that will ensure that the action- and futures-oriented goals of sustainability education can be met. This requires the development of sustainability education competencies. The notion of competency-based frameworks has sparked a wide and ongoing debate in education circles as to whether the complex skills involved in effective teaching can be broken down into separate and distinct competencies. Some argue that such an approach is simplistic and mechanized and may well lead to the further de-professionalization of the teaching profession. Others have argued that a more complex interpretation can facilitate planning of education provision and provide a platform upon which to ensure education for sustainability is addressed in a strategic and targeted manner. While competency-based approaches to teacher education have been criticized in the past due to crude behavioristic interpretations of skill in environmental education, recent approaches to competencies are more holistic and incorporate higher-order thinking skills, generic skills such as problem solving, and pedagogical and professional skills such as reflective practice.

The range of competencies advanced in the literature propose that in order to be proficient in sustainability education, trainee teachers should develop skill sets in both knowledge of the environment and the unique pedagogical approaches of sustainability education. Key competencies proposed are the following:

- Knowledge and understanding of environmental issues and problems
- Environmental action skills to enable action to take place
- Critical thinking skills
- An ability to effectively utilize a wide range of pedagogical approaches
- Teacher-as-professional skills such as reflective practice

Course Content

Trainee teachers from different contexts will, of course, require different content in their courses, and it would be inappropriate to provide a prescriptive set of course content for every student of education. However, certain general content central to sustainability education and also relevant to teacher education programs is outlined below:

- Environmental knowledge content that develops the ecological literacy of trainee teachers so that they are able to teach about a range of environmental and sustainability issues in a scientifically accurate manner is an essential component of sustainability education content.
- Environmental philosophy and ethics content would provide an important foundation for the development of teachers who are deep thinkers and reflective practitioners. It is important for trainee teachers to reflect on their educational and environmental philosophies and to evaluate how these affect their teaching practice.

- History of education content would encourage trainee teachers to critically examine the contexts and the history of contemporary educational systems, structures, and approaches. Such an examination would allow students to compare and contrast mainstream and alternative educational systems, curricula, and pedagogies and to reflect on those aspects that promote or hinder the teaching of sustainability education.
- Philosophy of education content would provide opportunities to examine the role of the modern scientific approach to knowledge, which emphasizes a single truth and reduces knowledge into separate components that, it has been argued, promote unsustainable development. An understanding of the ways in which new holistic approaches to science have the potential to integrate hard and social science approaches is central to sustainability education.
- Sociology of education content enables trainee teachers to explore critical theories such as reflexive modernization, which highlights the role of risk and uncertainty in post-industrial life and culture, and notions such as ecological citizenship require rethinking and reshaping perceptions of needs and aspirations in order to realize new kinds of wealth that focus on quality over quantity.
- Cultural studies and educational psychology content would enable trainee teachers to explore, for example, how marketing and popular culture shape young people's desires and identities and the impact of these on achieving a sustainable society.
- Citizenship education would enable trainee teachers to gain an understanding of the institutions and processes that regulate relations between people and between people and the environment, including the scope and power of governments, multinational corporations, and citizens.

Pedagogical Approaches

Sustainability education pedagogies, while having much in common with "good" general education approaches, are informed by a unique and particular set of educational philosophies. These philosophies pose a pedagogical challenge to teacher education because they consider current educational systems, because of their reductionist and utilitarian approaches to knowledge, to be contributors to the crisis of sustainability. Therefore, sustainability education represents a paradigm shift seeking to transform education, and as such requires pedagogical approaches that are very different from traditional teaching styles. UNESCO, the international body charged with implementing the education component of Agenda 21, has long called for a reorientation of teacher education to reflect this paradigm shift. Teacher education programs underpinned by this new paradigm would not only introduce systemic views of economy, environment, and education but would also model and promote pedagogical approaches that are interdisciplinary, holistic, experiential, and action oriented, as outlined below. Sustainability education pedagogy involves the following:

- *Inquiry-based* in order to encourage the adoption of a research stance to learning, teaching, and curriculum planning
- *Experiential*, that is, participatory and practice based to encourage trainee teachers to construct their own knowledge and to provide an opportunity for trainee teachers to reflect upon their own experiences
- *Critical and systemic* in its analysis of the environmental and educational values and assumptions underpinning education policies and practices to enable students to reflect on their role in achieving a sustainable society
- *Local*, that is, uses local issues and outdoor experiences to encourage community-based and real-world learning

- *Collaborative* to enable support and collegiality between trainee teachers and their communities
- *Reflective* in practice to encourage reflection and active engagement with teaching as a profession

Implementation Strategies

Despite growing support for sustainability teacher education, sustainability approaches have yet to be introduced consistently and coherently into teacher education programs anywhere in the world. It has been speculated that this is due to a number of factors, including lack of community and government demand for sustainability education; competition for time in already overcrowded curricula; poor dialogue between disciplines that makes interdisciplinary teaching and learning challenging; a shortage of teacher trainers with strong expertise in and passion for sustainability education; and a shortage of realistic models for implementing a sustainability teacher education program that reflects sustainability education principles and practices.

In conclusion, a sustainable society will only be achieved through a well-educated citizenry who have the requisite knowledge, attitudes, and skills. For schools to play a role, teachers are needed who are knowledgeable about both sustainability education content and pedagogies.

See Also: Agenda 21: "Promoting Education, Public Awareness, and Training" (Chapter 36); Declaration of Thessaloniki; United Nations Decade of Education for Sustainable Development 2005–2014.

Further Readings

Ferreira, Jo-Anne, Lisa Ryan, Julie Davis, Marian Cavanagh, and Janelle Thomas. "Mainstreaming Sustainability Into Pre-Service Teacher Education in Australia." Canberra: Australian Research Institute in Education for Sustainability for the Australian Government Department of the Environment, Water, Heritage and the Arts, 2009. http://www.aries.mq.edu.au/publications/aries (Accessed January 2010).

United Nations Educational, Scientific and Cultural Organization (UNESCO). "Guidelines and Recommendations for Reorienting Teacher Education to Address Sustainability." Rosalyn McKeown and Charles Hopkins, eds. Education for Sustainable Development in Action, Technical Paper No. 2 , 2005. http://www.allacademic.com/meta/p116506_index .html (Accessed January 2010).

United Nations Educational, Scientific and Cultural Organization (UNESCO). "Teaching and Learning for a Sustainable Future: A Multimedia Teacher Education Programme," 2002 and 2005. http://www.unesco.org/education/tlsf/index.htm (Accessed January 2010).

United Nations Educational, Scientific and Cultural Organization–United Nations Environment Programme (UNESCO-UNEP). "Environmentally Educated Teachers: The Priority of Priorities." *Connect*, XV/1 (1990).

Jo-Anne Ferreira
Griffith University

SUSTAINABILITY TOPICS CORRELATED TO STATE STANDARDS FOR K–12

Only 18 U.S. states currently have standards that address environmental issues, and of these states, only 12 have mandated some form of environmental education. However, many individual teachers and schools use sustainability curricula they have created themselves.

Source: U.S. Department of Agriculture, Natural Resources Conservation Service

Sustainability topics have become an increasingly relevant and important part of the curriculum in K–12 classrooms. Increased emphasis upon standards in the schools, however, has made it imperative that teachers and schools correlate sustainability curriculum to state-content standards. Fortunately for those interested in sustainability education, the underlying concepts that make rich and vibrant curricula that touch upon environmental concerns also are found in a variety of content standards, including those promulgated by national organizations and state boards of education. Teachers, curriculum coordinators, and other school leaders who desire to expose students to sustainability topics can do so while also adhering to state content standards that will assure curriculum that is rigorous, discipline based, and coherent.

Standards Movement

During the 1920s, the so-called scientific movement in education began to emphasize that teachers must state what students were expected to know in precise and defined terms. This movement reached its apex during the 1930s, when Ralph Tyler worked as an evaluator for the influential Eight-Year Study of the Progressive Education Association. Tyler's rationale explored the concepts of curriculum as defined both as what was taught in schools and as a scholarly body of knowledge. Tyler believed that curriculum must achieve the following:

- Identify educational experiences that can be provided that will assist in attaining these purposes
- Organize these educational experiences in an effective manner
- Evaluate educational experiences to determine whether the educational purposes are obtained

Tyler's approach linked teacher and student experiences through learning objectives, which could be clearly and precisely stated. Curriculum planning was transformed into a system where teachers would link the teacher-student-curriculum association by clear-cut and specific teaching objectives. Teaching objectives underscored what students were to know in detailed and explicit terms. Those evaluating student learning, be they teachers, administrators, or researchers, thus were able to compare student outcomes to the teaching objectives and to determine the success of the learning episode.

During the 1950s, Benjamin Bloom and his colleagues crafted the Taxonomy of Educational Objectives, which examined the cognitive domain as a taxonomic foundation necessary for student learning. Bloom cataloged six levels of understanding, ranging from the simple/concrete to complex/abstract. Bloom's taxonomy included knowledge, comprehension, application, analysis, synthesis, and evaluation. Knowledge involved those tasks that required students to recall specific elements of previously learned information. Comprehension included recalling bits of information, organizing these, transforming the form of information into a cohesive whole, and extrapolating from the information. Application referred to tasks asking students to use new information in different contexts from those in which it was initially learned. Analysis entailed identifying relationships between and among the parts of something. Synthesis necessitated putting together elements of different theories in a way that is novel and unique. Evaluation made judgments by using specific and credible criteria. Bloom's taxonomy provided those who promoted the use of clearly defined purposes for instruction with a new foundation for crafting educational objectives. An ever more precise relationship developed between what curriculum is to be learned, what instructional strategies the teacher uses, and what the student can expect to learn as a result of the instructional sequence.

In 1989, President George H. W. Bush called a national education summit, held in the Rotunda at the University of Virginia and attended by the governors from most states. The summit sought to establish academic achievement goals for U.S. schools and led to the National Educational Goals Panel (NEGP) and the National Council on Education Standards and Testing (NCEST). The NEGP and NCEST sought to establish the curriculum that schools should teach, the types of testing that should occur to assess that learning, and the student performance standards that should allow evaluation of this. These initiatives led to the standards movement, which was further supported by the No Child Left Behind Act of 2001. As a result of the standards movement, at least 50 different state content standards exist, each with a separate means of aligning curriculum, assessment, and evaluation. In addition to state content standards, a variety of organizations such as the National Council of Teachers of Mathematics (NCTM) and the National Research Council (NRC) have promulgated standards that cover a myriad of topics, including mathematics and science.

Standards-Based Instruction

Standards-based instruction is predicated on the notion that every child who receives a high school diploma is essentially warranted to know and understand certain key concepts and is able to perform specific tasks, especially those related to reading, writing, mathematics, science, and social studies, that will make him or her able to succeed in college or in the workforce. As much as possible, content standards attempt to emulate the practices and beliefs of members of the disciplines, therefore allowing students access to work that is engaging, meaningful, and appropriate. Content standards must be developmentally appropriate, written in a manner that supports both equity and excellence, and sensitive to the needs of both advanced and struggling learners so that all students receive an appropriate degree of challenge in the classroom. Because the proponents of standards-based instruction believe in accountability, students' mastery of particular standards is often measured by means of a criteria-referenced test. Criteria-referenced tests are those that translate test scores into an assessment of student mastery of certain behaviors. The objective of criteria-referenced tests is to see if the student has mastered certain material, not to determine how the student is performing in relation to his or her peers.

Although sustainability topics enjoy current popularity in K–12 classrooms, only 18 states currently have standards that address environmental issues. Of these states, only 12 have mandated some form of environmental education, and of these, most do not have the funding to enforce the instruction or testing of these topics. Nearly all states that have mandated sustainability topics incorporate them into the existing science curriculum rather than outline them as a separate subject. When a state has not designated sustainability as a topic to be included in the content standards, many teachers, schools, and school districts have created their own curriculum for teaching sustainability. Although sustainability instruction can be incorporated into any subject area, including English/language arts, mathematics, and social studies, many teachers, schools, and school districts that have created their own sustainability curriculum have chosen to do so through their science curriculum.

Science Standards Supporting Sustainability

The National Science Education Standards provide guidance regarding how to use what is understood about learning theory to best guide classroom learning experiences. Quality instruction related to sustainability topics thus should emulate practices that are aligned with these guidelines. Quality sustainability topics will address each of the following:

- Understanding sustainability is more than knowing facts.
- Students build new knowledge and understanding on what they already know and believe.
- Students formulate new knowledge through modifying and refining current concepts and adding new concepts to those already known.
- Learning is mediated through a social environment through which learners interact with others.
- Effective learning allows students to take control of their own learning.
- Transfer of learning, that is, the ability to apply existing knowledge to novel situations, is affected by the degree to which students learn with understanding.

Teachers, schools, and school districts desiring to build sustainability topics into their curriculum would be well served to allow for investigations of environmental issues that are pertinent to students and their lives.

Expertise is built upon a deep foundation of factual knowledge, understanding of these facts in a conceptual framework, and organizing knowledge in ways that allow for retrieval and applications. Thus, exposure to and use of the scientific inquiry process while exploring sustainability topics permits students to do more than know facts. Educational experiences where the teacher probes and discovers students' prior knowledge allow misconceptions to be addressed and new concepts to be acquired. Students who engage in hands-on work are more likely to make such a change, as students are willing to accept new ideas when they discover ideas that seem plausible and appear more useful than the ideas they previously held. Learners are more able to construct their own knowledge about sustainability when they have opportunities to articulate their ideas to others, challenge their peers' ideas, and in doing so, construct their own ideas. Allowing for project-based group investigations allows students grappling with sustainability topics to evaluate the type of evidence they need to support certain claims, build and test certain theories of phenomena, and utilize the meta-cognitive skills of monitoring and regulating their own thoughts and knowledge. Finally, students addressing sustainability topics must achieve an initial threshold of knowledge and then have opportunities to apply that knowledge to new and unique experiences.

A deep focus on inquiry thus allows classroom investigations of sustainability topics to address science standards. Building questioning and analysis into lessons strengthens students' ability to ask questions, make hypotheses, speculate, inquire, and develop answers. As part of the inquiry process, sustainability topics allow students to acquire knowledge of environmental processes and systems, allowing students to demonstrate an understanding of natural systems, habitats, and various changes that occur in the physical environment. Working with peers to investigate these issues provides students with both the learning environment best suited to acquire the inquiry and meta-cognitive skills that allow optimal learning as well as the social skills necessary to build a warm and vibrant classroom community. Thus, aligning sustainability topics with the operable science standards will build student mastery of content and process.

Math Standards Supporting Sustainability

Mathematics involves the interaction of knowledge, tools, and ways of doing and communicating about numeracy in both practical and theoretical settings. The National Council of Teachers of Mathematics (NCTM) has promulgated mathematics standards so that instruction in that subject can better address six overarching themes: equity, curriculum, teaching, learning, assessment, and technology. Each of these principles, although conceived to support mathematics instruction, also supports the introduction of sustainability topics in K–12 classrooms. Excellence in mathematics instruction and sustainability education requires equity, as manifested in high expectations, an adequate level of challenge, and strong support for all students. Similarly, a mathematics curriculum that addresses sustainability topics must be more than a collection of activities, instead focusing on important issues and well articulated across grade levels. Sustainability and mathematics teaching that is effective requires a strong grasp of what students must know, understand, and be able to do, followed by instruction that challenges and supports them to do it well. Students must learn mathematics and sustainability topics with understanding, combining prior knowledge with in-class experiences to actively build new knowledge. Assessment should furnish useful information about understanding of mathematics and sustainability topics so it can support learning by students. Finally, technology is essential to effective mathematics and sustainability teaching and learning insofar that it influences the material that is taught and enhances student learning.

Similarly, the NCTM content and process standards provide guidance and structure for educators interested in incorporating mathematics instruction with sustainability topics. The NCTM content standards focus on the content students should learn, whereas the process standards emphasize ways of acquiring, using, and exploring content knowledge. In all, there are five content standards and five process standards, which are the same from kindergarten through grade 12. The content standards—number and operations, algebra, geometry, measurement, and data analysis and probability—each comprise a small number of goals for each grade level, which build upon each other as a student progresses through school. Similarly, the process standards—problem solving, reasoning and proof, communication, connections, and representation—are also present across grade levels, although the precise emphasis varies over time.

Teachers desiring to introduce sustainability topics into their classrooms will often also utilize the NCTM content standards in their environmental curriculum. The NCTM process standards, however, provide multiple opportunities to recast mathematics instruction in a manner that explores sustainability topics as a guiding framework that crosses disciplines. As the NCTM process standards seek to build mathematical knowledge through problem solving, for example, teachers may wish to address sustainability by identifying environmental issues that affect the students. A guided investigation into ways of reducing car emissions in a neighborhood requires children to use mathematical problem-solving skills to explore the problem, define goals, and measure results of potential solutions. Numerical reasoning and proof may well prove necessary to defend a potential solution to the problem, as the recognition that investigating conjectures, developing and evaluating mathematical arguments and solutions, and selecting a defensible resolution are all vital to both mathematical and sustainability topics. The ability to communicate a solution, representative of coherent and clear thinking, is also something that students investigating a sustainable topic can use mathematics to do. Recognizing the connections between sustainability and mathematics allows students to build a coherent whole from two subjects sometimes treated separately. Finally, students can and should use mathematical ideas as representations that organize, record, and communicate their solutions to sustainable topics.

Other Subjects

A variety of other standards related to history, geography, economics, art, and other academic subjects have been developed by national organizations that can be used to address sustainable topics. Groups that have developed such standards include the National Council for the Social Studies (NCSS), the Center for Civic Education, the National Council on Economic Education (NCEE), the National Geographic Society, the National Center for History in the Schools, and the Consortium of National Arts Education Associations. These national standards as well as those disseminated by state and local education authorities provide a wealth of opportunities to explore sustainability issues while also providing students standards-based educational experiences. Creative, concerned, and conscientious teachers will align sustainable topics to standards to assure adherence to frameworks and improve student performance.

As sustainability topics become more prominent in public discourse, it is not surprising that educators seeking to respond to these issues, as well as to nourish student interests, will include them in their curriculum. To the extent possible, teachers who align sustainability topics with standards will assure their students access to both topics of relevance and a system that has been put in place to assure equity and excellence for all. As national organizations and state and local educational authorities continue to refine standards, the connections between sustainability and other academic subjects may become more explicit, but the current standards allow for many opportunities for student investigations and instructional strategies that support both goals.

See Also: Constructivism; Early Childhood Education; Environmental Education Debate; Experiential Education; Sustainability Topics for K–12.

Further Readings

Davis, J. M. *Young Children and the Environment: Early Education for Sustainability.* New York: Cambridge University Press, 2010.

Gray, D., L. Colucci-Gray, and E. Camino, eds. *Science, Society and Sustainability: Education and Empowerment for an Uncertain World.* New York: Routledge, 2009.

Hewitt, T. W. *Understanding and Shaping Curriculum: What We Teach and Why.* Thousand Oaks, CA: Sage, 2006.

National Council of Teachers of Mathematics (NCTM). *Principles and Standards for School Mathematics.* Reston, VA: NCTM, 2000.

National Research Council. *Inquiry and the National Science Education Standards: A Guide for Teaching and Learning.* Washington, DC: National Academy Press, 2000.

Stephen T. Schroth
Jason A. Helfer
Whitney L. Taylor
Knox College

Sustainability Topics for K–12

In the early grades, sustainability education may begin with the development of a general awareness of biodiversity and animal habitats. These children were bird-watching at the San Luis National Wildlife Refuge in California in 2008.

Source: U.S. Fish & Wildlife Service

As education is often a key element to the advancement and progression of ideas, so it is only natural that sustainability topics should be used by schools to address and improve the declining state of the environment. With the increasing awareness of the environmental damage caused by human actions, adults have made an effort to better the planet. As a response to this awareness of the importance of sustainability, "going green" has become a phenomenon among Americans. This change is reflected in many aspects of life, as sales of hybrid cars have skyrocketed, the percentage of Americans who recycle has steadily increased, and the demand for alternative energy has led to the construction of wind and solar farms across the nation. Not a great deal has been done, however, in terms of changing K–12 educational practices to reflect this new sensibility. Sustainability will soon be in the hands of a new generation, one that can be taught to be conscientious of the environment. Today's students require the knowledge and the tools to make the powerful changes that are necessary to reverse the damage that has been done to Earth and to prevent more harm from occurring. Countless topics can be covered and integrated into preexisting curricula in all grade levels, and separate units can be constructed that focus on sustainability. Sustainability education builds on previous and learned knowledge of the planet, starting with general awareness of biodiversity and habitats in the early grades to involvement with local environmental protection issues and a strong knowledge of sustainable practices upon graduation from high school.

Sustainability in K–12 Education

Sustainability, in essence, provides the tools and the setting for life to continue. More specifically, sustainability education focuses on twin goals of allowing life to continue in a manner that supports human and animal health and the possibility for future generations to be nourished and physically satisfied by the Earth. In this sense, sustainability is a measure of quality of life. Factors like air pollution, soil erosion, water contamination, and draught may decrease the quality of life of individuals around the world. In addition to quality of life considerations, sustainability represents the ability to meet the needs of the current population without endangering the ability for future generations to support themselves. Sustainability issues can be found on the local, national, and international level, and all of these levels are interconnected.

As an educational concept, sustainability and environmental education serve the purpose of increasing students' awareness of the issues surrounding environmental degradation and impressing that these problems are largely caused by human activity. This goal is supported for a variety of organizations such as the United States Partnership for Education for Sustainable Development. Advocates for sustainability education define it as a framework that engages students in all subjects by using the real-world context of the complex interconnections between the creation of vibrant communities, strong economies, and healthy ecosystems. Sustainability education can provide useful context for instruction in all areas, including the arts, social studies, English/language arts, mathematics, and science. When done properly, sustainability topics promote the development of critical and creative thinking skills, encourage systems thinking, and promote collaboration and communication.

State Mandates for Sustainability Education

Sustainability is a key component of environmental education, though an environmental education program is not required to introduce such topics. Indeed, there is little mandated curriculum for sustainability. Only 18 states have formal environmental education learner objectives and outcomes or are in the process of creating them. Of these states, 12 have mandated some form of environmental education. Many states with requirements for K–12 sustainability education do not have the funding to enforce the instruction or testing of these topics. Nearly all states that have mandated sustainability topics incorporate them into the existing science curriculum rather than outlining them as a separate subject. Even if a state has not designated sustainability as a topic to be included in the learning standards, school districts have the option of creating a curriculum for teaching sustainability. Districts that have done this have outlined their intentions to become more sustainable and have assembled resources for teachers to use when including these topics in the classroom.

Wisconsin, Maryland, Vermont, and Washington are a few states that have established environmental education and sustainability learning standards. Furthermore, nongovernmental organizations (NGOs) like the North American Association for Environmental Education and the U.S. Partnership for Education for Sustainable Development have published learning standards that provide educators with a framework for teaching sustainability topics. All of these sources vary in the number of principal learning standards that are set forth, ranging from two to six. They also differ in how the standards are organized; some dictate standards that should be reached by the completion of fourth, eighth, and 12th grade, and others offer standards for each grade level. Regardless of organization and slight variations in content, all of these sets of standards provide educators with the necessary framework to implement sustainability topics into the classroom.

Teacher-Created Sustainability Curriculum

Though sustainability education is not required in the vast majority of schools, topics pertaining to these issues can still be integrated into the extant curriculum using textbooks, trade books, realia, and other resources, including online sources. Since many subjects, including social studies and science classes, often examine current events in society, teachers frequently respond to environmental issues as part of their subject matter and explore individual and community efforts to improve sustainability. Furthermore, teachers with a particular interest in sustainability are able to teach about topics in relation to science and social studies and to investigate literature that condones sustainable behavior. A study conducted by the North American Association for Environmental Education (NAAEE) reported that 61.2 percent of K–12 educators include environmental studies topics in their curriculum in the hope that it will encourage students to become active about the environment. Some teachers also responded that they hope to show students that what they learn in the classroom is applicable to the outside world.

Community-Centered Learning

Sustainability education helps students apply knowledge learned about the environment to their lives, and it is also important to consider the environmental impact of one's actions on the lives of others around the world. Focusing on the planet as a whole, or as a "global community," provides students a different view of sustainability issues by allowing them to see how the actions of a small group can impact the well-being of millions. While many American students may have difficulty grasping some concepts of sustainability and linking cause and effect if they have not experienced the results of climate change, widening the scope of sustainability education can allow students to see the interconnectedness of the world and its many inhabitants. For example, students in the Pacific Northwest can more easily comprehend the implications of a drought after studying cases in sub-Saharan Africa. Likewise, students from the Midwest can have a better understanding of famine and crop failure after examining how China has been affected by flooding. All American students can conceptualize the importance of the rainforest on life all over the world by studying what products come from the rainforest, how the rainforest impacts the climate and weather patterns, and the rich biodiversity that exists there.

Green Curriculum

Broadening the scale on which students learn about the environment is important to consider when determining topics for sustainability. Practically all aspects of the curriculum can incorporate aspects of sustainability. Coupling sustainable curriculum with resources like the state standards or learning guidelines from the North American Association for Environmental Education (NAAEE) can help teachers and curriculum designers to decide which topics to approach at which grade levels. Furthermore, the standards serve as a practical guide and a helpful tool for educators to create assessments of students' knowledge of sustainability. Many of the standards coincide with science, social studies, math, and language arts instruction so teachers can introduce sustainability topics across the curriculum. Typically, sustainability standards outline learning goals for students by the time they have completed 4th, 8th, and 12th grades so there is flexibility in the topics that can be taught and in the order in which they are introduced.

The themes introduced in elementary school are vital to creating a foundation for further learning in the upper grades. As students grow older, their continued experiences with natural science both inside and outside the classroom lead to a better understanding of how habitats and various life forms interact. This essential knowledge is built upon with the continued appreciation of nature. The philosophy behind this direction in the curriculum is that students will want to protect nature if they learn about its diversity, rarity, and importance to human life. Hands-on activities in sustainability are popular at all grade levels because they help students take a stake in their learning. For example, growing plants in various conditions and charting their success allows students to be interactive and to control the variables. In this experiment and others, the focus is on the link between man-made conditions and the outcome—an imperative aspect of students' understanding of sustainability. Using these foundations, students in the upper grades can expand and experiment on more advanced levels to further their awareness of sustainability issues.

Objectives of Sustainability Curriculum

Using these foundations of sustainability knowledge, state standards, and guidelines set forth by NGOs, topics for grades K–12 can be outlined in four main objectives. The first is questioning and analysis, which strengthens students' ability to ask questions, make hypotheses, speculate, inquire, and develop answers. After fourth grade, students are expected to have obtained the basics of these skills and the ability to apply them simply to a variety of situations. Students also devise ways that they can investigate and come to conclusions and design simple experiments to reflect their mastery of this concept. In middle school, students will identify specific environmental issues that are of interest to them and pose clear questions and means of research to investigate sustainability. By the end of eighth grade, students can be expected to use experiments or simulations to explore sustainability topics and to communicate their findings from the initial hypotheses to the final results and to defend their logic. In high school, students use their skills in questioning and analysis to inquire about current issues in sustainability on a local, national, and global level and suggest possible inquiries and solutions. They also investigate all sides of the issues and learn to communicate their findings to groups that would be interested.

The next essential foundation concerns knowledge of environmental processes and systems, where students show an understanding of natural systems, habitats, and various changes that occur in the physical environment. Through this objective, students learn about the interconnectedness of the many facets of the Earth and come to understand the delicacy and balance of the planet's systems. This kind of knowledge is necessary for students to truly understand what is at stake and to take an interest in sustainability. These objectives are often found in existing science curriculum, meaning that the foundation for sustainability education is present and simply needs to be developed further. In the early grades, students can focus on energy in nature, the water cycle, and elements of ecosystems. Understanding how the Earth works leads to the examination of resources and how humans use them in their daily lives. In connection with this, students can examine the results of using these resources, including related jobs, environmental impact, and how to decrease dependence on nonrenewable resources. Middle school students learn more about the exchanges in natural processes, such as succession and evolution, and take a closer look at biodiversity and how it ranges around the world. Furthermore, students examine the human impact on biodiversity, how humans change their environment, and how these factors vary over cultures. Students also learn which natural resources can be found in their area and how other resources like gasoline and electricity reach their state. By the time students reach high school, they are prepared to mesh these ideas and create hypotheses of how populations and, more specifically, their generation will respond to environmental changes. This can be tailored to the students and to the community and can be detailed to expand into social studies and literature. For example, students can explore how environmental quality affects economic policies, how the risks of pollutants are shared with the populations at risk, how technology has changed the way humans harvest and use resources, and how natural resource trade impacts international relations. High school students can also benefit from exploring careers in resource management and other related fields.

These concepts are elaborated upon in the next essential foundation, investigation of sustainable practices. This set of objectives guides students to be able to identify environmental issues, the groups of people that are affected, and how society responds to these problems. In the early grades, students show an understanding of the relationship between humans and the environment by identifying proposals for sustainability and which groups of people are involved in making changes. Middle school students explore how human beliefs and values play a part in sustainability, with the goal of exploring both sides of an issue and improving their ability to determine credibility of sources. Students also revisit questioning and analysis. By high school, students have advanced to understanding the human activities that contribute to or support a sustainable environment and how societal values have changed or developed in response.

The decision and action skills foundation outlines how students can put their knowledge to use by identifying individual actions that impact sustainability, evaluating the reasons why people feel strongly about these issues, and communicating information in a variety of formats. Younger students can make posters to raise awareness of sustainability issues and learn how they can influence a sustainability issue individually or in a group. By the end of middle school, students will have a basic knowledge of the politics involved in sustainability, including the budgeting that is allotted for sustainability and the governmental organizations that are responsible for managing environmental issues. By high school, students can express their personal views on sustainability and create an action plan regarding the steps they will take to be sustainable adults. They will learn about the political and legal courses of action that are used to determine the outcome of an environmental issue and choose to take a stand by contacting government officials regarding sustainability.

Conclusion

When working to develop a curriculum that explores environmental issues, a key point across age groups is the identification of individual actions that impact sustainability and evaluation of the reasons why people feel strongly about these issues. There is an abundance of information pertaining to sustainable living. By introducing it in steps each year and building upon students' prior knowledge and experiences, educators can help shape a generation of young minds that think and work sustainably. Making sustainability a focus of all academic subjects and interweaving it with other curricular topics helps teachers and students to recognize the centrality of its value to our world.

See Also: Early Childhood Education; Green Math; Outdoor Education; Social Learning; Sustainability Teacher Training; Sustainability Topics Correlated to State Standards for K–12.

Further Readings

Davis, J. M. *Young Children and the Environment: Early Education for Sustainability.* New York: Cambridge University Press, 2010.

Gray, D., L. Colucci-Gray, and E. Camino, eds. *Science, Society and Sustainability: Education and Empowerment for an Uncertain World.* New York: Routledge, 2009.

Jones, Paula, David Selby, and Stephen Sterling, eds. *Sustainability Education: Perspectives and Practice Across Higher Education.* London: Earthscan Publications, 2010.

Stephen T. Schroth
Jason A. Helfer
Rose M. Van Grinsven
Knox College

SUSTAINABILITY WEBSITES

While there are an increasing number of print resources available for those education professionals interested in sustainability, the Internet has become a location for quick dissemination of information and for resources for administrators, facility personnel, and educators. This article briefly discusses a number of sustainability Websites available, including those from nonprofit organizations, sustainability associations, and federal agencies. These Websites, each a helpful resource for incorporating sustainability into education, also contain further resources both in print and online, expanding the knowledge base. It should be noted that while each of these Websites represents a tremendous resource, there is a lack of peer review for some of the articles and information presented in this format. Readers should also take the opportunity to receive e-mail updates or join listservs associated with each site.

Association for the Advancement of Sustainability in Higher Education (AASHE)

AASHE is an international association of institutions focused on spreading sustainability initiatives in higher education in campus operations, in the classroom, and in research. In addition to its biannual conference and weekly newsletter, the AASHE Resource Center is a great resource for those looking for case studies and for model programs across the country. The Resource Center is divided into five headings: AASHE Publications; General Resources; Education and Research; Campus Operations; and Planning, Administration and Engagement. Each heading is broken into further categories and, with the exception of the first two listed headings, provides links to programs at institutions around the world. Many of these areas are open to the public, and some do require a login and password, available to anyone associated with an AASHE member institution.

For example, under Education and Research, the viewer can choose between the following topics: co-curricular education, curriculum, research, curriculum integration workshops, and sustainability surveys. Choose "curriculum" and more subheadings are shown, including academic programs, courses on campus sustainability, course inventories, syllabi, and student research papers. Choose "curriculum" and search programs in sustainability by degree type, nondegree programs, or by discipline. Each of these then provides URL links directly to any related program's Website. In this way, the AASHE Resource Center serves as the umbrella for individual sustainability-themed Websites across higher education, almost completely user-generated.

The online resources created and published by AASHE include blogs, the AASHE Digest annual review, "how-to" guides, archives of the weekly newsletter, AASHE-produced survey results, and wiki books on a variety of topics. Also included within the Resource Center is a group of discussion forums on various sustainability topics, moderated by AASHE staff, which include the same topic headings as their general resources (Education and Research; Campus Operations; and Planning, Administration and Engagement), while allowing for conversation to occur between interested parties in real time without e-mail, listservs, or waiting until the next conference.

Campus Sustainability Case Studies

This resource, provided through the National Wildlife Federation (NWF) Campus Ecology Program, includes annual user-generated sustainability project case studies from campuses across the country as well as internationally. Viewers can sort projects by year (1997–2009), by school, by location, or by topic. Available topics include assessment, building design, climate action plan, composting, dining, energy, education, management, greenhouse gas inventories, purchasing, transportation, waste, and water.

Each topic includes a brief introduction, as well as a few additional general links to Websites and publications suggested by campus ecology staff. After the introduction, case studies are sorted alphabetically by school. Each case study includes a brief description and links to read more about the project or see additional projects from the same school. The full case study link opens a PDF created by each institution and organized by abstract, goals and outcomes, engagement and support (leaderships, funding, and outreach), and contact information for personnel involved as well as additional information about each school. The case studies site can provide valuable resources for someone interested in learning about schools that have already begun projects of interest.

The College Sustainability Report Card

The Sustainability Endowments Institute (SEI) releases a sustainability report card, which provides grades on a number of sustainability-related topics in higher education. SEI states that the Website can serve as a tool of prospective students, as well as researchers, while providing third-party evaluation and assessment. The 2010 Report Card includes 332 schools from the United States and Canada with endowments of $160 million or more. Each institution is sent four surveys to complete, and the results of each survey are included online as a case study.

Visitors to the Website are able to browse through schools, select schools to compare, or establish an account to save searches and profiles. Each school's site contains a good deal of information, including general demographics, as well as grades and summaries for each topic scored. Also included is a link to each school's complete survey. The areas evaluated for each school are administration, climate change and energy, food and recycling, green building, student involvement, transportation, endowment transparency, investment priorities, and shareholder engagement. The complete survey link includes the complete answers as stated by the institution's contact. As with the AASHE Resource Center and the NWF Case Studies, the College Sustainability Report Card provides an additional resource of campus projects.

American College and University Presidents Climate Commitment (ACUPCC)

The American College and University Presidents Climate Commitment (ACUPCC) is not just an association of schools pledging to address and to improve their environmental impact with the goal of carbon neutrality; their Website also provides signatories with a number of resources that aim to assist in reaching that goal. A number of topics are addressed for organizations, and they include climate action plans, energy, green building, emissions inventories, transportation, procurement, and waste. Each category includes a brief summary with overall goals that institutions need to address as ACUPCC signatories, a few descriptions of model programs (case studies), and external links related to the topic being addressed.

For example, under the energy topic, ACUPCC tells institutions that they should be prepared to address energy reduction, renewable energy generation (on-campus and purchased), and carbon offsets for any remaining emissions. The State University of New York at Buffalo and Carleton College were mentioned as model institutions, and external links included programs with the U.S. Environmental Protection Agency, the Alliance to Save Energy, and the U.S. Department of Energy. For those institutions aiming at carbon neutrality, the ACUPCC resource site provides a good deal of information under many sustainability topics.

Higher Education Associations Sustainability Consortium (HEASC)

The Higher Education Associations Sustainability Consortium (HEASC) is an organization consisting of higher education associations working together with their member institutions to address campus sustainability. At the central site, associations have submitted links to their individual Websites that include sustainability measures as well as resources created by the HEASC itself. Member institutions include, among others, the American Association of Community Colleges, the American Association of State College and Universities (AASCU), the Council of Christian Colleges and Universities, the National Association of Independent Colleges and Universities, and the Society for College and University Planning.

Under each association's heading, links are provided to Websites and publications created by that association. For example, the AASCU includes a document titled "State College Role in Advancing Environmental Sustainability: Policies, Programs, and Practices" as well as a link to a Website from the Grants Resource Center on funding sustainability projects. The Resource Center provided by the HEASC includes over 100 links that can assist virtually any type of institution or professional in finding ways to connect their area to sustainability.

National Council for Science and the Environment (NCSE)

The National Council for Science and the Environment (NCSE) is an organization that aims to assist policy makers with better information regarding environmental decisions. It accomplishes this through education, career development, a conference on policy and the environment, communicating environmental information, and promoting environmental science education.

Specifically for those institutions interested in sustainability, the NCSE has created an affiliate program of over 160 schools that network not just on science policy issues but on advancing environmental education, with a recent focus on sustainability. The organization's Programs and Initiatives site includes links to the affiliated schools and assists in communication between schools. For science educators, the greatest resource on the NCSE site is the Earth Portal.

The Earth Portal brings together an online community of over 1,000 international experts from over 50 organizations to share environmental information freely. The information gathered on the Earth Portal is assembled within the Encyclopedia of Earth, in addition to a stream of environmentally related news items and a page of moderated forms. Hot topics on the Earth Portal include many topics specifically related to sustainability, including climate change, energy, environmental health, and pollution. In addition, the Encyclopedia of Earth includes specific pages for educators that include publications and external links for pre-K through post-secondary instructors.

North American Association for Environmental Education (NAAEE)

The North American Association for Environmental Education (NAAEE) exists as a professional organization that supports environmental educators at all levels of education as well as government agencies and industry. Its Website includes a page that lists its programs for environmental educators, which can be very useful for educators, even those not currently members. The programs include a set of guidelines created by the NAAEE for environmental educators, guidelines for pre-service teachers, and a directory of higher education programs and researchers.

In addition, the NAAEE sponsors the EE-Link, which it describes as the most comprehensive search engine related to environmental education online. The EE-Link includes external resources, including related publications for K–12 students, higher education students, faculty, researchers, and teachers. Under each topic are resources provided by NAAEE, professional and classroom resources, and resources provided by educators themselves.

Federal Agencies

A number of federal agencies have created resources related to sustainability and green education that can provide a valuable source to environmental educators as they develop curricula. The U.S. Environmental Protection Agency's (EPA) Office of Research and Development is responsible for a sustainability-centered Website. This site specifically addresses the urban and built environment, water, energy, and human health. Each topic includes a brief description, EPA-sponsored policies and programs, a collection of external links of research areas and tools, and assessment and performance measures for each topic. Especially for those disciplines preparing students for careers in sustainability and the environment, this site can help develop real-world classroom assignments and assist in developing internship and community outreach opportunities.

The EPA also maintains a Website dedicated to information for colleges and universities. Included here are voluntary EPA programs for institutions, such as Energy Star, Smart Growth Network, and Labs 21. Also included are a number of external links to higher education sustainability associations and other links related to sustainability in higher education.

The National Aeronautics and Space Administration (NASA) also maintains a Website dedicated to climate change that can provide a valuable resource to educators. Included is information on NASA's role in evaluating climate change and developing solutions. In addition, there are separate pages for children and educators. NASA's Global Climate Change Education program is a grant program that has created a number of resources for K–12 and higher education teachers based on their awarded programs, utilizing NASA resources. Included here are links to additional NASA resources and data, as well as Websites of program awardees.

Finally, the U.S. Department of Energy (DOE) has a number of resources for educators related to climate change and sustainability. The Office of Policy and International Affairs is the central location for climate change at the DOE. Working along with this office are a number of related departments, including the U.S. Climate Change Technology Program, the Office of Energy Efficiency and Renewable Energy, and the Office of Science. Each provides educational background on its specific areas of interest and can be used by educators in developing curricula and finding resources.

A number of online resources are also available for those interested in green education: for finding relevant programs, resources, and data for curriculum development. Each resource listed below provides links to even more Websites as well as publications to assist the environmental educator.

See Also: American College and University Presidents Climate Commitment; Association for the Advancement of Sustainability in Higher Education; Higher Education Associations Sustainability Consortium; North American Association for Environmental Education; Second Nature: Education for Sustainability; Sustainable Endowments Institute.

Further Readings

American College and University Presidents Climate Commitment. "Resources and Events." http://www.presidentsclimatecommitment.org/resources (Accessed December 2009).
Association for the Advancement of Sustainability in Higher Education. "Campus Sustainability Resource Center." http://www.aashe.org/resources/resource_center.php (Accessed January 2010).

Earth Portal. http://earthportal.net (Accessed January 2010).

Higher Education Associations Sustainability Consortium. "Resource Center." http://www2
.aashe.org/heasc/resources.php (Accessed January 2010).

National Aeronautics and Space Administration. "Global Climate Change." http://climate
.nasa.gov (Accessed December 2009).

National Council for Science and the Environment. "NCSE Programs and Initiatives." http://
ncseonline.org/11programs (Accessed December 2009).

National Wildlife Federation Campus Ecology. "Campus Sustainability Case Studies." http://
www.nwf.org/campusEcology/resources/yearbook/dspYearbookbyYear.cfm?year=2009#
2009 (Accessed December 2009).

North American Association for Environmental Education. "Programs, Initiatives, Awards."
http://www.naaee.org/programs-and-initiatives (Accessed January 2010).

Sustainable Endowments Institute. "The College Sustainability Report Card." http://www
.greenreportcard.org (Accessed December 2009).

U.S. Department of Energy. "Climate Change." http://www.energy.gov/environment/
climatechange.htm (Accessed January 2010).

U.S. Environmental Protection Agency. "Colleges and Universities." http://www.epa.gov/
sectors/sectors/college.html (Accessed January 2010).

U.S. Environmental Protection Agency. "Sustainability." http://www.epa.gov/sustainability
(Accessed January 2010).

Justin Miller
Ball State University

SUSTAINABLE ENDOWMENTS INSTITUTE

The Sustainable Endowments Institute (SEI) is a nonprofit organization focused on campus sustainability practices, with a special emphasis on college and university endowments. Based in Cambridge, Massachusetts, SEI provides one response to a growing interest in assessing how institutions of higher education are performing with respect to sustainability. Its primary assessment tool is the College Sustainability Report Card, which SEI issues annually. Three hundred and thirty-two schools in the United States and Canada received grades in the 2010 report cards. These schools represent the 300 largest endowments in higher education, plus 32 additional schools that requested inclusion in the rating process.

Mark Orlowski founded SEI in 2005 after recognizing a need for campus sustainability evaluation that accounted for endowment practices. Because the investment policies of colleges and universities can have significant effects on the environment, SEI seeks to call attention to those issues. The report card SEI developed grades campuses in nine areas, three (one-third) of which relate to endowments: endowment transparency, investment priorities, and shareholder engagement. The remaining six categories examine policies and practices in administration, climate change and energy, food and recycling, green building, student involvement, and transportation.

The College Sustainability Report Card's rating system differs from other efforts to assess campus sustainability performance. Unlike programs such as the Association for the Advancement of Sustainability in Higher Education's (AASHE) Sustainability Tracking, Assessment & Rating System (STARS), with bronze, silver, gold, and platinum levels of accomplishment, the SEI report cards issue letter grades on a scale of A to F. In 2007, the first year of the report cards, only four schools earned the highest assigned grade of A–: Dartmouth College, Harvard University, Stanford University, and Williams College. A number of schools where extensive campus sustainability efforts have garnered wide recognition earned lower grades. Oberlin College, for example, which has pursued green building policies and which has signed the American College and University Presidents Climate Commitment (ACUPCC), earned a C+ on its 2007 report card, largely due to failing marks in the categories of Endowment Transparency and Shareholder Involvement.

The process SEI uses currently to assess campus sustainability draws on data collected from four surveys the institute sends to administrators and students on each campus. Three of the surveys contain questions on general campus practices, dining practices, and endowment practices. The fourth survey, completed by leaders of student sustainability and environmental campus groups, focuses on student-led efforts in these areas. In addition to information received through completed surveys, SEI seeks publicly available data on each campus from sources such as school Websites, AASHE, media coverage, and so on.

From the data it gathers on each campus, SEI uses 48 indicators to evaluate activity under each of the nine performance categories. Every indicator under each category has a weighted value to reflect its respective importance to overall campus sustainability. For example, the three indicators and their respective weights for the endowment transparency category are transparency of investment holdings (40 percent), availability of proxy voting records (30 percent), and accessibility of investment holding and proxy voting information (30 percent). Other categories have a wider distribution of indicators and weights. The administration category, for instance, contains six separate indicators with weights ranging from 5 percent for having a campus sustainability Website to 30 percent for the presence of sustainability policies and plans. In cases where a campus has achieved some but not all of what a particular indicator measures, SEI may assign partial credit.

Assessment of performance with respect to the various indicators leads to a full letter grade SEI assigns to each of the nine categories. To receive an A in a particular category, an institution must earn at least 70 percent of the credits available in that category. Grades of B, C, and D require 50 percent, 30 percent, and 10 percent of the available credits, respectively. Once it assigns grades in each category, SEI averages these scores to calculate the final letter grades for a campus's sustainability performance, using a standard 4.0 Grade Point Average (GPA) scale. For example, a university that earned six As and three Bs for the nine categories would have an overall score of 3.67, for an overall grade of A–. SEI recognizes that not every indicator may be relevant to every campus, in which case that element is excluded from the calculation of the final grades. Canadian colleges and universities, for example, are not able to sign on to the ACUPCC, as it is limited to U.S. campuses.

The 2010 report cards represent the fourth year that SEI has issued its campus grades. The evaluation instrument has changed to some extent in every version of the report card in order to reflect changing practices and emphases in campus sustainability. Transportation, for example, was not an evaluative category in the 2007 report cards. An addition to the most recent report cards is the potential for earning "extra credit" in cases where a campus has undertaken a noteworthy or unique sustainability initiative in a particular category.

One result of the SEI report cards and other similar assessment tools is that campuses wishing to participate in these evaluations must dedicate considerable time and resources to gathering and reporting the required data. Some of this data may be very difficult to obtain. Additionally, many institutions now recognize the difficulties involved in demonstrating exceptional campus sustainability performance, particularly when endowment practices are considered as an important part of that performance. While the number of schools earning an A– in 2010 has increased to 26 institutions, no campus has yet to earn a full A. The 4.0 scale would require a campus to earn As in each of the nine categories in order to earn this highest level of achievement. Nevertheless, many institutions have improved their grades substantially. Oberlin College, for example, which received a C+ on the 2007 report card, received an A– in 2010. The SEI Website makes it possible to compare an institution's performance over a number of years, so that one can determine the areas where improvement (or possibly decline) has occurred.

Accelerating interest in the sustainability efforts of colleges and universities has made the work of organizations like the Sustainable Endowments Institute prominent. SEI's focus on endowments, along with other more typical assessments of campus sustainability, has generated significant discussion as to the role of higher education's investment policies. As a dynamic tool that can respond to changing practices, the Campus Sustainability Report Cards are one way to assess the sustainability trajectory of higher education.

See Also: AASHE: STARS (Sustainability Tracking, Assessment & Rating System); Dartmouth College; Harvard University; Oberlin College; Stanford University (Global Ecology Research Center).

Further Readings

Gose, Ben. "Charity Urges Colleges to Focus on 'Sustainability' of Their Endowments." *Chronicle of Philanthropy,* 19/16 (2007).

Khadaroo, Stacy Teicher. "Now, A Green Report Card for U.S. Colleges." *Christian Science Monitor* (July 9, 2008).

Sustainable Endowments Institute. http://www.endowmentinstitute.org/index.html (Accessed January 2010).

Sustainable Endowments Institute. "The College Sustainability Report Card." http://www .greenreportcard.org (Accessed January 2010).

Stacey Swearingen White
University of Kansas

SWANSEA DECLARATION

The Swansea Declaration was a commitment made at the August 1993 Association of Commonwealth Universities (ACU) Conference in Swansea, Wales. The declaration was timely as it was made at the 15th quinquennial conference in Swansea, the theme of which was "People and the Environment—Preserving the Balance."

This was a key milestone in a series of international higher education conferences that were convened by university representatives in 1990. The series of declarations, of which the Swansea Declaration is one, is key in affirming the role that higher education can play in sustainable development, furthering the mission of universities to contribute to the improvement of society as a whole.

Drivers for the Swansea Declaration

The ACU meeting in Swansea was inspired by two previous similar declarations made by university associations in Talloires and Halifax; these were declarations that were made in an era of change and global recognition of man's impact on the environment. The Swansea Declaration was part of the higher education community's move to recognize and deal with the problem of man's impact on the environment and was notable—for although it was not the first declaration of its kind, it was one of the biggest in scale and scope.

The previous Talloires Declaration, made by University Leaders for a Sustainable Future (USCF) in Talloires, France, in October 1990, a conference of 22 university leaders, has since been signed by 413 college and university presidents. The Halifax Declaration followed the Talloires Declaration. It took place in Halifax, Canada, and was attended by representatives from 33 international universities from 10 different countries as well as senior representatives from university associations such as the International Association of Universities, the United Nations University, and the Association of Universities and Colleges of Canada. The Swansea Declaration was larger in scale and involved participants from over 400 universities from 47 different countries.

However, it should also be noted that there were a number of precedents for the Talloires and Halifax Declarations; the Belgrade Charter was introduced at the Belgrade Conference on Environmental Education, and the Tbilisi Declaration took place at the Intergovernmental Conference on Environmental Education.

There were a number of drivers for the Swansea meeting. It was felt that higher education was poorly represented at both the United Nations Conference on Environment and Development (commonly referred to as the Earth Summit) held at Rio de Janeiro June 3–14, 1992, and also at the United Nations Agenda 21 program (the full text of which was released at the Earth Summit).

The Swansea Declaration

Quoting directly from the Swansea Declaration, it establishes a series of seven actions to which universities should give consideration:

1. To urge universities of the ACU to seek, establish and disseminate a clearer understanding of sustainable development—"development which meets the needs of the present without compromising the needs of future generations"—and encourage more appropriate sustainable development principles and practices at the local, national and global levels, in ways consistent with their missions.

2. To utilize resources of the university to encourage a better understanding on the part of governments and the public at large of the inter-related physical, biological and social dangers facing the planet Earth, and to recognize the significant interdependence and international dimensions of sustainable development.

3. To emphasize the ethical obligation of the present generation to overcome those practices of resource utilization and those widespread circumstances of intolerable human disparity which lie at the root of environmental unsustainability.

4. To enhance the capacity of the university to teach and undertake research in sustainable development principles, to increase environmental literacy, and to enhance the understanding of environmental ethics within the university and with the public at large.

5. To cooperate with one another and with all segments of society in the pursuit of practical and policy measures to achieve sustainable development and thereby safeguard the interests of future generations.

6. To encourage universities to review their own operations to reflect best sustainable development practices.

7. To request the ACU Council urgently to consider and implement the ways and means to give life to this declaration in the mission of each of its members and through the common enterprise of the ACU.

Key Themes From the Swansea Declaration

One of the key themes to come out of the ACU conference is that all societies, developed and developing, are "mutually vulnerable" and therefore, there is a need for society to work together cooperatively to realize sustainable development. By stressing the equality among countries, this declaration was unique and added a new dimension to the dialogue surrounding higher education and sustainability.

Aftermath of the Swansea Declaration

Since the Swansea Declaration, there have been a number of other important university association declarations; it is important to consider the Swansea Declaration in the context of these in order to understand its importance as setting an early precedent in the higher education community. Not long after the Swansea Declaration, the Kyoto Declaration was signed in November of the same year at the 8th Round Table of the International Association of Universities.

The Swansea Declaration should be considered alongside the 1993 Copernicus University Charter made at the Conference of European Rectors; in the United States, the Campus Blueprint for a Sustainable Future founded at the Yale University Campus Earth Summit; the Thessaloniki Declaration in 1997; at the International Conference on Environment and Society: Education and Public Awareness for Sustainability; and the 2001 Lüneburg Declaration on Higher Education for Sustainable Development.

Conclusion

Declarations and partnerships such as the Swansea Declaration are an important tool to foster and galvanize attention toward promoting sustainable development. They reinforce the notion that universities and higher education institutions have a moral and ethical obligation to foster and nurture the concept of sustainability in society.

However, there are some caveats; while declarations are a statement of intention, they do not provide any guarantees of progress toward their objectives. Although they serve to guide and challenge institutions and provide objectives on which to focus, they do not offer any checks or guarantees.

See Also: Declaration of Thessaloniki; Halifax Declaration; Kyoto Declaration on Sustainable Development; Lüneburg Declaration; Tbilisi Declaration; United Nations Conference on Environment and Development 1992 (Earth Summit).

Further Readings

Calder, Wynn and Richard M. Clugston. "International Efforts to Promote Higher Education for Sustainable Development." *Planning for Higher Education* (2003). http://www.ulsf .org/pdf/International_SCUP_article.pdf (Accessed June 2010).

Corcoran, Peter Blaze and Arjan E. J. Wals. "The Evolution of Sustainability Declarations in Higher Education." In *Higher Education and the Challenge of Sustainability.* New York: Springer, 2004.

Herremans, Irene M. and Robin E. Reid. "Developing Awareness of the Sustainability Concept." *Journal of Environmental Education,* 34/1 (2002).

International Association of Universities. "Sustainable Development Declarations." http:// www.iau-aiu.net/sd/sd_declarations.html (Accessed June 2010).

Wright, Tarah. "Definitions and Frameworks for Environmental Sustainability in Higher Education." *International Journal of Sustainability in Higher Education,* 3/3 (2002).

Gavin Harper
Cardiff University

TBILISI DECLARATION

One of the significant documents in the history and development of environmental education on the international scene, the Tbilisi Declaration and its accompanying recommendations form part of the Final Report of the First Intergovernmental Conference on Environmental Education held in 1977 at Tbilisi, Georgia, in the former Soviet Union. At a time when scientific and public concern about the state of the environment was mounting, the Tbilisi Declaration underscored the crucial role that education has to play in enhancing awareness and understanding of the environmental issues faced by nations, in fostering positive behavioral change toward the environment, and in crafting solutions to environmental problems. The recommendations outlined goals, objectives, and guiding principles of environmental education; offered strategies for the development of environmental education; and stressed the need for regional and international cooperation to further progress in the field. The Tbilisi Declaration provided a foundation that would influence national, regional, and international initiatives in environmental education in the decades to follow.

Background to the Tbilisi Proceedings

Environmental education began on a widespread international scale during the 1960s and 1970s as scientific and public concern for both the environment and human quality of life increased. There was a concomitant emphasis on the value of education in increasing awareness about environmental issues and as a critical aspect of environmental problem solving.

At the United Nations (UN) Conference on the Human Environment held in Stockholm, Sweden, in 1972, environmental education gained currency, entering the official agendas of international organizations. Responding to Recommendation 96 emanating from this conference, the International Environmental Education Programme (IEEP) was established and launched by the United Nations Educational, Scientific and Cultural Organization (UNESCO) and the United Nations Environment Programme (UNEP). The International Workshop on Environmental Education, which subsequently took place in Belgrade in the former Yugoslavia in 1975, saw an overarching goal statement on environmental education, along with guiding principles, objectives, and recommendations, put forward in the Belgrade Charter: A Global Framework for Environmental Education. The preliminary framework outlined at Belgrade was reviewed in a series of follow-up regional meetings and taken forward two years later, in 1977, at the First Intergovernmental Conference on Environmental Education at Tbilisi, Georgia, the culminating event of the first phase of the IEEP.

The Tbilisi Conference was organized by UNESCO in collaboration with UNEP, taking place October 14–26, 1977. Over 300 participants were in attendance, composed of delegates from 66 UNESCO member states and observers from two nonmember states, along with representatives and observers from 20 international nongovernmental organizations, three intergovernmental organizations, and eight organizations and programs of the UN. With environmental education at the core of its agenda, the conference sought to raise awareness of environmental problems and their implications for society; emphasize the role of education in addressing environmental problems; assess national and international efforts and achievements in environmental education; pinpoint strategies for further development in the field; and underscore the global nature of environmental issues and the attendant need for cooperative efforts at regional and international levels to develop and implement environmental education.

The Final Conference Report included the Tbilisi Declaration and 41 recommendations. The declaration, affirmed by the delegates, enunciated the importance and nature of environmental education, stating: "Education utilizing the findings of science and technology should play a leading role in creating an awareness and a better understanding of environmental problems. It must foster positive patterns of conduct towards the environment and the nations' use of their resources." The recommendations endorsed a set of goals, objectives, and guiding principles for environmental education, suggested strategies for the development of environmental education, and highlighted the need for international and regional cooperation.

Goals, Objectives, and Guiding Principles of Environmental Education

The first five recommendations center on the role of environmental education. Three main goals of environmental education are identified: "to foster clear awareness of, and concern about, economic, social, political and ecological interdependence in urban and rural areas; to provide every person with opportunities to acquire the knowledge, values, attitudes, commitment and skills needed to protect and improve the environment; [and] to create new patterns of behavior of individuals, groups and society as a whole towards the environment."

Objectives in five specific areas are also highlighted: awareness, about the environment and its problems; knowledge, including experience in and understanding of the environment and its problems; attitudes, composed of both concern for the environment and the motivation to act on its behalf; skills, enabling groups and individuals to identify and solve environmental problems; and participation, that is, the active involvement by individuals and groups at all levels in resolving environmental problems.

Along with these goals and objectives, a set of principles that should underlie environmental education was outlined. These principles called for environmental education to be a lifelong process, incorporating historical, current, and future situations; accommodating local, national, regional, and international geographical perspectives; and recommending that it be interdisciplinary, holistic, and balanced in nature (taking into account the natural, built, technological, and social aspects of environments). The conference proposed that environmental education emphasize the root causes of environmental problems, their complexity, and the necessity for collaborative efforts to prevent and address environmental problems. Utilization of a range of teaching environments and approaches was also called for, along with educational approaches that allow learners to take an active role in their learning experiences and inculcate decision-making, critical-thinking, and problem-solving skills to address complex environmental problems. Of importance as well, the consideration of environmental issues in development plans was advocated.

Strategies for the Development of Environmental Education

The remaining series of recommendations identify issues relevant to and mechanisms for the advancement of environmental education at the national level. There are suggestions to establish new and/or strengthen existing organizational structures with multistakeholder membership to carry out specific functions, such as the coordination of environmental education initiatives, the assessment of research needs, the drafting of frameworks for the establishment of national environmental education committees, and the promotion of alliances among research, scientific, and education communities and other entities with an interest in environmental education. Other proposals focus on the target populations of environmental education (the general public, professionals and scientists, and specific occupational and social groups); the content and methods of environmental education (including issues of curricula development, the role of teachers, training and education institutions, potential of universities, and interdisciplinary approaches); the training of personnel; teaching and learning materials; dissemination of information; research; and the role of specific groups, such as youth, nongovernmental organizations, voluntary organizations, and the mass media.

Conference delegates also acknowledged the international and global dimensions of many environmental challenges, as well as the interdependence of the current global order. Consequently, the necessity for regional and international collaboration in order to improve the efficacy of environmental education initiatives was stressed, particularly with reference to areas such as curriculum development, resource centers, research and training facilities, and the exchange and sharing of expert and technical knowledge and experience.

Significance of Tbilisi

The declaration and recommendations of the Tbilisi Conference were pivotal outputs in the field of environmental education. Shaped by the philosophy of environmental education put forward at Belgrade, the Tbilisi Declaration built on this foundation, endorsing principles, goals, and objectives of environmental education that should characterize effective environmental education.

The declaration called for the incorporation of environmental education into formal and nonformal education, emphasizing, among other things, that it must be aimed at multiple audiences, be interdisciplinary in nature, and be a lifelong process. By outlining statements on the goals, objectives, and principles of environmental education, the declaration sought to broaden common understanding of the concept. Additionally, the conference called for sustained and enhanced efforts by UNESCO and UNEP to lead activity in the field and to further develop the IEEP.

Since Tbilisi, conferences have been held every 10 years in Moscow, Thessaloniki, and Ahmedabad. These forums, along with other major international meetings on environment and development issues, have built on the Tbilisi proceedings, reviewing, refining, and redefining understanding of environmental education. Notwithstanding the fact that the concept and field of environmental education would continue to evolve in subsequent decades, the proceedings at Tbilisi were of historic significance. The Tbilisi Declaration and its allied recommendations solidified environmental education as a critical item on the world agenda, signaled a commitment by the international community to advancing the cause of environmental education as a fundamental aspect of addressing environmental problems, represented a cooperative effort in the field, enhanced understanding of the role and nature of environmental education, and provided a foundation for ensuing initiatives in the field by offering guidelines for national, regional, and international activity.

See Also: Declaration of Thessaloniki; United Nations Conference on Environment and Development 1992 (Earth Summit); United Nations Conference on the Human Environment, Stockholm 1972.

Further Readings

Fensham, Peter J. "Stockholm to Tbilisi—The Evolution of Environmental Education." *Prospects,* 8/4 (1978).

McKeown, Rosalyn and Charles Hopkins. "EE ≠ ESD: Defusing the Worry." *Environmental Education Research,* 9/1 (2003).

Sterling, Stephen. *Coming of Age: A Short History of Environmental Education (to 1989).* Walsall, UK: National Association for Environmental Education, 1992.

United Nations Educational, Scientific and Cultural Organization (UNESCO). "Final Report of the Intergovernmental Conference on Environmental Education, Tbilisi." Paris: UNESCO, 1978.

Therese Ferguson
Independent Scholar

TUFTS UNIVERSITY

Tufts University is a private institution established in 1852 in Medford and Somerville, Massachusetts (near Boston), by the Universal Church on land donated by Boston businessman Charles Tufts. As of 2010, the university president is Lawrence S. Bacow and the provost and senior vice president is Jamshed Bharucha. Tufts has the following academic divisions: the School of Arts and Sciences, the School of Engineering, the Cummings School of Veterinary Medicine, the School of Dental Medicine, the School of Medicine, the Sackler School of Graduate Biomedical Sciences, the Gerald J. and Dorothy R. Friedman School of Nutrition Science and Policy, the Fletcher School of Law and Diplomacy (the oldest school of international relations in the United States), and the Jonathan M. Tisch College of Citizenship and Public Service. Tufts is a selective university with rankings in the top 40 for its undergraduate school and medical school, according to *US News & World Report* and international rankings in the second 200 in both the Times Higher Education University

The roof of Tufts University's Tisch Library, shown here, is planted with vegetation to improve insulation, reduce the urban heat island effect, and capture storm water before it enters the storm sewer system.

Source: Wikipedia

Rankings and Shanghai Jiao Tong University's Academic Ranking of World Universities. In 2008, Tufts enrolled 9,532 students (more than 5,000 undergraduates and 4,000 graduate and professional students), and employed 1,227 faculty and 2,982 staff members.

Tufts played a pioneering role in encouraging both research into sustainability and the practice of sustainable behavior by universities through the 1990 Talloires Declaration. This declaration was established at a convention of 22 university presidents and chancellors convened in Talloires, France, by Tufts President Jean Mayer and states that "Universities educate most of the people who develop and manage society's institutions. For this reason, universities bear profound responsibilities to increase the awareness, knowledge, technologies, and tools to create an environmentally sustainable future." Conference participants defined the responsibilities of universities in a 10-point program that included expanded research on the interaction of human activities and the environment, adoption of environmentally responsible behaviors on the university campus, and encouraging interdisciplinary focus on critical environmental issues. Originally signed by 31 university leaders from 15 countries, the Talloires Declaration (as of January 2008) was signed by over 360 university leaders in over 40 countries.

The Office of Sustainability

Tufts University first developed an environmental policy in 1990, making it the first university to do so. Today, the Office of Sustainability acts as a resource, catalyst, and advocate for sustainability at Tufts and also enhances the university's reputation as a leader in environmental sustainability. Tufts has many programs that encourage sustainable practices. The Eco-Ambassadors program, launched in fall 2008, trains staff members and provides small grants to encourage staff members to implement sustainable practices in their offices. Programs to conserve energy include an annual competition to reduce energy usage in campus dormitories for a month each year, a voluntary carbon offset purchasing program, a program initiated in 1990 to replace incandescent light bulbs with more efficient fluorescent bulbs (the goal is to have 90 percent of floor space lit by fluorescent bulbs), and free seminars for community members as well as Tufts affiliates on ways to save energy in the home.

Several measures have been taken to make Tufts buildings more energy efficient. Sophia Gordon Hall, completed in 2006 and the newest residence hall at Tufts, was awarded Gold certification by the LEED (Leadership in Energy and Environmental Design) rating system, a voluntary national standard for sustainable buildings. Among the environmentally friendly elements in this building—which was designed to reduce energy and water use by 30 percent and was constructed using 10 percent recycled or renewable materials—are the use of solar thermal and photovoltaic rooftop arrays for electricity generation and water heating, white Energy Star roofing that lowers cooling demand, use of insulated windows and louvered glass walls to reduce demand for heating and cooling, and landscape design that minimizes water use and light spillover. Tisch Library has a green roof (i.e., vegetation is planted on the roof), which improves insulation, reduces the urban heat island effect, and absorbs storm water that can then be released back into the atmosphere rather than being collected in the storm sewer system. Technologies to improve energy efficiency and reduce water use have been implemented in many Tufts buildings: these include installing high-efficiency lighting and motion sensors (a program begun in 2003), installing steam traps throughout the campus to improve the efficiency of the steam system, installing more water-efficient front-loading washing machines, and installing sensors to reduce lighting on vending machines when no one is near them. Tufts also runs a pilot organic turf management program to experiment with reducing the amount of herbicides and pesticides used while maintaining acceptable turf quality and appearance.

The FEAST (Food Education and Action for Sustainability at Tufts) program, a partnership among students, Tufts Dining Services, and several academic units, including the Fletcher School and the Tufts Institute of the Environment, is devoted to educating the Tufts community about environmental issues (including transportation costs) related to food production and consumption as well as promoting Free Trade, local farming, and organic growing methods. Beginning in fall 2010, all Tufts dining rooms instituted trayless dining, which was found in a pilot project to reduce food waste as well as water and energy use.

Tufts instituted a number of programs to reduce paper use and to encourage recycling. For instance, all office copiers have the default setting of two-sided printing, which reduces paper use on campus by an estimated 16 million pages per year. All university magazines are printed on recycled stock, and official Tufts stationery and envelopes are 30 percent postconsumer waste. Tufts developed a logo with the motto "Tufts Prints Green" that may be used on campus publications that meet at least three out of a selection of environmentally friendly practices, including use of vegetable-based ink, use of paper with at least 30 percent postconsumer waste content, use of chlorine-free paper, and limited, focused print runs.

In the area of transportation, Tufts has adopted several electric vehicles and hybrids, including an electronic tractor mower (which also reduces noise pollution) and encourages staff and students to bicycle and to use public transportation (some are eligible for a subsidized pass to use on area public transportation). Tufts students, faculty, and staff are eligible to join Zipcar, a car-sharing program, at a discount rate, and departments can establish a business account that entitles them to reduced rates. Tufts offers several shuttle bus services between campuses and facilitates use of car-sharing and vanpool services.

In 2008, the Sierra Club named Tufts University to its list of 10 "cool schools" notable for environmentally friendly and sustainable practices. Tufts was commended for overall commitment to sustainability and for creating a microfinance fund to invest in developing countries, but was criticized for not encouraging commuter students to use green forms of transportation. In 2010, the Northeast Energy Efficiency Partnerships recognized Tufts as a leader in energy efficiency and sustainable practices.

Academic Environmental Studies

In 2008, the National Wildlife Federation cited Tufts for offering an exemplary interdisciplinary program in environmental and sustainability studies. Tufts offered environmental studies classes beginning in 1962 and in 1984 established an Environmental Studies program that was one of the first multidisciplinary environmental programs in the United States. At Tufts, environmental studies may be taken as a second major along with any departmental major in the School of Arts and Sciences or the School of Engineering. The program, currently directed by Professor George Ellmore, offers three tracks: environmental science, which focuses on basic natural sciences, (e.g., biology, physics, chemistry, and geology) and their application to environmental problems; environment and technology, which emphasizes the applied science and engineering aspects of pollution abatement and prevention, resource management, and environmental impacts; and environment and society, which examines environmental problems with particular regard to their sociopolitical, humanistic, economic, and cultural aspects.

See Also: Association of University Leaders for a Sustainable Future; Social Learning; Sustainability Officers.

Further Readings

Association of University Leaders for a Sustainable Future. "Talloires Declaration." http://www.ulsf.org/programs_talloires.html (Accessed May 2010).
National Wildlife Federation. "Campus Environment 2008: A National Report Card on Sustainability in Higher Education." http://www.nwf.org/Global-Warming/Campus-Solutions/Resources/Reports/~/media/PDFs/Global%20Warming/CampusReportFinal.ashx (Accessed May 2010).
Sierra Club. "Cool Schools: The Third Annual List" (September/October 2008). http://www.sierraclub.org/sierra/200809/coolschools/ten (Accessed May 2010).
"Tufts Recognized for Energy Efficiency: Regional Nonprofit Names the University a Green Leader." *Tufts Journal* (April 7, 2010). http://tuftsjournal.tufts.edu/2010/04_1/briefs/03 (Accessed May 2010).
Tufts University, Institute of the Environment. http://environment.tufts.edu (Accessed May 2010).
Tufts University, Office of Sustainability. http://sustainability.tufts.edu (Accessed May 2010).

Sarah Boslaugh
Washington University in St. Louis

Tulane University

Sustainability is more than an academic concept for Tulane University. While its disaster preparation plans and academic focus had long considered the risks of hurricanes and wetland loss, no institution can fully prepare for the direct hit of a Hurricane Katrina. In late August 2005, Tulane's focus shifted from the start-of-fall-classes commotion to the protection of the lives of staff, students, and faculty, and preserving campus facilities. The uptown campus was evacuated while Tulane Medical Center downtown remained staffed under emergency conditions until an evacuation could be completed. The next phases—of immediate survival, a canceled semester, a dislocated student body and staff, and a long-term renewal plan—led

This Federal Emergency Management Agency photograph shows Tulane University and the I-10 highway in New Orleans on August 28, 2008, almost three years after Hurricane Katrina made landfall.

Source: FEMA/Jacinta Quesada

Tulane to reinvent itself as a university committed to service learning, urban sustainability and healthcare, and environmental research. As recovery progressed, Tulane went from being the largest private employer to "the" largest employer in New Orleans, Louisiana, making Tulane's recovery plans an integral aspect of the city's recovery. Today, Tulane's staff, faculty, students, and healthcare professionals support sustainability in an urban landscape that faces challenges from incomplete economic and infrastructure recovery, wetlands erosion, and the predicted impacts of climate change.

Since its 1834 founding as a medical college, Tulane has closed only twice—for four years during the U.S. Civil War, and for four months in 2005 after Hurricane Katrina. In March 2010, the university projected its first post–Katrina budget balance for the upcoming 2012 budget year. This fiscal recovery paralleled an academic revisioning, including active recruitment of students who wished to study while participating in the city's recovery. Before Katrina, service was a strong extra- and co-curricular component of student life; after Katrina, the undergraduate curriculum was redesigned to include a public service graduation requirement. Existing institutes and interdisciplinary study centers expanded to address the concepts of "Resilience, Recovery, and Sustainability," the catchphrase of the UrbanEco Initiative of the university's Katrina Environmental Research and Restoration Network, yet resiliency proves a faint concept when homes are uninhabitable and infrastructure in disrepair. In its 2008 application for Sustainability Leadership award from the Association for the Advancement of Sustainability in Higher Education (AASHE), Tulane officials noted that institutional recovery also supported civic rebuilding. Tulane crafted education initiatives in New Orleans and launched a charter school. Faculty, staff, and students from architecture, public health, medicine, the center for bioenvironmental research, and the environmental law clinic offered planning, leadership, and labor to help their city recover. Tulane was one of four AASHE award winners that year in the category of four-year and graduate institutions with 7,500+ full-time students, remarkable if only because three years earlier, this was an institution with no students on campus (although student progress continued as colleges across the country hosted Tulane students for the semester while allowing the school to retain the tuition fees and thus remain solvent).

History of Social and Environmental Advocacy

With its roots as a medical school, Tulane—along with Newcomb College, the women's college that is now a seamless element of Tulane—has become a leading educational institution in the South, in part due to its commitment to service and social advocacy. In 1914,

Tulane launched the first school in the Deep South to train social workers. Traditions of service evolved in the 1960s and 1970s as both the city and the south faced issues of racial, social, and environmental justice. With students drawn largely from outside the city known both for tolerance and social inequities, one response was the formation of the Community Action Council of Tulane University Students (CACTUS), based on "the belief that a college education should extend beyond the classroom" and should support the community. Today, CACTUS is considered the country's oldest and largest student-led community service organization; it provided both institutional tradition and a model for post-Katrina service initiatives, including the Center for Public Service.

In another reflection of the role of service, in 1987, the law school became the first in the country to require pro bono service as a graduation requirement. Like other programs at the college, the law school draws on its location to enrich the study and practice of legal principles concerning shipping and international trade, fishing and petrochemical industries, and a legal tradition in both the French and English languages. The law school's Environmental Clinic works on cases involving wetlands, the Clean Air and Clean Water Acts, and environmental justice concerns. The university also hosts a top-ranked Latin American Studies program, which drafted an environmental literacy curriculum in Costa Rica.

Pre-Katrina, Tulane was also known as a school of celebration, with classes being canceled for Mardi Gras, and frequent concerts on campus by the late Professor Longhair. After Katrina, the school's first celebration was bittersweet, with students welcomed back by the music of Wynton Marsalis, and the inauguration of a Renewal Plan that would support recovery of the school and its city. One element of this plan proved controversial: most of the Engineering School, along with numerous graduate programs and faculty, were cut due to financial exigency, which resulted in censure by the American Association of University Professionals. Other Gulf Coast colleges faced similar concerns after Katrina and other hurricanes, raising questions of how institutions prepare for and respond to natural disasters. With an increasing potential for future disruptions, Tulane and other affected colleges offer lessons learned regarding the role of academic governance during institutional emergencies. Likewise, Katrina highlighted the key role of information services in higher education, including backup of mission-critical data and the importance of immediate off-site replication of these services during a disaster.

Tulane Today: A Model of Engagement for the Post-Katrina World

Tulane's recovery is reflected in part by the response of students: in 2009, nearly 40,000 applicants competed for 1,500 seats in the freshman class. It is reflected in initiatives across the university to integrate academic studies with student service and civic recovery and by the leadership of Tulane's president, Scott Cowen. It is reflected by alumni, whose efforts include the pilot who flew Air Force One as President George W. Bush toured Katrina, and by the native New Orleanian Lisa Jackson, who was named the head of the U.S. Environmental Protection Agency by President Barack Obama. It is reflected in research funding that peaked at $157 million in 2007, a reflection of post-Katrina research and recovery efforts. It is reflected in unique interdisciplinary institutions like A Studio in the Woods, an artists' retreat managed by Tulane that also preserves a slice of river bottom hardwoods.

A range of research centers integrate the role of environmental studies and research, including the Katrina Environmental Research and Restoration Network and centers focused on applied environmental public health, bioenvironmental research, energy, architecture, and urban planning. Tulane hosts the Coastal Center for the National Institute for Climatic Change Research, with a charge that is integrally connected to the region: "to reduce scientific uncertainty about potential effects of climatic change on coastal ecosystems in the United States," with specific research on "sea-level rise, coastal subsidence, and the possibility of increased intensity and/or frequency of storms, including hurricanes." Additionally, Tulane has applied the very specific lesson learned in innovative programs, such as the Master's in Public Health with a concentration in Complex Emergency and Disaster Management.

As an institution, Tulane is not solely defined by its role as hurricane survivor, but each year post-Katrina sees an increasing synthesis of environmental and educational mission, one that is paralleled in colleges across the country. However, few colleges faced the impacts of a combined disaster of engineering and nature (and the implied role of climate change) so directly. In some cases, such as Tulane's recent improvement from C+ to B on the College Sustainability Report Card, the progress may seem slow compared to Tulane's past and goals, but rapid compared to comparable southern colleges and when considered in light of the fiscal and social hit of Katrina. Yet with energy awards for Tulane's dorms, greener dining practices, two LEED-certified buildings, and a car-sharing WeCar program and support for mass transit, it is apparent that sustainability is a motivating factor in transforming the institution.

Tulane's role as a center for student engagement with social and environmental issues may be best demonstrated by its selection to host the first Clinton Global Initiative University meeting held at Tulane in 2008, in which 700 national and international college students and 30 college presidents were convened on the Tulane campus (and off-campus at a service day in the Lower Ninth Ward) to brainstorm and practice solutions for energy conservation, climate change, and global health and human rights issues.

See Also: Association for the Advancement of Sustainability in Higher Education; Association of University Leaders for a Sustainable Future; Social Learning.

Further Readings

Taylor, Ian L. "Hurricane Katrina's Impact on Tulane's Teaching Hospitals." *Transactions of the American Clinical and Climatological Association*, 118 (2007).

The Tulane Hullabaloo. http://thehullabaloo.com/?s=sustainability (Accessed June 2010).

Tulane University. "Welcome to Green Tulane." http://green.tulane.edu (Accessed June 2010).

"Tulane University 2008 Campus Sustainability Leadership Award Application." Campus Sustainability Profiles, Association for the Advancement of Sustainability in Higher Education. http://www.aashe.org/resources/profiles/cat4_132.php (Accessed June 2010).

Ron Steffens
Green Mountain College

U

UNITED NATIONS CONFERENCE ON ENVIRONMENT AND DEVELOPMENT 1992 (EARTH SUMMIT)

The Forest Principles agreement made at the Earth Summit reflected an increased awareness of threats to the world's tropical forests, such as this partially cleared one in Indonesia, which had been found to be disappearing at a rate of about 11 million hectares per year.

Source: World Bank

The United Nations Conference on Environment and Development (UNCED), also known as the Earth Summit, was held in Rio de Janeiro on June 3–14, 1992. It is the largest international conference on the environment to be held in history, with the participation of 172 governments and 1,500 officially accredited representatives. Thirty thousand people descended on Rio for the summit in June 1992, including representation from 2,400 nongovernmental organizations (NGOs) and 7,000 journalists. An NGO forum, attended by 17,000 people, ran in parallel with the official summit. While on many fronts the Earth Summit was a great success—with the signing of five Earth Summit agreements—in other areas it was highly criticized, having been dubbed the "Summit of Hypocrisy" in the seemingly chaotic months of preparation just prior to its commencement.

Maurice Strong, the conference secretary general, writing after the summit, acknowledged that not all that had been hoped for was achieved, that some of the goals had been weakened, that there needed to be much stronger commitments on finance, and that targets and time-tables had to be set for the conventions to be effective. Nevertheless, he did suggest that one of the outcomes, Agenda 21, constituted the most comprehensive and far-reaching program of action ever approved by the world community. Alicia Bárcena, while recognizing its limitations, also points out that its achievements were significant in establishing new opportunities for partnership; a global, interdisciplinary framework; the integration of environment and development; the mobilization of many governmental and nongovernmental institutions; and a new way of working on sustainable development. UNCED was also, according to Mary Haney, a watershed in the recognition and involvement of women in international issues and provided a model for the involvement of women in every major United Nations (UN) conference since.

Twenty years prior to the Earth Summit, the United Nations Conference on Human Environment was held in Stockholm in June 1972. The setting up of this conference was a response to the growing awareness of the impact of human development on the planet, what Richard Stanbrook described as the "effluence of affluence." However, the Stockholm conference was beset with difficulties, what Stanbrook described as "a dialogue of the deaf between the rich and the poor." While the north wanted everyone to act together to clean up pollution, the south felt that the most pressing problem was poverty and, in order to tackle poverty, the south was prepared to adopt Western ways and to accept environmental problems as part of the package. Nevertheless, Stockholm did achieve some success by making its mark, and the Western world started to recognize that it could not continue in its current ways and began to introduce rules and regulations to govern its environmental problems, helped along by pressure from environmental groups.

Following the Stockholm Conference, the United Nations Environment Programme (UNEP) was established, which continues today to act as a forum and a catalyst for action to protect the environment. However, little was done following Stockholm to integrate environmental concerns into national economic planning and decision making, and the rate and scale of things that impacted the environment had increased dramatically in the intervening years, and many problems were exacerbated. In 1983, the UN set up the World Commission on Environment and Development, led by Gro Harlem Brundtland of Norway. In 1987, this commission published its report, known as the Brundtland Report, and put forward the concept of sustainable development as an alternative approach to one simply based on economic growth. It defined this as development "which meets the needs of the present without compromising the ability of future generations to meet their own needs."

Following the publication of the Brundtland Report, the UN General Assembly called for the UN Conference on Environment and Development. The goals of the conference were to arrive at an understanding of development that both supported socioeconomic development and prevented the continued deterioration of the environment. It was also intended to lay a foundation for a global partnership between the developing and more industrialized countries, recognizing mutual needs and common interests that would ensure a healthy future for the planet.

The Outcomes

Although, as reported, the conference had considerable difficulties in reconciling different perspectives between rich and poor nations, it did, nevertheless, result in five far-reaching agreements. Two of these agreements were legally binding conventions: the Framework Convention on Climate Change and the Convention on Biological Diversity. These two conventions had been negotiated independently of the main UNCED process in the two years prior to the summit, but on schedules devised to enable them to be signed at Rio.

The Convention on Climate Change was an attempt to address the threat posed by global warming. While it recognized the existence of the problem and the need to take a precautionary approach to the issue, it did not set any mandatory limits on greenhouse gas emissions for individual countries and contained no enforcement mechanisms. It was thus considered to be legally nonbinding. However, the framework did allow for further updates (called "protocols") that would set mandatory emission limits. The most notable of these was the Kyoto Protocol, which the United States signed but, under President George W. Bush, refused to ratify and still had not done so in 2010.

Biodiversity is the diversity of life on Earth, including its genetic diversity, and is often considered to be a measure of the health of an ecosystem. The Convention on Biological Diversity was the result of a growing awareness of the rapid acceleration of the loss of species and the increase in environmental degradation. The convention was negotiated under the auspices of UNEP and sought to conserve the biological diversity of the planet through protection of species and ecosystems and to establish terms for the associated uses of biological resources and technology. However, as mentioned by David Reid, key tensions in the convention came from the primary interest of the north, with its biotechnology industries, ensuring access to the diversity of species in the south. The framework asserted that individual states had sovereign rights over their own biological resources.

The Rio Declaration consists of 27 principles that reaffirm and build on the UNCHE Declaration adopted at Stockholm on June 16, 1972. As Michael Grubb pointed out, many of the principles address development concerns with recognition of the "right to development" and highlight the special needs and circumstances of developing countries. They also stress the need for development and poverty alleviation along with access to financial and technological resources.

The Forest Principles agreement could be said to be the result of a failed attempt at coming to an agreed convention on forestry, an attempt that had been initiated in the early 1990s as the culmination of growing awareness of the state of the world's forests. This had started in the 1980s when the UN Food and Agricultural Organization estimated that the world's tropical forests were disappearing at a rate of about 11 million hectares per year. The difficulties of arriving at an agreed convention prior to Rio reflected the tensions that existed between the north and the south. The southern countries, which contained the tropical forests, felt that the negotiations were an attempt by the northern countries to dictate the way in which they could use their forests, something that they resented given that the northern countries had been responsible for the destruction of much of their own native forests. The Forest Principles produced at Rio, as Grubb pointed out, "despite the lengthy build-up and preparatory process," were more the beginning of a process than a definitive agreement.

Agenda 21

Agenda 21 is a large document consisting of 40 chapters, which constitute what can be considered the action plan for sustainable development. The 40 chapters are clustered into four sections. Section I, chapters 1–8, deals with the social and economic dimensions of sustainable development. Section II, chapters 9–22, addresses the conservation and management of resources for development. Section III, chapters 23–32, deals with strengthening the role of major groups involved in achieving sustainable development; and section IV, chapters 33–40, addresses the means of implementation, including finances, technology, law, education, cooperative mechanisms, and institutional arrangements. While Agenda 21 is a large and comprehensive document, it is not a legal agreement and, as Grubb stated, it "is perhaps best seen as a collection of agreed negotiated wisdom as to the nature of the problems, relevant principles, and a sketch of the desirable and feasible paths towards solutions, taking into account national and other interests."

Agenda 21 is probably the outcome of the summit that has the greatest implication for education, containing as it does two chapters specifically dealing with children and youth (Chapter 26) and education (Chapter 36).

Despite the problems and lack of progress in a number of areas, the Earth Summit is considered to be a pivotal moment in international environmental politics. The UN General Assembly declared its broad support for UNCED's work and urged its members to implement the agreements.

See Also: Agenda 21: "Promoting Education, Public Awareness, and Training" (Chapter 36); Brundtland Report; Kyoto Declaration on Sustainable Development; United Nations Conference on the Human Environment, Stockholm 1972.

Further Readings

Bárcena, Alicia. "Marine Agenda of UNCED Role of the Earth Council." *Marine Policy,* 18/2 (1994).

Grubb, M., M. Koch, A. Munson, F. Sullivan, and K. Thomson. *The Earth Summit Agreements. A Guide and Assessment.* London: Earthscan, 1993.

Haney, Mary P. "Women's NGOs at UN Conferences: The 1992 Rio Conference on the Environment as a Watershed Event." *Journal of Women, Politics & Policy,* 27/1 (2005).

Lafferty, William and Katarina Eckerberg. *From the Earth Summit to Local Agenda 21.* London: Earthscan, 1998.

Quarrie, Joyce. *Earth Summit 1992.* London: Regency Press, 1992.

Reid, David. *Sustainable Development: An Introductory Guide.* London: Earthscan, 1995.

Stanbrook, Richard. "From Stockholm to Rio." In *Earth Summit 1992,* Joyce Quarrie, ed. London: Regency Press, 1992.

United Nations. "UN Conference on Environment and Development (1992)." In "UN Briefing Papers/The World Conferences: Developing Priorities for the 21st Century." http://www.un.org/geninfo/bp/enviro.html (Accessed January 2010).

Donald Gray
University of Aberdeen

United Nations Conference on the Human Environment, Stockholm 1972

The 1972 United Nations Conference on the Human Environment held in Stockholm, Sweden, (hereafter "Stockholm") represented the first major international environmental conference among officials representing a significant number of the world's nations.

Held over 12 days in June, the unofficial motto of Stockholm was "Only One Earth" from the book commissioned for the occasion by the United Nations Secretary-General Maurice Strong and written by Barbara Ward and René Dubos, who helped promote what became known as sustainable development. Stockholm signified the need for coordinated international action to deal with environmental issues that threatened each country and also the entire globe.

The negotiations were sometimes contentious, with concern voiced by various countries over topics such as the United States' participation in the Vietnam War, Palestine, colonialism, apartheid and racial oppression, commercial whaling, and the testing of nuclear weapons. There were also significant differences between the interests of developing and developed countries on issues such as the costs of environmental regulation and the responsibility for extant problems. The meetings moved forward, however, under what become known as the "Stockholm spirit of compromise."

Out of Stockholm came a document titled "Declaration of the United Nations Conference on the Human Environment," which included a proclamation and list of 26 principles and an action plan. A Governing Council of 54 member states and an environment secretariat were established, as was an Environment Fund seeded with voluntary contributions by countries to finance the activities supporting the agreements made. Other resolutions adopted included annually observing June 5 as World Environment Day and condemning nuclear weapons tests.

Stockholm reflected burgeoning environmental concern that began in the 1960s, highlighted by Rachel Carson's *Silent Spring,* published in 1962, which brought awareness of the dangers of the pesticide DDT and catalyzed the modern environmental movement. In 1968, the United Nations Economic and Social Council (ECOSOC) proposed holding a world conference on the environment, recognizing that environmental degradation posed threats to both economic and social development. The UN General Assembly endorsed the recommendations of ECOSOC that same year, emphasizing the importance of recognizing the special needs of developing nations in protecting their further development from any adverse effects of environmental regulation. A committee of delegates from 27 nations then met four times during 1970–1972 in preparation for Stockholm.

Delegates from 113 nations attended Stockholm. The preparatory and final meetings were boycotted by the Communist bloc countries—with the exceptions of Romania and Yugoslavia—because East Germany was not given voting rights. In addition, a separate Environment Forum was organized by the Swedish government for other interested parties, particularly nongovernmental organizations (NGOs); accredited NGOs were permitted to lobby delegates and to present statements to the official UN conference. An informal People's Forum and other meetings and demonstrations also occurred simultaneously.

Notably, the first component of the declaration and the first principle established during the meetings at Stockholm recognized the importance of environmental quality to human development and as a human right. Many of the other principles concerned the importance of development to the participants, including the presumed link between poverty and environmental degradation, but also the desire to prevent environmental protection from impeding development. Moreover, the need for development assistance for developing states was asserted, as was the claim that compensation was due to states that suffer from the resource use of other states. State sovereignty was also emphasized in statements that each nation sets its own environmental standards and population policies and also develops its own plans for resource use.

The action plan contained 109 recommendations, with six major themes: (1) Planning and Management of Human Settlements for Environmental Quality, (2) Environmental Aspects of Natural Resources Management and Conservation, (3) Identification and Control of Pollutants, (4) Educational, Informational, Social and Cultural Aspects, (5) Environment and Development, and (6) International Organizational Implications of Action Proposals. Few of the recommendations had specific targets for achievement, and some of those that did, such as a 10-year moratorium on commercial whaling, were not met. Most of the recommendations were fairly general suggestions; nonetheless, the Stockholm Conference must be judged a success if for no other reason than for focusing the world's attention on the need for environmental protection. It went beyond this, however, creating formal organizations and tools toward that end. Two enduring outcomes have been the creation of the World Heritage Trust to protect wilderness areas and sites of cultural and natural importance and the establishment of the United Nations Environmental Programme (UNEP), created by the UN General Assembly as the formal arm of the UN responsible for focusing international attention on the environment and safeguarding it from harm. Within UNEP, the Earthwatch program was established and continues to generate research on the environment and to monitor and disseminate information about it to states and the public. And UNEP has continued to organize international conferences on the environment, including the United Nations Conference on Environment and Development in Rio de Janeiro, Brazil, in 1992; the World Summit on Sustainable Development in Johannesburg, South Africa, in 2002; and the United Nations Conference on Climate Change held in Copenhagen, Denmark, in 2009.

See Also: United Nations Conference on Environment and Development 1992 (Earth Summit); United Nations Decade of Education for Sustainable Development 2005–2014; United States Partnership for Education for Sustainable Development.

Further Readings

Carson, Rachel. *Silent Spring.* Boston: Houghton Mifflin, 1962.

Clarke, Robin and Lloyd Timberlake. *Stockholm Plus Ten: Promises, Promises? The Decade Since the 1972 UN Environment Conference.* London: International Institute for Environment and Development, 1982.

Elliott, Lorraine. *The Global Politics of the Environment,* 2nd ed. New York: New York University Press, 2004.

Stone, Peter Bennett. *Did We Save the Earth at Stockholm?* London: Earth Island, 1973.

United Nations Environment Programme. "Stockholm 1972." http://www.unep.org/ Documents.Multilingual/Default.asp?documentID=97 (Accessed January 2010).

United States Government, Committee on Public Works. "Report on the United Nations Conference on the Human Environment." Washington, DC: U.S. Government Printing Office, 1972.

Ward, Barbara and René Dubos. *Only One Earth: The Care and Maintenance of a Small Planet.* New York: W. W. Norton, 1972.

Kirk S. Lawrence
University of California, Riverside

United Nations Decade of Education for Sustainable Development 2005–2014

In December 2002, the United Nations (UN) General Assembly adopted Resolution 57/254 that designated the period 2005 to 2014 as a UN Decade of Education for Sustainable Development (UNDESD). The Paris-based UN Educational, Scientific and Cultural Organization (UNESCO) was chosen to be the lead agency charged with promoting the decade and was requested to prepare a draft international implementation scheme (IIS).

The origins of the UNDESD can be traced back to Agenda 21, which had been approved by governments at the 1992 Rio Earth Summit (UNCED, UN Conference of Environment and Development). Chapter 36 of Agenda 21 stressed the need to reorient education to address sustainable development issues and to develop the capacity to deal with perceived social, environmental, and economic challenges. It also emphasized the need to raise public awareness and to provide training as a means to secure a more sustainable future. Agenda 21 was reaffirmed 10 years later at the World Summit on Sustainable Development (WSSD) that was held in South Africa. The Johannesburg Plan of Implementation reiterated the importance of education as a "key agent of change" and the need to integrate sustainable development issues into all levels and spheres of education. It explicitly supported the Millennium Development Goals and the education objectives set out in the Dakar Framework for Action (Education for All).

Following extensive consultation with other UN agencies, governmental and nongovernmental organizations, and others, UNESCO presented the draft ISS for the UNDESD to the UN General Assembly in October 2004. The draft ISS reflected ongoing debates surrounding the evolving concept of Education for Sustainable Development (ESD). While acknowledging the plurality of values across communities and different local and national priorities, the draft ISS judiciously sought to imbue ESD with a core set of values related to human dignity and rights, equity, and care for the environment. Education was regarded as vital to changing mindsets and behavior that was seen as essential for securing sustainable development. ESD, it was argued, had to address not only environmental issues such as biodiversity, water, or waste but also needed to deal with social and economic issues such as employment, human rights and security, gender equity, poverty reduction, and corporate responsibility and accountability. The ISS reaffirmed the need to improve access to quality basic education, reorient existing education programs, develop public understanding, and provide training. The draft was eventually published in 2006 as "Framework for the UNDESD International Implementation Scheme," which also outlined the numerous activities and events of the decade.

Assessing Effectiveness

Assessing the effectiveness of the UNDESD has been a core concern. In 2007, UNESCO established a Monitoring and Evaluation Expert Group (MEEG) to advise on progress in meeting objectives and to examine UNESCO's own contribution to the UNDESD. On the advice of MEEG, a Global Monitoring and Evaluation Framework (GMEF) was developed to determine what impact, if any, the UNDESD was having. Monitoring and evaluation work was divided into three phases. The first phase (2007–2009) focused on contexts and structures of work on ESD. The focus of the current second phase (2009–2011) is on processes and learning related to ESD. The final phase (2011–2014) will focus on impacts and outcomes of the DESD. As a consequence of this activity, a monitoring and evaluation report series was established titled "Learning for a Sustainable World." The first report was published in 2009, and two further reports related to the second and third phases are scheduled for 2011 and 2015.

The first report, titled "Review of Contexts and Structures for Education for Sustainable Development," highlighted certain key findings. To date, there remains no consensus surrounding the meaning of ESD. For example, the report noted that some countries focus on pressing issues such as combating poverty, dealing with human immunodeficiency virus and acquired immune deficiency syndrome (HIV/AIDS), or preventing loss of biodiversity, whereas others focus more on capacity building and how to change behavior in light of oil dependency and excess consumerism. Although UNESCO took the lead in coordinating ESD activity, the report found that it was too early to speak of a unified UN response. The rapid creation of national ESD coordinating bodies in the majority of countries that responded to a 2008 survey was notable, but it remained unclear what role each actually plays. Furthermore, many countries noted a significant presence of ESD in national policy documents, although most continue not to have a specific national ESD strategy.

Whereas the UNDESD aimed at reorienting the curricula, it was found that a more common response was to make minor adjustments by giving greater space in the curriculum to sustainability issues. The report noted that there had been further integration of sustainable development issues in training courses, teacher education, and higher education, but there remained negligible support of ESD in early childhood education. While ESD in nonformal education and learning spheres was increasingly on the agenda, there was little systematic evidence as to how this was being funded. In relation to global ESD research, it remains relatively undeveloped and, given the rise of ESD practice, the report called for improved quality assessment. Nevertheless, considerable networking activity was reported, with many examples of relevant multistakeholder networking taking place at local, national, and regional levels. Perhaps the most serious obstacle identified was the lack of funds or economic incentives available for ESD.

Areas for Action

In light of these findings, the report specified 10 areas for action for the remainder of the decade. For example, it was felt that there need to be more innovative approaches to curriculum development, further ESD-related professional development for educators and managers, and greater theoretical and critical thinking and evidence-based ESD research. The report called for the continued development of ESD indicators for purposes of monitoring and evaluation. It concluded that the success of ESD activities and projects relied on raising money. However, the specifics of funding remained vague in the action plan. The report merely suggested that multilateral and bilateral donors, as well as the private sector, could complement government budgets as potential sources for additional funds.

In March 2010, the "UNESCO Strategy for the Second Half of the United Nations Decade on Education for Sustainable Development" was published. It is noteworthy that the strategy acknowledged the prevailing global financial and economic crisis and, with the careful wording of a UN document, called into question "prevailing economic paradigms." A favorable light was cast on alternative ways out of the crisis, and reference was made to discussions within the UN system that led to the publication in June 2009 of the Interagency

Statement titled "Green Economy: A Transformation to Address Multiple Crises." The crisis itself raised concerns regarding continued commitment and funding of education and training programs. Consequently, it was felt necessary to stress the relevance of ESD by emphasizing education as an "investment in the future" as had been underlined in the April 2009 Bonn Declaration on Education for Sustainable Development.

Nevertheless, despite the changed economic circumstances, the strategic aims of UNESCO's strategy for the remainder of the UNDESD did not significantly differ from earlier broad visions. UNESCO outlined that it would coordinate a wide-ranging consultative process with multiple stakeholders with a view to securing action in four strategic areas. First, a key aim for the rest of the decade was to continue to build education partnerships and alliances at local, national, and regional levels. The second area for action was to further develop and strengthen capacities for ESD, including, for example, the integration of sustainability into teacher training programs. Sharing ESD knowledge and practice was a third area for strategic action. Finally, it was felt that the second half of the decade should support ESD advocacy and further raise public awareness.

What Remains

For the remainder of the decade, one can expect continued debates and development of the concept of ESD, such that it is not conflated with environmental education (EE). This will require more explicitness and evidence of the causal linkages between the social, economic, and environmental dimensions. This is not only relevant to raising public awareness and developing public policy, but to how "sustainability" is being integrated into the education curricula. ESD must continually stress more holistic inter- and multidisciplinary approaches and critical thinking. If ideas of "transformative education" are to be meaningful, it needs to be spelled out further, as it is central to developing capacities to live more sustainable livelihoods. More specifically, ESD in early childhood education needs greater support. Perhaps it is too ambitious to expect a unified UN approach to ESD, but certainly greater synergy is possible between UNESCO and the valuable work carried out in other UN agencies such as the United Nations Development Programme (UNDP), UNICEF, United Nations Environment Programme (UNEP), and the United Nations Framework Convention on Climate Change (UNFCCC).

In addition to constructive multistakeholder networks that continue to be developed, one area that is likely to be a key outcome of the UNDESD is the considerable work being done in relation to ESD indicators relevant for purposes of monitoring and evaluation. However, if the UNDESD is to be more than just a benchmarking exercise, it must develop critical thinking in relation to power in the world economy, to corporate accountability and responsibility, and to financial globalization. It would certainly make innovative ideas regarding "alternative economic paradigms," "transitions to a green economy," and more "sustainable societies" more credible. This should not detract from significant positive work that is being done under UNESCO's leadership and the UNDESD umbrella. Nevertheless, one fundamental obstacle remains—how to secure additional funding for ESD programs, particularly at a time of austerity in public budgets. The case for mainstreaming ESD into education and aid budgets remains as strong as ever.

See Also: Agenda 21: "Promoting Education, Public Awareness, and Training" (Chapter 36); Millennium Development Goals; United Nations Conference on Environment and Development 1992 (Earth Summit).

Further Readings

United Nations. "Plan of Implementation of the World Summit on Sustainable Development." Johannesburg: WSSD, 2002. http://www.un.org/esa/sustdev/documents/WSSD_POI_PD/English/WSSD_PlanImpl.pdf (Accessed June 2010).

United Nations Department for Economic and Social Affairs, Division of Sustainable Development. "Agenda 21." http://www.un.org/esa/dsd/agenda21/res_agenda21_00.shtml (Accessed June 2010).

United Nations Educational, Scientific and Cultural Organization (UNESCO). "Bonn Declaration on Education for Sustainable Development." Bonn: UNESCO/German Federal Ministry of Education, 2009. http://www.esd-world-conference-2009.org (Accessed June 2010).

United Nations Educational, Scientific and Cultural Organization (UNESCO). "Framework for the UNDESD International Implementation Scheme." ED/DESD/2006/PI/1. Paris: UNESCO, 2006.

United Nations Educational, Scientific and Cultural Organization (UNESCO). "Learning for a Sustainable World: Review of Contexts and Structures." Paris: UNESCO, 2009. http://www.unesco.org/education/justpublished_desd2009.pdf (Accessed June 2010).

United Nations Educational, Scientific and Cultural Organization (UNESCO). "Report by the Director-General on the United Nations Decade of Education for Sustainable Development: International Implementation Scheme and UNESCO's Contribution to the Implementation of the Decade." Paris: UNESCO, 2005.

United Nations Educational, Scientific and Cultural Organization (UNESCO). "UNESCO Strategy for the Second Half of the United Nations Decade of Education for Sustainable Development" 2010/ED/UNP/DESD/PI,1. Paris: UNESCO, 2010. http://unesdoc.unesco.org/images/0018/001873/187305e.pdf (Accessed June 2010).

United Nations General Assembly. "Resolution 57/254 United Nations Decade of Education for Sustainable Development" (2002). http://www.un-documents.net/a57r254.htm (Accessed June 2010).

United Nations System. "Green Economy: A Transformation to Address Multiple Crises. An Interagency Statement of the United Nations System." New York: UN System Chief Executive Board, 2009. http://www.unep.ch/etb/pdf/2009%20statement%20deliver%20as%20one/Interagency%20Joint%20Statement.%20E%20rev1.pdf (Accessed June 2010).

World Education Forum. "Dakar Framework for Action: Education for All: Meeting Our Collective Agreements." Paris: UNESCO, 2000. http://unesdoc.unesco.org/images/0012/001211/121147e.pdf (Accessed June 2010).

Frank Brouwer
University of Sussex
London RCE on ESD

UNITED STATES PARTNERSHIP FOR EDUCATION FOR SUSTAINABLE DEVELOPMENT

Education for sustainable development is a relatively new concept in K–12 and higher education systems, though environmental management and conservation have been rooted in practice

Shelburne Farms on Lake Champlain in Vermont is a partner in the United States Partnership for Education for Sustainable Development and serves as an innovative school for place-based sustainable education.

Source: Wikipedia

and teaching in the United States since the late 19th century, when conservation of national parks, forests, and natural resources was implemented. This movement is anchored in the work of Henry David Thoreau and the landmark text *Man and Nature* (1864) by George Perkins Marsh, who framed the necessity for sustainable human activities after observing long-term human impacts of deforestation in his native Vermont and in the Mediterranean. These concerns were updated in the land ethic of Aldo Leopold's *A Sand County Almana*c (1949). However, it has been within the past 50 years that environmental leaders and the general populace have expressed concern over the extravagant use of natural resources and the impact of such unsustainable economic development on the sustainable livelihoods of future generations. In response to public concerns and scientific research, the United Nations convened two Earth Summits, held in Rio de Janeiro in 1992 and in Johannesburg in 2002, which sought solutions to issues threatening sustainability. Outcomes of the summits resulted in the Johannesburg Declaration on Sustainable Development promoting the concept of sustainable development, including a call to integrate sustainable education in all levels of education and learning systems, since education is the key agent for change when promoting sustainable development. The forerunner of these was the First Intergovernmental Conference on Environmental Education, held in 1977, with the Tbilisi Declaration, which provided a foundation that would influence national, regional, and international initiatives in environmental education.

Emergence of the United States Partnership for Education for Sustainable Development

The United States Partnership for Education for Sustainable Development (USPESD) is one of many initiatives to implement the guidelines from the resolutions adopted by the Earth Summits. USPESD envisioned its role as a bridge to collaborate between environmental policy makers and schools and institutions. The organization was initiated by participants from more than 100 organizations gathering in Washington, D.C., in 2003 to discuss the newly adopted resolution by the United Nations (UN) General Assembly, which had declared the 10-year period beginning in 2005 as the Decade of Education for Sustainable Development. This decade also sought to implement sustainability education while supporting the Dakar Framework for Action on Education for All, which had been adopted during a previous UN conference.

Operational plans to establish the United States Partnership for Education for Sustainable Development (USPESD) were developed during the second gathering of the group at Gettysburg College in Pennsylvania. The participants agreed on a mission for the organization—"Sustainable development fully integrated into education and learning in the United States"—based on the UN resolution with a goal to develop sustainable development education programs in the United States by leveraging the initiatives of the UN Decade.

USPESD was launched as a nonprofit organization in 2005. After electing its first board of directors in 2006, the organization decided on its first and most crucial step. Instead of developing its own programs to introduce sustainable education in the nation's education system, the founders of the organization decided to forego authoring its own programs. It would support existing actions and emerging activities based on the framework of the UN Decade of Education for Sustainable Development by serving as a clearinghouse and catalyst to connect partners, foster collaboration, and build community. Two teams—Action and Sector—were formed to implement the organization's mission. The Action Team is responsible for supporting the operation of the organization, while the Sector Team is carrying out the collaboration work in order to meet the needs of different sectors.

In subsequent years, USPESD has worked with partner organizations to integrate education for sustainable development into the K–12 education and higher education systems. The USPESD embraced "the national trends to integrate sustainability in higher education" at all levels, including "mission and planning, curricula, research, student life, facilities, purchasing and operations."

USPESD partnerships are not limited to educational institutions, with collaboration occurring in a range of diverse sectors, including assistance for business owners to develop sustainable business plans and models, promotion of sustainable communities, and support for policy makers to encourage people to pursue sustainable lifestyle changes. Currently, the USPESD has partnerships with more than 300 organizations across the United States to collaborate for change by pursuing the partnership's mission to unite "individuals, organizations and institutions in the United States dedicated to education for sustainable development (ESD). It acts as a convener, catalyst, and communicator working across all sectors of American society."

Partners promoting sustainable education and lifestyles include the Association for American Colleges and Universities, Harvard Green Campus Initiative, Duke University, California State University, Georgetown University, George Washington University, Center for International Environment & Resource Policy, Coalition for Global Warming Solutions, Global Environmental Alliances, International Partner for Sustainable Agriculture, and the International Society for Ecological Economics.

One key result of USPESD's leadership is the ongoing development of Student Learning Standards for Education for Sustainability (ESF). The three core EFS standards suggest that K–12 students should (1) master the basic concepts and principles of sustainability, (2) gain the knowledge that sustainability is a dynamic condition of interconnected human and natural systems, and (3) apply multidisciplinary skills to envision and enact a sustainable world.

Establishing Sustainable Institutions

Through alliance with other institutions, USPESD supported the design and launch of several associations and initiatives that provide the foundation for sustainable education in the higher education system. USPESD listed five sustainable institutions in higher education whose emergence represents major efforts in the higher-learning system. One of these is the Association for the Advancement of Sustainability in Higher Education (AASHE), whose members initiate and assess projects for a sustainable future. Under AASHE's umbrella, branch associations such as Disciplinary Associations Network for Sustainability

(DANS) work to embed the value of sustainability in curricula, research, and professional development, while the Higher Education Association's Sustainability Consortium (HEASC), an association of college and university presidents, supports sustainable development plans on campuses.

Sustainability Academy at Lawrence Barnes

An example of USPESD's impact can be seen in the Sustainability Academy at Lawrence Barnes, opened in September 2009 with the partnership of Shelburne Farms, a partner farm of the USPESD. The K–5 school is one of two sustainable-themed elementary magnet schools in Burlington, Vermont, and is an international model using the tools of sustainability in its curriculum "for place-based education and service learning." The school built a partnership with a range of educators, professions, organizations, families, and the community to integrate sustainable concepts in the school's curriculum and its campus practices.

Shelburne Farms was selected as one partner so that its sustainability mission might guide students to become responsive citizens for their community and beyond. The farm's input helps students develop their commitment in shared responsibility to improve our interconnected lives, now and into the future. Shelburne Farms also serves as a classroom where students can conduct community studies outside their classrooms. The school noted that these real-life situations help students and teachers understand how skills in math, science, literacy, and social studies can be applied to problem solving.

The school's sustainable-themed curriculum encourages students to collaborate across grade levels and participate in learning programs outside the classrooms, with projects ranging from development of schoolyard habitats to interviews of neighbors to learn about immigration, and studies of local farming and food systems.

Consortium for Education in Renewable Energy Technology (CERET)

The USPESD and its partners made every effort to integrate sustainable themes in all education and learning platforms. An example is a training and research center based at the University of Wisconsin, the Consortium for Education in Renewable Energy Technology, which not only integrates sustainable education in its curriculum but also offers technical training and education about renewable energy technologies. The United States' energy consumption makes us dependent on other countries for fuel and energy supplies, which disturbs global environmental sustainability and undercuts long-term stability of source countries. Through the technical training programs on renewable energies, energy management, sustainability, and the technical skills to produce renewable energy, CERET and its partners offer sustainable solutions to energy dependency.

See Also: Center for Environmental Education; Cloud Institute for Sustainability Education; Integrating Sustainability Education Concepts Into K–12 Curriculum; Tbilisi Declaration; United Nations Decade of Education for Sustainable Development 2005–2014.

Further Readings

Cirillo, Jennifer. "Vt. Boasts First Sustainability-Themed Elementary Magnet School." *Rutland Herald* (January 3, 2010).

Cuesta, Camacho David E. *Environmental Injustices, Political Struggles: Race, Class, and the Environment.* Durham, NC: Duke University Press, 1998.

McMillan, Victoria and Amy Lyons Higgs. *Implementing Sustainability Education: Lessons From Four Innovative Schools.* Ann Arbor: University of Michigan Press, 2002.

Reid, Anna and Peter Petocz. "University Lecturers' Understanding of Sustainability." *Higher Education,* 51/1 (2006).

United Nations Conference on Environment and Development (UNCED). *Agenda 21, The United Nations Program of Action From Rio.* New York: United Nations, 1992.

United Nations Educational, Scientific and Cultural Organization (UNESCO). "Bonn Declaration." Bonn: United Nations World Conference on Education for Sustainable Development, 2009.

U.S. Partnership for Education for Sustainable Development. http://www.uspartnership.org/main/view_archive/1 (Accessed March 2010).

Wixom, Robert L. *Environmental Challenges for Higher Education: Integrating Sustainability Into Academic Programs.* Burlington, VT: Friends Committee on Unity with Nature, 1996.

Ron Steffens
Green Mountain College

UNITY COLLEGE

Unity College is a private liberal arts college in Unity, Maine, about 35 miles outside the coastal city of Bangor. Founded in 1965 by the local business community to boost Unity's economy, it has a curriculum that focuses strongly on the environment and sustainability issues. The small academic community of about 500 students and 35 faculty members leads to a close-knit community with strong ties to the town; the college's Dorothy W. Quimby Library serves as the public library for Unity and other nearby towns. The campus has been substantially rebuilt with energy efficiency and low emissions in mind, and all new buildings will be low or zero carbon; total carbon neutrality campus-wide is an actively pursued goal, and both solar and wind power are used on campus. Unity's sustainability coordinator is a member of the Master Planning Committee and works regularly with the college president. Furthermore, each campus department is required to have a sustainability plan. On the roof of the cafeteria building are 16 solar panels that provide hot water; the panels are the very same that were originally installed on the roof of the White House by President Jimmy Carter and that were removed by his successor.

All of Unity's degree programs incorporate a focus on environmental issues. There are 24 bachelor of science majors available: Adventure Education Leadership; Adventure Therapy; Agriculture, Food, and Sustainability; Aquaculture and Fisheries; Captive Wildlife Care and Education; Conservation Law Enforcement; Ecology; Environmental Analysis; Environmental Biology; Environmental Education; Environmental Policy; Environmental Humanities; Environmental Science; Environmental Writing; Forestry; General Studies; Landscape Horticulture; Liberal Studies; Marine Biology; Parks, Recreation, and Ecotourism; Sustainability, Design, and Technology; Wildlife; Wildlife Biology; and Wildlife Conservation. The Agriculture, Food, and Sustainability major, one of the newest, is conducted in association with the Maine Organic Farmers and Gardeners

Association (which also composts food waste from the campus to fertilize organic vegetables for low-income families), part of Unity's ongoing commitment to engaging with the off-campus world.

The school is actively involved in sustainability research and building, including community wind projects throughout central Maine. Other student activities include the Grease Car Collaborative, which retrofits engines to run on used vegetable oil from the cafeteria. Every fall, Unity students help to winterize homes for local low-income families and will begin conducting free energy audits for local homes to offer energy-efficiency suggestions. Unity College also purchases 250 carbon offsets a year to help the Maine State Housing Authority improve the energy efficiency of low-income family homes throughout the state.

In 2009, President Mitchell Thomashow laid out nine elements of a sustainable culture in a blog post, which has been used as a guiding document to address many of the campus's sustainability projects: energy, food, materials, governance, investment, wellness, curriculum, aesthetics, and interpretation. The Unity College Leadership Council—made up of representatives from the administration, faculty, staff, and student body—regularly assesses the college's sustainability in each of these areas. Thomashow's wife, Cindy Thomashow, serves as the executive director of Unity's Center for Environmental Education, and since 2008, the couple has lived in the Unity House, a campus residence designed from the ground up for sustainability, with solar panels, a concrete pad that retains heat in the winter and keeps the house cool in the summer, low-flow water fixtures and a one-shower-a-day limit, compact fluorescent lights, and high-efficiency appliances, insulation, and mechanical systems sufficient for Platinum-level LEED certification. In order to demonstrate the cost efficiency of sustainability, the house was built on a $200-per-square-foot budget, which forced the architect to cancel plans for composting toilets because of the cost of excavating the rock ledge on the property.

Students are heavily involved in the college's commitment to sustainability, and there are seven different student groups focused on sustainability and environmental issues, including the Primitive Skills and Naturalist Living Club, the Constructive Activists, and the Eco-Cottage. Junior-year students take a required course on sustainability,

In the summer, Unity offers summer programs to attendees of various ages, some with course credit and others without. All summer programs deal with sustainability, including the Education in a Changing Climate workshop for teachers, the Maine Arts Camp for school-age children, and the Sustainable Practices summer program for high school and college students, covering a variety of topics from beekeeping and greenhouse gardening to composting and photovoltaic systems.

See Also: Green Community-Based Learning; Green Mountain College; Leadership in Sustainability Education; Sustainability Officers; Whole-School Approaches to Sustainability.

Further Readings

Gough, Stephen and William Scott. *Higher Education and Sustainable Development*. London: Routledge, 2009.

Hargreaves, Andy and Dean Fink. *Sustainable Leadership*. Hoboken, NJ: Jossey-Bass, 2005.

Sandell, Klas, Johan Ohman and Leif Ostman. *Education for Sustainable Development*. New York: Studentlitteratur AB, 2005.

Silka, Linda and Robert Forrant, eds. *Inside and Out: Universities and Education for Sustainable Development*. Amityville, NY: Baywood Publishing, 2006.

Wiland, Harry and Dale Bell. *Going to Green: A Standards-Based Environmental Education Curriculum for Schools, Colleges, and Communities*. White River Junction, VT: Chelsea Green Publishing, 2009.

Bill Kte'pi
Independent Scholar

University Employee Training: Private Partnerships

By the early 21st century, sustainability efforts became a focus for many colleges and universities worldwide, influenced by the United Nations declaring 2005–2014 the Decade of Education for Sustainable Development. With this new emphasis, it became more important than ever to adequately train faculty and staff and provide them with the tools necessary for teaching sustainability and for operating campuses according to the dictates of the sustainability agenda. The need for training has often led colleges and universities to form partnerships with businesses and the local community.

Business and Foundation Partnerships

In the Puget Sound area, Sellen Construction has gained considerable experience in sustainable building projects, employing the use of environmentally friendly features such as operable windows and composting toilets. The roofs of Sellen structures are frequently filled with greenery to reduce energy consumption and costs while filtering rainwater. In 2009 alone, the company generated $500 million from projects earning the U.S. Green Building Council's Leadership in Energy and Environmental Design (LEED) certification. LEED projects comprised more than 90 percent of Sellen's total revenue. Using federal grant money, in 2010, the company created Sellen Sustainability and began partnering with the University of Washington and the Seattle Vocational Institute to create a training program for a "green collar workforce." The program is proving to be a valuable tool for employee training also. Sellen has also partnered with E2 Environmental Energy and Solutions of Pittsburgh (which also has offices in St. Louis, Missouri, Las Vegas, Nevada, and Orange County, California) to provide sustainability training. Sellen is in the process of extending sustainability training to Oregon, Alaska, and British Columbia in Canada.

In an effort to assist teachers across the United States in dealing with the issue of sustainability, the Ford Foundation's Difficult Dialogues initiative published *Start Talking* to be used as a tool for both learning and teaching. Using advice and experience gained from a partnership between the University of Alaska Anchorage, a large public university, and Alaska Pacific, a small private university, the book offers practical advice and assistance on both faculty development and classroom instruction. Only a limited number of hard copies have been published, but the Ford Foundation offers a downloadable version free of charge.

Just as educators look to private partnerships to incorporate sustainability into education, private companies depend on educators to provide essential instruction concerning sustainability. For instance, the National Center for Construction Education and Research of Gainesville, Florida, partners with the Myers-Lawson School of Construction of Virginia Tech University, located in Blacksburg, Virginia, to create "Your Role in the Green Environment" as a means of providing instruction on green building practices and dispensing knowledge needed to qualify for LEED certification in construction projects.

Beyond Grey Pinstripes

Throughout the world, colleges and universities have been charged with teaching students how to become more conscious of what sustainability means in their lives. As these students enter the workforce, they remain aware of the consequences of environmental irresponsibility. It is therefore in the interests of both educational establishments and businesses to share their knowledge through both employee training and the production of graduates who recognize the validity of sustainability in the workplace. As a result, large numbers of colleges and universities are incorporating sustainability into master of business programs, and trained faculty are needed to teach in these programs. In 2001, the think tank World Resources Institute and the Aspen Initiative for Social Innovation through Business (now known as the Aspen Institute Business and Society Program) first published *Beyond Grey Pinstripes,* a survey of MBA programs in colleges and universities around the world. While recognizing the efforts of schools such as the Asian Institute of Management in the Philippines, Harvard Business School in the United States, and York University's Schulich School of Business in Canada to address social, environmental, and sustainability topics, the survey concluded that in order to meet the needs of the global marketplace, there was a great need for more faculty trained in the field of sustainability and in incorporating sustainability across the curricula.

In 2010, 149 schools participated in the biennial *Beyond the Grey Pinstripe.* In that survey, 63 percent of the programs were located in the United States, and the rest were spread out over 24 countries. Using the criteria of how well institutions integrated social and environment issues into the curricula of MBA programs and how well trained and active faculty were in social, environmental, and sustainability issues, the top 10 schools in 2010 were York University's Schulich School of Business (Canada); the University of Michigan's York School of Business; Yale's School of Management (New Haven, Connecticut); Stanford Graduate School of Business (Palo Alto, California); Notre Dame University's Mendoza College of Business; University of California, Berkeley's Haas School of Business; Erasmus University's RSM Erasmus Business School (Rotterdam, the Netherlands); New York University's Stern Business School; IE Business School (Madrid, Spain); and Columbia Business School (New York). One of the overall conclusions of the survey was that while MBA students were being given more opportunities to take elective courses with social and environmental content, the rate of increase in requiring such courses had begun to decline, rising only 6 percent between 2007 and 2009 compared to 11 percent between 2001 and 2003. This decline was partly the result of a large number of schools already requiring courses in sustainability.

Globalization and Sustainability Training

A number of studies and reports have indicated that business education is essential to meeting the demands of globalization. At the insistence of employers as well as the alumni, faculty, and students of most business schools, there is mounting pressure to add depth to existing courses on sustainability and to step up the pace of integrating sustainability into the entire curriculum. Many students argue that the lack of such training is a contributing factor in many of the corporate scandals that fill the news around the world. In order to do this, faculty in all disciplines need to be adequately trained to teach their students about global issues that include sustainability and ethics. One way this is being accomplished is through institutions working and learning together to engage in sustainability education. When Presidio World College, which is affiliated with Alliant International University, Bainbridge Island Graduate Institute, located in Washington State, and the New College of California, based in Santa Rosa, enhanced faculty training by sharing expertise and engaging in faculty exchanges, the faculty and students of both schools benefitted greatly.

University staffs are constantly being asked to respond to calls for new forms of expertise concerning the teaching of sustainability to students of the 21st century. Innovative courses and new fields are being initiated in response to the increasing demands of globalization. Private foundations have sometimes joined in partnership with institutions of higher learning to expand their abilities to prepare business students to meet the needs of the rapidly changing business world. Largely in response to a recommendation by the International Commission on Education for Sustainable Development Practice, in 2009, the John D. and Catherine T. MacArthur Foundation donated an initial $7.6 million to nine universities in seven different countries for the purpose of founding Master's in Development Practice (MDP) programs. Once the projects were established, other universities were brought on board. The Earth Institute of New York's Columbia University was the first school to launch its MDP program.

There was a great deal of diversity in the programs created, and each imposed its own demands on the faculty involved in creating the MDP programs. Emory University in Atlanta, Georgia, was able to use its connections to the U.S. Centers for Disease Control and Prevention, CARE, and the Carter Center to create a program emphasizing the health and governmental aspects of sustainable development. New Delhi's Energy Resources Institute University, on the other hand, placed the emphasis on energy and climate science and green building. The faculties of Australia's James Cook University and Ireland's Trinity College Dublin and University College Dublin created programs that provided their students with field work in the Philippines and Indonesia and in Rwanda, respectively. China's Tsinghua University faculty determined to focus on generating developmental models for China. The faculty of some universities receiving MacArthur grants, including those at the University of Florida and the University of Botswana, created partnerships that enhanced their particular interests at the same time they advanced their knowledge of sustainable development. Nigeria's University of Ibadan applied for the grant with the ultimate intention of creating a Centre for Development Studies to serve as a program hub for English-speaking west African nations interested in developing their own MDP programs.

Innovation and Sustainability

Some schools have begun using all the tools at their disposal to teach faculty, staff, students, and other target groups about the advisability of sustainable behavior. In the summer of 2010, Arizona State University (ASU), which in 2007 became the first university in the United States to create a school of sustainability, partnered with *Grist* magazine to promote sustainability training and deliver both local and national sustainability news. ASU President Michael Crow declared that the partnership was a result of trying to expand the boundaries of the learning process and to encourage individuals to question the status quo.

In 1996, the United Nations Educational, Scientific and Cultural Organization (UNESCO) and EOLSS Publishers Ltd. formed a partnership to create *The Encyclopedia of Life Support Systems,* which they advertised as the world's largest Web-based online archive on subjects dealing with sustainable development. With advice from sustainability experts from Asia, Africa, Latin America, Russia, China, Japan, France, and the United States, the project was designed as a teaching tool as well as a learning resource. Initial offerings were around 25 million words, but the project grew to include approximately 120 million words at its completion, covering topics in natural science, biology, medicine, sociology, economics, the humanities, engineering, and technology.

Another online resource for faculty development is "Sorted: The Sustainability Online Resource and Toolkit for Education." It was commissioned in Great Britain by the Learning Skills Council and created through a partnership between Ecotec, the Environmental Association for Universities and Colleges, and Stephen and Maureen Martin. UNESCO also provides a toolkit for educators, "Training and Learning for a Sustainable Future," at www.esdtoolkit.org.

Other innovations may bypass technology, but they do offer in-depth faculty development by following the advice of the American Council on Education, which encourages administrators to promote travel abroad as an integral element in faculty sustainability training. Located in Winter Park, Florida, Rollins College is a liberal arts institution that has taken that message to heart. Rollins has a self-proclaimed mission of teaching students to develop global perspectives. In 2006, through funds donated by the parent of a former student, Rollins President Lewis M. Duncan established a program that awards $3,000 grants to faculty and staff members to allow them to travel abroad in faculty-led groups once every three years to learn about other cultures or to engage in relevant research. Destinations have included China, Ecuador, Tanzania, Europe, and India, and projects have ranged from trauma resolution in Turkey and Cyprus to a two-week program in Morocco led by an Arab-speaking anthropology professor. All participants in the program agree that it is always a transformative experience.

Other schools that operate sustainability faculty development programs that focus on traveling abroad include Rhodes College in Tennessee, Grinnell College in Iowa, Troy University in Alabama, and the University of Richmond in Virginia. Traveling abroad for faculty development is not always without risks, and one trip to Ghana in the summer of 2006 was cut short when half of the travel team was sent home for health reasons.

In West Virginia, where institutions of higher learning are generally too small to launch extensive efforts toward faculty development in sustainability, 20 institutions have formed the Consortium for Faculty and Course Development, which is housed in the Political Science Department of West Virginia University in Morgantown. Participation in FACDIS (Faculty and Course Development in International Studies) programs, which include annual workshops, the John A. Maxwell Scholar-Diplomat Program, and the International Studies Summer Institute for Teachers, is based on faculty involvement in either international studies or foreign languages. Nearly 400 teachers from more than 15 disciplines have met the consortium's criteria. The budget is underwritten by federal grant money, funds from state governing boards, institutional budgetary allotments, and private grants.

Conclusion

The trend toward sustainability at colleges and universities around the world has demonstrated that there is a distinct need for faculty and staff training, and private partnerships have been essential to the success of such efforts. The agenda established for the United

Nations Decade of Education for Sustainable Development is being realized on campuses at a much faster pace than in broader communities. Most experts agree that the greatest successes have been in the areas of curriculum revision and campus greening. They suggest that a greater emphasis on training faculty to develop international perspectives is needed to make them more amenable to promoting sustainable practices that protect the environment and make the world a safer place for both present and future generations.

See Also: Arizona State University; Association of University Leaders for a Sustainable Future; Cloud Institute for Sustainability Education; United Nations Decade of Education for Sustainable Development 2005–2014.

Further Readings

Aspen Institute Center for Business Education. *Beyond Grey Pinstripes 2010.* http://www .beyondgreypinstripes.org/pdf/2009–2010BGP_Brochure.pdf (Accessed July 2010).

Cloud Institute for Sustainability Education. http://www.sustainabilityed.org/ (Accessed July 2010).

Dernbach, John C. *Agenda for a Sustainable America.* Washington, DC: ELI Press, Environmental Law Institute, 2009.

FACDIS: The West Virginia Consortium for Faculty and Course Development in International Studies. http://www.wvu.edu/~facdis/index.htm (Accessed July 2010).

Fischer, Karin. "Professors Get Their Own Study-Abroad Program." *The Chronicle of Higher Education, 55/6* (2008).

Franson, Melissa. *The Impact of Classroom Exposure to Sustainability, Course Content, and Ecological Footprint Analysis of Student Attitudes and Projected Behaviors.* Unpublished Thesis. Auburn, AL: Auburn University, 2008.

Hignite, Karla, et al. *The Educational Facilities Professional's Practical Guide to Reducing the Campus Carbon Footprint.* Alexandria, VA: APPA, 2009.

Jones, Jeanne Lang. "Sellen, University of Washington, Vocation School Team Up on Training." http://www.bizjournals.com/washington/othercities/seattle/stories/2010/05/24/ story5.html?b=1274673600%5E3392711&s=industry&i=green (Accessed July 2010).

Landis, Kay, ed. *Start Talking: A Handbook for Engaging Difficult Dialogues in Higher Education* (2008). http://www.uaa.alaska.edu/cafe/difficultdialogues/upload/Start-Talking -Handbookcomplete-version.pdf (Accessed July 2010).

Lockerbie, Andrea. "Middlesex University to Offer New Industry Training Course" (2006). http://www.mrw.co.uk/home/middlesex-university-to-offer-new-industry-training -course/3002083.article (Accessed July 2010).

"MacArthur Funds New Master's Programs in Sustainable Development." http://www .sustainablebusiness.com/index.cfm/go/news.display/id/18483 (Accessed July 2010).

Rockwood, Larry, et al., eds. *Foundations of Environmental Sustainability: The Coevolution of Science and Policy.* New York: Oxford University Press, 2008.

Scott, William and Stephen Gough. "Sustainable Development Within the United Kingdom." *Journal of Geography in Higher Education, 30/2* (2006).

Weeks, Alison. "Business Education for Sustainability: Training a New Generation of Business Leaders." *GreenMoney Journal* (Summer 2010). http://www.greenmoneyjournal .com/article.mpl?newsletterid=29&articleid=309 (Accessed July 2010).

Weeks, Alison. "Grist for the Green Mill: An Econews Site Teams Up With a University."
    GreenMoney Journal (Summer 2010). http://www.greenmoneyjournal.com/article.mpl
    ?newsletterid=29&articleid=309 (Accessed July 2010).
World Resources Institute. "Beyond Grey Pinstripes 2001" (2002). http://www.greenbiz.com/
    business/research/report/2002/06/17/beyond-grey-pinstripes-2001 (Accessed July 2010).

Elizabeth Rholetter Purdy
Independent Scholar

UNIVERSITY OF BRITISH COLUMBIA

The University of British Columbia (UBC), a public university established in Canada in 1908, is the oldest university in British Columbia and enrolls over 50,000 students, including over 8,000 graduate students, on two campuses: the larger campus is in greater Vancouver, British Columbia, and a second campus was opened in 2005 in Kelowna, a city in the Okanagan Valley of British Columbia. UBC is well known as an educational and research institution and is regularly listed among the top 50 universities worldwide in both the Shanghai Jiao Tong University and the Times Higher Education rankings. Academic programs at UBC are organized into 24 faculties, schools, and colleges, including the Faculties of Arts and Sciences, Applied Science (Engineering), Business, Law, Medicine, and Dentistry. UBC is consistently ranked in the top three Canadian universities in terms of research funding, and in 2009 received over $475 million in research funds from government, industry, private, and other sources. The University-Industry Liaison Office at UBC facilitates technology transfer, an area where UBC ranks near the top among North American universities, and over 130 spin-off companies are located in British Columbia alone.

Sustainability Research

The University of British Columbia has many different research centers and institutes dedicated to sustainability research; information about these and other university sustainability resources can be found at www.sustain.ubc.ca. The University of British Columbia is a signatory to the Talloires Declaration of the Association of University Leaders for a Sustainable Future, which signifies support for making sustainability a critical focus of university teaching, research, operations, and outreach.

The Centre for Environmental Research in Minerals, Metals, and Materials (CERM3) is located within the Department of Mining Engineering and is focused on reducing the environmental harm caused by mining. CERM3 was founded in 2002 and consists of six research facilities: the Environmental Quality Laboratory, which provides analytical support; the Bioremediation and Reclamation Laboratory, which aims to address environmental pollution through research into biological systems; the Environmental Technology Laboratory, which aims to develop improved methods to deal with environmental problems; the Mine Health and Safety Laboratory, which studies issues of occupational health and safety; the Mine Automation and Environmental Simulation Laboratory, which studies remote mining and aims to improve processes to reduce processing cycle times; and the Energy and Mining Laboratory, which studies the feasibility of using alternative energy sources and reducing energy use in mining.

The Clean Energy Research Centre (CERC) officially opened its own physical plant in 2006, after being part of the Faculty of Applied Science since 2000. It includes over 40 faculty members and 100 graduate students involved in research into clean energy technologies and offers, in conjunction with the Faculty of Applied Science, the first Canadian master's degree program in Clean Energy Engineering. Research within CERC includes the use of biomass as an energy source, development of clean-burning engines, use of biocatalytics to create environmentally friendly fuels, and development of technology to separate hydrogen for use in fuel cells.

The Design Center for Sustainability (DCS) is located within the School of Architecture and Landscape Architecture and studies ways to apply sustainability concepts to development. DCS projects have won numerous awards for green design and sustainability, including an Honor Award for Research in 2010 from the American Society of Landscape Architects (for the report "Getting to Minus 80: Defining the Contribution of Urban Form to Achieving Greenhouse Gas Emission Targets"), a National Merit Award from the Canadian Society of Landscape Architects in 2009 (to professor Cynthia Girling for the report "Green Infrastructure in Calgary's Mobility Corridors"), and a Community Excellence Award for the report "100-Year Sustainability Vision: City of North Vancouver." The Greenskins Lab, located within DCS, monitors the performance of green roofs, green facades, green streets, and urban agriculture. The lab offers consulting and conducts expertise studies as well as conducting research into topics such as green-roof agriculture, rainwater harvesting, and storm water management.

The Centre for Interactive Research on Sustainability (CIRS) is devoted to accelerating sustainability in the urban environment, a process motivated by predictions that the urban population will double between 2000 and 2050. CIRS defines sustainability as requiring reconciliation of the ecological imperative to recognize the limited carrying capacity of Earth, the economic imperative to provide all people with a reasonable standard of living, and the social imperative to develop governance systems and methods to promote the values of communities and individuals. CIRS is dedicated to producing research on urban sustainability that is interdisciplinary, relevant to policy and decision makers, and involves nonacademic partners in the research process. The ultimate goal is to reduce the "performance gap" between innovations that are technologically and behaviorally feasible and those that are actually in use, with particular focus on the difference between predicted and actual performance on the built infrastructure, between stated environmental concern and actual behavior, and between the goals of environmental policy and actual outcomes.

The CIRS building, scheduled to open in 2011 in Vancouver, is intended to be a living laboratory to test and showcase methods of sustainable construction and building use. It embodies the principles of building construction that is smart (because it uses monitoring and feedback systems to get the greatest use from energy and material flows), green (because it is designed to have a positive effect on the environment), and humane (because it provides a healthy environment for those who work within the building). It is designed for a 100-year life cycle, including anticipated climate change, uses materials that are not toxic and that are able to be reused or recycled as much as possible, minimizes energy and water use, and is intended as a model that is economically feasible and replicable in other cities. Specific innovations in the CIRS building include a rainwater collection system to supply potable water, treatment and reuse of all wastewater on site, daylighting (designing a building so the building can be illuminated during the day by natural light), minimal use of mechanical cooling systems, and local procurement of products and services.

The Institute for Resources, Environment and Sustainability is an interdisciplinary research institute that focuses on a wide range of environmental and sustainability issues and also houses the interdisciplinary Resources Management and Environmental Studies (RMES) program that awards degrees at the master's and doctoral levels (with about 80 doctoral and 40 master's students in 2010) as well as a certificate in watershed management. The program integrates a political and biophysical approach to resource and environmental issues: specific topics included in RMES studies include land management, environmental assessment, policy analysis, coastal zone management, fisheries management, agroforestry, hydrology, risk perception and assessment, community development, and negotiation issues.

The Sustainable Forest Management Laboratory, established in 2000, conducts interdisciplinary research in sustainable forest management. Professor John Innes, chair of Forest Management, leads the program, and graduate students come from a variety of academic backgrounds and countries (including India, France, Switzerland, Romania, China, the United States, and Canada); areas of research include conservation, forest health, sustainable forest management, protection of forests, climate change, certification, First Nations' forestry, and international forest policy.

ISIS, founded in April 2009, is a research center within the Sauder School of Business at UBC that aims to encourage social innovation and support new ideas and ventures that can help solve environmental, social, cultural, economic, and political challenges. Social innovation is defined as including concepts such as social finance, social enterprise, and strategic social responsibility. ISIS has nine project areas: carbon management, sustainable transportation, clean technology, clean energy, social economy, food systems, digital media, development, and First Nations development. Currently, ISIS includes 2 postdoctoral fellows and 11 graduate fellows and has partnerships with several outside organizations, including Vancouver Economic Development, Mindset Social Innovation Foundation (focused on finding new solutions to drug development, access, and distribution), and Offsetters (focused on providing ways for individuals and organizations to reduce their climate impact.

See Also: Leadership in Sustainability Education; Sustainability Officers; Whole-School Approaches to Sustainability.

Further Readings

Pellow, James P. and Brij Anand. "The Greening of a University." *Change*, 41/5 (2009).

University of British Columbia, Centre for Interactive Research on Sustainability. http://www.cirs.ubc.ca (Accessed May 2010).

University of British Columbia, Design Centre for Sustainability. http://www.dcs.sala.ubc.ca (Accessed May 2010).

University of British Columbia, Institute of Resources, Environment and Sustainability. http://www.ires.ubc.ca (Accessed May 2010).

University of British Columbia. "Sustainability." http://www.sustain.ub.ca/research/centres-institutes-groups (Accessed May 2010).

Wixom, Robert L. *Environmental Challenges for Higher Education: Integrating Sustainability Into Academic Programs*. Burlington, VT: Friends Committee on Unity with Nature, 1996.

Sarah Boslaugh
Washington University in St. Louis

UNIVERSITY OF CHICAGO

The University of Chicago is a private institution with a main campus located in the South Side community of Hyde Park. The university created a Sustainability Council, Office of Sustainability, and campus director of sustainability to oversee the implementation of its sustainability initiatives. Campus sustainability actions incorporate a wide range of programs designed to reduce waste, to facilitate smart transportation and the use of environmentally friendly products, and to minimize the university community's collective impact on the environment. Students are involved in a number of sustainability-based campus groups and volunteer opportunities in the local community, facilitated by the University of Chicago Environmental Center (UCEC). The university offers an interdisciplinary Environmental Studies program located within the Program on the Global Environment and an Environmental Science program. The university also houses sustainability-related interdisciplinary research programs and centers.

Campus and Community Sustainability Initiatives

The University of Chicago places an emphasis on collaborative efforts across the university community to share the knowledge and skills necessary to implement sustainability throughout its on-campus and off-campus educational, research, operational, and outreach activities. Individual students, faculty, and staff members are also encouraged to practice sustainability in their everyday lives and receive guidelines and suggestions for specific steps they can incorporate both on and off campus. The university maintains a sustainability Website to facilitate the dissemination of sustainability information, policies, classes, research projects, and programs.

The Sustainability Council created a list of sustainability principles for university-wide adoption, including the mitigation of the school's carbon footprint, environmental awareness goals, and student, faculty, and staff involvement in the development of sustainable policies and practices that are in keeping with the university's present and future needs. The university also seeks to develop a sustainability initiative that reflects its role as a major research institution through the inclusion of intellectual debate and analysis. The University of Chicago's Office of Sustainability was founded in 2008 and is housed within the Facilities Services Department. Ilsa Flanagan was named the first campus director of sustainability.

The university's Facilities Services group has taken a leadership role in the implementation of many campus-based sustainability initiatives designed to reduce waste, facilitate sustainable transportation methods, introduce environmentally friendly products, and minimize the campus's environmental impact. Purchasing emphasizes green products and cleaning supplies in its acquisitions. Campus groundskeeping operations have limited the use of leaf blowers, pesticides, and synthetic fertilizers on campus and have switched to the use of integrated pest management and environmentally friendly Eco Salt made from sugar beets to remove surface ice during the winter months. They also purchase mulch made from recycled trees from local businesses. Dining Services purchases 20 percent of the food served on campus from local suppliers and contracts with a local dairy. They have introduced reusable dishware in campus dining halls and offer reusable cups, mugs, and shopping bags to students, faculty, and staff.

The University of Chicago seeks U.S. Green Building Council LEED certification. The campus features one LEED-registered project, with several other construction and renovation projects either in planning or currently under way that will seek LEED certification. The university not only seeks to integrate sustainable features into its building design, but to also ensure that all features meet the needs of the building's future occupants. Facilities Services works with outside consultants in the planning and design phases to meet these goals. Sustainability considerations include energy- and water-use reduction and waste minimization. Examples of projects implementing this process include the William Eckhardt Research Center, the Chicago Theological Seminary, the Reva and David Logan Center for the Creative and Performing Arts, and the Lab School renovation. The campus also features a new residence hall built with green and reflective roofs.

Students and faculty assisted the university in completing an inventory of campus greenhouse gas emissions using a computer program called Clean Air–Cool Planet to calculate carbon emissions. The Sustainability Office is developing a mitigation/reduction strategy and planning for its implementation. The university has partnered with the local business Blackstone Bicycle Works and the nonprofit community organization Experimental Station to offer a free bike-share program on campus. It has also entered into an agreement with the Chicago Transportation Authority to subsidize public bus routes. The university encourages biking, walking, and carpooling through a comprehensive university shuttle system and ride-sharing, carpooling, guaranteed ride home, and occasional parking programs. The university purchases green energy sources, such as the wind energy used to power some of the campus residence halls.

The university has implemented several recycling programs, including campus-wide one-stream recycling, with the exception of the student residence halls. One-stream recycling allows participants to place any recyclable items in any recycling container, which has been shown to increase participation rates. Recyclables are later sorted at off-campus recycling facilities. Real Estate Operations also includes one-stream recycling at all of its university rental residential apartments. Other recycling initiatives include the recycling of green landscaping waste, the composting of food waste from campus dining facilities, and the recycling of automotive repair fluids, car batteries, and tires.

Academic, Research, and Outreach Programs

The University of Chicago is known for its pioneering interdisciplinary education and research programs that provide courses, conferences, and programs to students, faculty, outside researchers, and the general public. The university publishes a guide with a listing and overview of the more than 100 sustainability-related courses that are offered to facilitate interest and enrollment. The university's academic, research, and outreach programs address critical global environmental issues, provide opportunities for new interdisciplinary collaborations, and help ensure a sustainable future for the campus, community, and world.

The University of Chicago's Social Sciences Collegiate Division offers a flexible interdisciplinary Environmental Studies program located within the Program on the Global Environment, designed to integrate undergraduate- and graduate-level research and teaching as well as public outreach efforts. The Program on the Global Environment includes the Environmental Studies bachelor of arts (B.A.) major and minor, and the graduate Workshop on the Global Environment as well as conferences, lecture series, and other events. The program hosts visiting researchers and postdoctoral fellows. The program serves as a liaison with student environmental groups and the campus Sustainability Council. The Program on the Global Environment is part of the Center for International Studies. The university also offers complementary bachelor of arts (B.A.) and bachelor of science (B.S.) programs in Environmental Science.

Undergraduate students majoring in Environmental Studies receive a basic foundation of knowledge in a variety of sustainability-related disciplines that provides them with the skills necessary to tackle current and future environmental challenges. The program offers students the choice of thematic tracks such as environmental economics and policy or socio-natural systems and frameworks. The program brings together faculty and courses from anthropology, biological sciences, chemistry, economics, English, geography, geophysical sciences, physics, political science, sociology, law, medicine, linguistics, and public policy studies. Basic coursework features a common introductory sequence of general education courses required of all majors as well as sustainability-related courses that cover the natural sciences and quantitative analyses, incorporating models and methods from the social sciences, humanities, and natural sciences. Other program requirements include a B.A. thesis and an internship or field studies component.

The Irving B. Harris Graduate School and the Division of the Physical Sciences offer a two-year master of science (M.S.) degree program in Environmental Science and Policy. The program provides students a background in environmental policy and science to examine the scientific impact of environmental policy decisions. Existing coursework in the Graduate School and Physical Sciences is combined with a required introductory course and independent study. Each student also works with his or her own advisory committee. The university also offers a master's degree in Environmental Studies, emphasizing the basic principles of ecology and environmental science. Students study environmental issues from scientific, social scientific, and humanistic approaches.

The university seeks to use its status as one of the nation's leading research institutions to foster the interdisciplinary intellectual dialogue and broad scientific understanding necessary for the development of global solutions to global environmental problems. The university's Climate Systems Center focuses on a broad range of climate problems, including global warming, and the scientific search for their solutions. The Climate Systems Center specializes in blending climate research with the development and implementation of state-of-the-art climate modeling simulation software such as atmosphere-ocean general circulation models. The center also offers advice and opportunities for small outside groups to put together or utilize computer-assisted climate modeling hardware and software that would otherwise be unavailable to them.

The interdisciplinary Human Rights Program combines academic coursework and research with real-world experience. The student curriculum centers on a set of core courses featuring disciplinary, thematic, and regional perspectives combined with a student internship program providing fellowships at both national and global host organizations that present practical experience. The program's community education mission is implemented through its conferences, workshops, lectures, and film series open to those outside as well as within the university community.

The University of Chicago Environmental Center (UCEC) provides a central location for sustainability-based student clubs and organizations. The UCEC contains a library of books, magazines, videos, and other sustainability resources. Interested students find a wealth of information on graduate school programs, internships, and jobs in sustainability-related fields as well as information on local, national, and international environmental organizations offering membership or employment opportunities.

Several student organizations promote campus sustainability initiatives. Student sustainability-based clubs and organizations include the Green Campus Initiative, the

Environmental Concerns Organization, the Vegan Society, Green Awareness in Action, Material Exchange, and the Outdoor Adventure Club. The Green Campus Initiative oversees the Battle of the Bulbs energy conservation competition. Student energy conservation initiatives also include the Uncommon Turbine Project. Students may also join the university's Sustainability Council, which offers an undergraduate-level student internship. The university also facilitates student involvement in a variety of off-campus projects and volunteer opportunities across the greater Chicago area and beyond.

The university's sustainability affiliations include Clean Air Counts, a Chicago-based coalition of businesses and educational institutions with the goal of reducing carbon emissions, and the Illinois Sustainable University Compact, which sets sustainability guidelines in the areas of product purchasing, renewable-energy purchasing, green building, transportation, water conservation, and dining practices. The Sustainable Endowments Institute awarded the University of Chicago an overall grade of C+ on its College Sustainability (Green) Report Card.

See Also: Leadership in Sustainability Education; Sustainability Officers; Sustainability Websites.

Further Readings

Barlett, Peggy F. and Geoffrey W. Chase. *Sustainability on Campus: Stories and Strategies for Change*. Cambridge, MA: MIT Press, 2004.

Koppes, Steve. "Advance Sustainability Planning Helps Make a Greener Campus." University of Chicago. http://news.uchicago.edu/news.php?asset_id=1712 (Accessed June 2010).

M'Gonigle, R. Michael and Justine Starke. *Planet U: Sustaining the World, Reinventing the University*. Gabriola Island, British Columbia, Canada: New Society Publishers, 2006.

Rappaport, Ann and Sarah Hammond Creighton. *Degrees That Matter: Climate Change and the University*. Cambridge, MA: MIT Press, 2007.

Sustainable Endowments Institute. "The College Sustainability (Green) Report Card." http://www.greenreportcard.org/report-card-2009/schools/university-of-chicago (Accessed June 2010).

Marcella Bush Trevino
Barry University

University of Copenhagen

The University of Copenhagen, an ancient research university located in Denmark, has emphasized steps that have made its operations an exemplar for sustainability. The University of Copenhagen has created a climate-friendly campus that uses sustainable energy whenever possible. Renovated and new buildings are designed to contribute to a green society, ensuring that structures on the campus are not only useful, but sustainable as well. Using a rigorous method of reporting and accountability, the University of Copenhagen uses data to make certain that its plans and systems regarding sustainability are in place and actually doing what they are intended to do. The University of Copenhagen has also worked with global partners to focus on sustainable development in a way that extends beyond its campus. The University of Copenhagen and the Danish government have continued to make significant investments to support sustainable operations at the institution, making it a model for others.

Background

The University of Copenhagen, founded in 1479, is the oldest university and research institution in Denmark, with a global reputation for excellence and innovation. Between 1675 and 1788, the university initiated degree examinations, with the first degree examination being held in theology, followed by law in 1736. Finally, by 1788, all faculties required examination before awarding degrees. In 1960, the university served approximately 6,000 students, but that number grew to over 26,000 by 1980. The University of Copenhagen's growth resulted from a variety of institutional changes such as in 1933, when law departments formed a Faculty of Law (the equivalent of a U.S. law school) or the 2007 merger of the Royal Veterinary and Agricultural University and the Danish University of Pharmaceutical Science. As a result, the University of Copenhagen now is the largest university in Denmark, with more than 37,000 students. The students, 59 percent of whom are female, are increasingly diverse, and the university has strived to implement ways to serve its student population better. While the majority of its classes are still taught in Danish, a growing number are being taught in other languages, predominantly English and German. The university is served by more than 7,000 employees, including faculty and staff. Since 1994, the university has regarded the Environmental Studies department as an area of special priority according to its long-term plan. All of these factors result in the university's reputation as a leading university in Scandinavia and its rank as one of the 50 best universities in the world.

Sustainable Initiatives

With the growing movement toward sustainability and ecofriendly societies, the University of Copenhagen has dedicated itself to building a climate-friendly campus that also encompasses sustainability efforts. Part of this movement has been undertaken by the university's board of directors, which has adopted the goal of reducing the university's energy consumption and carbon dioxide (CO_2) emissions by 20 percent between 2006 and 2013. As one would expect of a university founded in the 15th century, the University of Copenhagen is composed of many historical buildings that were constructed long before energy efficiency was a priority. Rather than creating new structures to completely replace the historic buildings already present, the university and its leadership have chosen to pursue initiatives that will update existing buildings that are not currently climate-friendly and sustainable. This has resulted in a sometimes massive overhauling of existing structures, sometimes resulting in essentially new structures that utilize the historic building's character and façade. For new buildings that are in the planning stages, there exists a rigorous set of criteria to be met. New buildings must utilize the most advanced environmental principles in their design and incorporate climate-friendly materials in their construction so that building efficiency is ensured from the beginning.

New buildings that are to be constructed for the university must be developed and constructed by the Danish University and Property Agency (DUPA). DUPA is an organization based on the goal of creating world-class public structures that are aesthetically pleasing and have optimal organizational, legal, physical, and economic frameworks; and, following these new commitments by participating universities and Denmark itself, seeks to create sustainable environments to provide for a safe and healthy global climate. The University of Copenhagen and DUPA have worked together extensively to plan and construct a series of green buildings. The relationship between DUPA and the University of Copenhagen has resulted in the Green Lighthouse Project. The Green Lighthouse Project demonstrates climate developments that many thought were futuristic in nature. For example, the Green Lighthouse Project has created a building that is CO_2 neutral and now houses faculty of the University of Copenhagen. The beliefs and practices of the Green Lighthouse Project mirror those of the university and its board of directors with regard to construction of new buildings. To that end, the environment and climate are of concern at the beginning of any construction, and energy-efficiency processes are designed during planning and implemented during construction. The process of creating buildings that are energy efficient from the beginning actually results in considerable cost savings, as energy expense is greatly reduced and the buildings serve as a model for other users. The University of Copenhagen is in the process of obtaining Leadership in Energy and Environmental Design (LEED) certification for all of its new and many of its historic buildings.

Green Campus

In October 2008, the University of Copenhagen received its first "green accounts," which are site-specific environmental reports that are generated to provide the public with information related to the amounts of energy, water, and raw material used at a particular site. This initial green account led to the University of Copenhagen's commitment to reducing its CO_2 emissions by 2013. The green accounts also led to the university's goal that by the end of 2009, 80 percent of all purchases made shall be based on purchase agreements run through the central procurement office and that by 2013, at least 75 percent of all procurements made via purchase agreements shall require sustainability standards to be met. This process will enable the University of Copenhagen to more carefully monitor green purchasing and to ensure that sustainability is at least considered for all purchases.

The commitment to utilize purchase agreements that ensure sustainability standards allows the University of Copenhagen to draw on its annual purchasing budget of $249 million to encourage sustainable practices from manufacturers and distributors. Approximately one-third of the purchasing budget is used to buy products that span a wide array of items that range from pencils and computers to freezers and other equipment used for research. The University of Copenhagen understands that many of its purchases use energy in their production, for transportation to campus, during their use, and for disposal when no longer useful. As a result, the commitment to change purchasing practices to encourage sustainable development will focus on creating standards that give special attention to decreasing energy consumption and CO_2 emissions during the life span of a product. The University of Copenhagen also hopes to use its considerable purchasing power to pressure manufacturers to produce more cost-effective and environmentally sustainable products and to create and sustain a market for such goods.

Through its green building practices and its revised purchasing standards, the University of Copenhagen has made a commitment to sustainability. As the university's sustainability profile has been raised, it has worked to make available and communicate outcomes of these initiatives and experiences to all interested parties. In addition, the university has sought to enter into cooperative agreements with partners who will contribute to the goals of the green campus movement.

Global Collaboration

As interest has grown in sustainable campus practices internationally, many universities have made it a priority to work with other institutions to promote climate-friendly behavior. The University of Copenhagen has begun collaborative efforts by joining with other universities to form the International Alliance of Research Universities (IARU). The IARU focuses on fostering sustainable development among its members and sharing this information through a variety of outlets. The IARU began in January 2006 as a network of 10 research-intensive universities, including Australia National University; the University of Tokyo in Japan; National University of Singapore; ETH Zurich in Switzerland; the University of Cambridge and the University of Oxford in the United Kingdom; and Yale University and the University of California, Berkeley in the United States. The IARU has taken a variety of steps to focus its members on sustainable development. These steps have included establishing common and binding goals related to sustainability; attending Climate Change, an international congress that took place in Copenhagen in March 2009; and issuing a statement and joint binding goals, endorsed by all member presidents, regarding reduction of CO_2 emissions. To demonstrate commitment to these goals, the IARU members have also developed a toolkit that provides a template for the development of campus sustainability programs and collaborated to devise greenhouse gas emission reduction strategies. Additional collaboration has emphasized developing key numbers, indicators, and tools that will assist in determining environmental benchmarks for individual universities.

In addition to the IARU, the University of Copenhagen also participates in the International Sustainable Campus Network (ISCN). The ISCN is composed of 22 institutions that seek to share best practices on building, transportation, and teaching to better equip others to uphold commitments to sustainability. To that end, the ISCN has developed a series of goals that will assist its members in creating more sustainable campuses. These goals include the following:

- Goal 1: sustainable construction, renovation, and operation
- Goal 2: sustainable master planning and development, mobility, and community integration
- Goal 3: linking facilities, research, and education for sustainable development

Through this collaboration, ISCN members hope to improve campus operations as they relate to building, renovating, and running their campus operations in a sustainable manner. The ISCN members also serve as a means to recognize exemplars of the sustainability efforts at the university level. The ISCN uses a variety of forums, including conferences, the designation of exemplary green practices through campus green building case studies, and other methods to disseminate useful information about sustainable efforts and research.

Energy and Climate

In 2008, the University of Copenhagen invested $2 million to improve its operations as related to energy and climate efficiency. This base contribution was made with the intention of increased amounts to follow in ensuing years. These funds were used to improve the energy climate screen through the installation of better insulation, the replacement of windows with energy-efficient alternatives, ensuring that all windows close tightly in order to prevent heat loss, and other such improvements. These efforts have become an integral step in ensuring that the university's operations support overall sustainability efforts, something that is important due to the age of many of the university's buildings.

Further improvements will include energy smart installations, which will update ventilation and lighting systems, heating and cooling sources, and other necessary infrastructure for buildings. This area is grounded in the university's desire for all installed systems to be efficient both during their use and after their life cycle has ended. A strong focus is placed on the human impact related to the use of materials. The installation and updating of materials within buildings is but one part of the creation of a sustainable university environment. This realization has helped the University of Copenhagen and its international affiliates establish energy-smart conduct. Facility users and managers operate in partnership with faculty and students to ensure that simple actions such as closing windows or turning off lights and equipment when leaving rooms are taken to further sustainability goals. Communication of sustainability goals to all stakeholders is seen as key to this effort.

See Also: American College and University Presidents Climate Commitment; Collaborative for High Performance Schools; International Alliance of Research Universities: Sustainability Partnership; NAEP Sustainability Institute.

Further Readings

Edwards, A. R. and D. W. Orr. *The Sustainability Revolution: Portrait of a Paradigm Shift.* Gabriola Island, British Columbia, Canada: New Society Publishers, 2005.
University of Copenhagen. "Green Campus." http://climate.ku.dk/green_campus (Accessed June 2010).
University of Copenhagen. "University of Copenhagen." http://www.ku.dk/english (Accessed June 2010).

Stephen T. Schroth
Jason A. Helfer
Jordan K. Lanfair
Knox College

University of Glasgow

The University of Glasgow, established in 1451, enrolls over 16,500 undergraduate students, 5,000 postgraduate students, about 5,000 adult learners, and employs over 6,000 staff members, including faculty and research staff. The main campus is in Glasgow, Scotland, with a second campus at Dumfries. Students at the University of Glasgow come from over 120 countries and enroll in one of eight faculties: Arts, Education, Engineering, Law, Business and Social Sciences, Medicine, Science, and Veterinary Medicine. It is among the top 10 universities in the United Kingdom (UK) in terms of research funding and is a member of the Russell Group of 20 top research universities in the UK, with £126 million in research grants and contracts in 2008–2009; it is a member of the International Research

A kayaker on Loch Lomond, where the University of Glasgow maintains a research field station for the Scottish Center for Ecology and the Natural Environment.

Source: iStockphoto.com

Universities Network and a founding member of Universitas 21, an international organization of universities dedicated to setting international standards for university education.

Research in Sustainability

Sustainability is a key research area at the University of Glasgow, and researchers at the university are investigating questions such as threats to biodiversity and the best ways to manage climate change.

The Scottish Centre for Ecology and the Natural Environment (SCENE), part of the Division of Ecology and Evolutionary Biology within the Faculty of Biomedical and Life Sciences, is a leading unit for training, teaching, and research in ecology and the environmental sciences. SCENE was established in 1946; it trains over 600 undergraduates each year and provides research facilities for professionals in many disciplines. The SCENE field station, a purpose-built structure that offers residential accommodation as well as training and research facilities, is located in Trossachs National Park on the shore of Loch Lomond in western Scotland and offers immediate access to a number of habitats, including moorland, oak woodlands, and upland and lowland terrestrial habitats and streams.

The current director of SCENE is Professor Colin E. Adams, a specialist in behavioral ecology whose research areas include the biology of the Arctic char, the impact of new species introduced into freshwater systems, evolutionary processes that lead to polymorphisms in fishes, and life-history strategies in salmonid fishes. Professor Felicity A. Huntingford is the deputy director of SCENE; she supervises research on the role of body reserves in reproductive strategies and the functional significance and biological consequences of aggression in fishes. Other topics being researched at SCENE include biogeochemical cycles, stoichiometry of natural systems, factors influencing nocturnalism in fishes, predator avoidance in fishes, and environmental constraints on parental investment in birds.

The Glasgow Centre for International Development is focused primarily on research in the global south, particularly in sub-Saharan Africa, and is focused on strengthening and developing the university's long-standing ties (dating back to the mid-19th century) with African universities and institutes. The center is built around four research themes: health; environmental management and infrastructure; education, lifelong learning and global citizenship; and economic development. In the area of health, the Boyd Orr Centre for Population and Ecosystem Health facilitates interchange among researchers focusing on the relationship between human and animal populations and the health of their ecosystems, with recognition of the links among human activity, ecological change, and health. A current (2010) research project that directly links ecosystem change and human health is a study led by Dr. Heather Ferguson that studies the impact of rapid population increase in the Kilombero Valley of southern Tanzania, which is accompanied by radical transformation of the landscape from wilderness to agricultural and urban uses. This research project focuses on how these changes affect disease transmission, particularly of malaria, which is transmitted by mosquitoes whose survival and breeding process as well as choice of host may be changing in response to the ecological changes.

In the area of environmental management and infrastructure, researchers from a number of fields, including chemistry, engineering, life sciences, and the geographical and earth sciences, are focused on how humans affect the natural and urban environment. Research projects in this area include a study of sustainable biodiversity management in northern Zambia (in partnership with the University of Zambia and the University of Aberdeen), a study of indigenous environmental knowledge among Bedouin livestock herders and its influence on sustainable development in arid and semi-arid environments, and two projects designed to provide training in research, environmental management, and sustainable development within Africa.

Work in the area of education, lifelong learning, and global citizenship is organized through the Centre for Research Development in Adult and Lifelong Learning, which has relationships with a number of African studies to facilitate research and development collaboration and student recruitment for postgraduate study. Current projects include building partnerships with universities in Nigeria, Botswana, and Malawi, supporting the creation of effective recordkeeping systems to facilitate good governance and the rule of law, and fostering female literacy and income generation in Bedouin countries.

Many other University of Glasgow researchers are engaged in studies that center on questions of sustainability. For instance, an interdisciplinary research team (including experts in biology, chemistry, and engineering) is working in the field of synthetic biology, which attempts to mimic natural biological processes for useful purposes. The Glasgow team is focused on producing clean fuel by harnessing the process of photosynthesis and thus converting solar energy into fuel, thus surmounting two difficulties with solar production of electricity, thus obviating the well-known problems of storing electricity as well as providing a more direct approach to energy needs (only about 70 percent of which is for fuels such as ethanol and methanol versus about 30 percent for electricity).

Alan Ervine, professor of water engineering, is the leader of a cross-disciplinary project initiated in January 2007 and including researchers from 16 universities worldwide (all members of Universitas 21, an international association of research universities) that addresses the problems of water use in urban environments, including consideration of the effects of population growth, rapid urbanization, increasing prosperity, and climate change. This project, titled Water Futures for Sustainable Cities, aims to develop a research portfolio including the following themes: water education, literacy, and citizenship: transition education for environmental sustainability; urban river corridors: multidisciplinary performance assessment for sustainability of restored river reaches; innovation strategies for city water systems: a critical examination of the new modeling frameworks for sustainability assessment; tracking the fate and transport of endocrine-disrupting chemicals and pharmaceutical and personal care products in ecosystems; sustainable waterfronts development and adaptation to climate change; recycled water; and water/energy relationships in urban water and wastewater management.

Martin Bees, lecturer in mathematics, is studying how to increase the yield of hydrogen from algae with the goal of enabling algae to be used as an energy source. Bees, who was awarded an Engineering and Physical Sciences Research Council Advanced Research Fellowship for his work, is conducting experiments in the effects of removing sulfur and oxygen from algae systems to increase hydrogen production, potentially creating a green source of energy that could be used to power vehicles in place of gasoline.

Education in Sustainability

The University of Glasgow offers several degree programs directly related to sustainability, and courses that include sustainability are included in many other degree programs as well. The Centre for Development Studies (CDS) at the University of Glasgow, based within the Department of Economics, is a leading center within the UK for research and postgraduate study in development. Faculty in many other departments, including Law, Business and Social Sciences, Accounting and Finance, Economic and Social History, Urban Studies, Sociology, Anthropology and Applied Social Sciences, and Central and Eastern European Studies, also teach courses in the master's programs of CDS. In 2009–2010, 102 students were enrolled in CDS, and since establishment of CDS in 1983, over 1,000 students have graduated and gone on to careers in many venues, including international organizations, nongovernmental organizations, government ministries, banks and financial institutions, and academic and research institutions.

The master of science (M.Sc.) in Carbon Management degree is a multidisciplinary program and the first of its kind offered in Scotland. This M.Sc. is offered jointly by CDS, the Crichton Carbon Centre (a not-for-profit charitable organization that conducts academic research and teaching in carbon management), and the Dumfries Campus of the University of Glasgow. The M.Sc. Carbon Management degree is intended to prepare students for careers in fields such as carbon auditing, carbon footprinting, environmental consultancy, protection or business, renewable energy technologies, and carbon-related study and applications within academic and research institutions and international agencies. All students take classes in theories and principles of sustainability, carbon auditing and management, climate, carbon and change, and either environmental and organization ethics or policies for sustainability and development, while electives include courses in sustainable buildings, sustainable energy technologies, and environmental economics. Students complete either a research dissertation or a work placement project that takes place within a sponsoring organization or company.

CDS offers an M.Sc. in Environment and Sustainable Development and Diploma in Environment and Sustainable Development: both are intended to prepare students for careers as managers of environmental projects, in international agencies, as part of the planning staff of government ministries, and in academic and research institutes; the diploma places less emphasis on research than does the M.Sc. All students in these programs study development policy, environmental economics, policies for sustainability and development, and the theory and principles of sustainability, with elective courses in topics such as econometrics, international trade, and globalization, trade, and economic growth.

The Faculty of Law, Business & Social Sciences offers postgraduate study leading to an M.Sc. or postgraduate diploma in Environment and Sustainable Development. This is a degree designed for students intending to work on environmental issues in academic and research institutions or international agencies, those intending to work as managers of environmental projects, and those intending to work in ministries as planning staff who design and implement environmental policies.

The University of Glasgow was named to the environmental magazine *Grist*'s top 15 green universities in 2007, one of only two UK universities so honored. Glasgow was cited for the use of energy assessments and energy-saving design in new buildings, for sustainable procurement practices, and for using energy from renewable sources for almost 45 percent of the university's needs. The university was also recognized for green practices by HEEPI (Higher Education—Environmental Performance Improvement), an organization financed by the Higher Education Funding Council for England. The University of Glasgow is a signatory to the Talloires Declaration of the Association of University Leaders for a Sustainable Future, which signifies support for making sustainability a critical focus of university teaching, research, operations, and outreach.

See Also: Association of University Leaders for a Sustainable Future; Leadership in Sustainability Education; Leeds University.

Further Readings

Grist: A Beacon in the Smog. "15 Green Colleges and Universities." (August 10, 2007). http://www.grist.org/article/colleges1 (Accessed May 2010).

The Times Higher Education Supplement: Sustainability. *Bright Ideas That Could Save the Planet: Green Gown Awards 2006* (April 14, 2006). http://www.heepi.org.uk/green%20 gowns%20GG%202005–6/thes_sustainability_small.pdf (Accessed May 2010).

Universitas 21: The Network for International Higher Education. "Water Futures for Sustainable Cities." http://www.universitas21.com/water.html (Accessed May 2010).

University of Glasgow, Glasgow Centre for International Development. http://www.gla.ac.uk/gcid (Accessed May 2010).

University of Glasgow, SCENE: Scottish Centre for Ecology and the Natural Environment. http://www.gla.ac.uk/scene (Accessed May 2010).

Sarah Boslaugh
Washington University in St. Louis

University of Maryland, College Park

The University of Maryland, College Park, (UM) is a public research university just outside Washington, D.C., in College Park, Maryland. Established in 1856, UM is the primary institution of Maryland's state university system and is the largest in the state and the District of Columbia metropolitan area. Unusual for a state flagship institution, UM has no satellite campuses; the other institutions of the state university system are autonomous. Because of its proximity to the nation's capital, the school has a long history of research partnerships with the federal government, including the National Center for Weather and Climate Prediction, constructed in conjunction with the National Oceanic and Atmospheric Administration; the 2005 launch of the Deep Impact spacecraft, a joint venture between UM and the U.S. National Aeronautics and Space Administration (NASA); and research funded by the U.S. National Science Foundation, the U.S. National Institutes of Health, and the U.S. National Security Agency. The UM Office of Sustainability was established in summer 2007, upon UM President Dan Mote's signing of the American College and University Presidents Climate Commitment, to further the university's goals of campus carbon neutrality. The Climate Action Plan Work Group finalized its Climate Action Plan in 2009, and the office has promoted a number of sustainability initiatives, with particular success in raising its recycling rate to more than half, thanks in part to a recycling program

Among the University of Maryland's many federal research partnerships was a collaboration with NASA on the *Deep Impact* spacecraft, shown here on a Delta II rocket at Cape Canaveral, which was used to study the composition of the comet Tempel 1.

Source: National Aeronautics and Space Administration

instituted at home football games, previously a significant source of unrecycled waste. The university's strategic plan, adopted in 2008, spells out the importance of sustainability in its goals:

> The University of Maryland will be widely recognized as a national model for a Green University. In 10 years' time, the University will have made substantial progress towards addressing energy issues. It will have slashed energy use, expanded green spaces, dramatically reduced its carbon footprint, and built and retrofitted buildings to strict environmental standards. The University will complement these concrete actions with its teaching, research, and development efforts in energy science and policy, smart growth, environmental mapping, sustainable agriculture, and other fields.

The University Sustainability Council was formed in the fall 2009 semester following the creation of the Climate Action Plan and advises the president, the Office of Sustainability, and the campus community. The 16-member council includes 10 permanent members from the administration, three faculty members serving two-year terms, two students serving one-year terms, and one at-large appointment. The council meets a minimum frequency of quarterly and oversees UM's mission to become a nationally preeminent green university by recommending long-term sustainability goals and policies and by suggesting cost-effective methods of meeting them. It monitors the implementation of these policies and of the Climate Action Plan, drafts policy and guidelines, provides cost-benefit analyses, recommends the metrics used to check progress, and makes recommendations to the Finance Committee, the Provost's Office, the vice president of Research, the Facilities Council, and other appropriate campus organizations. In the future, representatives of the graduate school will be included in the council, if a sustainability fee is assessed to graduate students, which it was not at the founding of the council.

Greening the Campus

The council helped to draft new policies implemented in 2010 to reduce energy usage on campus. The new building temperature policy set 68 to 78 degrees as the year-round target range for occupied-building temperatures. The lighting levels policy adopted the lighting levels recommended by the Illuminating Engineer Society of North America, and called for the use of the most energy-efficient technology available to provide lighting of 5- to 8-foot candles in restrooms, stairwells, and hallways, and 30- to 50-foot candles in offices, classrooms, and laboratories. Supplemental lighting is strongly discouraged but not forbidden. Finally, the environmentally preferable procurement policy dictates that UM will procure all supplies and services with consideration of the environmental impact of such procurement.

Every year beginning in 2009, the Chesapeake Project is held, a two-day workshop to help UM faculty integrate sustainability across all disciplines. Furthermore, student sustainability advisors present a lesson on sustainability to all incoming freshmen as part of a mandatory freshman seminar class. The sustainability conversation on campus is augmented by the Sustainability Speaker Series, which combines half-hour presentations and 10-minute discussions with lunch on Thursdays and has included presentations by sustainability staff members and other administrators.

The energy conservation project conducted in nine campus buildings is designed to reduce carbon dioxide (CO_2) emissions by 4,200 tons a year, and save nearly $2 million a year in energy costs. Energywise UM is a pilot project involving the Office of Sustainability, Facilities, and the Office of Fraternity and Sorority Life, attempting to meet both UM's goal of reducing electricity consumption by 5 percent through behavior modification strategies in buildings, and the Maryland governor's initiative to reduce energy consumption in state government buildings by 15 percent by 2015. Energywise both retrofits buildings to some degree and encourages better energy usage by their residents by advising occupants of strategies for more energy-efficient living.

The combined heat and power plant has been U.S. Environmental Protection Agency (EPA) Energy Star certified since 2003. All cleaning products used in residential facilities are Green Seal certified, and the products used in academic and administrative buildings (which have slightly different cleaning needs) are expected to follow. The entire campus is a certified Arboretum and Botanical Garden, and the golf course is a certified Audubon International Wildlife Sanctuary, planted with water-conserving grass varieties that have reduced its water usage by more than one-third. Furthermore, a 10,000-gallon cistern installed in the Washington Quad collects rainwater and uses a computer-controlled drip irrigation system to water the Quad. The dishwashers in Dining Services have been replaced with energy-efficient equipment using steam heat, and a gradual plan is being implemented to replace plumbing fixtures in the residence halls with low-flow toilets, showerheads, and faucets. Various green commuting options for students include free buses, the Terp Riders Carpool, improved bike lanes and parking spots to encourage bicycling, free bus service from area park-and-rides, and the Green Permit, which provides a 20 percent discount on parking permits for vehicles that meet the EPA's green standards.

The UM dining services have undergone significant greening since 2007. Bottled water has been eliminated from all dining halls, with complimentary triple-filtered water stations installed instead. Thousands of free, reusable cold beverage containers were distributed to students, and discounts are offered on refills of sodas and other fee beverages in order to encourage their use. In 2009, polystyrene containers were eliminated and replaced with biodegradable trays and recycled paper cups. Food waste from the Diner and the South Campus Dining Hall—about 20 tons per month—is composted off campus, while all used cooking oil is harvested for biodiesel. Recycling is taken seriously in the dining halls and includes recycling of toner cartridges from cash-register printers, fluorescent light tubes from the overhead lighting, paper, cardboard, bottles and cans, motor oil, and refrigerants.

Greening the Curriculum

Many of UM's degree programs incorporate sustainability and the study of the environment, including undergraduate degrees in Agriculture and Resource Economics, Agricultural Sciences, Animal and Avian Sciences, Anthropology, Architecture, Biochemistry, Biology, Civil and Environmental Engineering, Environmental Science and Policy, Environmental Science and Technology, Geography, Geology, Landscape Architecture, Natural Resource Management, and Natural Resource Sciences; and graduate programs in Agricultural and Resource Economics, Animal Sciences, Anthropology, Architecture, Atmospheric and Oceanic Science, Behavior, Ecology, Evolution, and Systematics, Biochemistry, Biology, Chemistry, Civil and Environmental Engineering, Engineering and Public Policy, Geography, Geology, Marine-Estuarine Environmental Sciences, Natural Resource Sciences, Public Policy, Public and Community Health, Sustainable Development and Conservation Biology, Urban Studies and Planning, and Urban and Regional Planning and Design. UM is the first U.S. school to offer a Master of Engineering in Sustainable Energy Engineering degree program, and offers it in its entirety online (though it is also available on campus). The program covers studies on various renewable energy applications, according to the student's focus in one of three areas: Energy Systems, Nuclear Engineering, or Reliability Engineering.

Sustainability clubs include the College Park Environmental Group, the Ecological Economics Student Group, Emerging Green Builders, Engineers Without Borders, the Maryland Food Collective, Quench, Rethink College Park, the Society for Green Business, the Terrapin Trail Club, the Solar Decathlon, St. Mary's Garden Club, and the Student Sustainability Committee.

Biodiesel University is a nonprofit renewable energy education organization run by UM's R. H. Smith School of Business Dingman Center for Entrepreneurship and known for its fleet of mobile laboratories. The greenest mobile labs in the world, the university's buses use biodegradable motor oil, hydraulic fluids, and greases.

Environmental research centers based at UM include the Center for Environmental Energy Engineering, the Center for Integrative Environmental Research, the Center for Social Value Creation, the Earth Science Interdisciplinary Center, the Environmental Finance Center, the Maryland Institute for Applied Environmental Health, the National Center for Smart Growth Research and Education, the University of Maryland Energy Research Center, and the Joint Global Change Research Institute.

The Joint Global Change Research Institute (JGCRI), founded in 2001, is a joint venture between UM and the Pacific Northwest National Laboratory. The interdisciplinary team at JGCRI, along with a network of international collaborators, conducts climate change research and trains university students in five areas: integrated assessment modeling, technology strategies to address climate change, natural resource modeling and assessment, vulnerability and adaptation studies, and local and global environment mitigation measures.

The institute's research activities are led by two groups. The Impacts, Adaptation, and Vulnerability Group focuses on potential future climate change impacts on natural resources, understanding the carbon cycle, and analyzing the vulnerability of the world to carbon change. The group's studies provide data that inform energy and economic models developed at the institute. One of the major initiatives of the group is the study of carbon sequestration in agricultural soil, which improves soil quality while removing carbon from the atmosphere. Several long-term agricultural experiment stations are operated as part of the initiative, while the historical carbon cycle of southeastern Arizona is studied in terms of its response to climate variability. The Climate Change Economics, Technology, and Policy Group assesses the possible future growth of greenhouse emissions and potential policies to mitigate those emissions and develops the energy and economics models used by the institute.

The main models used by the institute are the Second Generation Model, a computable general equilibrium model of the world with eight regions that projects energy consumption and greenhouse gas emissions in five-year time steps and investigates the impact of climate change policies and technologies for emissions mitigation; the Global Change Assessment Model, a partial equilibrium model of the world with 14 regions that projects energy consumption and greenhouse gas emissions in 15-year time steps and includes an agriculture land-use module and a reduced-form carbon cycle and climate module; and the Environmental Policy Integrated Climate model, a process-based agricultural systems model with simulation components for weather, hydrology, nutrient cycling, pesticide fate, tillage, crop growth, soil erosion, crop and soil management, and economics.

See Also: American College and University Presidents Climate Commitment; Outdoor Education; Whole-School Approaches to Sustainability.

Further Readings

Hargreaves, Andy and Dean Fink. *Sustainable Leadership*. Hoboken, NJ: Jossey-Bass, 2005.

Sandell, Klas, Johan Ohman, and Leif Ostman. *Education for Sustainable Development*. New York: Studentlitteratur AB, 2005.

Silka, Linda and Robert Forrant, eds. *Inside and Out: Universities and Education for Sustainable Development*. Amityville, NY: Baywood Publishing, 2006.

University of Maryland, Campus Sustainability. http://www.sustainability.umd.edu (Accessed August 2010).

Bill Kte'pi
Independent Scholar

University of Michigan (Erb Institute)

The University of Michigan (U-M) was founded in 1817 in Detroit (where it was the first public university in the Northwest Territories), moved its main campus to Ann Arbor in 1837 (the year Michigan became a state), and has since established regional campuses in Flint in 1956 and Dearborn in 1958. As of fall 2007, the university enrolled over 56,000 students on the three campuses and employed over 9,000 instructional staff members, including over 3,000 faculty members. The university comprises 19 schools and colleges on the Ann Arbor campus, including the College of Literature, Science and the Arts (founded 1841); the College of Engineering (founded 1854); the Stephen M. Ross School of Business (founded 1924); the School of Natural Resources and Environment (founded 1927); the A. Alfred Taubman College of Architecture and Urban Planning (founded 1931); and schools of medicine, dentistry, nursing, pharmacy, and public health.

The university president as of December 2010 was Mary Sue Coleman, Ph.D., and the vice president and secretary is Sally J. Churchill, M.A., J.D. U-M is a major research institution with over $875 million in research funding in 2007–2008, among the largest amount for any U.S. university, with almost 70 percent of this funding coming from the federal government. U-M is sometimes called a "public Ivy" in reference to the high quality of education and research provided and has been ranked highly on various lists of national and international universities, including a rank of 22 in the 2009 Academic Ranking of World Universities compiled by Shanghai Jiao Tong University and a rank of 19 in the 2009 Times Higher Education-QS World University Rankings.

The Erb Institute

The Frederick A. and Barbara M. Erb Institute for Global Sustainable Enterprise was created in 1996 as a partnership between the School of Natural Resources and Environment and the Stephen M. Ross School of Business at the University of Michigan. The focus of the Erb Institute is global sustainability, that is, finding ways to meet the needs of the human population in an equitable manner consistent with the carrying capacity of the natural world. The Erb Institute supports scholarship, outreach, and professional education in service of this goal, with an emphasis on an interdisciplinary approach that links knowledge generation with social action, advances understanding of the dynamics of human and natural systems, and relies strongly on collaboration both within and outside the institute. The Erb Institute was founded with a gift of $20 million from Frederick and Barbara Erb and also receives significant funding from corporations (including the Dow Chemical Company, the Holcim [U.S.] Corporation, the Weyerhaeuser Corporation, and the Ford Motor Company), foundations (including the McGraw Foundation, Molly Vincent Foundation, and Alcoa Foundation), and private individuals.

Thomas P. Lyon, Dow Chemical Professor of Sustainable Science, Technology, and Commerce, is the director of the Erb Institute. His research focuses on the interplay between corporate strategy and public policy, including innovation in the healthcare sector, corporate environmentalism, investment practices in electric utilities, and the introduction of competition to regulated industries. Thomas N. Gladwin, Max McGraw Professor of Sustainable Enterprise, is the associate director of the Erb Institute. His research focuses on establishing a science of sustainable enterprise through the study of relationships among social systems, ecosystems, economic systems, and organizations systems. Andrew J. Hoffman, Holcim Professor of Sustainable Enterprise, is an associate director. His research focuses on institutional chance, organizational culture, and corporate strategies related to social and environmental issues. In addition, a number of U-M faculty members are affiliated with the Erb Institute.

At the undergraduate level, the Erb Institute offers an undergraduate course on Global Enterprise and Sustainable Development and is involved in developing new curricula, learning experiences, and support opportunities for students interested in combining studies in business and sustainability. At the master's degree level, the Erb Institute offers a three-year M.B.A./M.S. program that combines coursework, research, and practical projects related to business, the environment, and sustainability, resulting in degrees from both the Ross School of Business (M.B.A.) and the School of Natural Resources and Environment (M.S.). Students in the M.B.A./M.S. program complete a master's project and a multidisciplinary action project in conjunction with a corporation or nonprofit organization, where they apply their academic skills to practical problems such as streamlining processes to reduce waste and developing models to test the effectiveness of organizational efforts at sustainability. These projects have been conducted in over 35 countries as of 2010. The Erb Institute does not grant a doctorate, but faculty work with Ph.D. students enrolled in U-M schools and colleges who are interested in doctoral work that includes a focus on issues related to global sustainable enterprise. The Erb Institute also offers funding at the postdoctoral level and in 2010, the Erb Institute hosted two postdoctoral fellows and one visiting scholar.

Research Themes

Research at the Erb Institute concentrates on five issues that are related to global sustainable enterprise: alternative energy and clean technology, climate strategy and carbon policy, mega-city mobility and accessibility, base of the pyramid and social enterprise, and green buildings and development. Clean technology products and services include alternative energy sources that are carbon neutral, cause minimal harm to the environment, and do not deplete natural resources; examples of alternative energy sources include wind, solar, thermal, photovoltaics, and geothermal. Clean technology ideally should provide superior performance at lower cost, reduce or eliminate ecological impact, and make more productive and responsible use of natural resources, thus offering advantages for investors and customers while also reducing the impact on the environment.

Businesses have recently become more concerned with climate change, and this is reflected in many ways, including changes in supplier relationships, procedures for procuring raw materials, and costs of shipping and transportation as well as the price of goods and services to the consumer. In part, this is due to regulations that will affect the costs of carbon, and therefore energy pricing and availability, at all levels. These changes impose a number of new risks on businesses that must be accommodated in the business model, including regulatory, legal, physical, and reputation as well as systemic risks. The University of Michigan Sustainable Mobility and Accessibility Research and Transformation (SMART) project brings together partners from many academic departments, including the Erb Institute as well as industry. The focus of SMART is on developing sustainable transportation and mobility systems in urban areas.

The area of base of the pyramid and social enterprise refers to two related concepts that integrate understanding of the current social and economic situation and organizations that use the business process for purposes beyond simple economic profit. The phrase *base of the pyramid* is used in the economic and social sciences to refer to that sector of the world's population that is lowest in income and largest in number. As defined by the Erb Institute, this sector includes about 4 billion people, who represent a largely untapped market for goods and services. "Social enterprises" refers to organizations that trade goods and services for some social purpose such as empowering women or achieving other social aims. Activities in these areas at the Erb Institute are conducted in partnership with the William Davidson Institute, a nonprofit research and educational institute established in 1992 at the University of Michigan. In the areas of green buildings and development, the Erb Institute studies ways to reduce the environmental impact of construction in areas such as energy systems, water use, construction processes, architectural design, and site planning. They are also concerned with establishing the cost-effectiveness of green construction in both the short and long term, studying issues such as creating healthier environments with natural light and clean air, increased productivity and satisfaction for those who live and work in green buildings, and monetary savings by reduced construction costs, reduced operation and maintenance costs, and reduced utility bills.

Conferences and Publications

As part of its goal of bringing together interested stakeholders, including academics, business executives, policy makers, and representatives of nongovernmental organizations, the Erb Institute regularly hosts conferences, often in collaboration with organizations such as the National Science Foundation or the World Resources Institute. These conferences not only help frame and inform research agendas but also help educate the public about the institute's activities. For instance, in May 2010, the Erb Institute sponsored a conference titled Constructing Green: Sustainability and the Places We Inhabit, which focused on the built environment's influence on human societies as well as on the environment, with a focus on the green building industry and how it is changing. In February 2010, Tom Lyon of the Erb Institute copresented (with Michael Toffel of the Harvard Business School and John Mayo of the Georgetown Center for Business and Social Policy) a one-day workshop at the National Press Club in Washington, D.C., that focused on information disclosure as a regulatory approach in various sectors including the environment.

The Erb Institute sponsors a biweekly research seminar with contributions from faculty members from the University of Michigan and other universities who work in the field of sustainable enterprise, broadly defined. Recent presentations have included "Vehicle Standards in a Climate Policy Framework" (John DeCicco, presenter; Mary Zimmerman Ross, discussant); "Benefits, Costs, and EPA/NHTSA's Light Duty, Greenhouse Gas-Fuel Economy Rule" (Gloria Helfand, presenter; John DeCicco, discussant); and "The Al Gore Effect: *An Inconvenient Truth* and Voluntary Carbon Offsets" (Grant Jacobson, presenter).

Faculty and students associated with the Erb Institute have produced a number of reports and teaching cases that are available for free download through the Internet. Topics covered in these reports include an analysis of the influence of organization culture on corporate sustainability initiatives, a plan to address sustainability issues at the Holy Cross Abbey, a study of the possibilities for developing solar energy resources on public land in the California desert, and a study of sustainable urban redevelopment in Detroit.

The Erb Institute Website also makes available for free download many faculty publications and working papers. Examples of topics covered in recent (as of 2010) publications include social and psychological barriers to green building, effects of disclosure on pollution in India, the influence of international trade and voluntary environmental standards (specifically ISO 14001 certification) on compliance with existing environmental regulations in China, and the influence of the CAFE (corporate average fuel economy) standard and incentive program on greenhouse gas emissions and alternative fuel vehicles.

See Also: Green Business Education; Leadership in Sustainability Education; Sustainability Officers; Sustainability Websites.

Further Readings

Deutsch, Claudia H. "A Threat So Big, Academics Try Collaboration." *New York Times* (December 25, 2007). http://www.nytimes.com/2007/12/25/business/25sustain.html?_ r=1&scp=2&sq=%22erb%20institute%22&st=cse (Accessed May 2010).

Erb Perspective Blog. http://erbsustainability.wordpress.com (Accessed May 2011).

University of Michigan, Erb Institute for Global Sustainable Enterprise. "Institute Reports." http://erb.umich.edu/research-initiatives/institute-reports (Accessed May 2011).

Sarah Boslaugh
Washington University in St. Louis

UNIVERSITY OF MINNESOTA

Founded in 1851, the University of Minnesota is a public research institution located in Minneapolis and Saint Paul and is the flagship of the state's education system. A land-grant university, the University of Minnesota has a long history of providing outreach and assistance to state residents on a variety of topics, including sustainability. With over 60,000 students, 3,000 faculty, and 13,500 staff members, the University of Minnesota is one of the largest institutions of higher education. In addition to Minneapolis and Saint Paul, the University of Minnesota has campuses in Crookston, Duluth, Morris, and Rochester, as well as numerous extension offices. The University of Minnesota has defined sustainability as meeting current needs without compromising future generations' ability to meet theirs. The University of Minnesota models the important principles of sustainability and undertakes tasks that will improve the world. The institution has taken an active role in its community to model best practices and partners with other organizations to develop this vision locally, across the state, and throughout the world. Through its operations, research, teaching, and outreach, the University of Minnesota is a leader in green education.

Guiding Principles to Support Sustainability

The University of Minnesota's board of regents has endorsed sustainability and adopted six guiding principles to support that goal. The guiding principles include leadership, modeling, operational improvements, energy efficiency, education and outreach, and research. The University Minnesota seeks to demonstrate leadership through excellence in environmental education, research, outreach, and stewardship, promoting and demonstrating the effectiveness of sustainability and energy efficiency, and producing informed citizens and leaders for the movement. Through meeting and exceeding all applicable regulatory requirements, preventing pollution, reducing emissions, and using life-cycle cost frameworks, the University of Minnesota models sustainability principles to guide its operations. A continuous improvement process that meets operation performance targets, goals, and objectives that support sustainability allows the University of Minnesota to make environmentally friendly operational improvements. Energy efficiency is supported by reducing dependence on nonrenewable energy and encouraging the development of alternative energy sources through research and innovation. Educational and outreach activities that are linked to operational improvements and innovation principles are promoted by the university. Research that promotes innovative, high-visibility projects focused on sustainability and energy efficiency are undertaken by the University of Minnesota, especially those initiatives undertaken with operations staff, students, public entities, community organizations, and industry. These principles allow the University of Minnesota to incorporate sustainability into its teaching, research, and outreach.

Leadership

The University of Minnesota, long known as a center for research on sustainability, builds on this background to provide leadership and outreach. In 2002, a report produced by the university's Commission on Environmental Science and Policy created the University-wide Sustainability Policy (USP). The USP helped coordinate pre-existing initiatives so that the University of Minnesota could prioritize significant projects that would have the most impact. In 2004, the University of Minnesota joined the Chicago Climate Exchange, a pilot program that seeks to reduce and trade greenhouse gas emissions in Canada, Mexico, and the United States by 4 percent below each user's average 1998–2001 baselines. The University of Minnesota also created the Institute on the Environment, which brings together scholars from design, engineering, health, law, the natural and social sciences, public policy, and other fields to collectively envision possible solutions to environmental needs and to deliver measurable results. The Initiative on Renewable Energy and the Environment (IREE), for example, part of the Institute on the Environment, has brought together and supported over 400 researchers focused on alternative energy sources, including bioenergy, biofuels, and bioproducts, solar and wind power, renewable hydrogen, as well as economic and policy implications of these innovations. Coupled with the University of Minnesota Extension Service, with offices in every county, these initiatives allow interested Minnesotans to be provided with university resources related to renewable energy and sustainable food choices. A more tangible result of this outreach is a university-funded community known as the University of Minnesota Outreach, Research, and Education Park (UMore Park). UMore Park is a planned community, where quality of life, innovation, sustainability, and connection with the environment are all promoted and explored by faculty, students, and residents.

Modeling

The University of Minnesota advocates green practices not only through its educational offerings and research but also through its practices on and around campus. All undergraduate students are required to take a course dealing with environmental issues before graduating. Classes work in conjunction with campus organizations and clubs to use natural energy and other environmentally friendly sources of fuel. In 2007, a forum hosted by the Initiative for Renewable Energy focused on renewable energy careers available, the skills necessary for those careers, and how students could best prepare for these careers.

The University of Minnesota believes that by becoming a sustainable institution, it will greatly assist the environmental movement. To that end, the University of Minnesota joined with Xcel Energy, a public utility company, and utilized Xcel's Energy Design Assistance Program (EDAP). The EDAP provides energy-efficient strategies for all new construction on campus. Frequent sustainability workshops assist students, faculty, staff, and members of the community to engage in policy discussion as well as to explore sustainability initiatives across the University of Minnesota. The university, in an effort to assist in toxic waste reduction, operates a chemical redistribution program that collects opened and unopened bottles of unwanted but usable chemicals and lab equipment and redistributes these to interested university departments or research laboratories. Similarly, electronic and organic waste are collected and disposed of appropriately, which includes refurbishing and collecting usable parts from unneeded or broken equipment and composting food and other wastes.

Operational Improvements

With nearly 80,000 people arriving on campus each day, the University of Minnesota is the third-largest traffic generator in the state. Even small reductions in energy expended for travel consequently have a large effect on the community. To that end, the University of Minnesota runs an intercampus bus system to transport students, employees, and others free of charge to assorted campus locations. Bicycle racks have been installed across campus, and a free 24-hour-per-day escort service reassures those who choose to walk on campus. Over 20 percent of the University of Minnesota's vehicle fleet use flexible fuel, which leads the nation.

Dining and housing services at the University of Minnesota also work to improve the connection between the land, the community, and local producers. A priority is placed on acquiring locally grown agricultural products, as shown by an ongoing partnership with the Midwest Food Alliance, a group that encourages socially responsible agricultural practices. A variety of organic foods are featured in university dining facilities, and cage-free eggs, Fair Trade coffee, and biodegradable packaging are used. As much as possible, the University of Minnesota recycles or composts waste from its dining facilities and uses recycled paper products exclusively. The University of Minnesota has built more dormitories and other student housing over the past decade, increasing the number served by over 40 percent. Housing students on campus reduces energy expended for transportation by reducing travel on Minnesota roadways by an estimated 25,000 miles per day.

Energy Efficiency

Efforts by the University of Minnesota to increase energy efficiency are an important component of its outreach activities, both in terms of research related to sustainable projects and as a working environment to provide models for innovations that can be embraced by private industry. Renovation of the Saint Paul Chiller Plant included the rehabilitation of a historic building and the consolidation of cooling systems in one central location. With funding provided by the Minnesota State Legislature, the university replaced aging and inefficient chillers from 16 buildings, resulting in annual cost savings of $1 million, increased energy efficiency by reducing carbon dioxide (CO_2) by over 3,300 metric tons annually, and increased the reliability of the system. Such benefits were repeated in the renovation of the Southeast Steam Plant, which now burns fuel that is at least 70 percent natural gas, unlike the mostly coal fuel used before. The plant also uses biomass fuels, such as oat hulls, significantly reducing the cost of fuel used to produce steam and greatly reducing greenhouse gas emissions. Burning an estimated 25,000 tons a year of oat hulls saves $2 million per year in heating costs. The University of Minnesota is also renovating many buildings on its historic campus to identify opportunities for efficiency, improvement, and energy conservation. Findings of these projects are shared through the University of Minnesota extension services offices.

Education and Outreach

Practically since its inception, the University of Minnesota has provided the community various extension services through offices located around the state. These extension offices have permitted communities and individuals to gain access to resources and services related to environmental findings and sustainable practices. Extension services help spread awareness of numerous green-related topics across the state.

Certain extension services have been designed specifically to enhance understanding of environmental issues. The Ecosystem Science and Sustainability Initiative (ESSI), for example, offers educational workshops specifically designed to assist teachers to become more aware of best practices related to sustainability so that they can revamp the curriculum used in their classrooms to reflect this. ESSI also provides workshops for journalists and media representatives who work on similar environmental topics. The Minnesota Institute for Sustainable Agriculture (MISA) works with farmers and provides information on effective and efficient alternative energy systems that might be used instead of more traditional methods. Students at the University of Minnesota are provided over 100 different environmentally oriented courses that allow interdisciplinary approaches to essential issues facing world leaders. Multiple departments and degree offerings enable those interested in acquiring a level of expertise on a host of environmental issues to pursue their passions.

Additionally, the University of Minnesota offers a variety of outreach opportunities designed to connect the campus with the community to enhance discussion related to seminal environmental issues. Some of the conferences the university has sponsored in the past decade include the E3 conference, which explored the intersection between innovative technologies, visionary policies, environmental benefits, and emerging market opportunities presented by renewable energy; the Biofuel Production and Wildlife Protection symposium, which examined the best management practices for the Great Plains; the Biotechnology and Bioenergy Workshop, held in conjunction with the Norwegian Consulate General, which explored biomass, bioenergy, commercialization issues, and student mobility; the Ecosystem Science and Sustainability Initiative, which explored the relationship between climate change and economic development; and the Minnesota Institute for Sustainable Agriculture, which assisted farmers interested in sustainable practices. These seminars, institutes, trade shows, and exhibitions assist various sectors of the state's population to access reliable and accurate information related to sustainability and to understand how it relates to their everyday lives.

A variety of student activities and organizations also provide outreach experiences that are fun, hands-on, and informative. These include Concerts for the Environment, promoted by a student group dedicated to raising student awareness of environmental issues through music; the Ecology, Evolution, and Behavior Club, which organizes volunteer opportunities and field trips and promotes the use of the Lake Itasca Biological Station; EcoWatch, a student organization that works to expand environmental dialogue and to engage students with campus environmental issues; Greenlight, which promotes a broad application of design and creative means as a way to implement environmental solutions; and the Solar Decathlon Student Workshop, which involves student teams from across the globe who compete to design, build, and operate the most attractive and energy-efficient solar-powered house.

Research

As one of the global leaders in research spending, the University of Minnesota provides a variety of outlets for high-profile projects focused on sustainability and energy efficiency. The Center for Science, Technology, and Public Policy (CSTPP), for example, provides outreach on energy-related issues, including the environment and climate change. Other research initiatives explore issues of benefit to many, including examinations of affordable housing alternatives, assessment of the economic value of the environment, and others.

Minnesota's mining, lumber, and manufacturing industries also benefit from work done at the university.

See Also: Education, Local Green Initiatives; Green Community-Based Learning; Social Learning; Whole-School Approaches to Sustainability.

Further Readings

Arnaud, B. S., L. Smarr, J. Sheehan, and T. DeFanti. "Campuses as Living Laboratories for the Greener." *Educause Review*, 44/6 (2009).

Edwards, A. *The Sustainability Revolution: Portrait of a Paradigm Shift*. Gabriola Island, British Columbia, Canada: New Society Publishers, 2005.

Elder, J. L. "Higher Education and the Clean Energy, Green Economy." *Educause Review*, 44/6 (2009).

University of Minnesota. *Sustainability and U*. Minneapolis: University of Minnesota, 2008.

Stephen T. Schroth
Jason A. Helfer
Sergio Ulloa
Knox College

UNIVERSITY OF NEW HAMPSHIRE

The University of New Hampshire (UNH) formed the University Office of Sustainability in 1997, making it the oldest endowed sustainability program in higher education in the United States. The Office of Sustainability was created to oversee campus-wide and community outreach sustainability initiatives in UNH's ongoing commitment to maintain its leadership role in the field. The office employs six full-time equivalent sustainability staff and reports to the Office of the Provost. UNH integrates sustainability principles in its curriculum, operations, research, and engagement and has held annual campus-wide dialogues on issues related to sustainability since 2005.

Campus and Community Sustainability Initiatives

The Office of Sustainability maintains multiple committees, each dedicated to a specific aspect of sustainability within four key sustainability initiatives. The Biodiversity Education Initiative focuses on campus biodiversity protection and ecological and public health. The Climate Education Initiative focuses on the future of sustainable energy through emissions reductions policies, practices, research, and education. The Food and Society Initiative focuses on healthy food systems from the farm to the table, and health and nutrition outcomes. The Culture and Sustainability Initiative focuses on the creation of a culture of sustainability through commitment to community, diversity, civic engagement, the public arts, and the conservation and development of cultural and natural resources.

UNH implemented the climate action plan WildCAP with the goal of reducing campus greenhouse gas emissions 50 percent by 2020 and 80 percent by 2050. WildCAP has set 2100 as the target date for the achievement of total carbon neutrality. The 2009 EcoLine project is a WildCAP initiative that will utilize landfill gas as UNH's primary campus energy source, making it the first university in the United States to do so. Funding for EcoLine and other campus energy-efficiency projects will come in part through the sale of renewable energy certificates.

UNH is also a leader in community, state, regional, and national climate initiatives. UNH is part of the grassroots New Hampshire Carbon Challenge, which seeks to reduce household greenhouse gas emissions by 10,000 pounds annually through educational and outreach assistance to state residents. The university is a member of the public-private partnership Carbon Solutions New England, which seeks regional carbon neutrality through a blend of scientific, technological, and policy initiatives. UNH partnered with Clean Air–Cool Planet in 2000 to develop the trademarked Campus Carbon Calculator, now in use by over 2,300 U.S. colleges and universities.

UNH mandates that all new construction and renovation projects meet the U.S. Green Building Council's LEED Silver certification or equivalent standards. The recently replaced DeMeritt Hall is LEED Silver equivalent, and the James Hall renovation is registered to seek LEED certification. Green building features adopted or under consideration include rainwater reclamation systems and low-flow toilets, urinals, faucets, and showers. The UNH campus features 11 Energy Star–rated buildings, including the nation's first Energy Star designation for residence halls.

The Transportation Department has implemented sustainable practices in its operations and awareness efforts. UNH fleet vehicles run on biodiesel and compressed natural gas. Pedestrian- and bicycle-friendly infrastructure changes include new bike lanes and bus shelters with solar-powered lights. Any member of the UNH community may sign out a bike through the Cat Cycles program. An increase in on-campus housing has reduced the number of vehicles driven to campus daily. UNH also promotes the community's extensive public transportation system with links to regional bus and train services. It is the largest public transit system in New Hampshire. UNH also supports the Amtrak Downeaster rail service, which has a newly renovated campus station, and the GoLoco ride-sharing program.

Environmental practice evaluations are a component of all purchasing considerations. The university's Green Cleaning Program incorporates the use of Green Seal cleaning products, energy- and water-saving processes and machines, and precise dispenser systems. Campus groundskeepers routinely replace synthetic landscaping products with organic versions. The UNH Dining Services Department offers discounts on beverages to customers with reusable mugs; serves local, organic, and Fair Trade food products, including those raised by the UNH Organic Garden Club; and sponsors an annual community dinner showcasing local food. Its pre- and postconsumer food waste composting program diverts over a half million pounds of food waste annually.

Academic, Research, and Student Sustainability Initiatives

The University of New Hampshire offers interdisciplinary campus and community educational and research experiences in many areas of ecological study and related social sciences.

The university's areas of emphasis include climate science and policy, fisheries management and restoration, marine sciences, environmental engineering, organic dairy management, and rural and sustainable community development. Coursework at both the undergraduate and graduate levels covers such diverse sustainability-related areas as climate change science and policy, marine sciences, sustainable engineering, environmental sociology, women's studies, art and theater, cultural heritage, organic agriculture, and public health.

UNH offers sustainability-related degrees through the Department of Natural Resources housed within the College of Life Sciences and Agriculture (COLSA). COLSA was recently reorganized to expand its leading role in the interdisciplinary fields of sustainable food systems, natural resources, and sustainable communities. Undergraduate students may pursue bachelor of science degrees in Community and Environmental Planning, Environmental Conservation Studies, Environmental and Resource Economics, Environmental Sciences, Forestry, Tourism Planning and Development, and Wildlife Ecology. Students in other disciplines may pursue a Sustainable Living minor.

The University of New Hampshire introduced the nation's first interdisciplinary, international dual major in EcoGastronomy in 2008. The EcoGastronomy major combines the fields of sustainable agriculture, hospitality management, and nutrition to prepare students for careers in the area of sustainable foods. The EcoGastronomy major also features an international component wherein all students spend a semester studying at the University of Gastronomic Sciences in Italy.

Sustainability courses often include projects that contribute to campus- and community-based sustainability initiatives. The 400-level Earth Sciences course titled Global Environmental Change requires students to interview university administrators and staff and engage in role-play in order to arrive at campus emission reduction strategies that can be presented to the UNH Energy Task Force. Other student research is a vital component of the Carbon Solutions New England program.

The Sustainability Internship Program matches students with employers like Pax World Funds, Clean Air–Cool Planet, the Mount Washington Resort, the U.S. Forest Service, and others to gain real-world experience in sustainability. Sustainability interns also hold paid positions in various university departments. Sustainability interns and others present their findings through blogs and presentations at the annual Undergraduate Research Conference. Past conference topics have included microbial fuel cells, sustainable community dinners, and wind turbines.

The University of New Hampshire also offers sustainability-oriented graduate programs. Graduate students may pursue a master of science (M.S.) degree in Natural Resources, Resource Administration and Management, Resource Economics, or the EcoQuest Study Abroad Program in New Zealand, offered through a partnership with Natural Resources and the Environment. The university also offers the interdepartmental Cooperative Doctoral Natural Resources and Earth System Science (NRESS) Program. UNH is committed to research in the field of sustainability; according to university administration, over 60 percent of the $99 million external research funding received in fiscal year 2008 was devoted to environmental research topics. UNH houses the nation's first land-grant-college organic dairy research farm.

The university also hosts a number of community outreach programs and centers. UNH offers training and networking opportunities for business professionals in a variety of fields through the Carsey Institute's Sustainable Microenterprise and Development Program (SMDP). The SMPD's professional development training is centered on sustainability program planning, implementation, and evaluation. The New Hampshire Farm to School Program is an outreach initiative helping local elementary and secondary schools provide more local foods in their cafeterias. The New Hampshire Center for a Food Secure Future promotes the creation of a comprehensive local and regional food system with an emphasis on nutrition. UNH also maintains a Cooperative Extension Service.

University of New Hampshire students participate in sustainability initiatives through a variety of campus and community outreach organizations and projects. Students may serve as campus ecological advocates, who raise awareness and encourage sustainable living practices among campus residents, or on the Student Environmental Action Council. Environmentally themed student organizations include the Organic Garden Club, the Slow Food Campus Convivium, Engineers Without Borders, and the UNH Energy Club. Students may also live in the campus Living Green community dedicated to the practice of a sustainable lifestyle. Students participate in the annual Thanksgiving power-down dorm energy challenge and the RecycleMania competition.

The University of New Hampshire showcased its commitment to all aspects of sustainability in its 2009 commencement ceremony. Guests received electronic rather than paper invitations. The ceremony implemented the sustainability theme through speeches and the awarding of honorary degrees, and programs were printed on recycled paper. Leftover food was composted. The university president launched the EcoLine project implementing landfill fuel as the university's primary energy source in his address.

Recognition

In 2007, UNH became the first New England land-grant university to sign the American College and University Presidents Climate Commitment, placing it in the early leadership circle of signatories. UNH has been a member of the Association for the Advancement of Sustainability in Higher Education (AASHE) since the group's inception in 2006 and was one of 90 pilot campuses for its Sustainability Tracking, Assessment & Rating System (STARS). The chief sustainability officer serves on AASHE's Advisory Council. UNH is also a founding member of the Northeast Campus Sustainability Consortium.

The University of New Hampshire has received recognition for its leading role in sustainability education, research, operations, and outreach. UNH received the nation's first Energy Star rating for residence halls in 2006, has been designated a "Best Workplace for Commuters" since 2004, and has won a Success in Enhancing Ridership Award from the Federal Transit Administration.

UNH scientists have served on the 2007 Nobel Prize–winning Intergovernmental Panel on Climate Change and received the 2009 Coastal America Partnership Award from the U.S. president. More than 60 faculty and staff across campus contributed their sustainable experiences to the book *The Sustainable Learning Community: One University's Journey to the Future*, published by the University Press of New England in 2009. The Sustainable Endowments Institute awarded the University of New Hampshire an overall grade of A– on its College Sustainability (Green) Report Card. UNH is nationally recognized as a sustainable learning community.

See Also: Leadership in Sustainability Education; Northeast Campus Sustainability Consortium; Sustainability Websites.

Further Readings

Aber, John, Tom Kelly, and Bruce Mallory, eds. *The Sustainable Learning Community: One University's Journey to the Future.* Lebanon, NH: University Press of New England, 2009.

Barlett, Peggy F. and Geoffrey W. Chase. *Sustainability on Campus: Strategies for Change.* Cambridge, MA: MIT Press, 2004.

M'Gonigle, R. Michael and Justine Starke. *Planet U: Sustaining the World, Reinventing the University.* Gabriola Island, British Columbia, Canada: New Society Publishers, 2006.

Rappaport, Ann and Sarah Hammond Creighton. *Degrees That Matter: Climate Change and the University.* Cambridge, MA: MIT Press, 2007.

Sustainable Endowments Institute. "The College Sustainability (Green) Report Card." http://www.greenreportcard.org/report-card-2010/schools/university-of-new-hampshire (Accessed June 2010).

University of New Hampshire. "University Office of Sustainability." http://www.sustainable unh.unh.edu (Accessed June 2010).

Marcella Bush Trevino
Barry University

UNIVERSITY OF OREGON

The design of the University of Oregon's Autzen Stadium includes sustainable features such as a grass parking lot, preserved wetlands on the grounds, and recycling of water runoff.

Source: Wikipedia

The University of Oregon (UO) is a public research university in Eugene, Oregon. The state's second-oldest public university, it was founded in 1876 and almost closed five years later due to its debts; today, much of its funding comes from the independent not-for-profit UO Foundation. UO is Oregon's only member of the Association of Research Libraries and hosts Scholars Bank, an open access digital repository for the creations of the faculty, students, and staff of the university. The university also operates the Riverfront Research Park on the banks of the Willamette River, across from the main campus. The research park, which has come under criticism by Oregonians because of the construction that displaced open civic space and wildlife habitats, is used for new technology research such as artificial intelligence.

The Sustainable Development Plan (SDP) was adopted in 2000, requiring that sustainable design principles be applied to all new development and remodeling projects on UO campuses. The SDP actually followed the sustainable design of several of UO's buildings, including Autzen Stadium, which was designed for energy efficiency and has a grass parking lot that reduces the amount of airborne dust and eliminates the need for asphalt. Runoff from the asphalt-paved areas of the parking lot—required for handicapped parking—is retained on-site, as is most runoff from the field and seating areas. Part of the stadium grounds have been delineated as a wetland and preserved accordingly. Both the Student Recreation Center and the Knight Law Center meet and exceed Oregon's State Energy Efficient Design requirements and were given the Energy Smart Award.

UO has been recycling all paper waste from UO Printing Services since 1975. It also has an award-winning Campus Recycling Program, one of the first in the United States, which began in 1989 as a student initiative. Elliptical workout machines in the student recreation center capture kinetic energy for the university's power grid—50 watts per 30-minute workout. The building's power usage is also offset by a student-funded photo-voltaic system on the gymnasium roof. The UO began a Bike Loan Program in 2008, and students ride free on the Lane Transit District buses, including a low-emission, hybrid-electric EmX line. Furthermore, all campus grounds are maintained using an integrated pest management program, with plant selection informed by shade provision, wildlife habitats, and a Campus Tree Plan that emphasizes the importance of trees on campus and sets forth specific policies for tree planting and for the impact of new construction on the campus tree population.

UO operates the Sustainable Cities Initiative (SCI), an interdisciplinary outreach program that promotes the design, development, and public support of sustainable cities, especially in Oregon. SCI approaches sustainability at scales from the individual building to the whole of a region and grounds its work in interdisciplinary discussion, especially among engineers, scientists, and policy makers. It seeks to expand research into sustainable cities and to promote the results of that research to scientists, policy makers, community leaders, and investors in order to shift sustainable ideas from discussion to reality. Along the way, SCI is involved in creating and promoting academic courses and certificates and in encouraging student and faculty interest in sustainability regardless of discipline. Specific SCI goals include conducting interdisciplinary research to meet goals at various scales of sustainable city design, from the local to the national; providing technical assistance on sustainable city design to the state of Oregon; attracting the nation's best students interested in sustainable cities in order to make Oregon a leader in sustainability; and engaging the nation's experts in an exchange of ideas and ongoing discourse on sustainable city design.

The SCI faculty is drawn from a variety of departments, including Architecture, PPPM (Planning, Public Policy, and Management), the law school, and the Lundquist College of Business.

Research Projects

SCI research projects include Overlooked Density: Rethinking Transportation Options in Suburbia, Linking Experiential Learning to Community Transportation Planning; and Increasing Capacity in Rural Communities: Planning for Alternative Transportation. The Healthy Communities and Urban Design project is a multidisciplinary national analysis of travel behavior, residential preference, and urban design studying the use of active transportation (walking and bicycling) according to neighborhood type and sociodemographic variables and the impact of active transportation choices on obesity, livability, carbon emissions, and other aspects of the local community. Active transportation and access to transportation options other than car ownership (including Zipcars, public buses and trains, and bicycle paths) are common areas of research conducted by SCI.

The Sustainable Cities Curriculum designed by SCI focuses on three complementary areas of study: ecological design, sustainable urban design, and sustainable land development. The curriculum is designed to integrate design, planning, public policy, and ecology in a way that transcends individual disciplines and is applicable at all scales. Interdisciplinary classes are offered in those three areas as well as public policy analysis, city planning, applied service learning, transportation and livability, and green architecture. Furthermore, a key component of SCI's educational initiative is its "professional reeducation" offerings through UO's Division of Continuing Education's Sustainability Leadership Academy, which offers workshops and residencies in Eugene and Portland for private and public sector officials and staff to help train them in sustainability issues. Launched in 2004, the Sustainability Leadership program focuses on the core competencies of sustainability, and rather than limiting itself to one or two one-size-fits-all courses, offers a range of courses, including a full track for local government officials. In collaboration with the League of Oregon Cities, Sustainability Leadership seminars are offered across the state for local government officials and city planners.

The Sustainable Cities Year is offered by SCI every year. Each year, the university partners with a different city in Oregon, with a number of the university's courses focusing on faculty and students working with officials and citizens from that city to identify, develop, and achieve sustainability goals. This hands-on learning provides students with a sense of how working toward sustainability goals works in the real world while also helping to better the Oregon community. The 2009–2010 Sustainable City was Gresham; the 2010–2011 city, Salem. Gresham-associated courses included Sustainable Suburbs, analyzing the suburban development surrounding Gresham (the fourth-largest city in Oregon), its surrounding residential areas, and the commercial development in the area; Crossings Site, focusing on the mixed-use development at the Crossings Site on Civic Drive in Gresham, with architecture students working in teams to identify synergies and conflicts between policies applicable in the area, with the goal of helping Gresham become its own city rather than a bedroom community; Specific Area Sustainable Development Concept Plan, a three-course PPPM project examining land use in the area; Sustainability Focused Data Analysis; Citizen Engagement, teaching students elements of community change by conducting surveys and observing the community; Gresham Quarry Site Analysis/Design, studying the soil, hydrology, history, and ecology of the Gresham quarry; Green Cities, studying city planning documents to make suggestions for net zero developments; Shaping Light, working with digital models and fabrication tools to retrofit the west-facing offices of city hall to enhance daylight; two different Gresham City Hall design courses developing proposals for a new, sustainably designed city hall building; and Public Folklore and Cultural Programming, studying the cultural groupings and community goals of Gresham. Over 70 students were involved in the first semester of the Sustainable Cities courses, meeting regularly with Gresham city planners and councilors.

The 2010–2011 Sustainable Cities program in Salem will address projects such as bicycle facilities, park development, inventory of energy data throughout the community, redeveloping certain areas of the city, restoring the Minto Island area, redeveloping downtown to manage parking needs, redeveloping the waterfront area, and addressing the energy efficiency needs of the Civic Center. The Architecture and Allied Arts (AAA) department operates a number of other similar service-learning programs through the

Community Service Center and designBridge, a UO student design organization that links AAA with the off-campus community through design development services, design projects, and weekend meetings. AAA is one of UO's six professional schools, with programs at both the Eugene and the Portland campus, and offers both undergraduate and graduate degrees in architecture, art, art history, arts administration, digital arts, historic preservation, interior architecture, landscape architecture, and PPPM, as well as graduate certificates in ecological design, nonprofit management, museum studies, and technical teaching in architecture.

The Institute for a Sustainable Environment (ISE) is located at UO in Eugene, and is a research center for innovative, interdisciplinary research on ecological, economic, and social sustainability. The university's oldest sustainability initiative, ISE is a center for collaborative research, policy education, and technical assistance, with the goal of developing sustainable communities throughout the Pacific Northwest and beyond. The information produced by the ISE is intended to assist in the resolution of difficult policy problems and private sector issues. ISE faculty includes political scientists, psychologists, environmental scientists, and others. Ongoing research programs include the Climate Leadership Initiative, which develops public education on climate change and strategies to promote public awareness of climate change issues and realities; the Social Acceptability of Forestry Practices, which assesses the public perception of various timber-harvesting methods; the Ecosystem Workforce Program, which aims to turn the rural Pacific Northwest into healthy communities with healthy environments; and the ISE Geographic Information Systems Lab, which analyzes patterns of landscape change to promote habitat conservation and water quality. Projects have included reducing wildfire risks in Curry County; creating jobs to restore the forest in northwest Oregon; examining the potential of oak savanna restoration to protect communities from wildfire; studying the rapid population growth impacts in the Willamette Valley; helping to create jobs in natural resource management in poor, rural communities; working with the U.S. Forest Service to implement tribal monitoring of government-to-government relationships; and assisting the city of Damascus with developing a comprehensive plan to conserve water quality and wildlife habitats.

See Also: Experiential Education; Sustainability Teacher Training; Whole-School Approaches to Sustainability.

Further Readings

Filho, Walter Leal. *Sustainability and University Life.* Ann Arbor: University of Michigan Press, 1999.

Hargreaves, Andy and Dean Fink. *Sustainable Leadership.* Hoboken, NJ: Jossey-Bass, 2005.

Sandell, Klas, Johan Ohman and Leif Ostman. *Education for Sustainable Development.* New York: Studentlitteratur AB, 2005.

Silka, Linda and Robert Forrant, eds. *Inside and Out: Universities and Education for Sustainable Development.* Amityville, NY: Baywood Publishing, 2006.

Bill Kte'pi
Independent Scholar

UNIVERSITY OF PENNSYLVANIA

The University of Pennsylvania (Penn), a private university located in Philadelphia, Pennsylvania, is one of the oldest universities in the United States. The organization of Penn was informed by the vision of Benjamin Franklin, one of Penn's founders, of a university that would prepare students for careers in business and public service rather than providing only education for the clergy. The predecessor of Penn, the Academy and Charity School of Philadelphia, began operation in 1751, and was the first American university to offer a modern liberal arts curriculum. In fall 2009, Penn enrolled 20,643 full-time students, including 10,206 graduate and professional students, and employed 4,127 faculty members, over 1,000 postdoctoral fellows, and over 5,400 academic support staff and graduate assistants. Dr. Amy Gutmann has been the president of the University of Pennsylvania since 2004 and is the third woman in succession to hold this position, after Dr. Clair Fagin (interim president, 1993–1994) and Dr. Judith Rodin (1994–2004; the first woman to hold the permanent position of president of an Ivy League university). Penn is a leading research university with an annual research budget of over $700 million, and Penn graduate schools are regularly ranked among the best in the United States; for instance, *US News & World Report* ranked Penn's medical school second in the nation, the Wharton School of Business fifth, the law school seventh, and many other programs in Education, Engineering, and Arts and Sciences in the top 20 nationally.

The Green Campus Partnership

In 2007, Penn's president, Amy Gutmann, signed the American College and University Presidents Climate Commitment, which committed Penn to developing plans to reduce its emission of greenhouse gases. The Green Campus Partnership at Penn was created in that same year as an umbrella organization to coordinate sustainability programs and initiatives and to facilitate campus planning and policy development relating to sustainability. The Green Campus Partnership includes Business Services, Facilities and Real Estate Services, the Environmental Sustainability Advisory Committee (ESAC), and several faculty and student groups.

The ESAC is an organization of students, staff, and faculty that advises the university president on issues relating to environmental sustainability. ESAC is composed of six subcommittees—Academics, Built Environment, Energy and Utilities, Waste Management and Recycling, Transportation, and Communications—whose recommendations formed the basis of Penn's Climate Action Plan.

The Green Campus Partnership offers many opportunities for Penn faculty, students, and staff to become involved in the campus sustainability efforts. The Eco-Reps program trains students, faculty, and staff to act as environmental leaders within their departments, offices, college houses (undergraduate residences), and sororities and fraternities for development campaigns, activities, and educational events. Several students groups are also involved in campus sustainability initiatives. The Penn Environmental Group (PEG) aims to increase campus awareness of global environmental issues, both at Penn and in the world at large. PennGreen provides an environmental orientation program to teach freshmen about campus recycling and LEED (Leadership in Energy and Environmental Design) building projects at Penn and offers the opportunity to learn about other opportunities, including participation in the sustainable food and living movement and volunteering at local urban farms. Farmecology is dedicated to raising awareness about the benefits of eating sustainably and coordinates a local food buying club as well as consulting with on-campus food outlets to encourage purchase of local food. CommuniTech is an outreach program run by students from the School of Engineering & Applied Sciences that provides computers to Philadelphia residents and offers education in computer use and information technology. Penn's two elected student bodies, the Undergraduate Assembly and the Graduate and Professional Student Assembly, also work on sustainability initiatives, and the Graduate and Professional Student Assembly representatives are part of the ESAC.

Climate Action Plan

The Penn Climate Action Plan, published in 2009, outlines the steps that the university will take to lower greenhouse gas emissions and to monitor progress toward this goal. The changes outlined fall into four categories: Utilities and Operations, Physical Environment, Transportation, and Waste Minimalization and Recycling. In the area of utilities and operations, Penn's primary goal is to reduce energy usage by 5 percent (using energy usage in 2007 as the baseline) by 2010 and 17 percent by 2014. Strategies to achieve this goal include education and management to reduce energy use in existing buildings; adopting conservation measures, including metering and incentives for better performance; and continuing to purchase renewable energy credits. In the area of physical environment, Penn's goals are to increase green space, decrease energy consumption, and increase aware-ness of sustainable design with the overall goal of creating and maintaining a sustainable campus. Specific strategies intended to help reach this goal include using LEED Silver certification as a standard for major renovations and new construction, training staff in sustainable design and construction practices, and implementing sustainable protocols for site planning and landscape maintenance.

In the area of transportation, Penn's overall goals are to provide a pedestrian-friendly campus environment, encourage use of public transportation, and provide efficient local transportation for the university community. Strategies to achieve these goals include investigating the possibility of partnering with SEPTA (Southeastern Pennsylvania Public Transportation Authority, an agency that provides bus and rail transportation services in the Philadelphia area), improving the environment for pedestrians and cyclists, and improving the efficiency of the university's vehicle fleet. In the area of waste minimization and recycling, Penn's goals are to reduce the overall waste stream and increase the diver-sion rate of paper, cardboard, and commingled recyclables to 40 percent by 2014. Strategies to reach this goal include instituting a comprehensive policy regarding recycling and waste minimization, providing education about recycling, and ensuring that campus buildings and public spaces have adequate recycling and waste bins.

Besides these concrete behaviors to reduce waste and increase sustainability, the Climate Action Plan includes goals in the areas of Academics and Communications. The primary goal for Academics is to make climate change and environmental sustainability part of the educational experience of all Penn students. Specific strategies include the creation of an undergraduate minor in Sustainability and Environmental Management beginning in fall 2009; provision of sustainability-related workshops and proseminar classes for faculty, staff, and students; and expansion of student participation in sustainability research. In the area of communications, Penn's goals are to provide clear, concise, and accurate informa-tion about Penn's commitment to sustainability and to encourage the Penn community to continue learning about sustainability. Strategies in support of these goals include delivering messages on the importance of individual behavior in meeting these goals; ensuring that all communications are accurate, accessible, valuable, and up to date; and creating events to publicize the sustainability campaign and to encourage the campus community to support these goals.

Green Campus Initiatives

In 2002, Penn was honored by both the U.S. Environmental Protection Agency and the Pennsylvania Department of Environmental Protection for its commitment to using energy from alternative sources, in recognition of Penn's commitment in 2001 to purchase 20 million kilowatt hours of wind-generated power annually for three years, a commitment extended to 10 years in 2003. The Operations Command Center at Penn controls air handling, chilled water, and steam utilities across campus and saves the university over $5 million annually in energy costs by implementing central monitoring and control, which enables it to conserve energy and avoid peak utility charges. Penn has a program, initiated in 2007, to replace incandescent light bulbs with more efficient compact fluorescent bulbs in on-campus residences; over 1,000 bulbs were exchanged in the first year alone.

Besides serving over 20,000 students, Penn is the largest private employer in greater Philadelphia and recognizes the importance of providing green methods for staff to commute to work. The Penn campus is compact and encourages walking; bicycle parking is also available on campus. Penn provides discounts and pre-tax benefits for faculty and staff members who use local public transportation, and students are offered a semester-based transit pass program. Some parking garages on campus offer charging stations for electric cars, and in 2008, Penn partnered with PhillyCarShare to become the largest car-share partnership in North America. This program allows students, faculty, and staff to rent cars by the hour and also allows Penn to rent the cars at a discounted rate, resulting in fewer cars purchased by individuals as well as a reduction in the size of the university fleet.

Penn Facilities and Real Estate Services is responsible for collecting traditional recyclables on campus; as of 2010, Penn recycles about 23 percent of its total waste and has diverted over 1,500 tons of waste annually from landfills since 2007. Penn has also installed solar-powered trash compactors at 10 high-traffic locations on the campus perimeter: each can hold four times the waste of a noncompacting bin, thus reducing the use of fuel by reducing the number of trips required for garbage pickup. Computers and electronic waste are disposed of through the local firm Elemental, which emphasizes reuse of materials as well as safe disposal of toxics; in 2009, Penn recycled over 122,000 pounds of electronic waste. Penn recently launched a program to recycle writing instruments (e.g., pens, markers, and highlighters) in partnership with Sandford, Office Depot, and Terracycle. Penn dining halls do not use trays, reducing food waste as well as eliminating the water and chemicals used to wash trays; other green measures in the dining halls include the use of compostable takeout containers and recycling cooking oil.

Penn offers several academic and research programs that focus on sustainability. An undergraduate minor in Sustainability and Environmental Management is available to students in any of the four undergraduate schools and provides a basis for interdisciplinary study in environmental issues. The Wharton School of Business also offers an undergraduate concentration in Environmental Policy and Management and an undergraduate minor in Organizations and Environmental Management. At the graduate level, Penn offers a Master of Environmental Studies degree through the College of Liberal and Professional

Studies, which offers concentrations in Environmental Advocacy and Education, Environmental Biology, Environmental Health, Environmental Policy, Resource Management, and Urban Environment as well as an individualized option for students whose interests do not fall into any of the available concentrations. The School of Medicine includes a Center of Excellence in Environmental Toxicology that aims to work with local communities and government decision makers to reduce environmental exposure to toxins and generally improve environmental health and justice. The Wharton School of Business offers a Master of Business Administration degree in Environmental and Risk Management, which focuses on designing and implementing strategies related to the health and safety impacts of business activities on the environment. The Penn-Tsinghua T. C. Chan Center, a cooperative program with Tsinghua University in Beijing, China, focuses on methods for developing energy-efficient buildings and sustainable environments.

See Also: Green Business Education; Leadership in Sustainability Education; Sustainability Officers.

Further Readings

American College and University Presidents Climate Commitment. http://www.presidents climatecommitment.org (Accessed May 2010).

Sierra Club. "Cool Schools: The Third Annual List." http://www.sierraclub.org/sierra/200909/ coolschools/allrankings.aspx (Accessed May 2010).

University of Pennsylvania, Climate Action Plan (September 15, 2009). http://www.upenn .edu/sustainability/Climate%20Action%20Plan%20FINAL.pdf (Accessed May 2010).

University of Pennsylvania, Penn Green Campus Partnership. http://ww.upenn.edu/ sustainability (Accessed May 2010).

Sarah Boslaugh
Washington University in St. Louis

University of Texas, Austin (Alley Flat Initiative)

The Alley Flat Initiative (AFI) is a collaborative project of the University of Texas, Austin (UT) Center for Sustainable Development, the Austin Community Design and Development Center, and the Guadalupe Neighborhood Development Corporation, which seeks to develop affordable housing alternatives for the city of Austin's low-income residents that also incorporate green and sustainable technologies. The AFI is an important component of the UT Campus Sustainability Policy that seeks to integrate sustainability throughout all components of the UT's mission. The AFI grew out of the University of Texas Sustainable Design and Development Workshop (SDDW). The program builds small, detached residential structures known as "alley flats" in Austin's underutilized alleys, targeting lots already owned by longtime neighborhood residents. The AFI completed two prototype alley flats and has 10 more in the planning, design, or construction phases.

A model depicting an Alley Flat house built behind another home in an Austin, Texas, alley. The Alley Flat Initiative, a collaboration of the University of Texas, Austin, and other local organizations, estimates that there may be 3,000 potential sites for the sustainable structures in the city.

Source: Flickr/Cubby_T_Bear

Program Goals

The Alley Flat Initiative seeks to develop affordable housing alternatives for Austin's low-income residents. The AFI believes its affordable housing alternative can thus help in the mitigation and reversal of long-term urban trends such as gentrification, in-filled neighborhoods, rising property costs and taxes, and environmental degradation that challenge the maintenance of the diversity and stability of Austin's urban neighborhoods as well as those across the nation and around the world. The project also seeks to create secondary housing that is environmentally sustainable, increases population density, and maintains the existing character of Austin's urban neighborhoods. Long-term goals include a voice in Austin's comprehensive planning process, public dialogue on affordable housing options, and the development of efficient housing designs that incorporate sustainable technologies and features and innovative financing methods.

AFI coordinator Sarah Gamble has stated that there are approximately 3,000 potential alley flat locations within the city. Gamble also noted that a 2008 City of Austin Neighborhood Housing and Community Development Office comprehensive housing market study found that there is a shortage of affordable housing options in Austin based on U.S. government guidelines stating that affordable housing should consume no more than 30 percent of income, including rent or mortgage payment and utilities. AFI also notes that many Austin residents in the project's targeted neighborhoods support the opportunity to obtain a secondary residential unit on their existing property lots. Target residents include students, the elderly living on fixed incomes, single parents, and low-income workers.

The AFI assists in the planning and construction of small, affordable, detached secondary residential units known as alley flats on existing but underutilized property lots with an existing primary residential structure that covers less than 25 percent of the lot. Benefits of the alley flat design include sustainable features, no additional burden to the existing primary residential structure, the distribution of land costs over two structures, and the maintenance of extended-family units. Partner organizations assist participating residents in the planning, design, and construction process and monitor the success of completed alley flats.

Program Management

The Alley Flat Initiative is part of the University of Texas Center for Sustainable Development (CSD). The CSD oversees all university sustainability-related research, education, and community outreach programs. The CSD assists the AFI through its research and policy work in the fields of affordable housing, secondary apartment development, and distributed infrastructure. Key Alley Flat Initiative participants include cofounder and AFI director Dr. Steven Moore, architect and AFI coordinator Sarah Gamble, AFI research and development member Barbara Wilson, AFI collaborators Mark Rogers and Michael Gatto, and AFI studio instructor Louise Harpman.

The CSD also facilitates connections between the AFI and the University of Texas School of Architecture and other members of the university community. The UT School of Architecture offers both undergraduate and graduate sustainable design degree programs, including the Master in Architecture I program designed for those graduate students without architecture degrees, the Master in Architecture II or postprofessional program designed for those with an undergraduate professional architecture degree, and the Master of Science in Sustainable Design program designed for students seeking Ph.D. degrees or employment in the areas of research, activism, or public policy.

The Center for Sustainable Development works in partnership with the Austin Community Design and Development Center (ACDDC) and the Guadalupe Neighborhood Development Corporation (GNDC) to run the AFI. The ACDDC provides sustainable design services to those families and neighborhoods considered low or moderate income. The GNDC provides both new and rehabilitated affordable housing alternatives or funding to low- and moderate-income residents of central East Austin neighborhoods. Both the ACDDC and the GNDC are 501(c)(3) nonprofit organizations.

Program Development

The Alley Flat Initiative began as a University of Texas workshop project. Professor Steven Moore recruited Professor Sergio Palleroni of the University of Washington to head a University of Texas Sustainable Design and Development Workshop (SDDW) funded through a grant from the Henry Luce Foundation. The SDDW partnered with the Blackland Community Development Corporation and the GNDC for a shared interest in the creation of affordable housing·alternatives in Austin. SDDW students conducted the necessary research and developed the project's first prototypes on an underutilized network of alleys with attached oversized residential lots in East Austin.

The SDDW students' first step, which was completed in the project's first year, consisted of interviewing Austin residents and exploring its neighborhoods in order to target suitable development locations for the project. The East Austin alleyways and attached lots were chosen due to Austin's neglect of the neighborhood due to budget constraints—the alleyways were overgrown and in an abandoned state—and for the large size of the unoccupied portions of the nearby residential lots and favorable zoning. They also learned that the Guadalupe Neighborhood Plan advocated the types of affordable housing the project envisioned. The students also explored the success of similar nationwide affordable housing initiatives that had been developed in cities such as Santa Monica, California; Boulder, Colorado; and Seattle and Portland, Washington.

The SDDW students' second step, begun the second year of the project, was the development of the preliminary concept of the alley flat and the design of several prototypes featuring state-of-the-art environmental systems. The students were able to utilize the research work of other design projects housed at the University of Texas, including the Civic Environmentalism Workshop, the East Austin Trapezoid study, and the UT Solar Decathlon house. The SDDW also found two residents of the targeted East Austin neighborhood willing to house the first alley flat prototypes on their lots.

Program Implementation and Outcomes

The Alley Flat Initiative's first goal was the creation of two prototype structures in East Austin neighborhoods on lots owned by longtime neighborhood residents. The AFI completed its first prototype building, known as Alley Flat Number One, in 2008 on an Austin lot located on Second Street that already contained a single-family residence. A community-wide housewarming celebration was held upon its June completion. The AFI completed its second prototype building, known as Alley Flat Number Two, in August the following year. Alley Flat Number Two was located on Lydia Street on the site of a property owned by the GNDC, one of AFI's partner community development organizations.

These two prototype residences served as showcases for the AFI's innovative design, which combined sustainable technologies with affordability and the ability to meet the needs of its occupants. The sustainable features used in the prototype residences included PV solar panel arrays, a tankless water heater, rainwater collection barrels to catch storm runoff, an energy-efficient heating and cooling system, and clerestory windows that provide cross-ventilation and natural daylight. The second home was also wheelchair accessible. Both homes were initially used to house extended-family members, although the residential owners of Alley Flat Number One later converted it into a rental property for an extra source of income.

There are 10 additional Alley Flats in various stages of the design, planning, and construction process as of 2010. They are Alley Flat Number Three, located on Willow Street; Alley Flat Number Four, located on East Cesar Chavez Street; Alley Flat Number Five, located on Clermont Street; Alley Flat Number Six, located on Gonzales Street; Alley Flat Number Seven, located on Providence Street; Alley Flat Number Eight, located on Johanna Street; Alley Flats Numbers Nine and 10, located on East Thirteenth Street; Alley Flat Number 11, located on Jobe Street; and Alley Flat Number 12, located on West Tenth Street. Construction on Alley Flats Numbers Three through Seven was in development as of 2010.

The ACDDC and the GNDC are long-standing partners in the Alley Flat Initiative. The ACDDC contributes project management, including such aid as site feasibility studies, the recruitment of designers willing to donate their time, completion of the SMART housing application, and assistance to lot owners in bidding and selecting a licensed general contractor for each alley flat project. The GNDC has been active in the targeted neighborhoods over a long period, establishing community relationships key to gaining neighborhood support for the project. The AFI can also draw upon the GNDC's prior experience in urban affordable housing development. The GNDC owns Alley Flat Number Two and is developing Alley Flats Numbers Three through Six.

The Alley Flat Initiative recruits new participants and funding sponsors through community outreach programs. An AFI exhibit for the general public was held at Austin City Hall from March 26 through April 9, 2010. The exhibit showcased the AFI's work as a viable affordable and sustainable housing alternative for Austin's urban neighborhoods. The Alley Flat Initiative maintains a Website containing an instructional guide for those interested in applying to become part of the program. The Website includes instructions on eligibility guidelines, the planning and design process, construction, and residing in the alley flat. The Alley Flat Initiative plays a key role in the development of affordable housing in the Austin area and serves as a model for similar initiatives. The AFI received the 2009 Envision Central Texas (ECT) Community Stewardship Award for Redevelopment.

See Also: Experiential Education; Social Learning; Sustainability Websites.

Further Readings

The Alley Flat Initiative. http://www.thealleyflatinitiative.org (Accessed May 2011).

Barlett, Peggy F. and Geoffrey W. Chase. *Sustainability on Campus: Stories and Strategies for Change.* Cambridge, MA: MIT Press, 2004.

Morgan, Sarah. "The Alley Flat Initiative: What Fits in Your Backyard?" http://rareaustin .com/?p=3945 (Accessed June 2010).

Rappaport, Ann and Sarah Hammond Creighton. *Degrees That Matter: Climate Change and the University.* Cambridge, MA: MIT Press, 2007.

Marcella Bush Trevino
Barry University

UNIVERSITY OF TOKYO

Gardens and a museum on the University of Tokyo campus. The campus produces about 136,000 tons of carbon dioxide every year, and the university has set a goal of reducing CO_2 emissions by 50 percent by 2030, largely by increasing the use of energy-saving equipment.

Source: Wikipedia

The University of Tokyo (also known as Todai, an abbreviation of *todai daigaku*) was established in 1877 by the Meiji government by merging two governmental schools that specialized in medicine and Western studies. The purpose of this merger was to create an elite educational institution to help Japan compete with Western countries. Todai is the most prestigious university in Japan, and has long been regarded as a gateway to leadership positions, particularly in the Japanese government. It also is highly ranked in surveys of international universities: Todai was ranked 19th in the Times Higher Education list in 2008, the highest ranking of any Asian university, and 20th in the 2009 Academic Ranking of World Universities produced by Shanghai Jiao Tong University. Originally admitting only male students, Todai became coeducational after World War II, also introducing the current system of undergraduate and graduate education at that time. Currently, Todai has about 30,000 students on three campuses in Komaga, Hongo, and Kashiwa; about 20 percent of the students are female and about 15 percent are foreign (most commonly from China and Europe). Junichi Hamada, Ph.D., is the president of the University of Tokyo.

Todai is organized into 10 faculties: law, medicine, engineering, letters, science, agriculture, economics, arts and sciences, education, and pharmaceutical sciences. The university offers graduate study in 15 fields, and is home to a number of research institutes, including the Institute of Medical Science, the Earthquake Research Institute, the Institute of Advanced Studies on Asia, the Ocean Research Institute, and the Research Center for Advanced Science and Technology. Several university-wide centers at Todai provide cross-disciplinary training, services, and research: these include the Environmental Science Center (responsible for implementing environmental safety measures, conducting research and offering education on environmental conversation, and supervising use and disposal of hazardous materials), the Asian Natural Environmental Science Center (responsible for international cooperative research in the fields of sustainability and environmental conservation), the Center for Climate System Research (responsible for developing climate models and conducting experiments to learn the mechanisms of climate variation) and the Biotechnology Research Center (responsible for promoting research and education in biotechnology).

The Todai Sustainable Campus Project (TSCP) is devoted to reducing greenhouse gas emissions and increasing sustainability on campus with the goals of not only reducing carbon output on the Todai campus, but also creating a model for this process that can be applied to other institutions. Currently, Todai produces about 136,000 tons of carbon dioxide (CO_2) per year, about 0.01 percent of the total for Japan, primarily through electricity usage (79 percent) and gas usage (19 percent). The university has set the goal of reducing CO_2 emissions by 15 percent in the first phase of this project (between fiscal year [FY] 2008 and FY 2010) by means such as installing power meters (to provide visual feedback on energy use) and switching to energy-saving equipment, including heat sources, lighting, air conditioners, and refrigerators. In the second phase, to take place between FY 2012 and FY 2030, the project aims to reduce carbon dioxide (CO_2) by a total of 50 percent from 2008 levels through further replacement of old equipment to more energy-efficient models, introducing new technologies, and creating energy (e.g., generating power from photovoltaic cells). The TSCP also includes financial support for research and practical activities to encourage faculty, staff, and students to become involved in the project, and collaboration with other universities in Japan and abroad to encourage widespread use of technologies and practices that will lead to a reduction in greenhouse gas emissions across society.

The Integrated Research System for Sustainability Science

The Integrated Research System for Sustainability Science (IR3S) is a new program at Todai in sustainability science, an academic field that is devoted to the study of linkages among global, social, and human systems, and seeks to understand the risks to human well-being and security caused by the interaction among those systems. IR3S consists of Planning and Administrative Headquarters, research centers located at five member universities, and research centers at six cooperating institutions. The Planning and Administrative Headquarters is charged with unifying and coordinating research from the participating universities and cooperating institutions and will decide basic research and education policy and implementation plans.

The five member universities involved in IR3S are the University of Tokyo, Osaka University, Hokkaido University, Kyoto University, and Ibaraki University; each has particular foci that contribute to the whole of IR3S efforts. The Transdisciplinary Initiative for Global Sustainability at the University of Tokyo aims to create a transdisciplinary sustainability science and conducts research in five areas: foundational studies incorporating satellite-based remote sensing, geographic information systems, and supercomputer modeling of climate and ecosystems; research into global warming, wide-area atmospheric pollution, and the relationship between resources and energy; research into sustainable water use and food production in the face of rapid population increase; research into the creation of optimal ecosystems in Asia's rapidly growing giant metropolises; and research into environmental risk management with consideration of cultural and lifestyle differences, international politics, and health issues.

The Kyoto Sustainability Initiative is focused on socioeconomic system reform, technology strategy, and creation of a flexible research and educational system. Particular goals for this initiative include developing optimal recycling systems, creating environmental policy based on economic and technical analysis, developing an ethical system to ensure cross-generational and intragenerational fairness in sustainability, and developing environmental governance that effectively manages risks. The Research Institution for Sustainability Science at Osaka University is a collaborative organization that focuses on designing systems for recycling and ultra-efficient production technologies. The Sustainability Governance Project at Hokkaido University focuses on issues specific to its northern location, including a subarctic waters environmental change system, and a systems model to cover the entire Northern Pacific Region. The Institute for Global Change Adaptation Science at Ibaraki University is a transdisciplinary research center that focuses on global warming and climate change, particularly in the Asia-Pacific Region.

The cooperating institutions, each of which has a particular focus or set of foci, are Toyo University, the National Institute for Environmental Studies, Tohoku University, the Chiba Association for Regional Sustainability Science, the Waseda Initiative on Sustainability Science for Political Decision-Making and Journalism, and the Ritsumeikan University. Toyo University's particular focus is on ecophilosophy, harmonious coexistence between people and the environment as well as between current and future generations, the philosophy of environmental design and ethics, and the application of eastern approaches to nature and ecology. The National Institute for Environmental Studies is focused on global warming and seeks to collaborate with researchers from China, India, South Korea, and Thailand to study the relationship between climate change and sustainability, and establish a global strategy to deal with environmental warming issues. Tohoku University is focused on health risk management and regional sustainability science, and aims to quantify benchmarks for human development, as well as proposing policies for sustainable development in Asia and Oceania. The Chiba Association for Regional Sustainability Science focuses on food resources and crop science and how sources of food and medicine may become more sustainable. The Waseda Initiative focuses on journalism and political decision making, including how journalism acts as an intermediary between politics and the public and how public opinion can affect political decision making. The Ritsumeikan University Research Center for Sustainability Science focuses on how East Asian countries can develop joint strategies to correct imbalances arising from economic development.

The IR3S has three flagship research projects: Sustainable Countermeasures for Global Warming, Development of an Asian Recycling-Oriented Society, and the Concept and Development of Global Sustainability—Reform of the Socioeconomic System and the Role of Science and Technology. Sustainable Countermeasures for Global Warming is aimed at developing a unified approach to predicting and measuring global warming as well as developing adaptation measures. This project requires interdisciplinary cooperation to create a comprehensive evaluation of global warming and to develop countermeasures and evaluate their effectiveness. Development of an Asian Recycling-Oriented Society takes the view that recycling is no longer simply a domestic issue for Japan alone, but must become an expectation throughout Asia, including China, which has become a major hub of international production. This project envisions recycling as consisting of three axes: one of production and consumption; one of the environment (including land, materials, and biology), which must support this production and consumption; and a third that incorporates geographic and temporal factors. The Concept and Development of Global Sustainability— Reform of the Socioeconomic System and the Role of Science and Technology recognizes that an environmental protection approach to global environmental problems is no longer sufficient, and that consideration must be given to the expected interactions in the future of factors, such as projected economic growth, expected development of science and technology, and natural disasters.

The Alliance for Global Sustainability

The Alliance for Global Sustainability (AGS) is an international partnership between the University of Tokyo, the Swiss Federal Institution of Technology in Zürich (ETH Zürich or the *Eidgenössische Technische Hochschule Zürich*), the Massachusetts Institute of Technology in the United States, and the Chalmers University of Technology in Sweden. AGS was founded in 1997 to contribute to developing capacity for integrated research, education, and outreach into sustainability, and to provide the global research agenda to create the scientific knowledge necessary for global sustainability. Each partner has a strong local program to build networks and encourage participation of faculty and staff in AGS activities, making AGS a sort of "network of networks" across the universities. Professor Akimasa Sumi of the University of Tokyo is the 2010 chair of the AGS Executive Board and Junichi Hamada, president of the University of Tokyo, is chair of the AGS Alliance Governing Board, the highest policy-making body within AGS.

The AGS has four research initiatives. Sustainability under Rapid Demographic Change addresses problems that arise with the changing demographic structure of many developed countries in which the population is living longer but having fewer children, and in which citizens must make changes in their current high consumption of natural resources. The Urban Futures Initiative addresses rapid increases in urbanization that are currently taking place around the world, and the implications for sustainability of this rapid shift from rural to urban populations. Energy Pathways attempts to identify energy pathways that bridge technologies, infrastructures, and markets currently available with those that will be required in the future to achieve sustainability, and to communicate this information to those who make policy decisions. The Food and Water program concentrates on creating food and water security necessary for human well-being, and conducting education and outreach to bring about the necessary changes in the current systems of food production and water access. The focus in this initiative is both on the immediate future (5 to 15 years) and the long term, and includes identifying development paths that can ensure sufficient food and water while also maintaining ecological balance.

The AGS 2010 annual meeting was held in Tokyo, March 16–19. Primary topics addressed at this meeting were mitigation and adaptation strategies for climate and demographic change, developing sustainable city regions, developing energy pathways for the future including smart grids and smart infrastructure, and facilitating information exchange and communication between academics and society. Plenary session and keynote lectures included "The Right to Sustainability? Possible Scope and Implications" by Junichi

Hamada of the University of Tokyo; "Nurturing Prosperity in Developing Countries With Sustainability" by George Hara of the Alliance Forum Foundation; "Sustainable Urban–Rural Systems and the Role of Simulation" by Gerhard Schmitt of the Swiss Federal Institution of Technology; "Sustainability Under Rapid Demographic Change" by Ralph Eichler, also from the Swiss Federal Institution of Technology; and "Looking for a Sustainable Society: Family Change in a Rapidly Aging Population," by Sawako Shirahase of the University of Tokyo.

Graduate Program in Sustainability Science

The Graduate Program in Sustainability Science at Todai is located within the Graduate School of Frontier Sciences, and educates students to become professionals and researchers who can take an active role in creating a sustainable society. The program accepts students from a variety of backgrounds and combines academic coursework, experiential learning, practical courses, and thesis work, leading to the degrees Master of Sustainability Science and Ph.D. in Sustainability Science. Professor Yukio Yanagisawa, a specialist in environment systems, is chair of the Managing Committee of the program and professor Takashi Mino, a specialist in sociocultural environmental studies, is the senior advisor. Research interests of faculty affiliated with the Graduate Program in Sustainability Science include institutional design for sustainability innovation, biological wastewater treatment and environmental engineering, sustainable forest management and timber production, landscape planning and urban agriculture, rural planning and development, desertification, biological diversity, ecological monitoring, and global climate change.

The Asian Program for Incubation of Environmental Leaders (APIEL) exists within the Graduate Program in Sustainability Science. APIEL, which is open to both Japanese and foreign students and is coordinated by Professor Takashi Mino, is a collaboration between the Graduate Program of Sustainability Science and the Department of Urban Engineering. The purpose of APIEL is to educate students to assume roles of leadership in environmental issues in Asia, to promote interdisciplinary research, and to establish a collaborative network of research and education with other Asian countries.

See Also: Green Community-Based Learning; Kyoto Declaration on Sustainable Development; Leadership in Sustainability Education.

Further Readings

Kato, Mariko. "Todai Still Beckons Nation's Best, Brightest but Goals Diversifying." *The Japan Times* (August 11, 2009). http://search.japantimes.co.jp/cgi-bin/nn20090811i1.html (Accessed May 2010).

University of Tokyo. "Alliance for Global Sustainability." http://en.ags.dir.u-tokyo.ac.jp (Accessed May 2010).

University of Tokyo. "Graduate Program in Sustainability Science." http://www.sustainability .k.u-tokyo.ac.jp (Accessed May 2010).

University of Tokyo. "Integrated Research System for Sustainability Science." http://en.ir3s.u -tokyo.ac.jp (Accessed May 2010).

Sarah Boslaugh
Washington University in St. Louis

UNIVERSITY OF VERMONT

The University of Vermont's Dudley H. Davis Center, pictured here in 2007, was the nation's first student center to attain LEED Gold certification.

Source: Wikipedia

One of the so-called Public Ivies, the University of Vermont (UVM) is a national public research university and Vermont's land-grant university. Located in Burlington, the university was established in 1791 as a private university, but merged with the public Vermont Agricultural College in 1865, a year after that school was created by the Morrill Land-Grant Colleges Act; the resulting entity was a public university that inherited Vermont Agricultural College's land grant. Today, UVM draws a considerable number of out-of-state students, who make up some 65 percent of the student body. This brings in greater amounts of tuition and prevents the university from needing to rely on public funding as a substantial source of revenue (as of 2010, about 10 percent of the budget), and helps UVM build its reputation as a school that combines the best of both the private and public academic worlds. This especially has helped during times of economic calamities—state budget cuts have less effect on the university than they would in other states. Reflecting the state's commitment to sustainability, UVM has long been greener than most public universities, and its Dudley H. Davis Center was the first student center in the country to be awarded LEED (Leadership in Energy and Environmental Design) Gold certification by the U.S. Green Building Council.

In the 21st century, UVM's commitment to sustainability has been foregrounded by the Leading By Design Project, which began in March 2007 when the Lewis Foundation offered the university $15 million if it could submit a plan to transform UVM into "a catalyst, a model, and an engine of change—a driving force that leads society to a sustainable and desirable future." Such a plan needed to address the full scope of the university's activities, including academics, financial operations, physical plant, and the local community. The university established a task force to develop a proposal, which as of 2010 was ongoing. In the meantime, the campus was already going greener, having converted to 100 percent recycled chlorine-free paper as of Earth Day 2006 and introduced buses running on natural gas in early 2007 as the result of a partnership between the university, Burlington, the Federal Transit Administration, and Vermont Gas Systems. The multimillion-dollar project created Vermont's first "fast-fill" natural gas refueling station, which pumps compressed natural gas into the clean-burning vehicles' rooftop tanks at the city's Department of Public Works. The previous fueling station used for the city-owned natural gas–burning vans had taken hours to refill, using older low-pressure refueling technology. Particulate pollution in natural gas vehicles is about 100 times lower than that of buses fueled with petroleum diesel, while nitrogen oxide emissions are cut in half, and engines operate at 15 decibels quieter.

The current draft of the plan proposes the formation of an Integrative Center for Sustainable Solutions, a nexus of open discussion for identifying ecological and social crises and developing responses, solutions, and prevention and mitigation strategies. The center will include research projects, community initiatives both on and off campus, and new sustainability courses, while promoting collaboration both interdisciplinarily and between UVM and off-campus institutions and communities. Existing UVM programs will be identified and linked with the new initiatives, and new financial and staff support will be available. Faculty members on the Leading By Design committee are drawn from a number of different departments and colleges, including political science, medicine, business administration, agriculture, and arts and sciences. Hundreds of students have attended discussions on the ongoing project, while nearly 1,000 have responded to surveys ranking vision statements generated by those discussions. The committee also maintains a blog in order to make its work transparent to the UVM community, has posted audio and video from various talks on its Web page, and has published a paper on its work in the *Journal for Sustainability in Higher Education*. The emphasis, first and foremost, has been on creating a shared vision for UVM's future, one built on fairness, mutual respect, and mutual concern for sustainability.

Student fees fund a $200,000 annual fund for renewable energy projects, and UVM carbon emissions have been tracked since 2002 with the goal of continuous reduction. Thirty percent of the university's food is purchased from local companies, and more than a third of the solid waste is recycled or composted. Dining services now use compostable to-go packaging, and used cooking oil is harvested for biodiesel production.

In 2010, the clean energy projects selected were the installation of a photovoltaic system on the roof of the Spear Street Equine Center; electricity-generation solar trackers on the roof of the Votey Building and near the Forest Service Building; a solar hot-water unit at one of the residence halls; a dashboard in four campus buildings to display energy use in real time; new courses to teach students to analyze campus energy use and study sustainable transportation systems; and an evaluation of the biomass energy potential of UVM's Trinity campus.

The Institute for Global Sustainability

UVM's Institute for Global Sustainability is an on-campus multidisciplinary institute that focuses on leadership programs, with a mission to educate students and citizens to lead a sustainable future. Founded in 2007 by the Continuing Education Department, its staff and advisory board are drawn from a number of academic and administrative departments throughout the university. One of its key programs is Sustainable Business, Management, and Entrepreneurship, a program of three courses (available to interested students in any combination): Sustainable Business, a five-day intensive summer program introducing students to principles of sustainable business practice; Collaborative Management, a five-day intensive summer program for organizational leaders, focusing on the relationships between organizations that must be formed for sustainability; and Introduction to Ecological Economics, an online course.

The institute's Food Systems and Agriculture program offers minors in Food Systems and Ecological Agriculture, as well as the Farmward Bound summer program, a three-day noncredit Cooking as Pedagogy workshop, and a 12-week intensive course on Integrated Large-Animal and Agricultural Operation Management, offered on a dairy farm. Courses offered include Vermont's Rural Food System, offered every summer; Environmental Cooking;

Exploring New York City's Urban Food System, which includes a weeklong trip to New York City; Entomology and Pest Management; Weed Ecology; Plant Pathology; Organic Farm Practicum; and Biomass to Biofuels.

There are a number of smaller programs available, especially for off-campus community members. The institute's Ecological Economics program offers a certificate of graduate study, for which all courses can be completed online apart from a short course during which students meet to create a community-based project. The Green Building and Community Design program includes certification and graduate study certification, as well as two-week, three-credit field studies devoted to clean energy in Vermont. The EMERGEneering program is an annual seminar for engineers, city planners, and policy makers, presenting holistic solutions to 21st-century engineering problems. The Greening a Business program is a five-day intensive summer program on carbon management, clean energy, and sustainable business practices and includes an optional four-week internship program for students. The Energy Systems program focuses on the issues of transitioning to a clean energy economy, addressing the problems of climate change and nonrenewable resources, and the potential of innovations in building design and new technologies. Coursework covers biofuels, energy policy, carbon-neutral energy solutions, and the aforementioned Greening a Business intensive seminar.

Other courses recommended by the institute for students interested in sustainability include Green Roof Design and Installation, Ecological Design, Green Remodeling, Introduction to Environmental Sciences, Redesigning Wastewater, Natural History of Vermont, Place and Placelessness, Earth System Science, Environmental Conflict Resolution, Introduction to Geographic Information Systems, Reading the Landscape, and the Diversified Agricultural Farm Residency.

Center for Sustainable Agriculture

The UVM Center for Sustainable Agriculture was founded in 1994 and promotes understanding of sustainable farming practices in the Vermont farming community. The center supports farmers and farming communities through its applied research and collaborative learning efforts; integrates the job experience and accumulated common wisdom of farmers with scientific expertise to single out ecologically sound farm management choices; and provides a forum for networking for farmers, both experienced and aspiring, as well as for consumers, community members, policy makers, and scientists.

The center has identified a number of specific goals that it pursues. It works on local food initiatives, including a local food pilot program teaching Vermonters to cook for themselves with local ingredients and the similar farm-to-plate program that encourages the market development of small-to-medium farms to increase Vermont's food security. It helps to market grass-fed products and addresses regional farm product distribution challenges. It offers technical workshops on consumer education and safe food production.

The Beginner Farmer and Land Access program helps new and aspiring farmers develop their plans and identify their options; it proactively supports all types of Vermont farms, including organic and conventional, livestock, dairy, and plant crops. Livestock grazing techniques are examined to maximize Vermont's short grazing season due to the severity of its winter, a project that also includes work with horse owners (who face challenges similar to those of livestock farmers) and the study of forage species.

Above all, the center is invested in promoting sustainable leadership and encouraging the involvement of beginning farmers, multigenerational farmers, students, professionals, and interested community members in keeping Vermont's lengthy history of farming alive, prosperous, and responsible.

Living Machines

One of the professors hired as part of UVM's attempt to become a leading environmental studies center was John Todd, who became a research professor and distinguished lecturer in 1999, his first teaching position since 1970. A leading biologist in the field of ecological design, Todd took the concept of the bioshelter (a solar greenhouse) that was developed at the New Alchemy Institute he'd founded, and extended it to the "living machines" concept while at UVM. Living machines are ecologically engineered systems that mimic the functions of wetlands for the purpose of wastewater treatment. They have since been trade-marked, branded, and further developed by Worrell Water Technologies.

See Also: Experiential Education; Green Mountain College; Leadership in Sustainability Education; Outdoor Education.

Further Readings

Filho, Walter Leal. *Sustainability and University Life.* Ann Arbor: University of Michigan Press, 1999.

Gough, Stephen and William Scott. *Higher Education and Sustainable Development.* London: Routledge, 2009.

Hargreaves, Andy and Dean Fink. *Sustainable Leadership.* Hoboken, NJ: Jossey-Bass, 2005.

Sandell, Klas, Johan Ohman, and Leif Ostman. *Education for Sustainable Development.* New York: Studentlitteratur AB, 2005.

Silka, Linda and Robert Forrant, eds. *Inside and Out: Universities and Education for Sustainable Development.* Amityville, NY: Baywood Publishing, 2006.

Wiland, Harry and Dale Bell. *Going to Green: A Standards-Based Environmental Education Curriculum for Schools, Colleges, and Communities.* White River Junction, VT: Chelsea Green Publishing, 2009.

Bill Kte'pi
Independent Scholar

University of Virginia (ecoMOD)

Established by Thomas Jefferson in 1817, the University of Virginia since its inception has focused on research that would improve the common good. In honor of that responsibility, the University of Virginia has initiated a far-reaching research initiative through its ecoMOD program, widely recognized for its excellence. Established in 2004, ecoMOD is a research project that focuses on designing, building, and evaluating sustainable residential designs, especially those for low- and moderate-income individuals and families. The project is conducted through the University of Virginia's School of Architecture and concentrates on projects that are both sustainable and affordable, thus making environmentally friendly housing available to a wide swath of the population. The School of Architecture teams with and is assisted by the School of Engineering and Applied Science to broaden the scope of ecoMOD from theory to implementation of the projects devised. The team that coordinates ecoMOD is persistent in challenging traditional housing industry practices and beliefs through the creation of prototypes that exemplify environmentally and economically potential prefabrication.

The prototypes ecoMOD endorses must excel in each of the following categories: affordability, modularity, and ecology. An extensive evaluation process allows ecoMOD to determine the efficacy of its practices and to refine procedures, material choices, and designs to better reflect consumer needs, efficiency, and effective practices. Many students and faculty from a broad variety of fields of specialization play an important role as eco-MOD uses these varying perspectives to enhance the designing, building, and evaluation of each prototype.

Goals of ecoMOD

Too often, sustainable housing initiatives have been exclusively the purview of the wealthy. This is largely because sustainable alternatives to traditional building practices are more costly than conventional alternatives. This reality deprives middle- and low-income individuals and families of the cost savings associated with sustainable housing. To change this, ecoMOD attempts both to provide alternatives and to change perceptions that hinder environmentally friendly alternatives in housing options, chiefly by revising the design, building, and evaluation phases of construction. As both an ongoing research project and an actual builder of homes, ecoMOD strives to be visionary in conceiving designs that are environmentally sustainable and practical in providing exemplars of viable green alternatives to standard building industry practices. In order to achieve these goals, ecoMOD focuses on three distinct but interrelated concepts that it feels can most greatly benefit the construction process: design, build, and evaluate. Each project ecoMOD undertakes thus focuses on these aspects of the construction process.

ecoMOD's integrated design teams focus on the socialization benefits that come from working with those from different backgrounds. Individuals from different fields of study bring different perspectives and learn to collaborate effectively and to form abstract opinions that have great impact upon the formulation of useful contributions to housing design. Project ecoMOD encourages spirited yet educated debates regarding the decision-making process related to both big and small ideas and concerns related to design. Unique and diverse perspectives contribute to the ecoMOD building stage as well. Prefabricated housing is investigated by ecoMOD as one means of improving affordability of ecologically friendly housing. Prefabrication can reduce both construction costs and utility bills. To that end, prefabricated modules and panels designed by ecoMOD are built in an airfield hangar the University of Virginia owns but that is no longer in use. These prefabricated parts are flown to designated construction sites after careful inspection; this is a critical way the ecoMOD team helps reduce environmental impact. Lastly, the complete house is evaluated. Everyone who took part in the designing and building phases of a given research project is encouraged to analyze his or her own efforts. Independent engineers, architects, and representatives of the housing industry later provide their own additional evaluations. Evaluations target environmental impact, energy operation and execution, affordability, occupant satisfaction, constructability, and site placement choices for the house. Analysis of these aspects of ecoMOD's projects allows the accumulation of evidence of effective and promising practices for others to emulate.

Design

As part of its ongoing research program, ecoMOD designs and develops several prototypes. Two prototypes, OUTin and Seam, are built for the Piedmont Housing Alliance (PDA) of Charlottesville, Virginia, and one, preHAB, for Habitat for Humanity. These prototypes focus on designing homes that feature durability and have increased value over time due to energy-saving features. These features are especially beneficial for low-income individuals, as they increase their chances to obtain comfortable housing and make it easier for them to maintain over time. Because PDA and Habitat for Humanity wanted housing that blends in with preexisting stock, the ecoMOD research team agreed to design the prototypes so that they resemble small houses. As a result, more time was devoted to creating multiple functions for things that would normally have a single use. This practice, according to ecoMOD's research, makes the houses seem much larger. Unlike the conventional housing industry, the ecoMOD research team also makes sure that houses are designed and built specifically to accommodate a particular site and its surroundings instead of merely being suitable for any environment. This focus on site-appropriate design practices allows the ecoMOD homes to be tailored to achieve maximum energy efficiency. Once complete, the houses will be evaluated and monitored not only to record the areas of success for future prototypes, but to assure high levels of occupant safety, health, and satisfaction.

All three prototype designs concentrate on sustainability in the hope of bringing more affordable homes to low-income families who would benefit most from reduced energy, water, and maintenance costs. Due to the results from its design experiments and research, ecoMOD provides housing that sustains the environment and meets the needs of occupants. Despite these successes, ecoMOD continually strives to improve its design practices, thus providing more benefits to the public. The ecoMOD research team puts particular design emphasis on devising ways that give one element or component of a house multiple functions and those methods that make energy systems used for heating and cooling reusable and more efficient. For example, whenever possible, ecoMOD uses sunlight as a major, if not the primary, source of lighting for its houses. Similarly, wind is utilized as a source of ventilation for the houses, and several efficient systems that provide reusable energy have been developed to reduce, if not eliminate, environmental impact. In addition to accommodating the occupants, the houses contain easily adjustable control systems for heating, cooling, and water. Also, the project demands that tools and anything that can be reused is reused, even if it is as small as carving and sanding a smooth doorstop out of a broken hammer.

Build

In the changing neighborhood of Fifeville/Castle Hill in Charlottesville, Virginia, stands the completed ecoMOD project OUTin house. OUTin is placed in a neighborhood that formerly was the home of predominantly wealthy Caucasians. Over time, the Fifeville/Castle Hill neighborhood has become far more diverse, with many low-income Latino and African-American residents who choose to live there due to relatively more affordable housing. The Fifeville/Castle Hill neighborhood also experienced an economic shift, which increased the need for providing low-cost housing options. The OUTin site features both a three-bedroom house and an accompanying single studio, thus making it the first two-unit condominium in the neighborhood. Both units have been sold with subsidized financing by the Piedmont Housing Alliance, allowing the purchasers to obtain their dwellings for low monthly payments.

Aside from its affordability and site adaptation, the OUTin house was built with many significant features that enhance the occupants' experience. It is considered both an "inside" and an "outside" house. The house was built to combine outside attributes to benefit the experience of those inside the house. Sunlight, breezes, and vegetation illuminate the OUTin house, creating a vibrant environment. The OUTin house contains a solar water heater that absorbs sunlight to accumulate hot water for the occupants, providing both practical comfort and energy efficiency. Further utilizing sunlight as an efficient source of energy, there are at least two windows of great size in each room as a way not only to provide a great deal of illumination but also to serve as a heating device. During warmer months, the abundance of sunlight can be controlled with curtains and blinds to adjust for immediate needs. The rainwater collection system is another environmentally friendly feature of the OUTin house. Rainwater collected from the roof is purified for the occupants' convenience, be it for taking a shower or drinking a glass of purified water. Other convenient features include easily operable windows that adjust wind ventilation and structural insulated panels that reduce air infiltration and thermal bridging. These innovative energy-efficient features bring considerable cost savings via reduced utility bills. The OUTin house also features 43 sensors that provide accurate readings of the humidity inside and outside the house; monitor the current, voltage, and water consumption; and measure the wind speed and other information.

The OUTin house was designed to be built in the Fifeville/Castle Hill neighborhood, but after further consideration by the ecoMOD team, the house was built in eight modules and transported to the site for construction. Each module was built at a different site and then transported to the OUTin site to reduce construction time and the environmental impact of the project. Each module was painted upon arrival at the OUTin house site and furnished with either low or no volatile organic compound (VOC) materials. VOC materials were minimized because they can harm the environment through air pollution and may endanger the health of the occupants over long periods of exposure. The OUTin house's floor is made of wood purchased from a sustainable forest grown in Virginia. Licensed builders and subcontractors worked with students and faculty from ecoMOD from start to finish, but the ecoMOD team was responsible for most of the actual building.

Evaluate

The evaluation aspect of ecoMOD expands traditional examinations of building practices, which focus primarily on cost effectiveness and durability, to include a spectrum of six considerations. These six considerations include a structure's aesthetic, environmental, social, technical, financial, and overall impact. Through this expanded evaluation process, ecoMOD is able to determine those aspects most significant to designing and building housing that is environmentally friendly as well as affordable and convenient to users. Through this process, ecoMOD discovered the cost savings associated with sustainable housing. Initial savings for each unit of the OUTin house, for example, were nearly $7,000 per unit, a sum that grows to over $20,000 over the course of a 30-year mortgage. Such savings can reduce the required annual income of qualified purchasers by roughly $3,000 per year, from $44,000 to $41,000. Findings such as these assist in making the case for sustainable practices in constructing dwellings and empowering sustainable building projects.

In general, the housing industry conducts few evaluations of projects, especially those of an experimental nature such as ecoMOD. Continual monitoring and evaluating of ecoMOD's projects has allowed it to make refinements and improvements in its designing and building processes. These improvements have facilitated the houses that ecoMOD has produced, but also assisted in understanding the prefabrication process. Although off-site construction has long been hailed as a cost-reduction tool, experience has demonstrated that prototype prefabrication efforts generally are more expensive than traditional alternatives. By focusing on what works during the construction process, both in terms of cost efficiency and consumer reaction to certain implementations, ecoMOD's evaluations provide the data necessary to support certain decisions, especially those that reduce the environmental impact of construction and improve occupant health and well-being.

See Also: Association of University Leaders for a Sustainable Future; Constructivism; Educating for Environmental Justice; Experiential Education; Green Community-Based Learning; United Nations Decade of Education for Sustainable Development 2005–2014.

Further Readings

Broadbent, G. and C. A. Brebbia. *Eco-Architecture: Harmonisation Between Architecture and Nature.* Southampton, UK: WIT Press, 2006.
ecoMOD. http://www.ecomod.virginia.edu (Accessed March 2010).
Moe, K. *Integrated Design in Contemporary Architecture.* New York: Princeton Architectural Press, 2008.

Stephen T. Schroth
Jason A. Helfer
Edel Vaca
Knox College

UNIVERSITY OF WASHINGTON

The University of Washington (UW) is a public research university founded in 1861 in Seattle, Washington, making it one of the oldest universities on the West Coast of the United States. Additional campuses were established at Bothell and Tacoma in 1990. In 2009, UW enrolled 27,785 undergraduate and 11,072 graduate students and employed over 25,000 faculty and staff members, including graduate assistants. UW is consistently at or near the top among public universities in terms of research funding, and in 2007, UW received over $1 billion in grant and contract research funding, nearly 80 percent of which came from the federal government.

UW has a long history of environmental and sustainability education and research as well as green operations dating from 1947, when the climatology-focused Department of Meteorology was founded and the Division of Health Services began providing on-campus environmental health and safety services. In 1973, the Institute for Environmental Studies was founded (superseded in 1997 by the Program on the Environment), and in 1977, the Joint Institute for Study of the Atmosphere and Ocean was founded in conjunction with the National Oceanic and Atmospheric Administration. In 1998, UW began admitting students to a new bachelor of arts program in Environmental Studies, and as of 2009, offered over 500 courses that relate to climate change and sustainability.

Sustainable Operations

The Office of Environmental Stewardship is the central location for information about sustainability and environmental projects and programs at the University of Washington. The Environmental Stewardship Advisory Committee, created in 2004, provides leadership in support of UW's commitment to becoming a sustainable campus. Many student organizations are also involved with environmental issues, including the UW Earth Club, Engineers Without Borders, the Sierra Student Coalition, the Student Environmental Health Association, the Coastal Society, and SEED (Students Expressing Environmental Dedication).

Housing and Food Services (HFS) at UW (which serves about 30,000 customers per day) works with the student organization SEED to promote environmentally sound practices. Many food products served in the UW dining halls are obtained locally, including eggs, milk, bagels and bakery items, bread and rolls, and produce (herbs are grown on campus). Coffee served at HFS facilities is organic, shade grown, and Fair Trade certified; hamburger beef is antibiotic and hormone free, and grab-and-go foods (e.g., sandwiches, fruit cups, sushi) are produced locally. Food waste is composted, and many products used in dining, including cups, straws, and food containers, are compostable as well. Cooking oil from HFS is recycled by a local producer of biodiesel fuel.

In 1987, UW's central utility plant was converted from coal to natural gas (reducing carbon emissions by 40,000 metric tons per year), and UW has since installed high-efficiency boilers and lowered building temperatures, reducing greenhouse gas emissions to 1980 levels. UW has also reduced electricity demand by committing to green building standards (as of 2009, three UW buildings had LEED Gold certification and three had Silver certification) and efficiency improvements in the data center of the Seattle campus, which achieved a 26 percent reduction in electricity demand. In April 2008, UW conducted the CFL Exchange Project, which replaced all incandescent bulbs on the Seattle campus with compact fluorescent bulbs that use 75 percent less energy.

In 1991, UW created the U-Pass program, which encourages alternatives to commuting in private cars, including subsidized transit use, discounted carpool parking, vanpool subsidies, and car-sharing discounts. The number of UW commuters using public transportation has almost doubled since introduction of U-Pass, while the number driving alone has dropped by about 25 percent; currently, 39 percent of commutes to the Seattle campus are by bus and 30 percent of trips are by foot or bicycle. A ride-share program on the Bothell campus has also reduced emissions due to automobile commuting. Almost half the university's vehicle fleet is composed of alternate-fuel vehicles, including biodiesel, hybrid, and electric.

Climate Action Plan

In January 2009, a team of over 100 UW faculty, students, staff, and administrators from the three campuses drafted the Climate Action Plan (CAP), which lays out strategies the university will follow to become climate neutral. The CAP was written as part of UW's obligations as a member of the American College and University Presidents Climate Commitment (ACUPCC) but also includes considerable information about the history of sustainable operations and environmental research, education, and outreach at UW. In the CAP, the UW sets a goal of achieving climate neutrality by 2050.

In the areas of research and scholarship, the CAP states UW's commitment to maintain its position as a leader in climate science, to foster undergraduate participation in research, to support junior faculty in environmental scholarship, and to include environmental topics in professional degree programs such as law, business, and public policy. UW also commits to developing environmental literacy programs that can be taken as part of any undergraduate program, establishing interdisciplinary centers at the College of the Environment, and creating new means of communication with the larger community as well as university affiliates.

The CAP divides plans for reducing greenhouse gas emissions into three approaches and five systems or sectors where reductions in emissions can be achieved. About 20 percent of the required reductions are expected to occur by changing behavior through education incentives, education, policies, and standards, and an additional 60 percent through technological changes such as reducing emissions and acquiring energy from alternative sources. The remaining 20 percent of reductions are expected to be met by purchasing allowances and offsets. The sectors or systems are campus energy supply, campus energy demand, computing strategies, information technology, and professional travel. The CAP then estimates which sectors are most amenable to each type of approach: for instance, campus energy supply and demand will probably be most successfully reduced through technological changes while commuting demands are considered most likely to be reduced through changes in individual behavior.

To take an example of proposed changes outlined in the CAP, the central utility plant on the Seattle campus is the largest single source of greenhouse emissions at UW. The ideal means of reduction in emissions would be a reduction in demand through behavioral change, thus enabling the use of only the most efficient boilers, as well as reducing the overall load carried by the plant. In terms of technological improvements, energy could be saved by replacing the electric motors that drive pumps with motors of greater efficiency, by creating a system to recover waste heat in the flue gas system so that it could be used to heat buildings, by reducing the amount of heat loss by increasing thermal pipe insulation and creating a preventative maintenance plan to reduce the frequency of steam leaks, and by installing new pressure-independent control valves in the chilled-water distribution system to improve hydraulic pumping efficiency.

To take another example, many strategies can reduce campus demand for energy used to operate buildings. These include the design of buildings so they use less operational energy per square foot, using integrated building design processes that include input from a variety of experts, and using life-cycle cost analysis and energy modeling to make informed decisions about energy efficiency and costs over the expected lifetime of a building.

In the area of information technology, a combination of behavioral and technological changes is proposed. The first is to "buy green," for instance, by purchasing flat LCD screens rather than CRTs because LCDs consume less power in operation and reduce the amount of lead and mercury that must be disposed of when the screens are discarded. Power management systems, which automatically turn off components when not in use, can also substantially reduce energy use and can be implemented automatically. Waste energy from data centers can be reused (a practice already being used on the Seattle campus), and consolidating computer equipment into data centers has proven to be the most efficient way to lower cooling demands. Virtualization and cloud computing are other techniques that can substantially reduce the energy demands of computer use without negatively impacting communications or workflow.

Memberships and Recognition

UW is a founding signatory to the American College and University Presidents Climate Commitment (ACUPCC), which signifies support for climate neutrality and sustainability in university operations as well as integrating education about sustainability into students' university experience. UW is also a member of the Association for the Advancement of Sustainability in Higher Education (AASHE), an association supporting sustainability in higher education. UW is a founding member of the Seattle Climate Partnership, a voluntary agreement among employers in the Seattle area to reduce greenhouse gas emissions.

UW's sustainability programs have received much recognition. For instance, UW ranked second among U.S. universities in the 2009 Sierra Club's listing of "cool schools" and was commended in the National Wildlife Federation's 2008 campus report card for having exemplary programs for transportation, on-campus clean energy sources, off-campus sources of renewable energy, and plans to do more with green landscaping. The *Princeton Review* cited UW on its 2009 "green honor roll" as one of 15 colleges and universities with exemplary environmental records; reasons include the commitment that all new buildings will meet the LEED Silver standard, the commitment to purchasing power from renewable sources, the emphasis on local organic foods in the dining halls, and composting of postconsumer waste, including paper cups and dishware. The UW alternative transportation strategy, which reduced automobile traffic while allowing the campus to expand, was featured as a case study in Will Toor and Spenser Havlick's *Transportation and Sustainable Campus Communities.*

See Also: American College and University Presidents Climate Commitment; Association for the Advancement of Sustainability in Higher Education; Whole-School Approaches to Sustainability.

Further Readings

National Wildlife Federation. "Campus Environment 2008: A National Report Card on Sustainability in Higher Education." http://www.nwf.org/Global-Warming/Campus -Solutions/Resources/Reports/~/media/PDFs/Global%20Warming/CampusReportFinal .ashx (Accessed May 2010).

Princeton Review. "Green Honor Roll." http://www.princetonreview.com/green-honor-roll .aspx (Accessed May 2010).

Seattle Climate Partnership. http://seattleclimatepartnership.org (Accessed May 2010).

Sierra Club. "Honor Roll: Our Top 20 Schools." http://www.sierraclub.org/sierra/200909/ coolschools/top20.aspx (Accessed May 2010).

Toor, Will and Spenser Havlick. *Transportation and Sustainable Campus Communities: Issues, Examples, Solutions.* Washington, DC: Island Press, 2004.

University of Washington. "Climate Action Plan" (September 2009). http://f2.washington .edu/oess/uw-climate-action-plan (Accessed May 2010).

University of Washington. "Environmental Stewardship & Sustainability." http://f2 .washington.edu/oesss (Accessed May 2010).

Sarah Boslaugh
Washington University in St. Louis

V

VOCATIONAL EDUCATION AND TRAINING

The Green Jobs Act of 2007 and the American Recovery and Reinvestment Act of 2009 have mandated training for green jobs, such as installing solar panels. This woman is prepping and sanding energy-efficient windows for installation on a home addition.

Source: iStockphoto.com

Vocational education and training prepares students for specific trades, occupations, or vocations. Traditionally less academic in orientation than other subjects, vocational education and training is usually provided to students in secondary (grades 9 through 12) and postsecondary settings. Because its focus is on training students for jobs, vocational education and training concentrates on practical and hands-on activities so that students will become familiar with and adept in particular techniques or practices that are common in certain occupations. As interest has grown in building a green economy, vocational education and training has been seen as a way to prepare workers to meet the needs of sustainable development. New programs have been started to build workforce capacity for green initiatives, although some believe the potential for employment in such fields is overstated.

Background

Traditionally, many workers acquired the skills necessary for their occupations by serving an apprenticeship during which the apprentice learned a chosen trade from a master craftsman. Many trades were learned in this manner, including baking, blacksmithing, plumbing, masonry, and the like. During the 20th century, high schools began to offer vocational courses for students interested in going directly into the workforce. As a result, coursework in home economics, wood shop and metal shop, typing, drafting, business, and auto repair were offered in some schools. Vocational schools also became available for students who had graduated from high school, although offerings varied widely from state to state. This postsecondary vocational education and training was offered in a wide variety of venues that varied by state, including community colleges, state-sponsored technical colleges and institutes, and for-profit and proprietary career schools. These offered training in a wide variety of fields, including traditional trades but also practical nurse, computer technician, and other careers.

Government Involvement

The Office of Vocational and Adult Education (OVAE), an agency of the U.S. Department of Education, oversees policies and procedures pertaining to vocational and adult education as well as postsecondary offerings. Because the OVAE oversees financial assistance to students, it has tremendous influence over how programs operate and the types of programs that qualify for federal assistance. Beginning in 2007, the U.S. Department of Labor also became involved in job training initiatives related to environmentally sustainable employment. Pursuant to the Green Jobs Act of 2007, the U.S. Congress authorized $125 million to establish state and national training programs to address job shortages that are negatively impacting the growth of sustainable industries. The American Recovery and Reinvestment Act of 2009 also included a mandate for training that will allow more Americans to be ready to work in green jobs.

Green jobs include those constructing and renovating energy-efficient buildings, generating renewable electrical power, building and maintaining energy-efficient vehicles, and developing biofuels. As interest in green jobs has grown, many employers have reported difficulty hiring vocational or trade-skilled workers for sustainable development projects. Especially needed have been electricians who install solar panels, plumbers who work with solar water heaters, workers who install energy-efficient windows, and construction crews who build power-generating wind farms. The push for green jobs needs workers who can deliver the necessary skills and expertise to make sustainable economic development possible.

As interest in sustainable development has grown, spurred both by consumers and government incentives, attention has also been directed to vocational education and training programs. Community colleges, vocational schools, technical institutes, and career centers are looked to for guidance and training. As programs have been developed to meet the need for green-collar workers, no single model has emerged to train people for these jobs. Students can be trained to retrofit old houses with more environmentally friendly technologies in as little as three months, and several programs have been developed that focus on one month of classroom instruction and two months of on-the-job training. Programs that train plumbers or electricians for green jobs tend to be more time-intensive, as the training process for such trades is intense and extensive.

The push for vocational education and training programs to focus on green jobs has its critics. Some argue that well-trained plumbers, electricians, and those from other highly skilled trades are able to work on environmental development without special training. Others suggest that the majority of green jobs are low-paying, menial positions that offer little chance for advancement. For-profit education providers have been accused of overselling the opportunities for employment once training in their programs is completed, while publicly funded community colleges and trade schools have found it difficult to respond to the sudden demand for green jobs training. As interest in sustainable development grows, however, interest in vocational education and training to assist in meeting the needs of this burgeoning field will also grow.

See Also: Eastern Iowa Community College; Environmental Education Debate; Experiential Education; Green Business Education; North American Association for Environmental Education.

Further Readings

Finch, Curtis R. and John R. Crunkilton. *Curriculum Development in Vocational and Technical Education: Planning, Content, and Implementation,* 5th ed. London: Allyn & Bacon, 1998.

Gordon, H. R. D. *The History and Growth of Technical Education in America,* 3rd ed. Long Grove, IL: Waveland Press, 2008.

Gray, Kenneth C. and Edwin L. Herr. *Workforce Education: The Basics.* London: Allyn & Bacon, 1997.

Sherlock, Jim and Nicky Perry. *Quality Improvement in Adult Vocational Education and Training: Transforming Skills for the Global Economy.* London: Kogan Page, 2008.

Stephen T. Schroth
Cale T. Dahm
Knox College

Whole-School Approaches to Sustainability

Schools have long been seen as a vehicle for green education, and school-based approaches to environmental education have varied significantly since the 1970s, when content and field studies approaches to environmental education dominated the field. Since the mid-1990s, whole-school approaches to sustainability—targeting all aspects of school operations, not just the school curriculum—have become increasingly popular across many countries of the world. This approach not only has environmental education embedded in the school curriculum, but it also models sustainability policies and practices within the whole-school context.

A whole-school approach to sustainability would therefore consider, for example, the school's waste management systems; how students are transported to and from school; where food is sourced and how it is grown; how much energy the school uses and how energy needs are powered; the volume of water used and the way water is consumed; the consumption of other natural resources such as paper; the amount of habitat and green space provided for in the school grounds; and the role of the school within the broader community. Whole-school approaches to sustainability view the school as a microcosm of society and promote the take-up of sustainable lifestyles through the modeling of natural resource conservation and efficiency, and by adopting whole-school policies aimed at reducing the school's ecological footprint. Such an approach is a direct response to international calls to reform and reorient education toward a new global sustainability agenda.

History

Whole-school approaches to sustainability represent the amalgamation of a range of initiatives in the early to mid-1990s, including the United Kingdom's Learning Through Landscapes project, Canada's Evergreen project, the Organisation for Economic Co-operation and Development's (OECD's) Environment and Schools Initiative Learnscapes, and Alice Waters' Edible Schoolyard Project in the United States. These all focused mainly on greening school grounds, including utilizing school vegetable gardens as a way of connecting students to food production systems and healthy eating habits, and maximizing the potential of school landscaping and environs as a tool for quality environmental and educational experiences.

The whole-school approach was built upon the idea of school grounds as sites for learning, and added a commitment to a range of other complex education and sustainability issues throughout the full workings of the school including school governance, learning, and teaching approaches, natural resource conservation, curriculum integration, and community outreach. Because of these complex concerns, the whole-school approach differs from more conventional approaches to education for sustainability (EFS) in that support is required by the whole-school community, rather than by an individual teacher who usually initiates and maintains sustainability interventions. In this way, the whole-school model attempts to mitigate some of the weaknesses inherent in more traditional approaches to environmental education that place enormous pressure on individual teachers that all too frequently lead to fatigue and "burnout." A whole-school approach is systemic, involving the whole-school community, and the workload burden is shared across the community.

The whole-school approach is usually viewed as being a democratic and participatory program that provides opportunities for students to develop active citizenship skills through participation in school decision-making processes around a range of natural resource management issues. Students may, for example, be involved in monitoring their school water consumption practices by conducting school water audits that measure the amount of water consumed per day by the school community. Students could measure tap flow rates, count dripping taps, estimate flush volumes in toilets, and read water meters to undertake the audit. After collecting this baseline data, students could discuss the issue and come up with a variety of decisions aimed at reducing water consumption in the school community. These decisions may be aimed at the policy level, for example, creating a school policy about checking and fixing leaks and drips on a regular basis, or may involve infrastructure decisions, perhaps about installing rainwater tanks, or even making decisions about educating the school community about water use in the school and in the home. Such an approach ensures that learning is student-centered, meaningful, and relevant to students.

Geographical Reach

Internationally, the whole-school model is driven by a few key organizations. The largest of these, for example, is the Foundation of Environmental Education's (FEE) Eco-Schools Program, which has been in operation since 1992. This organization provides support through a seven-step process for becoming an Eco-School, finally culminating in the Green Flag award, an accreditation scheme for schools that meet specific environmental targets and benchmarks. According to FEE's 2008 annual report, there are now over 27,000 schools across 44 different countries in North and South America, Europe, Asia, Oceania, the Pacific, and Africa—reaching some 6.3 million students—that have formally committed to becoming an Eco-School. There are also a variety of other initiatives in other countries with a similar philosophy: for example, in Australia, the initiative is called Sustainable Schools; in New Zealand, Enviroschools; and in China, Green Schools.

Although the process for becoming a sustainable school varies across contexts, there are certain guiding principles driving the uptake of the approach across countries and contexts. In their international review of whole-school approaches to sustainability programs, Daniella Tilbury and Kate Henderson identified some common practices across countries and programs. They found, for example, that such approaches were not only participatory in involving the whole-school community, but that they also involved reciprocal learning partnerships between the family, the community, and the school. They also found that whole-school approaches enhanced the integration of education for sustainability across all key learning areas of the curriculum and that such approaches also assisted the school in ensuring that there was consistency between the schools explicit and hidden curricula, that is, between what is explicitly taught in the classroom and the informal messages promoted by the school's values and actions.

Benefits of Whole-School Approaches

There are also a number of other widely agreed-upon benefits attributed to the whole-school approach. Schools participating in such programs have achieved success in many areas although natural resource management outcomes are most prevalent. Others report financial savings; an increase in environmental awareness of students and teachers; improved school environment; greater access to green space; development of active citizenship and participation in global, and local, sustainability issues; and enhanced learning outcomes for school students through real-life, meaningful, and relevant participation in sustainability issues.

Challenges

Although there are enormous benefits to be gained by implementing whole-school approaches to sustainability, there are also significant challenges. Such an approach demands a wide variety of additional skills, often beyond those of a traditional classroom teacher, and many teachers feel they do not have requisite technical skills. Pre-classroom teacher education is therefore needed to ensure that new graduates are equipped with the specialist skills required.

Another significant issue militating against a whole-school approach to sustainability is the perceived lack of time and/or space in an increasingly overcrowded curriculum, which is often compounded by difficulties in leveraging administrative level support, especially when there are competing national agendas, such as a return-to-basics focus on literacy and numeracy. Recent research on the educational outcomes of whole-school approaches conducted by the United Kingdom's Office for Standards in Education, Children's Services and Skills (Ofsted) indicates, however, that whole-school approaches to sustainability, when conducted well, can lead to improved academic performance through extending students' critical and creative thinking skills and by motivating learning through engagement in relevant, real-world issues leading to change.

While these two issues offer the most significant challenges to the successful implementation of whole-school approaches to sustainability, some schools find there may be significant costs involved in implementing whole-school approaches to sustainability, such as setting up school gardens or buying infrastructure such as solar panels and water tanks. While these initial start-up costs may be high, many schools find that they actually save money in the long term through a reduction in costs of utilities, such as water and electricity, by using their improved natural resource efficiency measures.

See Also: Education, Federal Green Initiatives; Integrating Sustainability Education Concepts Into K–12 Curriculum; United Nations Decade of Education for Sustainable Development 2005–2014.

Further Readings

Eames, C., B. Cowie and R. Bolstad. "An Evaluation of Characteristics of Environmental Education Practice in New Zealand Schools." *Environmental Education Research,* 14/1 (2008).

Foundation for Environmental Education. "About Eco-Schools." http://www.fee-international .org/en/Menu/Programmes/Eco-Schools (Accessed March 2010).

Gough, A. "Sustainable Schools in the UN Decade of Education for Sustainable Development." *Southern African Journal of Environmental Education,* 23 (2006).

Henderson, K. and D. Tilbury. *Whole-School Approaches to Sustainability: An International Review of Sustainable School Programs.* Report prepared by the Australian Research Institute in Education for Sustainability (ARIES) for the Department of the Environment and Heritage, Australian Government, 2004.

Mogensen, F. and M. Mayer. *Eco-Schools—Trends and Divergences.* Vienna: Austrian Federal Ministry of Education, Science and Culture, 2005.

Ofsted. "Education for Sustainable Development: Improving Schools, Improving Lives." http://www.ofsted.gov.uk/publications/070173 (Accessed August 2010).

Scott, W. "Judging the Effectiveness of a Sustainable School: A Brief Exploration of Issues." *Journal of Education for Sustainable Development,* 3/1 (2009).

Elizabeth Ryan
University of the Sunshine Coast

WILDERNESS-BASED EDUCATION

These young science camp participants setting up tents in a birch forest in the Innoko National Wildlife Refuge in Alaska in 2006 were taking part in a structured form of wilderness-based education.

Source: U.S. Fish & Wildlife Service

Wilderness-based education can be defined as learning that occurs as a result of intentional immersion in a natural setting, outside the conventional classroom. It encompasses both learning about and through wilderness. The objectives of wilderness-based education include acquisition of a wide variety of knowledge and skills, including technical or hard skills and interpersonal and social skills. The former includes wilderness survival, crisis management, and environmental awareness, whereas the latter consists of communication and aspects such as self-discovery and social attributes, including compassion and personal responsibility. The mechanisms through which learning takes place span a wider range than classroom learning. Apart from instructor- and student-centered approaches, the interaction with the environment, both natural and social, expands the learning opportunities, often in unforeseen ways. The fluidity of the multifaceted encounter between the student, physical environments, peers, and group leaders can create a dynamic that is full of possibilities for deep learning.

The philosophical underpinnings of wilderness-based education are often derived from John Dewey's ideas on education. Education is not seen as simply an individual attribute, but as a repertoire of values, dispositions, problem-solving skills, and critical thinking tools that will enable the learner to be an active and productive citizen in a democracy. One of the key features of this educational philosophy is to break down the wall between life and the classroom. Education is thus not limited to what happens in the classroom; rather, it is through experiencing life as education that deep learning can take place. Wilderness-based education is thus premised on the efficacy of learning by doing. However, learning involves more than experience; it is reflecting on experience that makes true learning possible.

Outdoor learning experiences are often described as being "intense," where intensity refers to the extraordinary nature of the wilderness experience, especially as it is distinct from everyday life and connected to the individual's inner life. Specifically, the intensity inherent in outdoor learning arises from the capacity of nature and natural occurrences to facilitate reflection. The teaching moments occur with an unpredictability that arises both from the novelty of the setting and from the dynamic interaction between the teacher and the student in a radically different context. The learning context is an unstructured nature that engages multiple senses, leading to a richer and potentially transformative educational experience.

Wilderness-based education has specifically been used to promote leadership skills and to improve self-esteem. Due to both the challenging nature of the physical environment, as well as the emphasis on cooperation while engaged in collective problem solving in groups, participants feel motivated to exercise leadership. The experience of autonomy engenders a spirit of inquiry and discovery that creates possibilities for personal growth usually unavailable in a classroom. Wilderness therapy presents a relatively formalized approach to using wilderness to facilitate psychological and interpersonal functioning. The underlying principles are similar to those of wilderness-based education, although relatively more elaborately conceptualized. The therapeutic effects of wilderness are theorized as a product of the positive effects produced independently through wilderness, physical self, and social self. Wilderness allows an escape from cognitive overload produced by everyday life and pressures related to urban existence. The natural features of the landscape, such as clouds, streams, and mountains, engage the attention, but in a nondirected way, producing a relaxing state of "soft fascination." The beneficial effects of the physical self are mediated through the relationship between physical health and mental health. One of the key effects is the enhancement of the perception of self-control, resulting from the increasing control over the body. Wilderness experience, in groups, promotes the development of the social self. This happens through a variety of different but related pathways. Individuals need to cooperate to accomplish the tasks associated with typical activities, such as cooking and camping. Cooperation is also facilitated by the relative equality of the individuals within a group and the diminishing of the differences in social status. Additionally, learning by observation also takes place through the inspirational effect of the group leaders.

Educational institutions, including universities, have incorporated wilderness education in the curriculum under the ambit of "service learning." The concept of service learning involves a multistep process, including the establishment of a formal relationship between the university and a community organization. Students gain experience that includes a service component, in a mutually beneficial arrangement, and reflect on their experience through a variety of means such as journals and role-playing. The students' experience and outcomes are guided by appropriate course content and lesson plans developed by the instructor.

Organizations active in the area of wilderness-based education include Outward Bound, National Outdoor Leadership School, and Wilderness Education Association (WEA). WEA developed the well-known 18-point curriculum for wilderness. There has been concern that wilderness-based education, in recent times, has become unstructured and haphazard, where the curricular elements are neglected. Intentional programming is advocated as a rigorous approach to deliver concrete, measurable benefits to participants as well as to compensate for the lack of seasoned leadership that is becoming increasingly common with the explosion in the number of wilderness programs.

In the final analysis, the big question about the utility of wilderness-based education pertains to the transferability of its benefits to other domains of life. There is increasing evidence, mainly based on subjective measures, including self-perception and self-reported enhancement in self-esteem, self-confidence, and interpersonal relationships that indicates that transference indeed takes place. The students participating in wilderness-based programs demonstrate changed values and improvement in their self-perception of social functioning and other psychological attributes.

See Also: Experiential Education; National Wildlife Federation; Outdoor Education.

Further Readings

Freeman, Patti, Douglas Nelson and Stacy Taniguchi. "Philosophy and Practice of Wilderness-Based Experiential Learning." *Journal of Physical Education, Recreation and Dance,* 74/8 (2003).

Goldenberg, M., D. Klenosky, J. O'Leary and T. Templin. "A Mean Send Investigation of Ropes Course Experiences." *Journal of Leisure Research,* 32/2 (2000).

Hendricks, Williams and Barbara Miranda. "A Service Learning Approach to Wilderness Education." *Journal of Physical Education, Recreation and Dance,* 74/7 (2003).

Paisley, Karen, Nathan Furman, Jim Sibthorp and John Gookin. "Student Learning in Outdoor Education: A Case Study From the National Outdoor Leadership School." *Journal of Experiential Education,* 30/3 (2008).

Russell, Keith and Jen Farnum. "A Concurrent Model of the Wilderness Therapy Process." *Journal of Adventure Education and Outdoor Learning,* 55/4 (2004).

Sibthorp, Jim, Karen Paisley and Eddie Hill. "Intentional Programming in Wilderness Education: Revisiting Its Roots." *Journal of Physical Education, Recreation and Dance,* 74/8 (2003).

Neeraj Vedwan
Montclair State University

Yale University

Yale University, established in 1701, is a private research university in New Haven, Connecticut, serving over 11,500 students. Yale's dedication to sustainability, spearheaded by its Office of Sustainability, has helped it become an international leader in sustainability within the higher education sector. Along with strong connections to University curriculum and research, through such organizations as the Yale Center for Environmental Law and Policy and the Yale Project on Climate Change, the university has set itself apart in University Operations.

Yale's Office of Sustainability was established in 2005 to help increase, focus, and centralize sustainability efforts at Yale and in the surrounding community. To accomplish its mission, Yale integrates best practices and innovative models to create appropriate policies, assist with campus operations, campus engagement, integration into the academic community, and partnerships. Five full-time staff members, together with 15 other Yale staff members who are responsible for sustainability on campus (e.g., transportation options, waste and recycling management, and sustainable food) and 20 part-time research assistants coordinate this effort. As of 2009, the Yale Office of Sustainability reports directly to the vice president.

In addition to the central office, Yale has a number of advisory committees, including the following:

- Biodiversity
- Land and Water
- Building Design and Construction (university-wide policies for campus projects)
- Yale Sustainability Working Group (which focuses on educational initiatives and communication)
- Green Cleaning
- Sustainability Task Force (which oversaw the development of the Strategic Plan)

In the fall of 2010 Yale released the Sustainability Strategic Plan—a comprehensive framework that sets a combination of quantitative and conceptual goals that spans all campus systems. This plan distinguishes itself from other such sustainability strategies in that it was developed via committee with input via a stakeholder engagement process and has a system of accountability built into the plan. The following categories are highlights from the Yale Plan.

Purchasing

Yale's Purchasing Policy includes Environmentally Preferable Purchasing standards that cover office supplies to carpeting to university catering. These standards encourage purchasing of recycled copy paper, energy-efficient office equipment, recycled toner, and office supplies and furniture with recycled content. Whenever possible, the university promotes cleaners that meet Green Seal standards, including organic pesticides, and all electronic purchases are held to Electronic Product Environmental Assessment Tool standards. In addition, the university provides a Website to educate staff on Environmentally Preferable Purchasing.

Climate Change, Energy, Facilities, and Planning

Yale employs a number of renewable energy strategies across campus to help meet its commitment to reduce greenhouse gas emissions by 43 percent by 2020. Between 2005 (the baseline year) and 2008, the university achieved a 7 percent reduction in emissions. The Central Power Plant has been turned into a cogeneration plant, while most buildings feature recommissioned HVAC systems, programmable thermostats when possible, and window replacements.

The Class of 1954 Chemical Research Building uses a 250 kilowatt (kW) fuel cell that produces 40 to 50 percent of the building's electricity and has been recognized by the U.S. Environmental Protection Agency for leadership in combined heat and power. In addition, Fisher Hall (Yale Divinity School) features a 40 kW solar system of 255 panels, which generate 45,200 kW-hours of electricity, approximately 17 percent of the building's energy needs. A project currently in progress would put a 20 kW thin-film solar array on the roof of a selected dorm building, providing up to 5 percent of that building's needs.

Kroon Hall (School of Forestry and Environmental Studies) utilizes a variety of renewable energy systems to work toward its goal of carbon neutrality, including a 100 kW solar system, solar hot water tubes, and a ground source heat-pump geothermal system. A recent project on campus focuses on using small wind turbines, 1 kW each, that generate power from the wind that travels up the side of a building and require a breeze of only 7mph to start. Yale also purchases renewable energy credits to offset electricity usage for the School of Forestry and Environmental Studies, up to 1,175,000 kWh.

Yale utilizes sustainable building practices to reach its goal of reducing greenhouse gas emissions. Per university mandate, all new construction on campus must meet LEED Gold standards for certification. Examples on campus include the Class of 1954 Chemical Research Building (LEED Silver), the Sterling Hall of Medicine C3 Laboratory (LEED-CI Gold) and the School of Forestry and Environmental Studies (pending LEED Platinum status).

All sustainable buildings and labs feature low-flow bathroom and kitchen fixtures, as well as greywater systems, which use runoff and wastewater in toilets and irrigation. Locally procured building materials are used whenever possible, including those with a high percentage of recycled content. Natural light is used as much as possible, and buildings are positioned to reduce heat trapping, while room sensors control lighting and ventilation.

A Building Design and Construction Committee oversees these guidelines and continues to develop new processes on campus. In addition, 75 percent of Yale's nonhazardous construction and demolition waste is diverted from landfills.

Within student living, Yale has addressed energy conservation through a number of methods. A "Light 4 Light" campaign exchanged over 5,000 incandescent bulbs for compact fluorescent lamps (CFLs). Project Lux and the Unplugged Campaign encourage students to unplug appliances when not in use, such as during semester breaks. In addition, a sustainability pledge can be taken by faculty, staff, and students to commit to energy-saving behavior in dorms, offices, homes, and classrooms.

Recycling and Compost

All types of materials are recycled at Yale, including aluminum, cardboard, glass, paper, and plastics, all totaling a diversion rate of 22 percent. In addition, electronic waste is regularly reducing, totaling 2 tons of batteries, 11 tons of light bulbs, 7 tons of printer cartridges, and 114 tons of additional e-waste, such as televisions, computers, DVD players, and cell phones. Yale also has a commitment to composting or mulching 90 percent of its landscape waste.

Food and Dining

One exciting program within university dining is Yale's Sustainable Food Project. Founded in 2001 as a partnership between students, faculty, and staff, the project was led by Yale President Richard Levin and chef Alice Waters. The Sustainable Food Project addresses a number of goals on campus. First, it ensures that each college serves sustainable food, including fully sustainable meals for four days of the week, as well as having a sustainable entrée and side for every lunch and dinner, with the goal of having each meal (1.8 million annually) be local, seasonal, and sustainable.

A second aspect of the project involves overseeing the Yale Farm, a one-acre plot within New Haven. Focusing on sustainable agriculture, the farm holds workshops, tours, and seminars for students and the community. As a result of the farm and dining initiatives, sustainable agriculture has become a curriculum focus at Yale, including the Sustainable Agriculture concentration for the Environmental Studies major.

Transportation

Yale's campus bus fleet runs on ULSD/B20 biodiesel, which reduces sulfur content from 500 parts per million to 50 parts per million, and reduces its petroleum usage by 20 percent (20,000 gallons). The 440-vehicle transportation fleet at Yale also includes 13 hybrid and 4 electric vehicles. A bicycle program provides 30 commuter bicycles to departments for university business; this free service also includes a biannual repair service.

For those wishing to carpool, Yale Transportation Options provides a parking permit discount for two-person commuter cars, and full waiver for those with three or more. The program is open to all faculty and staff, and approximately 5 percent of Yale employees participate. For those taking public transportation, the university provides an Online Commute Benefits Program that can pay up to $230, pre-tax, per month for transit passes.

Yale also began participating in a Zipcar car-sharing program in 2007, which includes 17 cars, of which 4 are hybrids. Yale faculty, staff, and students receive a reduced membership that is applied to car usage, which is $8 per hour. Approximately 63 percent of faculty and staff commute to campus in environmentally friendly ways, including biking, telecommuting, walking, carpooling, and public transportation.

See Also: Green Community-Based Learning; Leadership in Sustainability Education; Sustainability Websites.

Further Readings

Sustainable Endowments Institute. "The College Sustainability Report Card: Yale University." http://www.greenreportcard.org/report-card-2010/schools/yale-university (Accessed May 2010).

Yale University. "Green Purchasing." http://www.yale.edu/procurement/green_purchase.html (Accessed July 2010).

Yale University. "Office of Sustainability." http://sustainability.yale.edu (Accessed June 2010).

Yale University. "Transportation Options." http://www.yale.edu/transportationoptions (Accessed June 2010).

Yale University. "Yale Recycling." http://www.yale.edu/recycle (Accessed June 2010).

Yale University. "Yale Sustainable Food Project." http://www.yale.edu/sustainablefood (Accessed July 2010).

Justin Miller
Ball State University

Green Education Glossary

A

AASHE: Association for the Advancement of Sustainability in Higher Education, an association of colleges and universities founded in 2005 to help coordinate and assist sustainability efforts in higher education.

Active Solar: As an energy source, energy from the sun collected and stored using mechanical pumps or fans to circulate heat-laden fluids or air between solar collectors and a building (contrasted with passive solar).

ACUPCC: The American College and University Presidents Climate Commitment, an agreement signed by over 660 presidents and chancellors at U.S. colleges and universities as of 2010, committing them to developing a plan for their institutions to achieve zero net greenhouse gas emissions.

Alternative Energy: Energy derived from nontraditional sources (e.g., compressed natural gas, solar, hydroelectric, wind).

B

Base Period: The period of time for which data used as the base of an index number, or other ratio, have been collected: For instance, reductions in greenhouse gases or energy consumption are generally calculated relative to some base period before the reforms intended to cause the reductions began.

Behavioral Change: As it affects energy efficiency, behavioral change is a change in energy-consuming activity originated by, and under control of, a person or organization. An example of behavioral change is adjusting a thermostat setting, or changing driving habits.

Biodiesel: A fuel typically made from soybean, canola, or other vegetable oils; animal fats; and recycled grease that can serve as a substitute for petroleum-derived diesel or distillate fuel.

Biofuels: Liquid fuels and blending components produced from biomass feedstocks, used primarily for transportation.

Biomass: Organic nonfossil material of biological origin constituting a renewable energy source.

C

Campus Ecology: A program of the National Wildlife Federation to promote sustainability programs at colleges and universities.

Center for Ecoliteracy: An organization founded in Berkeley, California, in 1995 by Fritjof Capra, Peter Buckley, and Zenobia Barlow to promote environmental education.

Climate Change: Climate change refers to any significant change in measures of climate (such as temperature, precipitation, or wind) lasting for an extended period (decades or longer), resulting from natural factors, such as changes in the sun's intensity or slow changes in the Earth's orbit around the sun; natural processes within the climate system (e.g., changes in ocean circulation) or human activities that change the atmosphere's composition (e.g., through burning fossil fuels) and the land surface (e.g., deforestation, reforestation, urbanization, desertification).

Climate Neutrality: The state in which an institution such as a university has no net greenhouse gas emissions, achieved by some combination of reducing greenhouse gas emissions and creating or purchasing carbon offsets (e.g., by planting trees) to compensate for those which remain.

Co-Benefit: The benefits of policies that are implemented for various reasons at the same time—including climate change mitigation—acknowledging that most policies designed to address greenhouse gas mitigation also have other, often at least equally important, rationales (e.g., related to objectives of development, sustainability, and equity).

Cogenerator: A generating facility that produces electricity and another form of useful thermal energy (such as heat or steam), used for industrial, commercial, heating, or cooling purposes.

Compact Fluorescent Bulbs: Also known as "screw-in fluorescent replacements for incandescent" or "screw-ins," compact fluorescent bulbs combine the efficiency of fluorescent lighting with the convenience of a standard incandescent bulb.

Conference on University Action for Sustainable Development: A conference held in Halifax, Nova Scotia, in December 1991, during which 16 Canadian universities adopted the Halifax Declaration committing them to taking a leadership role in promoting environmental sustainability and restructuring their own policies and practices to become more sustainable.

Constructivism: A philosophy of education based on the ideas of Jean Piaget, Lev Vygotsky, John Dewey, and others that argues that individuals learn best when allowed to formulate meaning based on their own experiences rather than by memorizing information provided by an authority figure.

Conventionally Fueled Vehicle: A vehicle that runs on petroleum-based fuels such as motor gasoline or diesel fuel.

D

Daylighting: The practice of designing buildings to take advantage of natural lighting during the daytime through siting and placement of windows (including skylights) to reduce the need for artificial illumination.

Daylighting Controls: A system of sensors, sometimes referred to as photocells, that assesses the amount of daylight and controls lighting or shading devices to maintain a specified lighting level.

Dual Fuel Vehicle: A motor vehicle that is capable of operating on an alternative fuel and on gasoline or diesel fuel. This term is meant to represent all such vehicles whether they operate on the alternative fuel and gasoline/diesel simultaneously (e.g., flexible-fuel vehicles) or can be switched to operate on gasoline/diesel or an alternative fuel (e.g., bi-fuel vehicles).

E

EETAP: The Environmental Education and Training Partnership, a consortium funded by the Environmental Education Division of the U.S Environmental Protection Agency to promote environmental education at the state and local levels.

Electric Hybrid Vehicle: An electric vehicle that either (1) operates solely on electricity, but contains an internal combustion motor that generates additional electricity (series hybrid) or (2) contains an electric system and an internal combustion system and is capable of operating on either system (parallel hybrid).

Emissions: Anthropogenic releases of gases to the atmosphere. In the context of global climate change, they consist of radiatively important greenhouse gases (e.g., the release of carbon dioxide during fuel combustion).

Environmentally Preferable: A term used to describe products and services that have less negative effect on the environment and/or human health as compared to competing products and services that serve the same purposes.

Experiential Education: A philosophy of education, drawing on the work of John Dewey and others, which argues that education should not be considered simply a process of accumulating knowledge by a process of growth through reflection on the individual's own experiences.

Externality: The positive or negative unintentional result of an action that affects someone other than the primary actor and for which the affected parties are not charged or compensated.

F

Fairchild Challenge, The: A program of the Fairchild Tropical Botanic Garden in Coral Gables, Florida, which offers free environmental education programs to schoolchildren and conducts an annual competition among schools participating in the program.

Fleet Vehicle: Any motor vehicle a university owns or leases as part of its normal operations, including gasoline/diesel powered vehicles and alternative-fuel vehicles.

Flexible-Fuel Vehicle: Also known as a variable fuel vehicle, a flexible-fuel vehicle has a single fuel system that can operate on alternative fuels, petroleum-based fuels, or any mixture of an alternative fuel (or fuels) and a petroleum-based fuel.

Fossil Fuel: Any naturally occurring organic fuel, such as petroleum, coal, and natural gas.

G

Geothermal Energy: Hot water or steam extracted from geothermal reservoirs in the Earth's crust that can be used for geothermal heat pumps, water heating, or electricity generation.

Global Warming: An average increase in the temperature of the atmosphere near the Earth's surface and in the troposphere, which can contribute to changes in global climate patterns. Global warming can occur from a variety of causes, both natural and human induced.

Greenhouse Effect: The result of water vapor, carbon dioxide, and other atmospheric gases trapping radiant (infrared) energy, thereby keeping the Earth's surface warmer than it would otherwise be. Greenhouse gases within the lower levels of the atmosphere trap this radiation, which would otherwise escape into space, and subsequent re-radiation of some of this energy back to the Earth maintains higher surface temperatures than would occur if the gases were absent.

Greenhouse Gases: Those gases, such as water vapor, carbon dioxide, nitrous oxide, methane, hydrofluorocarbons (HFCs), perfluorocarbons (PFCs), and sulfur hexafluoride that are transparent to solar (short-wave) radiation but opaque to long-wave (infrared) radiation, thus preventing long-wave radiant energy from leaving Earth's atmosphere. The net effect is a trapping of absorbed radiation and a tendency to warm the planet's surface.

Green Roof: A building roof that incorporates living plants such as grass in order to provide insulation, reduce rain runoff, reduce the heat island effect, and possibly provide recreational space or wildlife habitat.

H

Harvest of Shame: A documentary television program by Edward R. Murrow, broadcast on CBS in 1960, which focused on the lives and working conditions of migrant farm workers in the United States, which was among the first to highlight not only environmental justice but also the disconnect between those producing America's food and those consuming it.

Heat Pump: Heating and/or cooling equipment that, during the heating season, draws heat into a building from outside, and during the cooling season, ejects heat from the building to the outside. Heat pumps are vapor-compression refrigeration systems whose indoor/outdoor coils are used reversibly as condensers or evaporators, depending on the need for heating or cooling.

I

Incandescent Lamp: A glass enclosure in which light is produced when a tungsten filament is electrically heated so that it glows, including the familiar screw-in light bulbs. Much of the energy is converted into heat, making this class of lamp a relatively inefficient source of light.

Inner City Outings: A community outreach program of the Sierra Club that conducts trips to wilderness areas for low-income young people living in urban areas of the United States.

International Alliance of Research Universities: Sustainability Partnership: An organization founded in 2006 by 10 leading research universities (located in Asia, Europe, Australia, and the United States) to promote sustainability education and their own campuses and also to cooperate in this endeavor through processes such as student exchange programs, summer courses, and conferences.

J

John Muir Youth Award: A program of the Sierra Club, launched in 1996, which encourages students to learn about the environment and become stewards of wild areas.

L

LEED: Leadership in Energy and Environmental Design, a program of the United States Green Building Council that encourages green building practices and provides rating and certification at different levels.

N

NAAEE: The North American Association for Environmental Education, a professional association founded in 1971 of professionals, students, and others interested in environmental education.

NAEP: The National Association of Education Procurement, an association of professionals (primarily in the United States and Canada) in charge of purchasing for institutions of higher education. NAEP held its first sustainability institute in 2007 to assist colleges and universities in making purchasing decisions that are environmentally sustainable.

National Environmental Education Act: An act passed by the U.S. Congress in 1990 that called for improved environmental education in the face of increasing threats to human health and the environment including global warming, species extinction, and increasing pollution.

National Network for Environmental Management Studies: A program founded by the U.S. Environmental Protection Agency in 1986 to create more interest in environmental careers; it awards fellowships annually to graduate and undergraduate students to allow them to complete an environmental project related to their field of study.

National Project for Excellence in Environmental Education, The: A program of the NAAEE in the late 1990s to develop resources for educators and establish standards for environmental education materials, student learning, professional development, and nonformal programs; these standards have been adopted by other organizations as well, including, for instance, the National Wildlife Federation.

Net-Zero Energy: Characteristic of a building that produces as much energy as it consumes on an annual basis, usually through incorporation of energy production from renewable sources such as wind or solar.

Nicodemus Wilderness Project: A nonprofit organization located in Albuquerque, New Mexico, which promotes environmental education and youth leadership in conservation and ecological restoration projects.

Nonrenewable Fuels: Fuels that cannot be easily made, grown, or otherwise replaced at a rate comparable to their consumption; examples include oil, natural gas, and coal.

O

Occupancy Sensors: Also known as "ultrasonic switchers," these are energy-saving devices that turn lights on when movement is detected and off following a period during which no motion is detected (considered a marker that the room or area is unoccupied).

Off-Hours Equipment Reduction: A conservation feature where there is a change in the temperature setting or reduction in the use of heating, cooling, domestic hot water heating, lighting, or any other equipment either manually or automatically.

Organic Food: Food that is produced without using most conventional pesticides, fertilizers made with synthetic ingredients or sewage sludge, bioengineering, or ionizing radiation, and by farmers who emphasize the use of renewable resources and the conservation of soil and water. Organic meat, poultry, eggs, and dairy products come from animals that are given no antibiotics or growth hormones.

Our Vanishing Wilderness: The first environmental television series aired in the United States, it was created by National Educational Television and aired in 1970 with episodes looking at issues such as the effects of pesticides on the food chain, environmental damage from the 1969 offshore oil leak near Santa Barbara, and the destructive effects of development on the Florida Everglades.

Ozone: A molecule made up of three atoms of oxygen. Occurs naturally in the stratosphere and provides a protective layer shielding the Earth from harmful ultraviolet radiation. In the troposphere, it is a chemical oxidant, a greenhouse gas, and a major component of photochemical smog.

P

Passive Solar: Methods for using the energy contained in sunlight (for instance, to heat air or water) without requiring the using of mechanical systems (contrasted with active solar).

Poisoned Air, The: A 1967 CBS television documentary that helped educate the American public on the dangers posed to human health and the environment by air pollution.

R

Recycling: Collecting and reprocessing a resource so it can be used again. An example is collecting aluminum cans, melting them down, and using the aluminum to make new cans or other aluminum products.

Rio Earth Summit: Formally known as the United Nations Conference on Environment and Development, a United Nations conference held in Rio de Janeiro, Brazil, in June 1992, that resulted in several major documents including the Framework Convention on Climate Change (which led to the Kyoto Protocol) and the Convention on Biological Diversity and that led to increased emphasis on the roles played by education, training, and public awareness in fostering sustainable development.

S

Solar Energy: The radiant energy of the sun, which can be converted into other forms of energy, such as heat or electricity.

STARS: Sustainability Tracking, Assessment & Rating System, a voluntary program of the Association for the Advancement of Sustainability in Higher Education for colleges to rate themselves on sustainability in three categories: Education and Research; Operations; and Planning, Administration, and Engagement.

Sustainability: A state in which the needs of the current population in a specific location can be met without compromising the ability of other populations (in future times or in other locations) from being able to meet their own needs.

W

Waste Heat Recovery: Any conservation system whereby some space heating or water heating is done by actively capturing byproduct heat that would otherwise be ejected into the environment. Sources may include refrigeration/air-conditioner compressors, manufacturing or other processes, data processing centers, lighting fixtures, ventilation exhaust air, and the occupants themselves.

Wind Energy: Kinetic energy present in wind motion that can be converted to mechanical energy for driving pumps, mills, and electric power generators.

X

Xeriscaping: Landscaping technique particularly popular in desert or drought-prone regions that incorporates plants (usually those native to the region) that require little or no water beyond that provided by rainfall.

Sarah Boslaugh
Washington University in St. Louis

Sources: U.S. Environmental Protection Agency (http://www.epa.gov/OCEPAterms), U.S. Energy Information Administration (http://www.eia.doe.gov/tools/glossary)

Green Education
Resource Guide

Books

Allen, Thomas B. *Guardian of the Wild: The Story of the National Wildlife Federation, 1936–1986.* Bloomington: Indiana University Press, 1987.

Barlett, Peggy F. and Geoffrey W. Chase. *Sustainability on Campus: Stories and Strategies for Change.* Cambridge, MA: MIT Press, 2004.

Boardman, Philip. *Patrick Geddes, Maker of the Future.* Chapel Hill, NC: University of North Carolina Press, 1944.

Boud, David, Ruth Cohen, and David Walker. *Using Experience for Learning.* Buckingham: Open University Press, 1993.

Carter, Neil. *The Politics of the Environment: Ideas, Activism, Policy.* New York: Cambridge University Press, 2007.

Corcoran, P. B., Mirian Vilela and Alide Roerink. *Earth Charter in Action: Towards a Sustainable World.* Amsterdam: KIT Publishers, 2005.

Corcoran, Peter Blaze and Arjen E. J. Wals. *Higher Education and the Challenge of Sustainability. Problematics, Promise and Practice.* Amsterdam: Springer, 2004.

Davis, Brent. *Inventions of Teaching: A Genealogy.* Mahwah, NJ: Erlbaum, 2003.

Davis, J. M. *Young Children and the Environment: Early Education for Sustainability.* New York: Cambridge University Press, 2010.

Dewey, John. *Democracy and Education.* New York: Simon and Schuster, 1916.

Dewey, John. *Experience and Education: The 60th Anniversary Edition.* Indianapolis: Kappa Delta Pi Editions, 1998.

DeWitt, Calvin B. and Ghillean T. Prance. *Missionary Earthkeeping.* Macon, GA: Mercer University Press, 1992.

Edwards, A. *The Sustainability Revolution: Portrait of a Paradigm Shift.* Gabriola Island, British Columbia, Canada: New Society Publishers, 2005.

Filho, Walter Leal. *Sustainability and University Life.* Ann Arbor, MI: University of Michigan Press, 1999.

Gough, Stephen and William Scott. *Higher Education and Sustainable Development.* New York: Routledge, 2009.

Gray, D., L. Colucci-Gray and E. Camino. *Science, Society and Sustainability: Education and Empowerment for an Uncertain World.* New York: Routledge, 2009.

Hargreaves, Andy and Dean Fink. *Sustainable Leadership.* Hoboken, NJ: Jossey-Bass, 2005.

Hewitt, T. W. *Understanding and Shaping Curriculum: What We Teach and Why.* Thousand Oaks, CA: Sage, 2006.

Louv, Richard. *Last Child in the Woods: Saving Our Children From Nature-Deficit Disorder.* Chapel Hill, NC: Algonquin Books, 1992.

McMillan, Victoria and Amy Lyons Higgs. *Implementing Sustainability Education: Lessons From Four Innovative Schools.* Ann Arbor, MI: University of Michigan, 2002.

M'Gonigle, R. Michael. *Planet U: Sustaining the World, Reinventing the University.* Gabriola Island, British Columbia, Canada: New Society Publishers, 2006.

Myers, N. J. and C. Roffensperger. *Precautionary Tools for Reshaping Environmental Policy.* Cambridge, MA: MIT Press, 2006.

National Research Council. *Inquiry and the National Science Education Standards: A Guide for Teaching and Learning.* Washington, DC: National Academy Press, 2000.

Palmer, Joy A. *Environmental Education in the 21st Century. Theory, Practice, Progress and Promise.* London and New York: Routledge, 1998.

Pinsoneault, Eric. *The Environmental Impact of Green Mountain College: A Study of Resource Use and Waste Creation.* Poultney, VT: Green Mountain College, 2003.

Pramling Samuelsson, I. and Y. Koga. *The Contribution of Early Childhood to a Sustainable Society.* Paris: UNESCO, 2008.

Rappaport, Ann and Sarah Hammond Creighton. *Degrees That Matter: Climate Change and the University.* Cambridge, MA: MIT Press, 2007.

Reid, David. *Sustainable Development. An Introductory Guide.* London: Earthscan Publications, 1995.

Rhodes, Edwardo Lao. *Environmental Justice in America.* Bloomington, IN: Indiana University Press, 2003.

Russo, Rosemarie. *Jumping From the Ivory Tower: Weaving Environmental Leadership and Sustainable Communities.* Lanham, MD: University Press of America, 2010.

Sandell, Klas, Johan Ohman and Leif Ostman. *Education for Sustainable Development.* New York: Studentlitteratur AB, 2005.

Silka, Linda and Robert Forrant. *Inside and Out: Universities and Education for Sustainable Development.* Amityville, NY: Baywood Publishing, 2006.

Sterling, Stephen. *Sustainable Education: Re-Visioning Learning and Change.* Totnes, UK: Green Books, 2001.

Ward, Harold. *Acting Locally: Concepts and Models for Service-Learning in Environmental Studies.* Washington, DC: American Association for Higher Education, 1999.

Wiland, Harry and Dale Bell. *Going to Green: A Standards-Based Environmental Education Curriculum for Schools, Colleges, and Communities.* White River Junction, VT: Chelsea Green Publishing, 2009.

Wixom, Robert L. *Environmental Challenges for Higher Education: Integrating Sustainability Into Academic Programs.* Burlington, VT: Friends Committee on Unity with Nature, 1996.

Journals

American Educational Research Journal

Canadian Journal of Environmental Education

Diverse Issues in Higher Education

Early Childhood Education Journal
Educause Review
Environmental Education Research
Equity & Excellence in Education

Higher Education
History of Education

International Journal of Education & the Arts
International Journal of Environmental & Science Education
International Journal of Sustainability in Higher Education

Journal of Adventure Education & Outdoor Learning
Journal of Geoscience Education
Journal of Teacher Education

National Wildlife
Nature News

Science & Education
Studies in the Education of Adults
Support for Learning
Sustainability: The Journal of Record

Websites

American College and University Presidents Climate Commitment
www.presidentsclimatecommitment.org

Association for the Advancement of Sustainability in Higher Education
www.aashe.org

Australian National University Sustainable Learning Community
slc.anu.edu.au

Brandeis University Climate Action Plan
www.brandeis.edu/campussustainability/climate/actionplan.html

Campaign for Environmental Literacy
www.fundee.org/campaigns/hesa

Center for Environmental Education
www.ceeonline.org

The Cloud Institute for Sustainability Education
www.sustainabilityed.org

The College Sustainability Report Card
www.greenreportcard.org

Earth Portal
www.earthportal.net

15 Green Colleges and Universities
 www.grist.org/article/colleges1

GovTrack: Higher Education Sustainability Act of 2007
 www.govtrack.us/congress/bill.xpd?bill=h110-3637

Green Awakenings: Students Caring for Creation
 www.renewingcreation.org

Green Design Institute at Carnegie Mellon University
 gdi.ce.cmu.edu

Green Education Foundation
 www.greeneducationfoundation.org

The Greens: A Site for Kids Looking After the Planet
 meetthegreens.pbskids.org

Green.Tulane.Edu
 green.tulane.edu

Guidelines and Recommendations for Reorienting Teacher Education to Address
Sustainability
 www.allacademic.com/meta/p116506_index.html

Institute for Global Sustainable Enterprise at the University of Michigan
 www.erb.umich.edu

Institute of the Environment at Tufts University
 environment.tufts.edu

National Religious Partnership for the Environment
 www.nrpe.org

National Wildlife Federation Campus Ecology
 www.nwf.org/campusEcology

North American Association for Environmental Education
 www.naaee.org

Northeast Campus Sustainability Consortium
 sustainability.yale.edu/necsc

Office of Sustainability at Dalhousie University
 office.sustainability.dal.ca

Office of Sustainability at Tufts University
 sustainability.tufts.edu

Our Sustainable Future: The Institute for Sustainable Development, California State
University, Chico
 www.csuchico.edu/sustainablefuture

Sierra Club Top 20 Environmental Schools
 www.sierraclub.org/sierra/200909/coolschools/top20.aspx

Sustainable Endowments Institute
www.endowmentinstitute.org

Teaching and Learning for a Sustainable Future: A Multimedia Teacher Education Programme
www.unesco.org/education/tlsf

United States Environmental Protection Agency, "Colleges and Universities."
www.epa.gov/sectors/sectors/college.html

Universidad EARTH
www.earth.ac.cr/ing/index.php

University of British Columbia—Center for Interactive Research on Sustainability
www.cirs.ubc.ca

University of British Columbia—Design Centre for Sustainability
www.dcs.sala.ubc.ca

University of British Columbia—Institute of Resources, Environment & Sustainability
www.ires.ubc.ca

Green Education Appendix

Educational Institutions

Amherst College is committed to sustainability in operations from building practices to food services to transportation. Several campaigns on campus promote ecological awareness, including the Eco-Rep program, which trains students to educate others about environmental issues; a light bulb exchange program; and a competition among dormitories to lower energy usage. The college has an aggressive recycling program that includes materials such as clothing, furniture, and electronics as well as traditional materials such as paper, cardboard, and glass. Dining services emphasize purchase of locally grown food and have been composting food waste since 2009. The college makes its annual "fuel footprint" and an inventory of greenhouse gas emissions (from 2007) available through its Website. Amherst offers a major in Environmental Studies that includes courses from many different disciplines that examine the interactions of man and nature. (www.amherst.edu/cam puslife/greenamherst)

Ball State University has been a leader in sustainable operations for years and included sustainability goals in its *Strategic Plan, 2007–2012.* Most notable among these goals is the construction of a large-scale geothermal energy system on campus, unique among American universities and the largest such system in the United States. Construction began in 2009 and when completed, the system, which involves 45 campus buildings, will replace four coal boilers and reduce the university's carbon footprint by about half. Ball State is also committed to green construction, and all new campus buildings and renovations to existing buildings must meet LEED certification guidelines. Ball State's Council on the Environment, established in 1991, works to raise awareness of sustainability issues and increase sustainable practices on campus. Ball State offers several majors related to the environment and sustainability, including Environmental Management, Environmental Design, and Environmental Communication/Interpretation—Natural Resources and Environmental Management. The interdisciplinary Center for Energy Research/Education/Service focuses on research and education related to energy and resource uses in the local and state communities as well as on campus. (cms.bsu.edu/about/rankings/commitment.aspx)

Bates College has a number of programs to reduce environmental impact and increase sustainability in operations. In dining services, 30 percent of the food budget is spent on local foods, and 82 percent of food waste is composted, recycled, or donated to a food bank or farmer. Bates has an action plan to become climate neutral, which includes converting its main steam plan to a biomass cogeneration facility and offsetting current emissions. Bates was the first college in Maine to participate in the Zipcar (car-sharing) program and many other initiatives promote conservation among students and staff, including a bicycle co-op, a van-pool program, and environmental-themed houses for students. The Environmental Studies program at Bates includes courses in many disciplines, including science, ethics and economics, and the humanities, to give students a broad-based awareness of the issues in the field. (bates.edu/x217211.xml)

Bowdoin College has an Environmental Mission Statement that states its commitment to increasing sustainability efforts on campus, integrating environmental awareness into students' daily lives as well as their education, and providing leadership to the larger community on environmental issues. The Environmental Studies Program offers over 25 courses and aims to provide students with the analytical tools necessary to investigate environmental issues; many other departments offer courses related to sustainability as well. Bowdoin offers the only undergraduate program in the United States in Arctic Studies and also has two coastal research stations that engage in environmental research. Bowdoin is committed to becoming carbon neutral by 2020 through initiatives that include increasing fuel efficiency in transportation, improving the power grid, reducing energy consumption, and increasing use of renewable sources of energy. (www.bowdoin.edu/sustainability/statements/mission-statement-statement.shtml)

Bowling Green State University is a center for education and research into sustainability with about $17 million in research funding (including programs in alternative energy, environmental monitoring, and global change) over the past three years. The School of Earth, Environment and Society was formed in 2007 to strengthen multidisciplinary research and to integrate the fields of geology, geography, and environmental studies. The Department of the Environment and Sustainability offers three major fields of study: Environmental Health, Environmental Science, and Environmental Policy and Analysis. The Center for Environmental Programs at Bowling Green, founded in 1968, promotes campus sustainability efforts and environmental education for students and the community. (www.bgsu.edu/departments/envp/pages/center_for_environmental_programs.htm)

California Polytechnic State University recently established the Center for Sustainability in the College of Agriculture, Food and Environmental Sciences to support research and education in food systems, resource management, and sustainable agriculture. Cal Poly also offers minors in Environmental Studies, Sustainable Environments, and Sustainable Agriculture. The Empower Poly Coalition, founded in 2006, is a coalition of student clubs and organizations interested in sustainability, with 27 members as of 2010. The Sustainability Advisory Committee studies resource use on campus and makes recommendations to the university regarding land use, physical projects, and resource utilization. The primary metric used for tracking conservation efforts at Cal Poly is energy use per square foot, and the university has achieved a reduction of over 15 percent in this measure since 2006. (www.sustainability.calpoly.edu)

Chatham University is committed to achieving carbon neutrality in operation by 2025 and conducted its baseline greenhouse gas inventory in 2007. In order to achieve this goal, the university has committed to steps including requiring all new campus buildings to meet the LEED Silver standard, requiring all new appliances purchased to be energy-efficient (Energy Star certified, if possible); providing discounted public transportation passes for students, faculty, and staff; purchasing at least 15 percent of electricity from renewable sources; and participating in a number of recycling and waste reduction programs. The Climate Committee, made up of students, faculty, and staff, meets biweekly to plan initiatives to help Chatham achieve carbon neutrality. Chatham offers undergraduate majors in Environmental Science and Environmental Studies. (www.chatham.edu/outreach/sustainability)

Clarkson University (New York) offers undergraduate degree programs in Environmental Engineering, Environmental Health Science, Environmental Science & Policy, and a professional concentration in Environmental Engineering and graduate programs in Environmental Science and Engineering (M.A. and Ph.D.). Clarkson is also currently developing a cross-disciplinary minor in Sustainable Energy Systems through the School of Engineering and is looking into developing a similar minor that would be available to all Clarkson students. The Clarkson Center for the Environment is home to Clarkson's environmental activities in research (including the Center for Air Resources Engineering and Science, the Center for Sustainable Energy Systems, the Clarkson BioMass Group, the Great Rivers Center, and the Rivers and Estuary Observatory Network), education and outreach (including a K–12 learning partnership program and a Hazardous Waste Site Operation and Emergency Response Course). Clarkson has numerous campus programs to reduce environmental impact and increase sustainability: For instance, all new buildings are built to LEED certification standards, all washers and dryers have been replaced with Energy Star equipment, and the campus has an extensive recycling plan including electronic waste, tires, used oil, and antifreeze. (ww.clarkson.edu/green/index.html)

Clemson University has an Environmental Committee to formalize its commitment to environmental stewardship and sustainability; coordinate and consolidate efforts on campus toward this end; provide environmental education for students, staff, and faculty; and promote environmental learning and research. Clemson conducted an environmental audit in 2006 that is available from the campus Website, which evaluated the university's current energy use and waste production and proposed options to reduce both. The Solid Green program on campus promotes environmentally conscious behavior such as recycling waste from tailgating parties and football games (which draw 80,000+ fans). Among the environmentally relevant degree programs at Clemson are undergraduate programs in Environmental Science and Policy and in Environmental Engineering, and graduate programs in Plant and Environmental Sciences and in Environmental Toxicology. (www.clemson.edu/cuec)

Colby College coordinates environmental activities through the Environmental Advisory Group (EAG), formed in 2000 to advise the president and college community. The mission of the EAG is to encourage teaching about environmental issues, disseminate environmental information to the community, promote conservation of resources and reduction of waste and pollution, make recommendations about campus and community environmental issues including consideration of social justice, prioritize projects and assess them quantitatively whenever possible, and encourage members of the community to take personal responsibility for their effects on the environment. Colby has many campus programs to increase sustainability and has committed to reduce greenhouse gas emissions 9 percent by

2010 (from a 2005 baseline) and 20 percent by 2020. Colby has an Environmental Studies program that offers majors in Environmental Studies: Interdisciplinary Computation, Environmental Studies: Policy, and Environmental Studies: Science, while the chemistry and biology departments also offer degrees focusing on Environmental Science. (www.colby .edu/wag/index.shtml)

The **College of William & Mary** has been recognized by the *Princeton Review* as among the top green colleges in the United States. William & Mary's Committee on Sustainability, created in 2008, coordinates campus efforts to achieve greater sustainability as part of a campus-wide initiative begun that same year to improve environmental sustainability. Students elected to charge a "green fee" whose funds would go to improve campus facilities (it raises more than $200,000 annually) and many on-campus initiatives help to reduce resource use and waste, including creation of a solar cell testing station that will ultimately result in a solar power facility on a campus building; recycling programs for many items, including batteries, chemicals, and fluorescent light bulbs; use of only Fair Trade organic coffee on campus; composting of organic waste from the dining halls; and recycling of cooking oil for biofuel. (www.wm.edu/sites/sustainability)

Colorado College included sustainability as one of seven core values in its Vision 2010 road map document, created in 2003. The Office of Sustainability coordinates campus efforts to increase sustainability, and many campus programs are already in place. The Russell T. Tutt Science Center, completed in 2005, is the first LEED-certified science center in the United States and the first LEED-certified building in southern Colorado. The Cornerstone Arts Center, opened in 2008, received LEED Gold certification. Campus landscaping uses regionally appropriate plants and avoids the use of pesticides, and uses environmentally friendly compounds for snow removal. Campus dining services use local and organic produce wherever possible, compost food waste, and use trayless dining, which reduces food and water waste. Colby offers an interdisciplinary Environmental Program, which includes tracks in Environmental Science, Environmental Physics, Environmental Chemistry and Environmental Policy. (sustainability.coloradocollege.edu)

The Center for Sustainable Communities and Civic Engagement at **Daemen College** (New York), founded in 2001 with a grant from the John R. Oishei Foundation, has as its primary purpose to establish long-term collaborations with several communities in Buffalo (Seneca Babcock, the West Side, and the Fruit Belt) to support community-driven projects and provide education about economic and sustainability issues. Daemen is a member of the Consortium for North American Sustainability, an exchange program among six North American universities that allows students academic and experiential opportunities to learn about Civil Society and Sustainable Communities. Daemen offers an undergraduate major in Environmental Studies through either the Natural Sciences or Interdisciplinary Studies division. (www.daemen.edu/academics/centersinitiatives/cscce/pages/default.aspx)

Duke University created an Environmental Policy in 2005 that commits the university to becoming a leader in environmental research and education, environmentally responsible operations, and environmental stewardship in the community. Duke publishes several journals related to sustainability, including the *Duke Environmental Law & Policy Forum* (established in 1991), the *International Journal of Sustainability in Higher Education, Sustainability: The Journal of Record* (founded 2008), and the open access e-journal

Sustainability: Science, Practice & Policy. Duke offers undergraduate majors in Environmental Science and Policy, Environmental Sciences, Earth and Ocean Sciences, and Civil and Environmental Engineering; an undergraduate minor in Environmental Science & Policy; and a Master of Environmental Management. Sustainable activities on campus include use of locally grown food in campus dining facilities, retrofitting campus steam plants to improve efficiency, exchanging incandescent light bulbs for energy-saving compact fluorescents, recycling of 17 different items, and using compressed natural gas facilities for a high percentage of the university's fleet. (sustainability.duke.edu)

The Ad Hoc Committee on Environmental Stewardship at **Emory University**, formed in 1999, supports sustainability efforts on campus. The committee drafted a Campus Environmental Mission Statement that states the university's commitment to conservation, sustainability, and biodiversity. Particular emphasis is placed on preserving the ecosystems of the natural areas on campus (including Lullwater, Hahn Woods, and Baker Woodlands) and in restoring the balance of building and environment envisioned by Henry Hornbostel, first architect of Emory. In 2001, Emory appointed its first campus environmental officer, and in 2002, Emory began its first Energy Conservation Project. As of 2004, Emory has sought LEED certification for each new building or building renovation, has proposed a storm water management and forest management plan, and has prioritized bicycle and pedestrian access to campus. (www.environment.emory.edu)

The David E. Shi Center for Sustainability at **Furman University**, founded in 2008 with support from the Andrew W. Mellon Foundation, coordinates Furman's efforts to support sustainability through academic study, research, leadership, and community service. Since 2002, all new buildings and renovations at Furman have met at least the LEED Silver standard, and in 2005, the Sustainability Planning Group was created to champion sustainability on campus. Furman includes several sustainability Living/Learning Laboratories on campus, which facilitate teaching and research related to sustainability, including Cliffs Cottage (which houses the Shi Center), the Furman Farm, and the Charles H. Townes Science Center. Furman offers majors in Earth and Environmental Sciences and a concentration in Environmental Studies. (www.furman.edu/sustain)

George Washington University (GWU) President Steven Knapp created the Presidential Task Force on Sustainability in 2007 and signed the American College and University Presidents Climate Commitment in 2008. GWU sustainability activities, including creating an inventory of greenhouse gas emissions and a plan to reduce them, are guided by the Office of Sustainability. GWU is currently converting its oil-burning equipment to natural gas, requiring all new buildings to receive at least 16 LEED points, requiring all new appliances and electronic equipment to be Energy Star products if possible, and recycling about 30 percent of its waste, including cooking oil, which is processed into biofuel, and food scraps, which are composted. GWU participates in the Zipcar program, provides discounted parking for carpools, and encourages public transportation use by employees through the Metrocheck program. GWU has highly ranked programs in Environmental Law and offers an Environmental Focus in the School of Business. The Institute for the Analysis of Solar Energy is a multidisciplinary program that conducts research into economic, technical, and public policy issues surrounding solar power while the Sustainable Landscape Design Program teaches the principles of landscape conservation and sustainability. (sustainability.gwu.edu)

Hampshire College has a Sustainable Campus Plan that states its commitment to using the campus and surrounding community as a laboratory for demonstration of sustainable development. The college is committed to global environmental issues and social justice both in terms of campus operations and in preparing its students to take leadership roles on these issues. The college has an active recycling program, a green purchasing plan (for instance, white copy paper purchased for campus use must be 100 percent recycled content and chlorine free), and new buildings are constructed to LEED certification standards. Hampshire's Environmental Studies and Sustainability Program is an interdisciplinary program including environmental science, ecology, sustainable agriculture, conservation, technology and design, and sustainable communities. (www.hampshire.edu/offices/5230.htm)

Harvard University has made four major commitments to increasing sustainability: a 30 percent reduction in greenhouse gases by 2016 (from a FY 2006 baseline), a commitment to fostering research and education into environmental issues, a campus-wide energy policy that sets temperature guidelines to reduce energy costs and greenhouse gas emission, and a policy of establishing green building guidelines that include a minimum of LEED Silver certification and use of the Integrated Design approach and Life Cycle Costing. The Office for Sustainability was established in 2008 to coordinate sustainability efforts across the university. Many academic areas at Harvard are involved with teaching and research related to the environment (including the fields of ecology/biodiversity, public health, climate, and economics and policy), and in 2009, the Graduate Consortium on Energy and Environment was formed to facilitate the exchange of ideas across disciplinary boundaries. (green.harvard.edu)

Illinois State University (ISU) was named a "green college" by the *Princeton Review* in recognition of its Green Team sustainability committee, active recycling and energy efficiency programs, provision of transportation alternatives, and for having a full-time sustainability coordinator. ISU's Center for Renewable Energy, which combines college faculty with business partners, fosters applied research, provides public education, and strengthens the interdisciplinary Renewable Energy major at ISU (the first such program in the United States). ISU also offers an undergraduate major in Environmental Health, undergraduate minors in Environmental Health and Environmental Studies, and master's programs in Conservation Biology and Applied Community and Economic Development. (sustainability.illinoisstate.edu)

Indiana University (IU) at Bloomington has a Sustainability Office to promote and coordinate sustainability efforts on campus. A Campus Sustainability Advisory Board is working to address issues identified in IU's 2008 Campus Sustainability Report and has seven working groups: Academic Initiatives, Energy & Built Environment, Environmental Quality & Land Use, Food, Resource Use & Recycling, Sustainable Computing, and Transportation. The campus has a number of Green Teams that promote environmentally friendly practices in their offices, departments, and living areas. The Student Sustainability Council is composed of a number of campus organizations that are working to increase sustainability and promote communication and cooperation among them. IU held its third annual Energy Challenge in March–April 2010, in which campus dorms, Greek houses, and academic buildings competed to reduce energy use as much as possible (from the previous year's baseline) for one month. The IU Department of Environment Science offers undergraduate and graduate study drawing on courses from the College of Arts and Sciences and the School of Public and Environmental Affairs. (www.indiana.edu/~sustain)

Ithaca College established a Comprehensive Environmental Policy in 2001 and has two staff members working to coordinate sustainability efforts on campus. The college is committed to carbon neutrality by 2050 and has reduced emissions by 5 percent from 2007 levels. The campus has a 50 percent waste diversion rate and composts food waste. New buildings are expected to be at least LEED Silver certified, and the Park Center for Business and Sustainable Enterprise has LEED Platinum certification. The college has a car-sharing program and subsidizes public transportation for students, faculty, and staff. Ithaca has a Department of Environmental Studies and Science that offers undergraduate degrees in Environmental Science and Environmental Studies.

Johns Hopkins University (JHU) launched its Sustainability Initiative in 2006 as an outgrowth of several programs already in place, including sustainable dining efforts and a recycling program. The Sustainability Committee was formed at the same time to facilitate reducing the university's environmental impact. JHU has had a recycling program for more than 10 years but has updated it over the years to include not only traditional recyclables (paper, cans) but also items such as computers and other electronic equipment. In March 2010, the university completed the Implementation Plan for Advancing Sustainability and Climate Stewardship, which commits it to reducing carbon dioxide emissions by half over the next 10 years. JHU offers undergraduate studies in Global Environmental Change & Sustainability, Environmental Engineering, and Geography & Environmental Engineering while the Energy, Resources and Environment major offers an interdisciplinary curriculum integrating economics, political science, law, and other disciplines. (www.sustainability.jhu.edu)

The **Lewis & Clark College** Sustainability Council was formed in the late 1990s to promote environmental stewardship on campus and is made up of interested students, faculty, and staff members. The college established a major in Environmental Studies in 1995, which aims to address environmental issues comprehensively through interdisciplinary studies drawing on science, ethics, economics, law, and other fields. The college has been tracking its greenhouse gas emissions since 2000 and became compliant with the Kyoto Protocol in 2003 through purchase of carbon offsets. The Sustainability Council offers the Evan T. Williams Sustainability Prize for campus projects that promote or increase sustainability. The college offers an interdisciplinary undergraduate major and minor in Environmental Studies, and the Law School offers a summer program in Environmental and Natural Resources Law. (legacy.lclark.edu/dept/lcsc)

Luther College (Iowa) employs two full-time staff members working on issues of sustainability and created a Campus Sustainability Council in 2008 that works on issues including energy and water use, waste, land use, and education. Luther is committed to reducing greenhouse gases by half by 2013 (using 2003 to 2004 as a baseline) and has already achieved a 15.5 percent reduction, primarily through upgrades to lighting and HVAC systems and use of geothermal heating and cooling in two campus buildings. Dining services purchase produce and meat from local farms and serve Fair Trade coffee, and the college composts landscaping and food waste. New buildings must meet LEED Silver standards, and the college fleet includes hybrid, electric, and compressed natural gas vehicles. Luther offers a major in Environmental Studies and is participating in an initiative to integrate sustainability into the undergraduate curriculum. (www.luther.edu/sustainability)

Sustainability efforts at **Macalester College** are coordinated through the Sustainability Advisory Committee, whose primary mission is to oversee implementation of the Talloires Declaration and the American College and University Presidents Climate Commitment. The college has made public an inventory of its greenhouse gas emissions from 1990 to 2006. Macalester offers an interdisciplinary Environmental Studies program that integrates knowledge from the humanities, natural sciences, and social sciences and looks at global studies on the local, national, and global level. The Three Rivers Center at Macalester was established in 2007 with support from the Andrew W. Mellon Foundation to develop a curriculum that takes advantage of the confluence of the Mississippi, Minnesota, and St. Croix Rivers to create student–faculty research collaborations, and to upgrade instructional research and instruction facilities. The EcoHouse on the Macalester campus gives students a chance to test the effectiveness of green technologies and practice environmentally sustainable lifestyles. (www.macalester.edu/environmentalstudies/ceic/envissues.htm)

The **Massachusetts Institute of Technology** (MIT) has a number of environmental initiatives to manage resource consumption and reduce environmental impact. The Green Building Task Force at MIT was formed in 2000 to identify long-term goals for environmental sustainability and to develop green performance specifications for buildings. The Environmental Programs Task Force is composed of faculty, staff, and students who seek concrete ways to make the campus more environmentally friendly. The task force's accomplishments include expansion of the campus recycling program and increasing from 5 percent to 30 percent the amount of waste recycled, starting programs to compost food and yard waste, and increasing the use of recycled copy paper from less than 1 percent to 100 percent. The Environment, Health and Safety Management System is concerned with protecting the health and safety of the MIT community as well as overseeing the management and disposal of hazardous materials. Many departments at MIT are involved in environmental research and teaching, including the Department of Earth, Atmospheric and Planetary Sciences; the Department of Civil and Environmental Engineering; the Department of Urban Studies and Planning; and interdisciplinary labs and centers such as the Earth System Initiative, the Laboratory for Energy and the Environment, and the Joint Program on the Science and Policy of Global Change. (web.mit.edu/environment/commitment/e_initiatives.html)

The Environmental Action Coalition (EAC) at **Mt. Holyoke College** has the goal of educating the campus community in ecological responsibility and encouraging members of this community to adopt sustainable approaches to their activities. The EAC has implemented a number of energy-saving initiatives, including using biodiesel to power groundskeeping equipment, promoting the use of LEED certification for new buildings, and joining the Million Monitor drive, which encourages students to use the power save mode on their computers to put the hard drive to sleep during long periods of inactivity. The EAC sponsors a Kill-a-Watt competition among campus dormitories, offering a prize each month to the dorm that reduces energy use the most from the previous year. (www.mtholyoke.edu/org/ccc/website/homepage.htm)

New York University (NYU) has a comprehensive Climate Action Plan that details its current greenhouse gas emissions and strategies to reduce them by 30 percent by 2017. Currently, NYU recycles more than 30 percent of its waste stream and is the largest university purchaser of wind energy in the United States. NYU is also upgrading its Cogeneration

Plant (which provides both thermal energy and electricity) to improve output and efficiency: When complete, the improved plan will emit 75 percent fewer regulated pollutants. The dining program is committed to including more locally grown, organic, and Fair Trade products and sustainable seafood in dining halls and retail outlets, and to increasing the use of biodegradable containers for takeout meals. Many NYU academic programs focus on the environment, including the undergraduate major in Environmental Studies, the Environmental Health Clinic, the graduate Certificate in Sustainable Design, Construction, and Development (through the SCPS Schack Institute of Real Estate), the Bioethics program, the Center on Environmental and Land Use Law (through the School of Law), the Wallerstein Collaborative for Urban Environmental Education (through the Steinhardt School of Culture, Education and Human Development), the Environmental Conservation Education program (through the Steinhardt School), and the Master of Urban Planning degree (through the Wagner Graduate School of Public Service). (www .nyu.edu.sustainability)

The Office of Sustainability at **North Carolina State University** (NCSU) is the clearinghouse for campus efforts to balance environmental, social, and economic sustainability. In 2001, NCSU created an Office of Waste Reduction and Recycling to further develop existing programs, and in 2003, created the Office of Energy Management to support conservation efforts. Since 2008, all new NCSU buildings have been built to the LEED Silver standards, and NCSU has been an Energy Star partner since 2008. University dining halls have been trayless since 2008, reducing water use and food waste, donate used cooking oil to a company that converts it to biodiesel fuel, and regularly feature organic, locally grown, and Fair Trade products in their meals. Ten of the colleges within NCSU offer environmental degrees, and the university is developing an interdisciplinary curriculum in Environmental Science Research. (www.ncsu.edu/sustainability)

Northern Arizona University was selected by the *Princeton Review* as a "green college" and has the goal of becoming carbon neutral by 2020. The university's Strategic Plan includes the goal of increasing sustainability and environmental stewardship through scholarship, campus practices, and practical engagement. Many campus operations have already adopted green practices: For instance, food services composts food waste and buys local produce and milk, used cooking oil is converted to biodiesel to power campus vehicles, and the university has a free campus bike program for students, faculty, and staff. The College of Earth Sciences and Environmental Sustainability offers undergraduate degree programs with an interdisciplinary focus (including natural sciences, social sciences, and the humanities) in Environmental Sciences and Environmental Studies, and graduate degree programs in Environmental Sciences and Policy and Quaternary Sciences. The Institute for Tribal Environmental Professionals, established in 1992, brings together tribal governments, university personnel, government agencies, and the private sector to help safeguard natural resources on Native American lands. The Western Regional Center of the National Institute for Climate Change Research, covering 13 states, is centered at Northern Arizona University. (green.nau.edu)

Occidental College created the Sustainable Oxy/Eco-LA program in 1999 to enhance sustainable practices on campus and to link the college with community partners to further urban greening programs. Occidental is a signatory to the Talloires Declaration, and in 2000 created the Council for a Livable Campus to increase recycling and other environmentally friendly practices on campus. Occidental offers a major in Urban & Environmental Policy that combines academic studies with practical experience in civic action and governmental affairs. The Urban & Environmental Policy Institute at Occidental College is an academic and research center linked with the Urban & Environmental Policy program but that also acts as a community development organization. (departments.oxy.edu/uepi)

Ohio University created an Office of Sustainability in 2006 (then called the Office of Resource Conservation) that coordinates campus efforts to reduce consumption, waste, and environmental impact on campus. Environmental stewardship and sustainability are among the goals listed in *Vision Ohio,* the university's statement of goals first formalized in 2004. Ohio University has numerous programs to reduce environmental impact, including alternative energy sources, conversion of waste cooking oil to biodiesel, holding a Residence Challenge for dormitories to reduce their energy use, composting food waste, and conducting a biodegradable service war in dining halls. Ohio University offers many academic programs with an environmental focus, including (at the undergraduate level) Environmental Biology; Marine, Freshwater, and Environmental Biology; Environmental Chemistry; Environmental Geography; Environmental Geology; and (at the graduate level) Conservation Biology, Environmental and Plant Biology, Environmental Studies, and Environmental Sustainability. (www.ohio.edu/sustainability)

Pacific Lutheran University signed the Talloires Declaration in 2004, and in 2007, President Loren Anderson joined the leadership of the American College and University Presidents Climate Commitment. Pacific Lutheran has numerous campus efforts to reduce environmental impact and increase sustainability, which are overseen by the Sustainability Committee. The university uses geothermal heating and cooling, purchases Energy Star appliances when available and computers with a Gold EPEAT rating, and requires that new construction and renovations achieve at least the LEED Silver standard. Pacific Lutheran subsidizes transit passes for faculty, staff, and commuter students; offers bicycles for rent; and uses alternative-fuel vehicles for one-fourth of the university fleet. Pacific Lutheran offers an interdisciplinary program in Environmental Studies that prepares students for many careers, including laboratory science, consulting, corporate and government environmental regulation offices, and nonprofit organizations focused on environmental concerns. (www.plu.edu/sustainability)

Penn State (Pennsylvania State University) has a Center for Sustainability, founded in 1995, which promotes environmental stewardship and sustainability at the university, and promotes education, research, and outreach, which advance sustainability. The Green Destiny Council produced the *Penn State Indicators Report* in 2000, which calculated resource use at Penn State, and the *Ecological Mission for Penn State* in 2001, which suggested ways Penn State could reduce its environmental impact in eight areas: energy, water, materials, food, land, building, transport, and community. Penn State is a leader in developing affordable and sustainable housing: Demonstration projects include the Maple Point development in Philadelphia, which uses 80 percent less energy than comparable traditionally built homes; Hundredfold Farm in Gettysburg, Pennsylvania, which includes multiple sustainable technologies, including passive solar design; and the MorningStar Solar Home on the Penn State campus, which acts as a teaching and research facility. Penn State offers a number of environmentally focused academic programs, including Community, Environment, and Development; Environmental and Renewable Resource Economics; Environmental

Resource Management; Environmental Soil Science; Environmental Studies; and Energy and Sustainability Policy. (www.cfs.psu.edu)

Pomona College has a full-time sustainability coordinator and is committed to reducing its emissions through a sustainability audit and measures such as the use of solar power, sleep settings for all computers and occupancy sensors to reduce power use in unoccupied space, and replacement of incandescent bulbs with fluorescents. Currently, three academic buildings have LEED Gold or Silver certification. The purchasing department is committed to buying from environmentally responsible vendors, and dining services uses biodegradable utensils and purchases some locally grown, Fair Trade, and organic products. Food and landscaping waste is composted and used as mulch for an on-campus organic farm. Pomona subsidizes public transportation and rideshare programs and has a bicycle-renting program for students. Pomona has an interdisciplinary Environmental Analysis program with courses in four broad areas: Experiential Relation to the Environment, Natural Science Analysis of the Environment, Social Science Approaches to the Environment, and Environmental Design and Architecture. (www.pomona.edu/administration/sustainability/index.aspx)

Ecological awareness, concern for social justice, and experiential learning and field studies are integrated into all degree programs at **Prescott College** but the college also offers undergraduate studies in Environmental Studies with a choice of nine emphasis areas (including agroecology, conservation biology, environmental education, and marine studies); a master's degree program in Environmental Education with concentrations in Environmental Education, Conservation Ecology and Planning, Sustainability Science and Practice and Social Ecology; and a Ph.D. in Education with a concentration in Sustainability Education that emphasizes global citizenship and environmental responsibility. Research centers and institutes at Prescott, which emphasize ecological awareness, include the Global Change and Sustainability Institute, the Ecological Research Institute, the Green Recreation Institute, and the Center for Children and Nature. Prescott publishes the *Journal of Sustainability Education;* the first issue appeared in May 2010. (www.prescott.edu/seed)

Sustainability activities at **Princeton University** are organized through the Office of Sustainability, and the university committed in 2008 to a Sustainability Plan that includes goals for research, education, and outreach; conservation of resources (including reduction of water use by 25 percent of 2007 levels by 2020); and reduction of greenhouse gases to 1990 levels by 2020. University efforts to attain these goals include a purchasing plan that favors recycled products and uses 100 percent recycled paper for all regular printing and copying, programs to reduce electricity use, including lighting retrofits, and requirements that all new building projects and major renovations meet at least LEED Silver standards. Several green building programs are under way, including a chemistry building that uses rainwater collection and a residential college with green roofs. The university has a bike repair and lending program, offers incentives to use public transportation and carpooling, and owns several electric and hybrid vehicles. Dining services uses biodegradable cups and 100 percent recycled paper plates, offers organic and locally grown food and Fair Trade coffee, and has set a goal of reducing waste by 50 percent through composting and recycling. (www.princeton.edu/sustainability)

Rensselaer Polytechnic Institute has several well-established research centers that do research related to sustainability, including the Center for Future Energy Systems (founded in 2005), the Lighting Research Center (founded in 1998), and the Darrin Freshwater Institute (founded in 1970). In addition, the Center for Architecture Science and Ecology, recently founded by Anna Dyson, does research to improve solar generation of electricity. Rensselaer offers majors in several disciplines that include work relevant to sustainability, including Environment and Society, Environmental Economics, Science of the Environment, Environmental Impact Analysis, and Introduction to Environmental Studies. Efforts to increase sustainable practices in university operations began with the Greening of Rensselaer Institute program in 1995. It was most recently renewed when Rensselaer held its first Sustainability Charrette in 2009, in which a team of students, faculty, and staff generated ideas for ways to increase sustainability at Rensselaer. Current activities to reduce environmental impact include plans to construct a facility to convert cooking oil to fuel and a second initiative to use food waste to create methane gas, encouraging trayless dining, instituting single-stream recycling, and building a green roof on the student union. (www .rpi.edu/about/sustainability)

Several programs within the Biotechnology Center for Agriculture and the Environment at **Rutgers University** focus on environmental and sustainability concerns. The Renewable and Sustainable Fuel Solutions for the 21st Century is a research and educational program funded by the National Science Foundation. Over 40 faculty members are involved in this program, which has four main research areas: development and optimization of biofuels, innovative catalysts and engineering systems for synthetic fuels; land use; sustainability and environmental impact; and fuels deployment logistics, economics, and policy. Other research and training projects within the Biotech Center include studies in environmental and health genomics, environmental bioremediation, and the Botanical Research Center. The Rutgers Center for Green Building, located within the Edward J. Bloustein School of Planning and Social Policy, focuses on research, education, and training related to green building and forming partnerships with industry, government, and not-for-profit agencies. Rutgers offers an undergraduate program in Environmental Sciences and graduate programs in Environmental Sciences and Ecology and Evolution. Rutgers has had a green purchasing program since 1988 and has numerous policies in place to cut waste, encourage recycling, and increase sustainable practices in university operations. (biotech.rutgers.edu)

The Office of Environmental Sustainability at **Smith College** is the central organizing body to promote sustainable practices on the Smith campus. It helps develop campus policy in conjunction with the Committee on Sustainability, oversees the Earth Rep program, coordinates the Green Team, coordinates with other campus organizations including Facilities Management, and works with the Environmental Science and Policy program and the Center for the Environment, Ecological Design and Sustainability to connect student learning about sustainability with campus operations. Smith has reduced electricity consumption by 7 percent between 2004 and 2009 through a number of initiatives, including replacing incandescent light bulbs with compact fluorescent bulbs and encouraging faculty, students, and staff to implement energy-saving procedures on their computers. Smith installed a cogeneration power plant that is expected to reduce carbon emissions by 238 metric tons over the next 20 years. Several different curricular pathways at Smith are available to study the environment, including the department of Environmental Science and Policy. The Center for the Environment forms a framework for integrating environmentally relevant studies from various departments, including the natural sciences, engineering, design, landscape studies and architecture, the social sciences, and philosophy. (www.smith.edu/green)

The **State University of New York (SUNY) at Binghamton** incorporates environmentally sound principles in its vision statement and has been twice named one of the top 15 universities in the United States for its environmental friendly practices. President Lois B. DeFleur was a charter signatory in 2007 to the American College and University Presidents Climate Commitment, and SUNY has a number of sustainability efforts on campus, including the first LEED-certified buildings in the SUNY system, an organic gardening and composting program for the dining halls, a solar hot water program, and use of water-saving faucets and toilets. Binghamton offers about 40 courses related to sustainability or the environment in a given year as well as majors including Environmental and Resource Management and Environmental Studies. The E.W. Heier Teaching & Research Greenhouse is a living laboratory for ecology- and plant-related courses, with over 6,000 exotic plants maintained in the greenhouse, and the University Nature Preserve provides students with opportunities to study over 200 species of birds as well as diverse mammal, amphibian, and reptilian populations. Several other research initiatives at Binghamton relate to sustainability, including the Small Scale Systems Integration and Packaging Center, the Center for Advanced Sensors and Environmental Systems, the Center for Integrated Watershed Studies, and the Institute for Material Research. (www2.binghamton.edu/think/our-vision)

The **State University of New York (SUNY) at Buffalo** includes Environmental Stewardship in its UB2020 Strategic Plan. In 2007, Buffalo created an Environmental Stewardship Committee to support the American College and University Presidents Climate Commitment that commits the university to creating a plan to become carbon neutral. The university has a number of programs already in place to promote this goal, including a nationally recognized energy conservation program (credited with saving the university more than $9 million annually), a recycling program that currently recycles over 30 percent of the university's solid waste (with a goal of increasing this to at least 50 percent), use of recycled paper for printing and copying (about 50 percent of which is 100 percent postconsumer recycled paper and processed without chlorine), and use of alternate fuel vehicles (including electric, hybrid, and compressed natural gas) in the campus fleet. Most university diesel vehicles use B-20 fuel, which is 20 percent biodiesel. Buffalo offers a major in Environmental Geosciences, and undergraduate and graduate degrees in Civil, Structural, and Environmental Engineering, as well as graduate study in the Division of Environmental Health Sciences within the School of Public Health and Health Professions. (www.ubgreenoffice.com)

The **State University of New York's College of Environmental Science and Forestry** (ESF) is devoted to education and research related to natural resources and the natural and designed environments. Academic departments include Environmental and Forest Biology, Environmental Resources and Forest Engineering, Environmental Science, Environmental Studies, Forest and Natural Resources Management, and Landscape Architecture. ESF is a signatory to the American College and University Presidents Climate Commitment and a member of the Association for the Advancement of Sustainability in Higher Education; ESF has committed to becoming carbon neutral by 2015. The SUNY Center for Sustainable and Renewable Energy, located on the ESF campus, is a research and development clearinghouse for information about energy efficiency and sustainability. The Adirondack Ecological Center, established in 1971, is engaged in ongoing research monitoring trends and changes in the Adirondack ecosystem. The Center for the Urban Environment is dedicated to finding ways to mitigate urban environmental problems and working with communities, business, and governmental and nongovernmental organizations to produce a more sustainable program. ESF began a biodiesel program in 2006 and currently uses a BioPro 190 reactor to produce 50 gallons per week of biodiesel (from waste cooking oil) to power student transport and maintenance vehicles on campus. (www.esf.edu/sustainability)

The Sustainability Commission at **St. Bonaventure University** ties ecological and sustainability issues to the university's mission as a Franciscan institution, helps the university form a correct relationship with the environment that is consistent with that mission and creates opportunities for education, discussion, and dialogue, which can contribute toward this mission. In October 2008, St. Bonaventure held its first Sustainable Bona's Day in which staff, students, and faculty attempted for one day to reduce carbon dioxide emissions through means such as lowering thermostats, reducing motor vehicle use, serving local foods in the dining halls, and recycling soy containers and paper products. Electricity use dropped 24 percent on the first Sustainable Bona's Day as opposed to the same day one year earlier, and natural gas use was reduced by 50 percent. In 2009, St. Bonaventure initiated a trayless dining program, which reduces food waste and reduces the amount of heated water and detergent used by dining services. (www.sbu.edu/About_SBU.aspx?id= 21450&terms=environment)

Stetson University has committed to incorporating principles of sustainability on its campus. In 1998, Stetson created the Environmental Responsibility Council to promote sustainability practices in the university's curriculum, policies, operations, and external ties. The university is committed to environmental education, environmentally responsible purchasing, efficient use of resources, minimization of solid waste production and hazardous materials use, and planning and campus design that is environmentally responsible and centered on the use of Florida native plants. Stetson had the first LEED-certified green building in Florida, and the Rinker Environmental Learning Center boasts several ecological features, including a rainwater collection system, a geothermal heating system, and recycled metal roofing. The Stetson University Native Plant Initiative commits the university to using only Florida native plants on its main campus, both in order to preserve the natural ecosystem and also because such plants require less water, fertilizer, and pesticides than nonnative plants. Stetson has a broad recycling program and placed fifth nationally in the 2010 RecycleMania competition. Stetson offers a major and minor in Environmental Science as an interdisciplinary program through the geography department. (www.stetson .edu/other/erc)

The Sustainability Task Force at **St. Olaf College** has created a report that identifies the values of the college with regard to the environment, identifies the current state of sustainability at the college, and sets goals for the future. Current initiatives to reduce St. Olaf's environmental impact include purchasing a composter that turns food waste, paper, and cardboard into mulch, using recycled paper for copying and printing, planting trees on farmland, restoring wetlands on campus, restoring an area of prairie, and considering adding "green screens" to lessen the environmental impact of investment opportunities for the college endowment funds. St. Olaf's began offering an interdisciplinary concentration in Environmental Studies in 1987–1988 and an Environmental Studies major beginning in 1999–2000. The major offers three tracks: Environmental Science, Social Sciences, and Arts and Humanities. (www.stolaf.edu/green/report/principles)

The **University of California, Berkeley (UCB),** along with the rest of the University of California system, adopted a Green Building Policy and Clean Energy Standard, since renamed the Policy on Sustainable Practices. The Berkeley campus formed the Chancellor's Advisory Committee on Sustainability to assess the current state of sustainability on campus and promote measures to protect human and environmental health. In 2007, Berkeley committed to reducing greenhouse gas emissions to 1990 levels by 2014 through a series of energy efficiency projects, increased use of renewable energy, and reductions in fuel usage by the campus fleet and commuters. Berkeley has reduced water usage by over 10 percent since 1990, currently diverts 57 percent of its solid waste, uses recycled paper for 74 percent of copying, and 24 percent of all food and beverage products bought on campus in 2008–2009 were sustainable. Berkeley produces not only traditional annual campus sustainability reports (including measures like recycling and energy and water use) but also a report focusing on social and economic measures of sustainability. Berkeley offers many undergraduate and graduate programs that include a focus on sustainability, including Agricultural and Environmental Chemistry, Bioengineering, City and Regional Planning, Ecosystem Science, Environmental and Science Journalism, Environmental Health Sciences, Environmental Law, Environmental Science, Landscape Architecture and Environmental Planning, Molecular and Environmental Biology, Society and Environment, and Transportation Engineering. (sustainability.berkeley.edu)

The **University of California, San Diego** (UCSD) has implemented a number of measures to reduce environmental impact. Students serve as "Econauts" to provide peer-to-peer education with the goal of achieving zero waste and a 20 percent reduction in water use by 2012. Dining services have replaced Styrofoam with compostable plastic containers and permanent dishware that students may borrow. A variety of local and organic produce and Fair Trade coffee, tea, and sugar is used in the campus dining services. Landscaping has been adapted to use native plants and minimize the need for additional water. Solar panels have been installed on the roofs of two parking garages; each generates more than 17,000 hours of energy per year. Green Campus is a student program at UCSD funded by the Facilities Management Department and the Alliance to Save Energy, which works to build campus awareness about environmental issues, include sustainability concerns in course content, and implement projects targeting energy use. USCD offers a number of courses relevant to the environment as well as a major in Environmental Systems and a minor in Environmental Studies. (greencampus.ucsd.edu)

The **University of California, Santa Barbara** (UCSB) approved a campus sustainability plan in February 2008 that commits the university to achieving 2000 emissions levels by 2014, 1990 emissions levels by 2020, and carbon neutrality by 2050. This document also commits UCSB to environmental leadership in academics and research and to a sustainable program of building, operating, and retrofitting campus buildings. All buildings programmed after 2004 must meet at least LEED Silver standards, and UCSB has a number of environmentally friendly programs in place, including use of reclaimed water for irrigation and to flush toilets, various transportation alternatives to traveling by automobile, an Environmentally Preferable Purchasing policy that requires purchasers to consider sustainability issues, and composting of food waste and gardening waste. USCB gets 16 percent of its electricity from renewable sources, and several buildings use solar panels to create energy and/or to heat water. Several departments at UCBS offer coursework relevant to sustainability, including Geography, Environmental Studies, Global Studies and the Bren School;

and the Academic Senate Sustainability Work Group is currently developing a Ph.D. program in Environment and Sustainability and creating an environment and sustainability general education requirement. The Green Campus Program is a student-run effort to save energy by building awareness, collaborating with university administration, implementing projects that will reduce energy use, and incorporating energy conservation into course curricula. (sustainability.ucsb.edu/gcp)

The **University of Colorado** (UC) was the first university in the United States to have an Environmental Center: The CU Center, then called the Eco-Center, was founded in 1970 as part of the inaugural Earth Day celebrations. The center provides leadership and education in environmental issues and has helped implement many sustainable operations on campus, including establishing the first campus recycling program in the United States (in 1976) and establishing the first prepaid bus pass program for students. The CU Environmental Justice Project was founded to address the fact that poor and nonwhite communities are often unduly burdened with pollution and environmental degradation; principal activities include education, outreach, and dissemination of information regarding environmental justice. Earth Education is a volunteer and intern program that provides environmental education to schools in the greater Boulder community. CU offers an interdisciplinary Environmental Studies program that awards B.A., M.S., and Ph.D. degrees. The Cooperative Institute for Research in Environmental Sciences is home to five research centers, including the National Center for Atmospheric Research, the National Institute for Standards and Technology, and the National Renewable Energy Laboratory, providing many opportunities for collaboration in the environmental sciences. (ecenter.colorado.edu/energy)

The **University of Connecticut** adopted an Environmental Policy in April 2004 that commits the university to using the best available environmental practices and continually monitoring and improving its environmental performance, designing and constructing buildings and grounds in a sustainable manner, conducting outreach and embracing environmental initiatives in surrounding communities, and advancing environmental understanding through its academic programs. In 2007, the university established an additional Sustainable Design and Construction Policy relating to design, construction, renovation, and maintenance of sustainable and efficient building. The Office of Environmental Policy was created in 2002 to focus on sustainability and environmental performance on campus. Connecticut offers majors in Environmental Science, Environmental Health and Safety, and Environmental Engineering; and the Center for Environmental Sciences and Engineering promotes multidisciplinary research in engineering and sustainability. (www .ecohusky.uconn.edu)

The **University of Georgia** (UGA) has a Sustainability Task Force to promote energy efficient practices on campus and provides free bus passes to students, faculty, and staff, using a bus fleet powered by biodiesel. The GoGreen program at UGA encourages environmentally sustainable practices within the university's physical operations: Measures include composing plant waste (100 percent of leaf and limb debris has been composed since 1983); recycling about 1,000 tons of paper, cardboard, bottles, and cans annually; composing animal bedding from the Veterinary School and research facilities; and using green cleaning procedures to reduce exposure to harmful chemicals. The Odum School of Ecology fosters interdisciplinary studies in the relationships among organisms and their environment and provides undergraduate and graduate education in Ecology and Conservation Ecology and

Sustainable Development. The Georgia Plant Conservation Alliance was formed in 1995 to coordinate efforts to protect natural habitats and preserve Georgia's endangered fauna. The Center for Invasive Species and Ecosystem Health studies forest health, natural research, and agricultural management and invasive species from the state to international levels. (gogreen.uga.edu)

The **University of La Verne** Sustainable Campus Consortium, founded in 2002, promotes environmentally responsible practices on campus. The Sustainable Campus Task Force does an environmental audit, with results available on the campus Website, including seven categories: energy efficiency, water efficiency, landscaping, transportation, recycling and waste reduction, reducing toxins, and environmental lessons. In 2003, LaVerne was awarded a Waste Reduction Awards Program (WRAP) award from the California Integrated Waste Management Board for its recycling program—it was one of only five schools in California to receive this award. (www.laverne.edu/resources-services/facilities-manage ment/sustainability)

The **University of Maryland** (UMD) Office of Sustainability facilitates sustainability on campus by promoting education in sustainability, developing sustainability programs, coordinating efforts among departments and with external resources, and measuring and reporting on campus sustainability efforts. In 2008, UMD adopted a 10-year strategic plan that included becoming a national model for a Green University by reducing energy use and the university's carbon footprint, retrofitting buildings to strict environmental standards, expanding green spaces, and promoting teaching and research in energy science, policy, and related fields. UMD is committed to achieving carbon neutrality by 2050 and has taken a number of measures in this direction. Polystyrene foam has been replaced by biodegradable products in dining services, food waste is composted, a $20 million energy project was implemented in 2009 that should reduce CO_2 emission by over 4,200 tons per year, and all new construction and major renovations must meet at least the LEED Silver standard. The rate of recycling on campus is currently 54 percent, up from 17 percent in 2003. Water use has been reduced through installation of low-flow toilets, showers, and faucets; improved dishwashing machines; and use of a 10,000 gallon rainwater cistern and computer controlled drip irrigation system to water the campus's native landscaping. Several campus centers conduct environmental research, including the Center for Environmental Energy Engineering, the Center for Integrative Environmental Research, the Center for Social Value Creation, the Earth System Science Interdisciplinary Center, the Environmental Finance Center, the Joint Global Change Research Institute, the Maryland Institute for Applied Environmental Health, the National Center for Smart Growth Research and Education, and the University of Maryland Energy Research Center. (www.sustainability.umd.edu)

Sustainability programs at the **University of North Carolina at Chapel Hill** (UNC) are coordinated through the Vice Chancellor's Sustainability Advisory Committee. UNC is committed to reducing carbon dioxide emissions by 60 percent over the next 45 years through means such as constructing a state-of-the-art cogeneration facility and constructing energy-efficient buildings. The Center for Sustainable Enterprise at UNC, located in the Kenan-Flagler Business School, teaches business and industry leaders how to achieve a positive "triple bottom line" of financial profitability, ecological integrity, and social equity. The Center for Sustainable Energy, Environment and Economic Development within the Institute for the Environment coordinates campus activities related to energy and the environment, including the areas of energy sciences, environment and health, policy, planning and economic development, and efforts to green the UNC campus. The Institute for the Environment is currently developing a minor in Sustainability to be offered at UNC. (www.ie.unc.edu/cseeed/greening_the_campus)

The Mascaro Center for Sustainable Innovation within the Swanson School of Engineering at the **University of Pittsburgh** (Pitt) focuses on designing sustainable neighborhoods that have a positive effect on residents' quality of life as well as the environment. Current programs include research and development in green building design and construction, infrastructure technology and networks, and green materials. The Center for Sustainable Landscapes, designed to exceed the LEED Platinum standard, will generate all its own energy with renewable resources and capture all water used on site. Pitt has offered an interdisciplinary undergraduate program in Environmental Studies since 1995 and offers several relevant graduate programs, including the Environmental Law, Science and Policy Certificate Program, and several programs in environmental health through the School of Public Health. (www.pitt.edu/~esweb)

The **University of Puget Sound** Sustainability Advisory Committee was formed in 2005 after President Ron Thomas signed the Talloires Declaration on behalf of the university. The committee works to reduce environmental impact and increase sustainable practices on campus, working in four key areas: transportation, waste reduction, energy, and water conservation. The university guidelines for green building and design include using native plants, using natural ventilation and daylighting, maximizing renewable energy use, reducing waste, and using local materials. The university offers an undergraduate major in Environmental Policy and Decision Making, and faculty are working to integrate sustainability into the curriculum and are working with the Washington Center at Evergreen State College to develop goals and approaches to promote environmental literacy. (www.puget sound.edu/about/sustainability-at-puget-sound)

The Sustainable Communities Initiative at the **University of Rhode Island** (URI) developed as a response to the need to address issues of sustainability in human activity and to take advantage of several existing URI programs that were related to this issue, including the Coastal Institute, the Community Planning and Landscape Architecture Department, and programs associated with the Sea Grant, Land Grant, and Urban Grant programs. Currently, URI is developing a Sustainability minor, and all freshmen in the College of the Environment and Life Science are required to participate in a community service learning project related to sustainability. Sustainable campus practices currently in use at URI include installation of low-energy lighting, a green purchasing program, a demonstration project using solar shingles on a campus building, and a community bike program. (www .uri.edu/sustainability/curriculum.html)

The **University of South Carolina** (USC) has instituted a number of programs to reduce the environmental impact of university activities. The West Quad Residence Hall, a.k.a. the Green Quad, which earned a LEED Silver rating, was the largest green residence hall in the world when it opened in 2004. Most university shuttle buses use biodiesel, and an experimental hydrogen-powered bus is also included in the campus fleet. A biomass facility uses wood chips to provide electricity for the campus, and dining facilities have replaced Styrofoam containers with biodegradable takeout boxes. Water usage was reduced by over

40 percent by installing low-flow faucets, showers, toilets, and urinals. USC has an innovative Green Learning Community of 25 to 30 students who live on the same floor of the Green Quad and design projects and community activities that will lead to a more eco-friendly society. (www.sc.edu/green)

The "Make Orange Green" program at the **University of Tennessee at Knoxville** is a campus-wide program to reduce environmental impact and build a culture of sustainability on campus. The program offers a light bulb exchange for residence halls and holds an annual POWER challenge in which the residence halls compete to reduce water and energy use and increase recycling. The university is one of the largest purchasers of green power in the Southeast, purchasing 5,000 blocks (750,000 kWh) per month. Buildings constructed since 2007 must meet minimum LEED standards. (environment.tennessee.edu)

The Center for Integrated Agricultural Systems at the **University of Wisconsin–Madison** conducts research and offers education relating to sustainable agricultural practices. The Business, Environment & Social Responsibility Program, housed within the Wisconsin School of Business, offers education for business leaders and students interested in sustainable enterprise. The IGERT China program, funded by the National Science Foundation, trains students to address problems in biodiversity, conservation, and sustainable development. The Nelson Institute for Environmental Studies consists of four research centers—the Center for Climatic Research; the Center for Culture, History and Environment; the Center for Sustainability; and the Global Environment and the Land Tenure Center—that conduct research, offer nine degree and certificate programs, and conduct community programs and public events. In 2006, Wisconsin committed to an aggressive program to reduce environmental impact, including achieving a reduction of 20 percent in overall energy consumption and the campus environmental footprint by 2010. (www.conserve.wisc.edu)

The Global Environmental Management (GEM) program at the **University of Wisconsin–Stevens Point**, established in 2000 and affiliated with the College of Natural Resources, is a center for education and outreach in natural resources and environmental management. Programs at GEM include Sustainable Agriculture and Forestry, Permaculture Design, Sustainable Communities & Campuses, Sustainable Energy Systems, Land Use Planning and Rural Leadership, and Community Development. The Wisconsin Center for Environmental Education, created in 1990, develops environmental educational programs for K–12 teachers and students. The National Environmental Education Advancement Project promotes the creation and expansion of environmental education programs across the nation. (www.uwsp.edu/cnr/gem)

Sustainability is a central value at **Warren Wilson College** (WWC): Environmental stewardship is included in the campus mission statement, and students are expected to make sustainable choices and advocate best practices in their work and community service as well as learning about issues in the classroom. WWC began offering environmental studies courses in 1977, and in 1979, established an Environmental Studies major; it is currently the largest major at the college. The Orr Cottage at WWC, completed in 2005, was the first LEED Gold standard building in North Carolina. In 2006, WWC became the first college in the Southeast to offset 100 percent of its energy use with renewable energy credits. WWC includes biodiesel, hybrid, and solar-powered vehicles in its fleet and promotes carpooling, bicycling, and public transportation use. (www.warren-wilson.edu/~elc/sustainability)

Washington University in St. Louis has a Sustainability Office to coordinate campus efforts to reduce environmental impact. In 2008, the university created a Sustainability Plan with the primary goal of reducing greenhouse gas emissions. Major achievements include switching from coal to natural gas for steam generation, reducing energy usage per square foot by 31.3 percent (Danforth Campus) and 51.0 percent (Medical Campus), conducting a greenhouse gas emissions inventory from 1990 to 2009, and constructing the Living Learning Center at Tyson Research Center, a building that provides its energy and water needs on site and has a zero carbon construction footprint. Twelve campus buildings have or are awaiting LEED certification, and the university provides free public transportation passes and is a partner in the WeCar car-sharing program. (www.wustl.edu/sustain)

Williams College set a goal of reducing greenhouse gas emissions by 10 percent from 1990 levels by 2020 by measures such as installing solar hot water systems, using a cogeneration plant for 20 percent of campus electricity, upgrading light, installing steam and electrical meters, and retrocommissioning HVAC systems. More than 10 percent of the dining hall food budget is spent on local and organic foods, and food waste is composted. Two campus buildings have LEED Gold certification, most dining halls are trayless, most buildings have daylight sensors, and residence halls and locker rooms have low-flow plumbing fixtures. Williams has a car-sharing program, and the college fleet includes electric and hybrid vehicles. (www.williams.edu/resources/sustainability/index.php)

General References

American College and University Presidents Climate Commitment. http://www.presidents climatecommitment.org (Accessed June 10, 2010).

Association of University Leaders for a Sustainable Future. "University and College Sustainability Websites." http://www.ulsf.org/resources_campus_sites.htm (Accessed June 10, 2010).

National Wildlife Federation. "Campus Environment 2008: A National Report Card." http:// www.nwf.org/campusEcology/docs/CampusReportFinal.pdf (Accessed June 10, 2010).

Princeton Review. "College Sustainability Report Card." http://www.greenreportcard.org/ report-card-2010/awards (Accessed June 10, 2010).

Sierra Club. "Cool Schools: The Third Annual List." http://www.sierraclub.org/sierra/200909/ coolschools/allrankings.aspx (Accessed May 24, 2010).

Sarah Boslaugh
Washington University in St. Louis

Index

Article titles and their page numbers are in **bold**.